ELECTRON MICROSCOPY

ELECTRON MICROSCOPY

Pushpa Viswanathan
Department of Electron Microscopy
Cancer Institute
Chennai, Tamil Nadu

MJP Publishers

Cataloguing-in-Publication Data

Pushpa Viswanathan (1951-).
Electron Microscopy / by Pushpa Viswanathan. –
Chennai : MJP Publishers, 2023
xxxviii, 532 p.; 24 cm.
Includes Appendix, Glossary, References and Index.
ISBN 978-81-8094-075-0 (Hb.)
1. Microscopy, Electron I. Title.
502.825 dc 22 PVS MJP 093

ISBN 978-81-8094-075-0

MJP PUBLISHERS
47, Nallathambi Street
Triplicane

Publisher : J.C. Pillai
Managing Editor : C. Sajeesh Kumar
Marketing Manager : S.Y. Sekar
Project Editor : P. Parvath Radha
Acquisitions Editor : C. Janarthanan
Editorial Team : B. Ramalakshmi, M. Gnanasoundari, Lissy John
CIP Data : Prof. K. Hariharan, Librarian
RKM Vivekananda College, Chennai.

To

Our most revered Advisor Dr. S. Krishnamurthi, who gave me the opportunity to work with this wonderful instrument.

Our most beloved Chairman Dr. V. Shanta, who had been my role model from my MSc. days at JIPMER.

My mother Mrs. Padmambal Subrahmanyam who taught me hardwork and sincerity.

PREFACE

Irrespective of all the modern, sophisticated molecular techniques which enable one to look into the atomic substructures, the ultrastructural morphology has its own charm. The morphological evidence offered by Electron Microscope (EM) to substantiate and support other findings is definitely a highly rewarding exercise. Hence, it is not surprising that for scientific publications wherever feasible EM study is demanded. Even for nanosciences, often, EM has become an additional mandatory technique.

Many students approach me for doing ultrastructural studies for their samples. Their non-exposure to this field leads sometimes to non-optimal preservation of the tissue specimens jeopardizing the final research findings. Because improperly and inadequately preserved samples would produce "artifacts" which may mislead the observations, the conclusions/inferences derived from them may not be reliable. Similarly even if the study focuses on a particular organelle, say for example, mitochondria, the most popular target and which many students insist on to look at, at high magnifications, it is essential to scan the whole tissue or cells to rule out effects of artifacts introduced during processing. Rarely students have approached me with a few micrographs taken elsewhere and request for an explanation and inference from them. Most often these micrographs exhibit a few organelles with some aesthetic appeal not giving a holistic view of the specimen which is absolutely essential to draw any valid conclusion.

Thus, we may realize that this sophisticated technique is not one where you feed the sample and get back some values or graphs or data for further interpretation. In EM, subjective observations and interpretations with an objective viewpoint produce genuine results. Therefore it is imperative that the students have an idea about what EM is before proceeding further. And, my main intention in writing this book is to create a "feel for this subject" particularly among those who want to use this technique for their research work.

When I started preparing for the book, I read books which narrate the historical background and the evolutionary processes of this technique. I was surprised to learn that before the discovery of use of interference colours for determining the thickness of the floating sections in order to select ultrathin sections suitable for EM work, the operator had to scan grid after grid in the EM before ascertaining ultrathin sections useful for ultrastructural study. Those who have worked with EM will undoubtedly agree that this is an extremely time-consuming process, taxing one's patience.

But now, the technology is so advanced that both sectioning and screening in EM have become easy routines.

Applications of EM are very vast and varied. It is not possible to include all aspects of applications in one book. So, I have concentrated mainly on the aspects I have done in my laboratory. I have purposely included different kinds of samples to introduce the reader to the several possible roles of EM in the biological research arena. Numerous photographs, rather, electron micrographs are illustrated to acquaint the reader with the appearances of highly magnified features in the EM. This is important because without this experience it is impossible to infer any details from this technique.

I have omitted a few important topics in this book. One is stereology and 3D tomography, by which the 3D structures can be reconstructed. Quantification of structures including immunogold labelling is another aspect not touched upon. Plant specimens are also not included. SEM is also not elaborated.

I would like to point out here that hands-on training is necessary for users of EM. AIIMS, New Delhi and B.R. Doshi School of Biosciences, Sardar Patel University, Vallabh Vidhyanagar, Gujarat, offer yearly workshops to train technicians, students and scientists in this field.

The very first day I was left alone with the Siemens Elmiskop IA Transmission Electron Microscope at IIT, Madras, is still fresh in my memory. When I increased the magnification from 1000× to 25,000× at a stretch, the beam suddenly disappeared. I got terribly scared, switched off the beam and even started praying. Then Dr. Jagasivamani and Dr. Kesavan Nair, Metallurgy Department, explained the reasons for the disappearance of the beam, assured me that nothing can go wrong with the instrument by operating it and encouraged me to make use of the fullest potentials of the instrument. This is exactly what I want to convey to the readers—with the training and knowledge of the technique, just go ahead boldly using the instrument and travel into the magic world of ultrastructure.

Finally, I wish to share the joyful experience I felt during the preparation of this book with the readers.

Pushpa Viswanathan

ACKNOWLEDGEMENTS

I express my deep sense of gratitude to our Advisor (late) Dr. S. Krishnamurthi, who appointed me to work with this instrument at the Cancer Institute. I am ever indebted to him as well as to our Chairman, Dr. V. Shanta, who, have been a constant source of motivation and encouragement to undertake research activities in the Institute.

Our then Director and Scientific Director, Dr. T. Rajkumar, a rare combination of clinician-turned-researcher whom I always admire for his "scientific spirit", generously granted me the permission to undertake this venture and encouraged me to take up external samples. I am indeed very thankful to him.

I am grateful to our present Director and Dean, Dr. T.G. Sagar for providing me all facilities required for the working of the instrument and also permitting me to accept samples from students for EM study.

I owe a lot to the large number of students who have given their samples to me for the fine structural analysis. My exposure to different kinds of specimens is due to the variety of samples received from these students, which was an enjoyable experience and I could learn so much from this exposure and experience. Most of the illustrations included in the text are from these samples. My special thanks to each one of them.

My emphatic expression of gratitude and appreciation goes to Dr. V. Vibhu Krishnan and Dr. Sakthivel Rajan for writing the part on "Bone and soft tissue tumours" in Chapter 20 (Electron Microscopy in Cancer Diagnosis). The potential of EM in this area is neither fully explored nor utilized. These tumours have both benign and malignant components. This area being quite complicated, I had taken the help of these two doctors for picking up the salient features necessary for this book. They have indeed done a good job and I thank them for that.

I sincerely acknowledge and thank Shri K. Seshasayee of Toshniwal (SR) Brothers Private Limited, Bangalore, for sparing time to go through Chapter 4 (Operation and Maintenance of TEM) and offer corrections and additions. From the time of installation he has been helping us in maintaining the TEM at our institute in good working condition. He patiently explains the intricacies and nuances of the instrument and lends a helping hand at all times. He and his company deserve a special mention and thanks.

I extend my most sincere thanks to Shri N. Murali, National Manager (Service), Labindia Instruments Private Limited, Chennai for correcting Chapter 9 (Sectioning) as well as for providing the pictures of the ultramicrotome and the knife maker. I am grateful to him for also clarifying the controversy over the positioning of the upper and lower surfaces of the trimmed block in the ultramicrotome for ultrathin sectioning. He and Shri Natarajan of Labindia Instruments always help us instantaneously to keep our ultramicrotome and knife maker in purposeful condition.

I am highly thankful to Mrs. Varalakshmi Vijayakumar, Senior Microbiologist, Cancer Institute, Chennai for correcting and editing Chapter 22 (Electron Microscopy in Microbiology) and also for being a rare kind of friend sharing the same wavelength of thoughts and ideals.

I greatly appreciate and acknowledge the neat work done by Shri K. Vadivel, Laboratory Technician, Department of Electron Microscopy, Cancer Institute, Chennai in processing the tissue samples including ultrathin sectioning.

I also appreciate the help rendered by Ms. R. Latha, Telemedicine System Administrator, Cancer Institute, Chennai, in scanning the photographic negatives and Mrs. B. Thilagavathy, Typist, Scanning Section, Cancer Institute, Chennai, in scanning the photographic prints during the preparation of the manuscript.

I wish to thank Dr. Sushma Viswanathan for motivating me into work and also helping me in the corrections of the manuscript.

Last but not the least, I am deeply indebted to the publishers, M/s MJP Publishers, Chennai, for offering me the task of writing a book on electron microscopy. I gratefully acknowledge the constant support and freedom given by Mr. J.C. Pillai, Director, Mr. C. Sajeesh Kumar, Managing Editor, Mr. C. Janarthanan, Acquisitions Editor. Yes, there was no restriction regarding the contents or volume or duration/timeframe. They waited for me to complete the book to my reasonable satisfaction. The excellent editing done by Ms. Parvath Radha, Project Editor and Ms. B. Ramalakshmi, Assistant Editor and the neat and clear presentation of around 400 illustrations by the design editors Ms. Gnana Soundari and Ms. Lissy John are highly commendable.

Pushpa Viswanathan

CONTENTS

FIGURES

TABLES

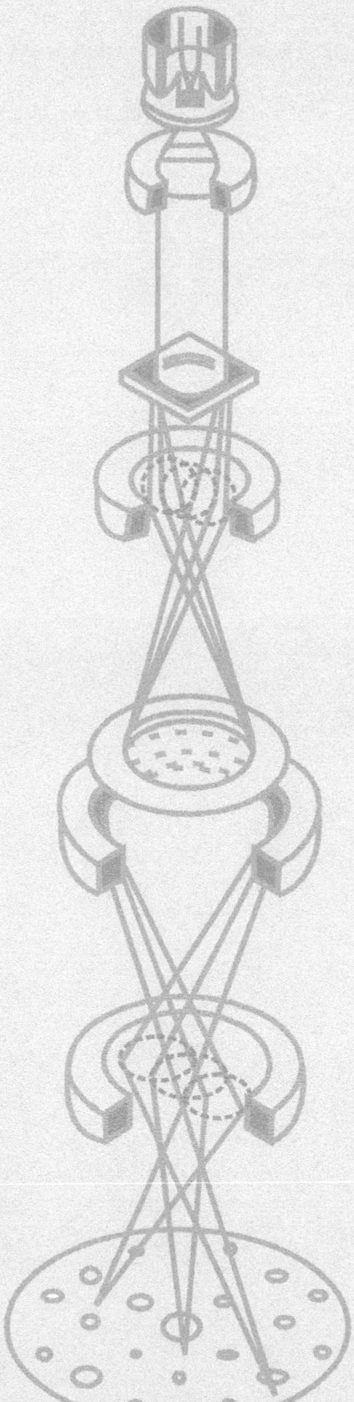

PART I

INSTRUMENTATION

- Electron Microscope—An Introduction
- Electron Microscope—Principles, Theories and Comparison with Optical Microscope
- Parts of the Transmission Electron Microscope
- Operation and Maintenance of TEM
- Scanning Electron Microscopy

1

ELECTRON MICROSCOPE—AN INTRODUCTION

What is a microscope? When I put this question to my students, the answer invariably I get is, "it's an instrument which magnifies the microorganisms." Does this mean the objects or rather microorganisms put in the microscope get magnified and become macro-sized to be visualized by the naked eye directly? No. The microscope only magnifies the image of the specimen and the object as such does not undergo any size variation. The image is formed with the help of the lenses and the illuminating source. In the routine optical microscopes, the illuminating source is a beam of light and the lenses are optical lenses, whereas in the electron microscope, the source of illumination is a beam of electrons, and the lenses are electromagnetic lenses.

A microscope basically consists of a source of illumination, lens systems for formation as well as magnification of the image of the specimen, a stage to mount the specimen and finally a provision to visualize and store the image. The main purpose of a microscope is magnification. But mere magnification, or increase in size, is meaningless until and unless the inner details, for which the magnification is attempted, are separated or resolved out. The resolving power of a microscope is defined as the shortest distance between two points or two lines, which can be recognized as two different images. The resolving power of the human unaided eye is 0.2 mm which means the eye cannot visualize two points which are closer than 0.2 mm as two distinct points. This is under normal viewing conditions, i.e., with an object at the distance of distinct vision, 25 cm. The resolving power of the optical microscope is 0.2 μm. Even the advanced optical microscopes could not aid direct visualization of biological matter smaller than about 0.2 μm. Light microscopic resolution is limited by illumination used and hence an improved resolution dependent on the use of a much smaller wavelength was considered. This created the necessity for inventing other types of microscopes. Figure 1.1 gives the scale of lengths/limit of resolution of various microscopes.

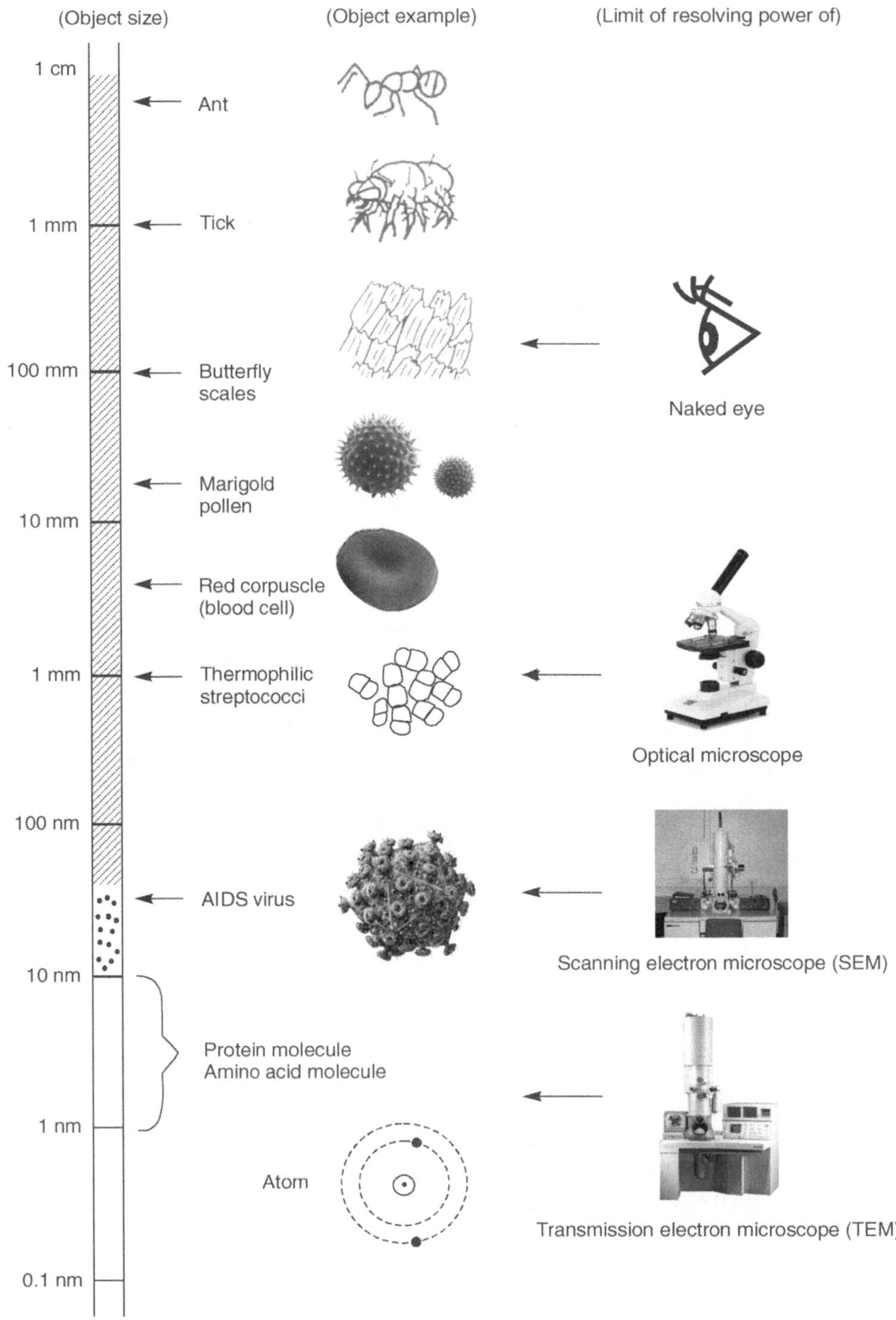

Figure 1.1 Scale of lengths (*Source: JEOL SEM Manual*, p.1)

Two essential developments in Physics led directly to the development of the electron microscope (EM). The first was the de Broglie theory (1924) which implied that a moving electron could be considered to have the properties of a light wave. The second was the demonstration by Busch in 1926, that suitably shaped magnetic or electrostatic fields could be used as true lenses to focus an electron beam to produce an image, which essentially means that a suitably shaped magnetic field could operate as an electron lens.

In 1932, Knoll and Ruska built the first transmission electron microscope (TEM) which can be considered as a prototype of modern magnetic instruments. Their instrument consisted of an electron source and the magnifying lenses. No condenser lens was used. The resolution obtained with this instrument was below that obtained with optical microscope. Nevertheless they secured the first electron micrographs of an illuminated specimen. Ruska added a condenser lens to this which improved the resolving power which was higher than that of the optical microscope. In 1938 von Borries and Ruska constructed an EM of advanced design which was capable of resolving up to 100 Å.

The commercial production of EM began in Germany by Siemens in 1939 and in North America in 1941. Many subsequent developments consisting of more stable power supplies, shorter focal length objectives, greater mechanical and thermal stability and multiple conveniences led to routine instruments with higher resolution of less than 10 Å. The latest reported is the near atomic level resolution. Roger and Hendrix using the emerging technique of single particle electron cryomicroscopy captured a 3D image of a virus at a resolution of 4.5 Å which is the largest resolution ever achieved for a biological organism.

So what is an electron microscope? In essence it is a microscope in which a beam of electrons is used as a source of illumination. Electrons have very low wavelengths and hence very high magnification and resolution can be achieved in the EM. Being particles of negligible mass, the electrons can be easily deflected away by the air molecules. Therefore, the whole microscope column is maintained in high vacuum. The lenses used are electrostatic or electromagnetic lenses, because electrons can be deflected only by electric or magnetic fields. The specimen preparation is a highly critical procedure. As electrons have very low penetrating power, the specimen has to be ultrathin, particularly in the case of transmission electron microscopy. The preparatory techniques vary with the nature of the specimen to be viewed. The high vacuum in the microscope column and the nature of the electrons render it impossible for the human eye to capture the image directly as in the case of light microscope where the image directly falls on the eye through the eyepiece. Hence in electron microscopes, the image falls on a fluorescent screen which can be viewed through a glass window. This image can be photographed for permanent storage as in optical microscopy.

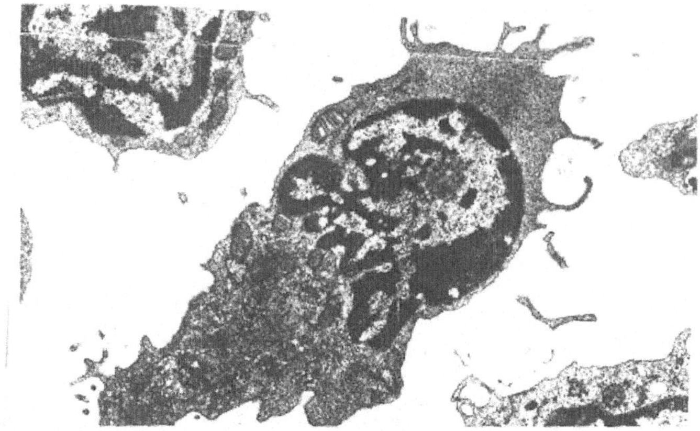

Figure 1.2 Transmission electron micrograph of a leukaemic cell (Cell morphology). (*Source:* Douglas, W. Henderson, John, M. Papadrimitriou & Mark Coleman. (1986). *Ultrastructural Appearances of Tumours,* 2nd edn. Churchill Livingstone. p. 308, Fig. 28.10.)

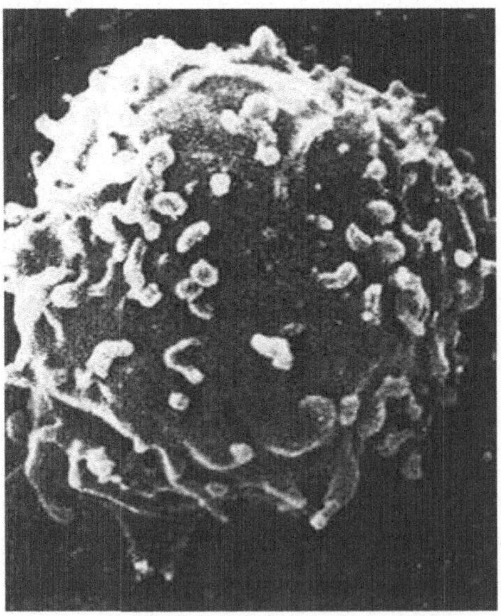

Figure 1.3 Scanning electron micrograph of a leukaemic cell (Surface topography). (*Source:* Douglas, W. Henderson, John, M. Papadrimitriou & Mark Coleman. (1986). *Ultrastructural Appearances of Tumours,* 2nd edn. Churchill Livingstone. p. 308, Fig. 28.9.)

Basically, there are two types of electron microscopes: transmission electron microscope (TEM) and scanning electron microscope (SEM). In the former, the electrons, as the name implies, are transmitted through the specimen, while in scanning, they are allowed to scan the surface of the specimen. Thus, in the transmission type, the internal details of the specimen can be viewed as a two-dimensional (2D) image. In the scanning type, a three-dimensional (3D) image of the surface topography of the specimen can be seen. But by using replica techniques, surface topography can be imaged in the transmission type, and by using ultrathin sections, inner details can be visualized in the scanning type.

Figures 1.2 and 1.3 show the TEM and SEM pictures respectively of a leukaemic cell. The nucleus and other internal details can be visualized in the TEM along with surface projections as a 2D image while in SEM only the surface features are seen as a 3D image.

In the world of biology, many cellular features were identified only after the advent of EM. A few of them are mentioned below:

- Cellular membrane
- Golgi complex which was originally thought to be a stain artifact. This controversy was solved by high-resolution electron microscopy.
- Presence of DNA in mitochondria was first described using EM.
- Lysosomes were distinguished from mitochondria only by EM in isolated preparations.
- Viruses identified only with EM.

2

ELECTRON MICROSCOPE—PRINCIPLES, THEORIES AND COMPARISON WITH OPTICAL MICROSCOPE

Electron Optics is a science on the motion of free electrons under the influence of electrostatic and electromagnetic fields. It utilizes the fact that electrostatic and electromagnetic fields act on the path of an electron in a way similar to the action of a glass lens in conventional light optics. This science is indispensable for designing electron microscopes, CRT, vacuum tubes, photomultiplier tubes (PMT), and electron accelerators.

For actual operation of the electron microscope, knowledge about the principles and theories of electron optics is not a prerequisite. But an understanding of the basic principles involved would contribute to gaining maximum benefits from this highly sophisticated technique.

GENERAL PRINCIPLES

The electron microscope is a highly sophisticated and highly specialized instrument. It has wide applications in various fields including biology, medicine, material science, metallurgy, thin films technology, forensic sciences, archaeology, pharmaceuticals, nanotechnology, etc.

EM is used to examine biological materials such as, microorganisms, cells and a variety of large molecules, tissue biopsy samples, plant materials, metals and crystalline structures and characteristics of various surfaces. It is also used for inspection, quality assurance and failure analysis applications in industry, including, in particular, semiconductor device fabrications.

Fundamentally and functionally the electron microscope and the optical microscope are identical. They serve to magnify the image of minute objects normally invisible to the naked eye. The basic difference is in the source of illumination, which is a beam of electrons in the electron microscope and a beam of light in the optical microscope. As we have already seen in the previous chapter, there are two types of electron microscopes, transmission electron microscope and scanning electron microscope. In the transmission type, the electrons are allowed to be transmitted through the specimen while in the scanning type the incident electrons scan the surface to render the surface topography of the specimen. Table 2.1 shows the differences between the light microscope (LM) and the transmission electron microscope (TEM).

Table 2.1 Differences between LM and TEM

Description	LM	EM
Source of illumination	Light beam	Electron beam
Wavelength	750 nm (visible light 400–800 nm)	0.006 nm at 40 kV
Medium	Air (atmosphere)	Vacuum
Lenses	Optical lens	Electron lens (magnetic or static lens)
Aperture angle	~70 degrees	~35 degrees
Resolving power	200 nm (visible) 100 nm (UV)	0.5 nm or even up to 0.2 nm
Magnification	10× – 2000×	100× – 2,00,000×and more
Focusing	Mechanically	Electrically
Alteration in magnification	By changing the objective lens	By altering the current to the projector lens and adjusting that of the objective lens
Contrast	By absorption and reflection	By scattering, absorption, diffraction and phase
Thickness of specimen	In microns	In angstroms
Image visibility	Directly with the eye	Image falls on a fluorescent screen

Photography is possible in both, but in EM, the photographic attachment is a part of the microscope while in LM it is only optional.

The light microscope relies upon the refractive property of the boundary between air and a transparent substance. The light diffracted by the specimen enters the lens and is bent sharply from the original direction and focused accurately such that the resulting image is a true representation of the original object. Electrons entering the electric or magnetic field experience a change of direction similar to light in a glass lens except that the bending is a continuous process rather than a sharp refraction. Owing to the smaller wavelength of the electron compared to light, much finer details can be seen in the electron microscope because resolution depends on the wavelength of the illumination used. Figure 2.1 gives the schematic representation of image formation in LM and TEM.

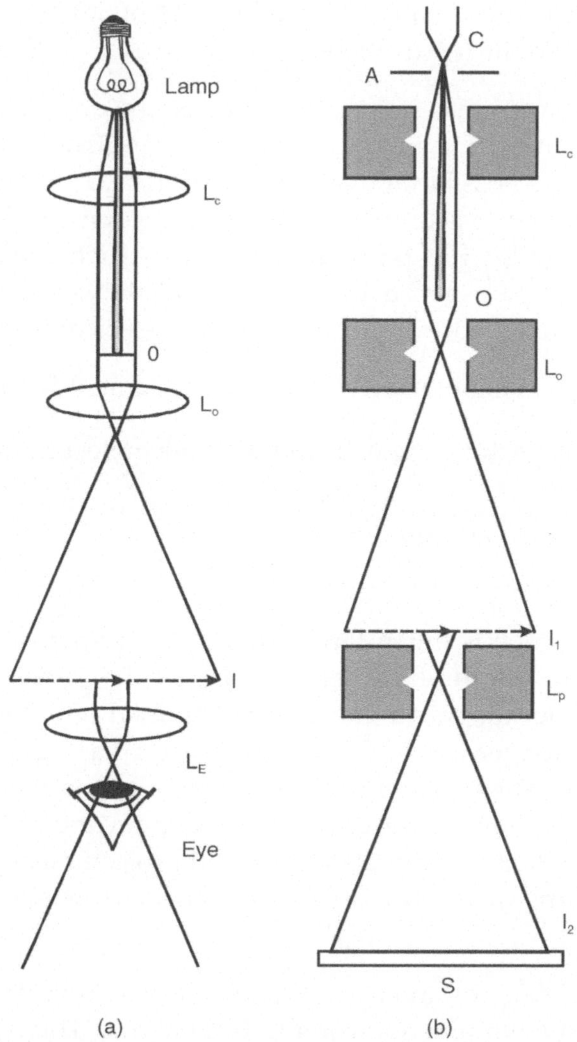

Figure 2.1 Schematic diagram of image formation using a) light microscope and b) electron microscope. A—Anode, C—Cathode, L_c—Condenser lens, O—Object, L_o—Objective lens, I—Image, L_E—Eye lens/eyepiece, I_1—Intermediate image, I_2—Final image, L_p—Projector lens, S—Screen. (*Source*: Cosslett, V.E. (1951). *Practical Electron Microscopy.* Butterworths Scientific Publications. p.5, Fig. 3.)

The source of electrons, which is the illuminating source, is a hot cathode filament from which a stream of electrons is drawn by the high voltage applied between it and the anode which has a small hole for admitting the electron beam. This beam is converged on the object by the condenser lens. The objective lens, close to the object, focuses the image on the intermediate screen which is further magnified by the magnifying lenses, intermediate and projector lenses. The sharp, enlarged image falls on a screen coated with fluorescent material which emits light under electron impact. The fluorescent screen is placed across the axis of the instrument.

Because our eyes do not have the power of perceiving an electron and also the entire column is in high vacuum, the image cannot be visualized directly as in the light microscope. Since no life is possible at that high vacuum, live specimens cannot be viewed in the EM. Recently the low-vacuum scanning microscope or environmental scanning electron microscope (ESEM) has been designed, in which the specimen chamber is maintained at a relatively lower vacuum so as to enable viewing live specimens. Very recently this is being attempted even in transmission electron microscopy.

A very high vacuum is maintained within the instrument for avoiding the following.

- loss of energy of electrons by collision with air molecules owing to the negligible mass of electrons
- oxidation of cathode at high temperatures
- ionization of air molecules at high voltages

The electron microscope is much larger than a light microscope because of higher magnification. The focal length of EM lenses being of the same order as that of optical lenses (a few mm), higher magnification can be obtained only by increasing the length of the tube, giving the longer image distance, or by increasing the number of stages of magnification. The three-stage magnification available in commercial instruments gives a more compact form. It allows a wider range of magnification to be covered simply by varying the current supplied to the intermediate and projector lenses (magnifying lenses). At higher magnifications the field of view is very small being of the order of a few microns only.

The general arrangement of an electron microscope is inverted, compared with that of a light microscope, the illumination coming from above. This arrangement is adopted owing to the need to keep the high voltage terminal of the electron source well out of reach of the operator.

A photograph of the transmission electron microscope is given in Figure 2.2. The inverted form can be clearly seen with the electron gun housing the source (the cathode) at the top and the screen for viewing the image at the bottom.

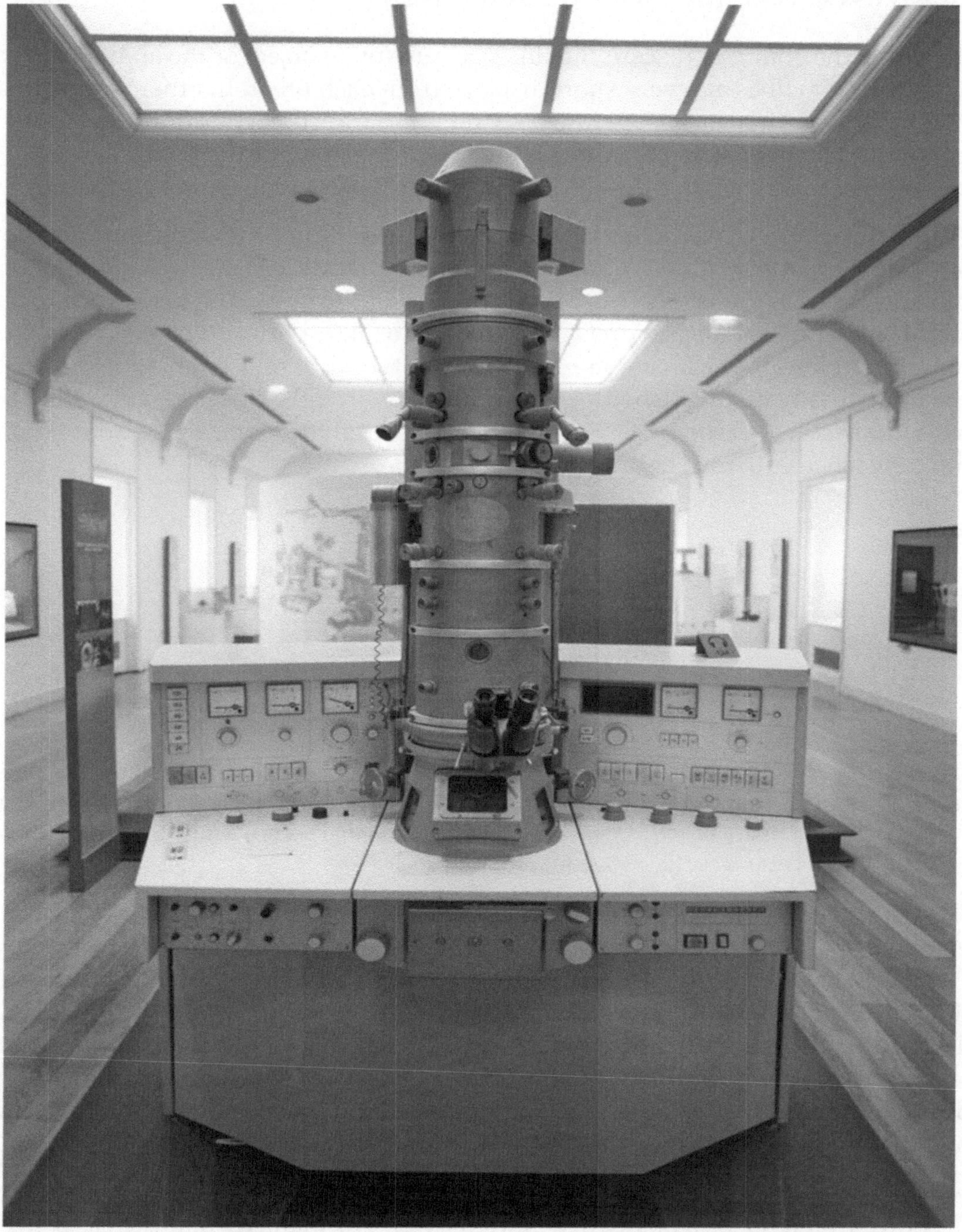

Figure 2.2 Photograph of the TEM (inverted form)

RESOLVING POWER AND RESOLUTION

Resolution is the ability to discern the fine structure of an object or the ability to identify two points or two lines at a very short distance from each other. In other words, it is the smallest distance that may separate two points on an object and still permit their observation as distinct separate points. The resolving power of the naked eye is 0.2 mm which means that we cannot differentiate any two points that are closer than this.

Any optical system (electron or light) has a natural limit set to the smallest details which can be made visible in an object. This depends on the wavelength of the illumination and on the geometry and dimensions of the system. It is referred to as the resolving power of the microscope and is defined as the shortest distance between two points (or two lines) which can be recognized as two different images, such that they can be individually distinguished. As the points are brought closer together they begin to fuse but can still be recognized as separate points until their centres are approximately within a distance equal to their radii (Figure 2.3).

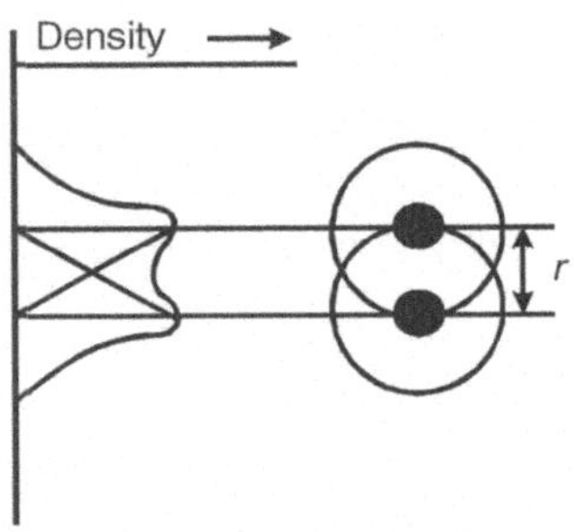

Figure 2.3 Fusion of circles (*r*—distance between the centres of the two adjacent images). (*Source*: *V.E.* Cosslett. (1951). *Practical Electron Microscopy.* Butterworths Scientific Publications. p.10, Fig. 4.)

The resolving power depends on the wavelength of the illumination, object size and angular size of the objective lens as viewed from the object. Therefore, by decreasing the wavelength, the resolving power can be improved. The resolving power of the unaided eye is 0.1–0.2 mm under normal conditions of vision, i.e., object at a distance of distinct vision which is normally 25 cm. By using a lens, this resolution can be improved. Further improvement of resolution is possible by immersing the object in a medium of high refractive index, like oil. But very high resolution can be achieved by using an electron beam instead of light.

Abbe's theory Even if the perfect lens of aperture angle 2α were used, the image of a point source would not be a point but a blurred circle. The diameter of the circle is given by

$$d = \frac{0.61\lambda}{n \cdot \sin\alpha}$$

where,

λ is the wavelength of the light,

n is the refractive index of the medium, and

α is half of the aperture angle of the lens.

This equation is called Abbe's equation and gives the theoretical limit of resolution.

This is the case with optical microscopes where *'d'* is determined by diffraction aberration, as spherical and chromatic aberration, can be eliminated in them. But in electron microscope, *'d'* is influenced mainly by spherical and diffraction aberrations. This is because, removal of aberrations from electron microscopes is difficult when compared with light microscopes. But a high resolving power is possible in electron microscopes because the wavelength of the electron beam is very short.

The limit of resolving power will be usually between ½ or $^1/_3$ of the wavelength of incident beam. In the case of light, the wavelength is around 5000 Å and so the limit of resolving power is around 2000 to 2500 Å only. When UV light of wavelength 2000 Å is used, the resolving power can be improved to 1000 Å. But, in the case of EM, when the electron wavelength is 1/100,000 of light, a resolution of 2 Å can be attained. But lens aberrations limit it to 10–20 Å.

The resolving power of a microscope and the resolution of a micrograph are different. The photograph taken with the EM is generally referred to as **electron micrograph**. The resolution of a micrograph is inferior to the resolving power of microscope because of the specimen, and also microscopic and photographic conditions. So to obtain best resolution, special attention has to be given to specimen preparation, microscope operation, maintenance of microscope, photographing and photographic processing.

MAGNIFICATION

Magnification is the increase in size of the image over the size of the object viewed. But increased image without improved resolution only results in a large blurred image. The human eye has no power of magnification. A suitable magnification is required to determine the resolving power *d* by naked eye. The maximum useful magnification of a microscope is that which will enlarge features of a size equal to the resolution up to the limit of the eye itself. If this is taken as 0.2 mm, then for optical microscope,

$$\text{Magnification } M = d_1/d_2$$

$$\text{i.e., } 0.2/(0.2 \times 10^{-3}) = 1000\times,$$

where,

d_1 is the resolution of the eye (0.2 mm) and

d_2 is the resolution of LM. (0.2 μm or 0.2×10^{-3} mm)

Beyond this value, higher magnification simply enlarges further the blurred images without making them any clearer.

When UV light is used as the source, the useful magnification is 3500×. For EM with high resolution, the magnification can be even in lakhs. With resolving power of 200 nm, the effective magnification can be 2,50,000× or more in EM.

PRINCIPLES OF ELECTRON LENSES

Electron lenses can be of magnetic type or static type. But mostly electromagnetic lenses are used. All of the parallel electron beam incident to the curved magnetic field produced by the pole pieces, converge at one point. The converging action of an electron lens is similar to that of an optical lens. Thus, an electromagnetic lens uses a magnetic field to focus an electron beam.

Focal Length

The effective focal length of an electron lens will be determined by the shape and intensity of its electric or magnetic fields. The focal length is directly proportional to the acceleration voltage or speed of the electrons and inversely proportional to the field strength. The change in focal length with change in wavelength gives rise to chromatic aberration which is more severe in electron lenses than optical lenses. Therefore extensive stabilization of the high voltage supply to produce a beam of great uniformity of velocity is essential in an electron microscope. Here, the focal length is not fixed as in optical lenses but can take a range of values. For focusing, it is only necessary to adjust the strength of a lens instead of raising or lowering the microscope column.

Lens Aberrations

Similar to optical lenses, electron lenses also suffer from lens defects but they possess a few additional aberrations also. The aberrations which set the limit to the performance of an EM are:

- **Spherical aberration** This arises due to differences between the axial and peripheral rays. This is the property of a lens whereby it focuses the beam strongly from its outer zones than from the regions near the axis. The effect is that a point object can only be imagined as a disc. The blurring of images increases very rapidly with increasing lens aperture. Thus a practical limit on resolution of EM of 2 to 3 Å is set due to this aberration.
- **Chromatic aberration** This arises due to variation in wavelength of the incident electrons. This expresses the inability of a lens to focus illumination of different

wavelengths at the same point. This is a defect not readily corrected in an electron lens as in the optical lenses but it is possible to obtain a reasonably monochromatic beam of electrons by applying a highly stabilized accelerating voltage. The degree of stabilization required is 1 or 2 volts in 1,00,000 V which is severe but attainable.

- **Astigmatism** Apart from the above two, the only lens defect which seriously limits the resolution of EM objective is astigmatism which arises due to defects in the construction of the lens. This arises from inaccuracies in machining the pole pieces of the magnetic circuit and from magnetic inhomogeneities in the iron itself. The mechanical tolerances from making a lens sufficiently free of astigmatism for a resolution of 3 Å or even of 10 Å are beyond attainable limits. But a corrector lens for astigmatic defect can easily be made and is a feature of EM. A limit of 5 Å due to astigmatism can easily be achieved. Mechanical vibration of the column may impose a similar limit.

In the optical microscopes, the lens aberrations can be eliminated almost completely, so that the image can be regarded as a faithful reproduction of the object. The optical lenses mainly suffer from diffraction limitations. But in the electron microscopes, the resolution is limited by the lens aberrations. The removal of aberrations from the lenses in EM is difficult when compared with the optical microscope, but its resolving power is high since the wavelength of the electron beam is very short (1/100,000 of the wavelength of the light rays).

The electron microscopes are influenced by spherical aberration which cannot be corrected. Since this aberration arises due to difference between axial and peripheral electrons, a narrow beam of electrons arising near the axis is used.

The non-uniformity of wavelength of the electrons contributes to chromatic aberration. Though there is no colour in EM unlike light microscope, the aberration arising out of wavelength differences in electrons is still referred to as chromatic aberration. The wavelength of an electron beam is determined by its voltage just as the wavelength of a light beam being determined by its colour. The chromatic aberration in EM can be contained by maintaining the high stability of the accelerating voltage, since the wavelength of the incident electrons depends on the accelerating voltage.

The third significant aberration, astigmatism, is caused by the minute variations in lenses—their pole pieces, alignment and magnetic inhomogeneity. If the bore of pole piece of an electromagnetic lens is not a perfect circle or if the material of the pole piece is magnetically non-uniform, the resulting magnetic field will not be axially symmetrical and also the refractive power will be different in the two mutually orthogonal planes passing through the axis. As a result, the electron beam which should normally be focused at a point will appear as an ellipse. This phenomenon is called astigmatism. If there is astigmatism in the objective lens, a sharp image cannot be obtained even at the in-focus condition, and the image blurs in one direction at the out-of-focus condition. This can be removed by adjusting the two knobs X and Y of the stigmator.

As astigmatism is caused by an axially asymmetrical magnetic field in the lens pole piece, it is possible to correct it by placing another asymmetrical magnetic field in the lens and adjusting its strength and direction. This correction lens is called a **stigmator**. Generally, it consists of a pair of quadrupole electromagnetic coils. This correction is applied routinely to both the condenser and objective lenses.

Depth of Field and Depth of Focus

The aberrations in EM limit its resolving power. On the other hand, they give a valuable advantage over LM in respect of depth of field and depth of focus.

Depth of focus refers to the range within which the fine focus control produces no visible departure from best focus in the image. In other words, it means that the object point could be moved over a small axial distance above and below its true position, without changing the size of the corresponding image point. For a given degree of resolution, the depth of this focal range depends on the angular aperture of the illumination.

In the case of LM, owing to the large illuminating apertures used, the depth of specimen in focus is only of the same order as the resolving power and focusing is extremely sharp carried out by the axial shift of the microscope tube. But in EM, lens defects necessitate the usage of a beam with small aperture, which has the positive advantage of giving a large focal length. Thus the focal depth is absolutely and relatively greater than that of LM of high power. This makes focusing less critical than might have been expected.

The extent of the object along the axis which is in sufficiently sharp focus in the image is usually several microns in LM as compared with a fraction of a micron in EM. As a result, the whole of a specimen is in focus at once in EM since its thickness has to be less than a micron on account of poor penetration of electrons, while in LM, the microscope has to be focused on successive layers in the object. Therefore EM can be used to examine the replicas of rugged surfaces such as fractures in metal which are not accessible to direct observation with light.

An added advantage of the large depth of focus in EM is stereophotography. It is possible to obtain a direct reconstruction of the 3D structure of any object which is not too dense with respect to electrons, from a pair of pictures taken from two points of view. Most of the electron microscopes have a facility of tilting the object with respect to the illuminating beam, in order to take stereophotographs.

Depth of field refers to the corresponding effect in the image space, i.e., the range over which the recording plate can be displaced without producing a loss of focus, within prescribed limits. These limits involve the grain size of the photographic emulsion as well as considerations of lens defects. Therefore, placing of photographic plates is not critical in EM. The camera chamber which houses the photographic plates is situated below the fluorescent screen, but it is not necessary to refocus the image on the position of the plate after focusing on the screen.

CONTRAST

It will not be possible to confirm the presence of the object, leave alone studying the configuration, if image contrast is not sufficient, even if the microscope has high resolving power. In other words the actual resolution depends greatly on the image contrast. The various features in the specimen can be clearly identified with their outlines only when there is a contrast between the features and the surroundings.

In light microscope, image contrast is determined by differences in the absorption coefficients between different portions of the specimen and partly by the differences in reflectivity. But in EM, image contrast is by scattering, absorption coefficient, diffraction contrast and phase contrast. Contrast is increased by increasing the scattering by using heavy metals either by staining or shadow casting. Thus in EM, contrast arises mainly by differential scattering between different parts of the specimen.

Image contrast is less in exact focus than in slight underfocus. This is due to interference effect. This effect, in fact, gives good indication of true focus when operating at high magnification and low intensity of illumination. Instrumental contrast is higher at lower voltages and hence for biological specimens, low voltages are usually preferred.

ELECTRON DIFFRACTION

Like an X-ray, an electron beam is reflected (diffracted) at the net plane of a crystal lattice, which can be used for the study of crystal structures. The advantage is that the structure of the specimen can be examined by correlating the diffraction pattern and the microscope images.

Generally, the identification of a substance in an electron diffraction pattern is accomplished through a comparison with the X-ray diffraction data. Electron diffraction is used to estimate the orientation of a crystal and the size and shape of a particle from the fine structure of the pattern. Since electrons are charged particles, their paths can be easily altered by the effect of magnetic or electrostatic fields. This facility provides a variety of diffraction methods for the selection of high resolution, selected area and high dispersion. Figure 2.4 gives the schematic diagram of image formation and electron diffraction in TEM.

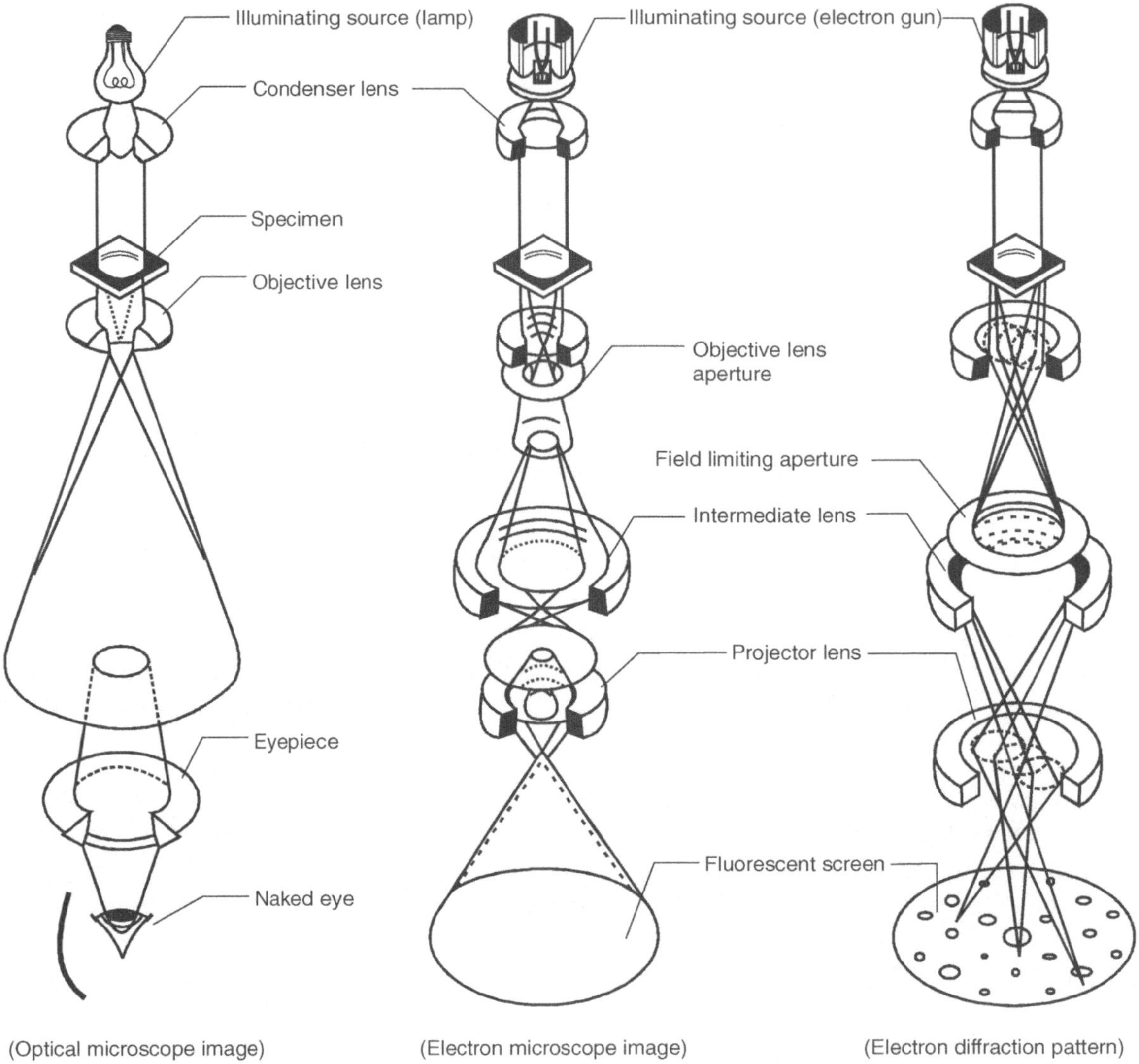

Figure 2.4 Schematic diagram of electron image and electron diffraction in TEM. (*Source: TEM Instructions Manual* for JEM 100 SX JEOL Transmission Electron Microscope, JEOL Ltd., Tokyo, Japan. pp. 1–5, Fig. 1.2.)

INTERACTION BETWEEN ELECTRON BEAM AND SPECIMEN

When an electron beam passes through a substance, due to coulombic interactions, the beam is scattered, the direction of movement of electrons is changed with or without partial energy loss.

When electrons of a given energy impinge on a substance, depending on the nature of atoms they encounter in the specimen, the following are generated (Figure 2.5).

1. transmitted electrons
2. backscattered (reflected) electrons
3. secondary electrons
4. Auger electrons
5. X-rays
6. cathodoluminescence

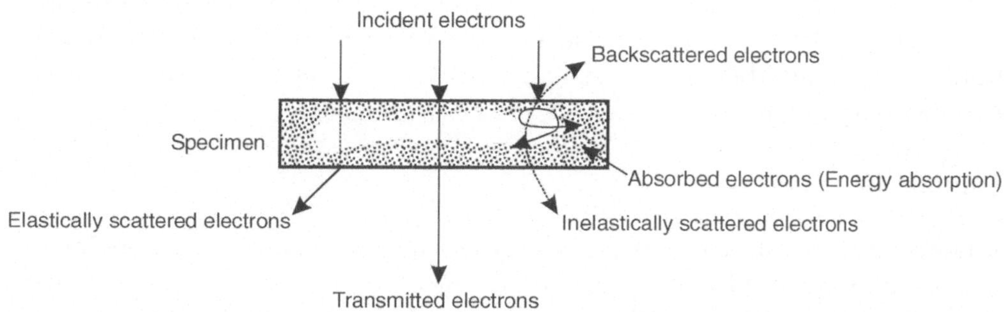

Figure 2.5 Interaction between the specimen and the electron beam. (*Source: TEM Instructions Manual* for JEM 100 SX JEOL Transmission Electron Microscope, JEOL Ltd., Tokyo, Japan. pp. 1–13, Fig. 1.10.)

1. Transmitted electrons As the name implies, these electrons are those which pass through the specimen and get transmitted. These can be classified into:

i. Transmitted electrons which pass through the specimen with little change of direction or energy.
ii. Elastically scattered electrons which undergo a directional change due to atomic collision but retain their energy.
iii. Inelastically scattered electrons which undergo both directional change and energy loss.

2. Backscattered electrons The electrons that are scattered back (reflected) are with an energy comparable to the incident beam. The tilt angle of the specimen and the atomic number of the elements constituting the specimen determine the quantity of these electrons.

3. Secondary electrons These electrons are those which are produced by the specimen as a result of absorption of the primary electrons from the source. Absorbed energy is the energy which the electrons have lost while passing through the specimen, and by which the atoms and the atomic nuclei get excited. Some of the absorbed energy is liberated as secondary electrons which arise from the specimen and different from the incident primary electrons.

The transmitted and scattered (elastically and inelastically) electrons contribute to the image formation and contrast in the transmission electron microscope (TEM), while the secondary electrons or the backscattered electrons contribute to the formation of the image in the scanning electron microscope.

DAMAGE AND CONTAMINATION OF THE SPECIMEN DUE TO ELECTRON BOMBARDMENT

Damage

The energy absorbed by the specimen due to the bombardment of electrons also causes heating, magnetization, ionization, secondary radiations, etc. Heat generation and ionization, in particular, damage the specimen but they can be reduced by increasing the energy of incident electrons.

Contamination

The contamination in the column mainly arises due to electron bombardment on the organic molecules released from sources like diffusion oil, vacuum grease, specimens, particularly, biological, etc. Even the fingerprints of the operator or the service engineer while opening the column will lead to contamination and also deterioration of the vacuum. The column is periodically opened for changing the filament and also, of course, for loading and removing the photographic films. The life of the filament usually ranges from 60 to 100 hours depending on the filament and operating conditions of the microscope. The sources and effects of contamination in the column are given in Figure 2.6.

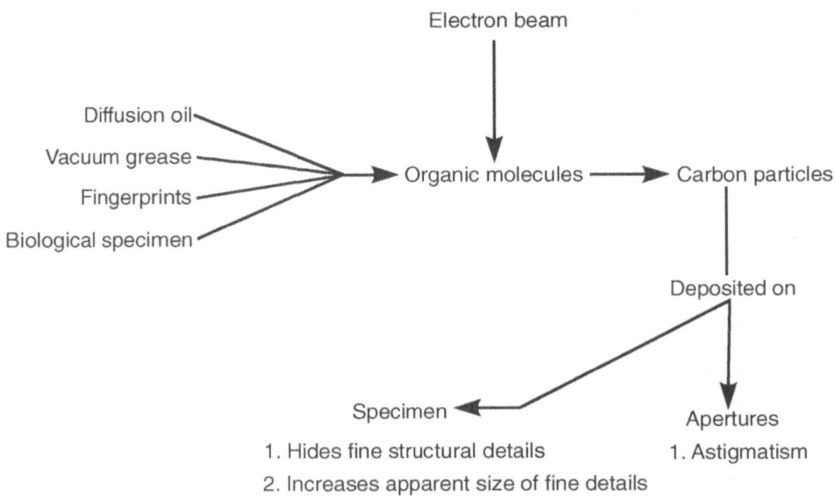

Figure 2.6 Contamination in TEM column

ADVANTAGES AND LIMITATIONS OF EM

Advantages

- The primary and most important advantage is its high magnification and high resolving power, the latter of course being limited by lens aberrations.
- Its lenses have continuously variable focal length over a wide range which simplifies operation and renders the instrument more flexible such that a great range of magnification (for example 100 to 2,00,000×) can be achieved without hindrance.
- Small aperture of illumination imposed by lens defects has a positive compensation of giving a great depth of focus facilitating stereophotography.
- Because of high magnification and resolution it finds wide application in various fields of science.
- In the field of biology, the cell organelles were identified only after the advent of electron microscope.
- EM has established itself as an indispensable tool in the diagnosis of certain diseases. Its uses in various facets of biology is discussed in Part III.

Limitations

- The chief inherent limitation is the poor penetration of the electron beam which necessitates only ultrathin specimens to be investigated.
- Another major disadvantage is maintenance of high vacuum.
- There is a need for high stability of electrical supplies, particularly, high voltage generation and stability to reduce chromatic aberration.
- Spherical aberration is high which acts as the chief limitation on resolution. Its correction is more difficult than in LM.
- Real electron microscope images do not carry any colour information.
- Electron microscopes are expensive to build and maintain.
- They are dynamic rather than static in their operation, requiring extremely stable high-voltage supplies, extremely stable currents to each electromagnetic coil/lens, continuously pumped high- or ultrahigh-vacuum systems, and a cooling water circulation system through the lenses and pumps.
- As they are very sensitive to vibration and external magnetic fields, microscopes aimed at high resolutions must be housed in buildings (sometimes underground) with special services.

- ✳ Some desktop low-voltage electron microscopes have TEM capabilities at very low voltages (around 5 kV) without stringent voltage supply, lens coil current, cooling water or vibration isolation requirements and as such are much less expensive and far easier to install and maintain, but do not have the same ultrahigh (atomic scale) resolution capabilities as the larger instruments.
- ✳ The specimen has to be dried before introducing into vacuum which itself may produce artifacts as wet, live biological samples cannot be viewed directly in the EM because of high vacuum. One exception is the ESEM, (environmental scanning electron microscope), which allows hydrated samples to be viewed in a low-pressure, wet environment.
- ✳ The biological samples require an elaborate processing both for SEM and TEM. These processing techniques create various artifacts. Even the conductive coating for SEM has the potential to disturb delicate samples.

In spite of all the above limitations, and also the advancements in molecular techniques, the ultrastructural study has its own charm and advantages. There is no substitute for EM in fine structural morphology, and technological innovations have revolutionized to the level of atomic resolution with this technique.

3

PARTS OF THE TRANSMISSION ELECTRON MICROSCOPE

COMPONENTS OF TEM

The following are the main parts of TEM.

- Electron gun with cathode and anode
- Condenser lenses
- Specimen stage
- Objective lens
- Magnifying lenses (Intermediate and projector lenses)
- Fluorescent screen
- Camera chamber

Apart from the above, supporting systems like the following also form the essential components of the microscope.

i. vacuum system,

ii. high-voltage generating system,

iii. cold water circulating system,

iv. electrical and electronics system, and

v. air compressor which opens and closes the valves at various positions in the column

Figure 3.1 shows the external view of the TEM column, while Figure 3.2 depicts the cross-sectional view of the same.

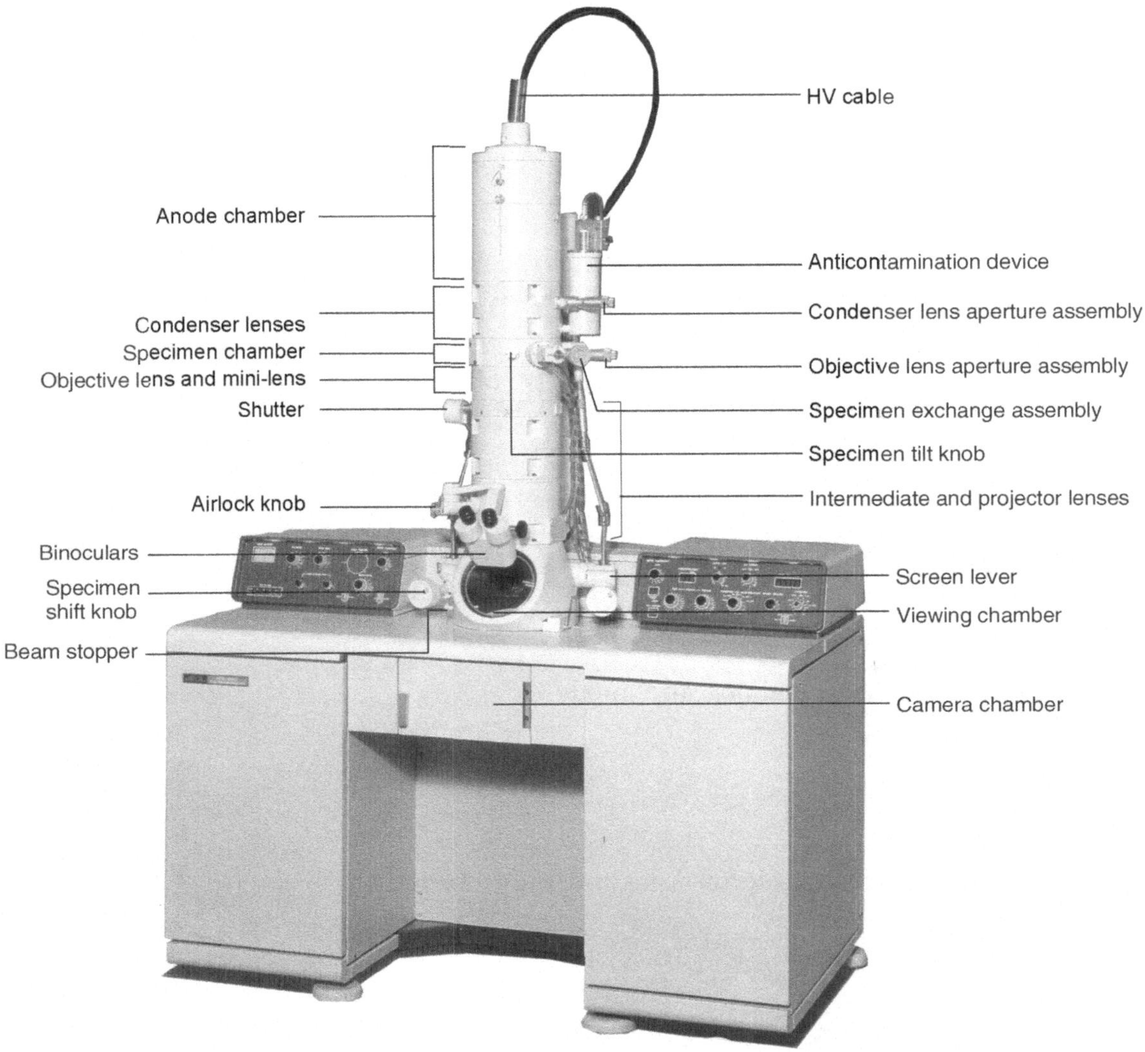

Figure 3.1 External view of the column of a TEM. (*Source: TEM Instructions Manual* for JEM 100 SX JEOL Transmission Electron Microscope, JEOL Ltd., Tokyo, Japan. p. 2.)

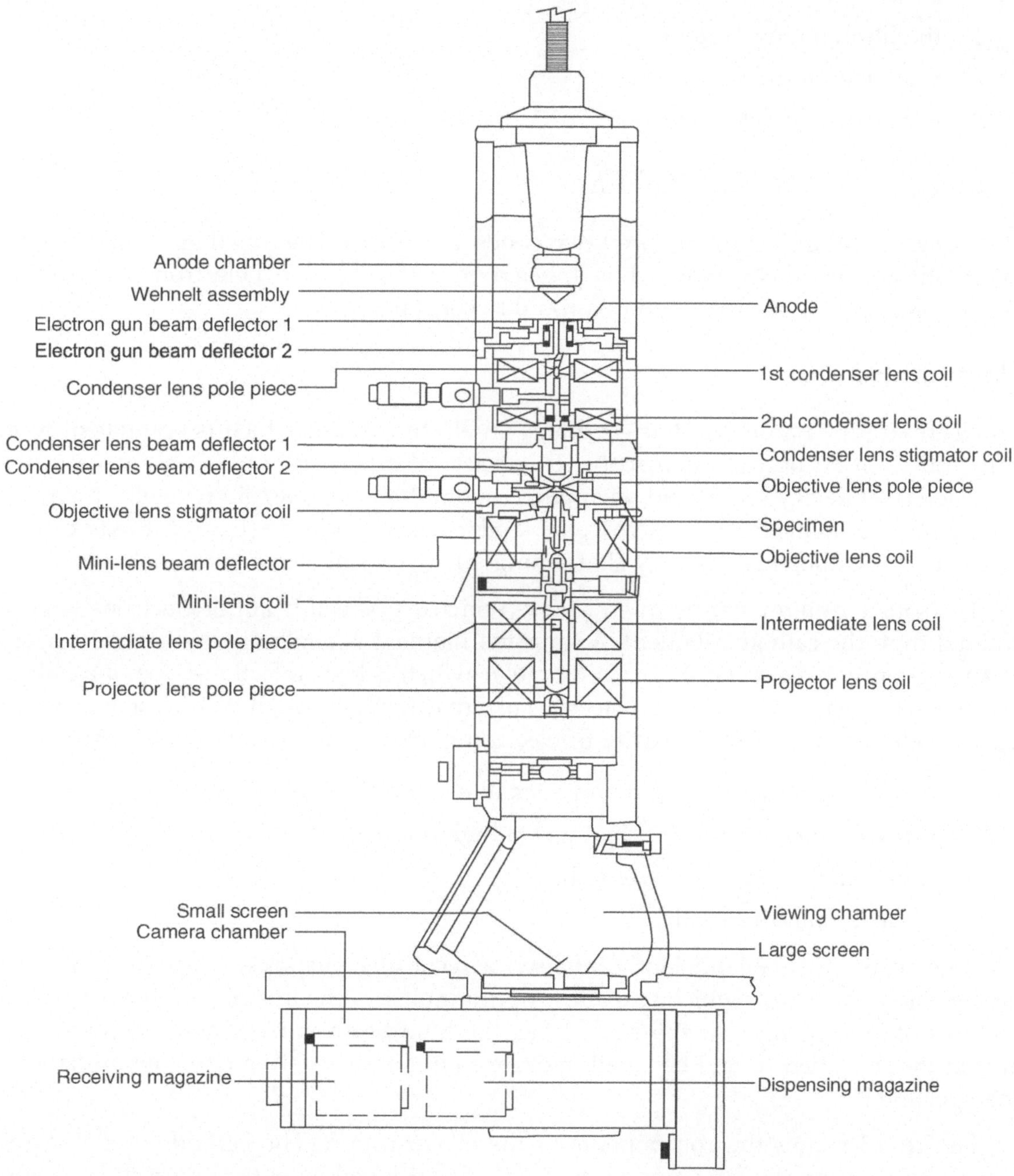

Figure 3.2 Cross-sectional view of TEM. (*Source: TEM Instructions Manual* for JEM 100 SX JEOL Transmission Electron Microscope, JEOL Ltd., Tokyo, Japan. p. 3.)

Functionally, the EM consists of three major components. They are

1. the illuminating system
2. the image forming system
3. the image recording system

THE ILLUMINATING SYSTEM

This consists of the electron gun and two condenser lenses. The function of this system is to provide an electron beam which is directed on to the object. An electron beam is a fine flow of electrons in one direction at a roughly constant velocity.

Electron Gun

An electron gun is a device that generates an electron beam which is equivalent to the light source of an optical microscope. It consists of a hot cathode (filament), a control electrode (Wehnelt cylinder) and an anode. Though tungsten hairpin type electron gun is generally used in an electron microscope, lanthanum hexaboride (LaB_6) cathode can also be used. This cathode is made by sharpening a LaB_6 crystal into a conical shape.

The source of illumination in an electron microscope is a beam of electrons which is emitted from the cathode filament. A tungsten filament is generally used which consists of a tungsten wire of about 0.12 mm diameter which is bent into the shape of a hairpin, and heated to about 2800°C by an electric current directly through it. Tungsten is used as the cathode material of the majority of electron guns for the following reasons.

1. It has a high melting point and a small evaporation rate.
2. It does not react with residual gas at high temperature.
3. It has high mechanical strength.
4. It can be handled easily.

The beam must be sufficiently intense for recording purposes and should converge on the object at an angle which will allow optimum use of the imaging system to produce pictures of high quality. At the same time, the total beam current and the illuminated area at the specimen should be small in order to minimize heating and contamination of the specimen.

Figure 3.3 shows the components of the electron gun. The cathode is a V-shaped hairpin (F) of tungsten surrounded by a cylindrical metal shield (S) pierced by a small aperture opposite the point of the V. Anode (A) is a similar metal cylinder with an aperture in the face exposed to the cathode shield. The shield is referred to as Wehnelt cylinder. It is an electron beam control electrode which is made of cylindrically shaped stainless steel with 1 to 2 mm diameter hole for an electron beam to pass through it. It functions to control and stabilize the electron beam emitted from the cathode when a bias voltage is applied to it. It is sometimes called a "grid".

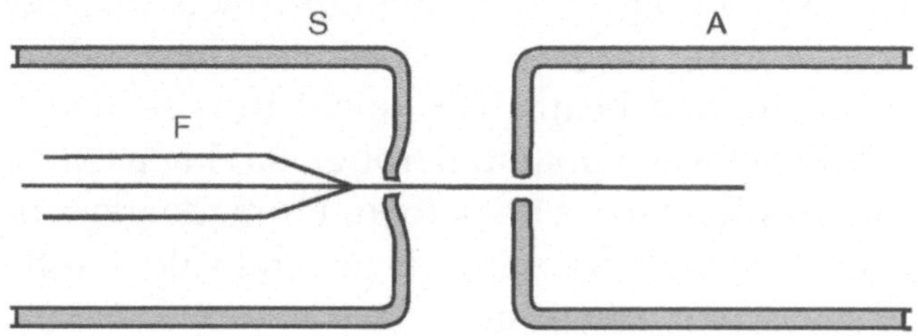

Figure 3.3 Gun assembly. (*Source*: V.E. Cosslett. (1951). *Practical Electron Microscopy.* Butterworths Scientific Publications. p.71, Fig. 36.)

The characteristics of the source are determined by the shape and position of F, G and A and also on the relation between their electric potentials.

The filament wire is about 0.1 mm in diameter and is heated directly by a current of a few amperes which is called **beam current**. When a high voltage of the order of kilovolts (40 to 400 or even more) is applied to the cathode, electrons are emitted from it. This voltage applied is called **accelerating voltage** which is defined as the voltage applied to accelerate the electrons emitted from a cathode in order to illuminate a specimen. The emitted electrons are attracted through the shield aperture towards the anode by the very strong electric field existing between the two electrodes. The anode is an accelerating electrode which gives the specified energy to the thermoelectrons that are emitted from the filament. In practice, it is convenient to maintain the anode and microscope body at earth and the cathode at high negative potential, well out of way of the operator. The beam emerging from the anode is focused on to the specimen by the condenser lens which is placed about midway between the cathode and specimen.

A range of accelerating voltages is fixed for each microscope. Though the lowest for TEM is 40 kV, in many microscopes, particularly those used for biological materials, a maximum of 400 kV is easily attainable now. Million-volt electron microscopes are also in existence. Generally for biological materials, lower voltages of 60 to 80 kV are preferred for the following reasons.

1. The biological materials cannot usually withstand the high intense beam generated by the high accelerating voltages.
2. The instrumental contrast (i.e., contrast introduced by virtue of construction of the instrument) will be low at higher voltages.

But at lower voltages, the energy of the emitted electrons and also their penetrating power are low, and thus the specimen thickness becomes critical. Ultrathin sections are mandatory though some authors have used one-micron semi-thin sections for TEM employing higher voltages, 125 kV, for biological materials.

The electron gun constitutes an electron lens of the immersion type, i.e., the lens action is dependent on the electron source being immersed to a greater or lesser degree, in the field from the accelerating electrode, the anode. The anode aperture by itself has a diverging effect on the emitted beam. The shield thus seems to have an important function, as its aperture introduces a potential trough having positive focusing properties even when it is at cathode potential. As its potential is made more negative, the focusing action becomes stronger in accordance with the general rule that the slower an electron beam, the greater is the effect of a field upon it.

When a negative potential is applied to the shield, it becomes a **biased gun**. The bias potential required is of the order of hundreds of volts, for the usual accelerating voltage of 50 to 100 kV. An increase in the distance between filament tip and shield aperture will lead to a decrease in bias needed. The negative field which it sets up may be pictured as a throttle, controlling the extent to which the positive field of the anode can penetrate through the aperture towards the filament to extract electrons from it. It acts in the same way as the control grid in a normal triode valve.

The biased electron gun gives a greater intensity of illumination and a smaller spot at the object, allowing, more favourable electron optical imaging condition, and at the same time making the adjustment of the filament position less critical. On the other hand, the greater beam intensity causes heating and contamination of the object and of all apertures to be much greater and the filament life is shorter than in the unbiased gun. These disadvantages can be minimized by using a variable bias control.

When a high negative potential from the accelerating power source is applied to the filament and a current from the heating power source is sent to the filament, thermionic electrons are accelerated by the electric potential difference between the filament and anode. At the same time, voltage drop by the bias resistance supplies a bias potential (self bias) for the Wehnelt. The electrons are concentrated to a point and emitted from that point. This electron-converging point is called the **crossover point** and is taken as the size of the source (Figure 3.4).

When the filament is heated gradually, the quantity of the emitted electrons reaches its limit (saturation state) in accordance with space charge. This heat condition must be secured to provide the most stable source. Overheating shortens the filament service life. Thus, the life of the filament may be prolonged if the filament is kept at the temperature needed for adequate brightness with maximum bias potential.

The character of illumination depends critically on the height of the filament. Small errors of adjustment can be compensated by variation of the bias voltage but the range available is not great. If the filament is badly displaced, it gives rise to double images. An efficient gun creates minimum of X-rays and delivers a beam of maximum intensity on to the object.

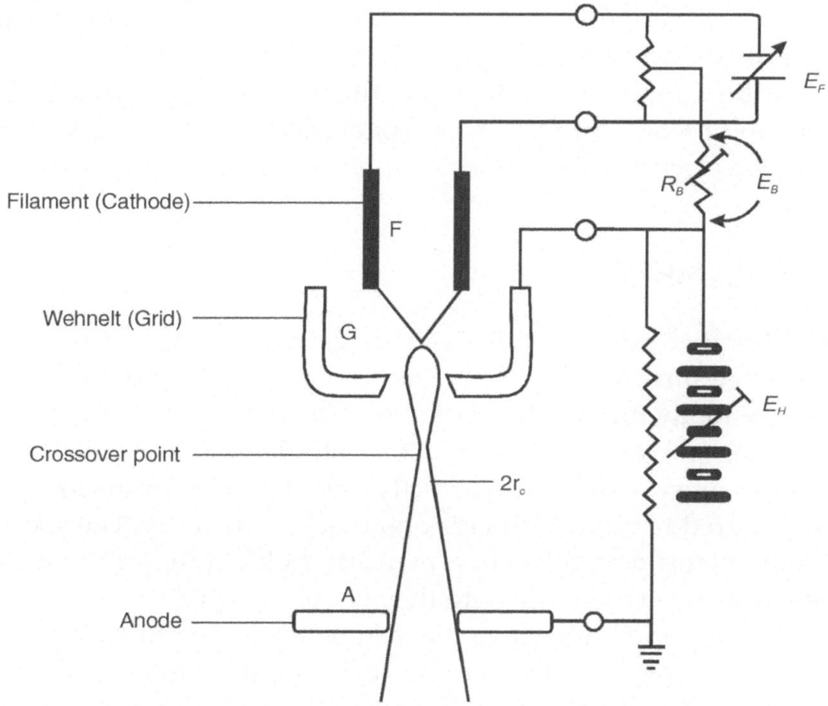

E_H—accelerating voltage applied to the filament

E_F—current from the heating power source sent to the filament

R_B—bias resistance which supplies a bias potential E_B for the Wehnelt

Figure 3.4 Structure of electron gun. (*Source*: Cosslett, V.E. (1951). *Practical Electron Microscopy.* Butterworths Scientific Publications. p.74, Fig. 39.)

Condenser Lens

This is an electromagnetic lens located just below the electron gun in order to make the electron beam emitted from the electron gun into a fine beam. It plays an important role in controlling the electron beam intensity and also the diameter of the electron probe.

A condenser lens is required to converge the electron beam from the electron gun and to illuminate the specimen as desired. Varying the condenser power very radically alters the aperture and intensity of illumination and the spot size. Since the field of view in high-magnification microscopy is limited to a very small area (approximately 1 μm^2 at 1,00,000×), the illuminated area must be small. If a wide area is illuminated, adverse effects will result, e.g., a temperature rise in the specimen.

To achieve a narrow illumination area, a double condenser lens is used. The focus is generally formed near the specimen plane to illuminate the specimen evenly. Shortening the focal length of condenser 1 (CL 1) (the first condenser lens in a double condenser system), makes the source image smaller, thus narrowing the illumination area. By changing the focal length of the second condenser lens (CL 2), the illuminated area will expand and the electron beam intensity will become low, thereby making the image darker.

SPECIMEN CHAMBER

The specimen chamber must contain a stage capable of easy and quick specimen exchange. This stage must hold a specimen or specimen holder in a stable condition and must move smoothly for the selection of field of view. The external vibration which affects the resolving power adversely must be eliminated. The microfield of view at high magnification must be selected correctly and the mechanism for quick exchange of specimen is required to improve image observation efficiency. The specimen suitably prepared and mounted is placed in a non-magnetic holder (Figure 3.5) which is inserted directly into the objective pole piece with the help of the specimen exchange rod shown in Figure 3.6. It is supported in a stage that can be translated in two directions at right angles by exterior controls in order to bring different areas into the field of view. At a total magnification of 10,000× the field of view will only be 5 micron square. The stage motion usually allows the searching of at least several spaces in each direction. The controls employ a fine screw and some form of reducing mechanisms to ensure that the motion is adequately fine. Also the control should be free of backlash and the two directions of motion are constant without side slip. The object holder should sit firmly in the stage and the whole assembly should not be subjected to vibration. Absence of vibration in the object seating is mainly a matter of design but can be further achieved by firmly pressing the holder into position when inserting it in the stage. This good contact between the specimen holder and its stage helps in rapid conduction of the heat absorbed by the object from the electron beam so as to minimize temperature rise in the specimen. Care should be taken to avoid settling of dust particles or object debris on the bearing surfaces of the holder or on which it rests in the microscope. To reduce specimen contamination,

- a good vacuum condition is essential, and
- a cooling trap to hold liquid nitrogen is installed near the specimen which is often referred to as **anti-contamination device**.

The specimen exchange device usually contains an airlock mechanism such that the specimen can be removed or introduced without disturbing the column vacuum. In the airlock method, the sample is exchanged through a small chamber (pre-evacuation chamber) which is separated from the overall electron optical column by means of an isolation valve. Air is introduced only to the pre-evacuation chamber when exchanging the specimen, hence the inside of the electron optical column is not exposed to the atmosphere.

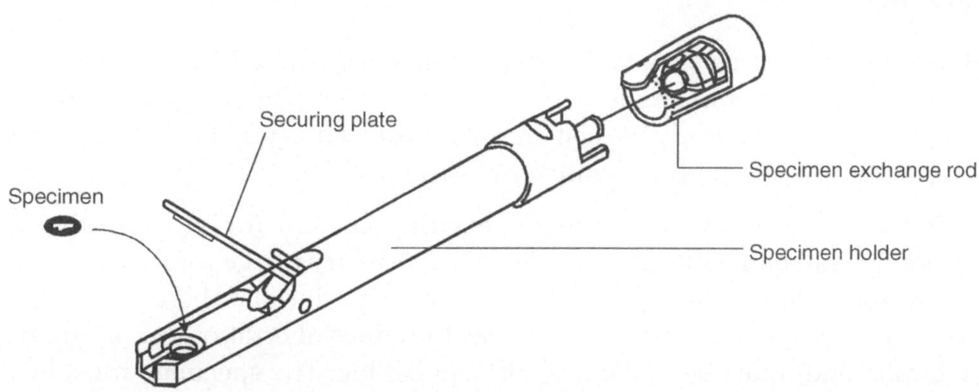

Figure 3.5 Specimen holder. (*Source: TEM Instructions Manual* for JEM 100 SX JEOL Transmission Electron Microscope, JEOL Ltd., Tokyo, Japan. p.5.11, Fig. 5.8.)

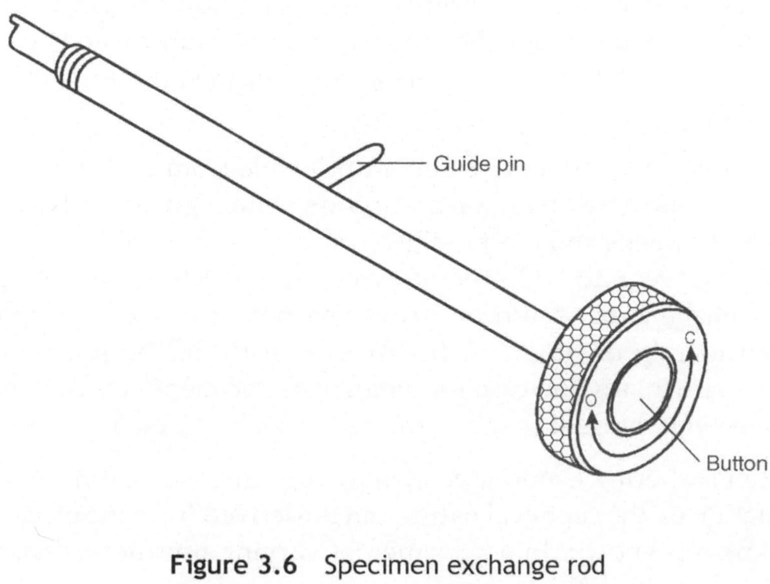

Figure 3.6 Specimen exchange rod

IMAGE-FORMING LENS SYSTEM

This is composed of the following lenses:

- Objective lens
- Intermediate lens
- Projector lens

Objective lens

In a high magnification lens system a first-stage image is formed by the objective lens (of small aberration) located immediately under the specimen. The image is further enlarged by the intermediate lens and the last image is formed on a fluorescent screen or a photographic film by the high magnification projector lens.

Focusing is achieved by changing the exciting current for the objective lens and magnification by that of the intermediate lens. The primary image formed by the objective is affected by aberrations arising from defects of the objective and therefore the conditions of construction and operation are so chosen as to reduce aberrations to a minimum. The objective focal length must be made as short as possible. The specimen must be situated close to the focal plane of the objective in order to obtain a high magnification at the given image distance which is fixed by the column between this lens and the other magnifying lenses.

A physical aperture introduced near the back focal plane of the objective lens will eliminate the widely scattered rays, clearing up the background and thus greatly increasing the clarity and contrast of the image. The resolution also is improved because the effect of spherical aberration is greatly reduced. The location of the aperture is best in the centre of the objective pole gap.

Thus the image-forming process differs in principle from that in optical microscope. In the optical microscope the differential absorption of light gives rise to visible differences in various parts of the image and it is possible to apply stains to the specimen to accentuate different structures as required. On the other hand, in electron microscopy, the electron density over the image depends on the scattering power of the corresponding parts of the object which mainly depends on the mass density of the matter composing the specimen. This is particularly the case for biological specimens where the atomic number of most of the elements present is much the same (O, N, C, etc.).

The amount of scattering is not dependent on the chemical nature of the specimen so that little information on the chemical nature can be derived from the electron micrographs even if the thickness is known. In a specimen of varying thickness, dense regions in the picture may represent either thick parts or presence of elements of high atomic number.

Intermediate and Projector Lenses (Magnifying Lenses)

The intermediate image formed by the objective lens is further magnified by the intermediate and projector lenses. This system consisting of three lenses (including the objective) offering three-stage magnification has the following advantages:

- compactness of the microscope column
- high range of magnification
- freedom from distortion

The projector lens is so called because it projects the final magnified image on the screen. The effect of aberrations in the projector lens is very different and much less serious than in the objective. Projector pole piece may be made symmetrical, since the specimen does not have to be accommodated (contrary to the objective lens) with the result that it can be made of shorter focal length. This is desirable to minimize **distortion effects**. Thus the main magnification takes place in the projector stage.

RECORDING THE IMAGE

This system consists of the following parts:

- Viewing chamber and camera chamber (Figure 3.7)
- Camera chamber (Figures 3.7 and 3.8)

Viewing Chamber

Since the images formed by an electron beam cannot be directly viewed by the naked eye, they must be converted into light images for selection of the field of view and focusing. (The high vacuum in the column also does not allow direct viewing). Therefore, the image is made to fall on a fluorescent screen which can be viewed through a glass window. The screen is coated with a fluorescent material which can convert electron energy into visible light energy. The most useful fluorescent materials are mixtures of zinc and cadmium sulphides, coated on a backing plate with a fixative such as gum arabic or collodion. They are usually made for high light output and good contrast rather than for high resolution. As they are to be viewed directly by the eye their spectral response to the beam is arranged to be of the colour to which the eye is most sensitive—yellow or green—rather than blue. For easy observation of images on the fluorescent screen, the light-emitting efficiency and resolving power of the screen must be superior. The images on the screen can be viewed through a glass window. For better and easier observation, a binocular with a clear field of view are installed to enlarge the image formed on the screen, especially while focusing for photography. The magnification offered by this binocular may be 3 to 10 times which is not limited by the grain size of the screen.

Electron microscopes are fitted with a screen of high contrast and efficiency and with a resolution well beyond that of the unaided eye. The screen may deteriorate slowly on prolonged exposure to the electron beam and also due to the mechanical shock of raising it for exposing the photographic plate. In such cases, it is possible to either re-coat or replace the screen.

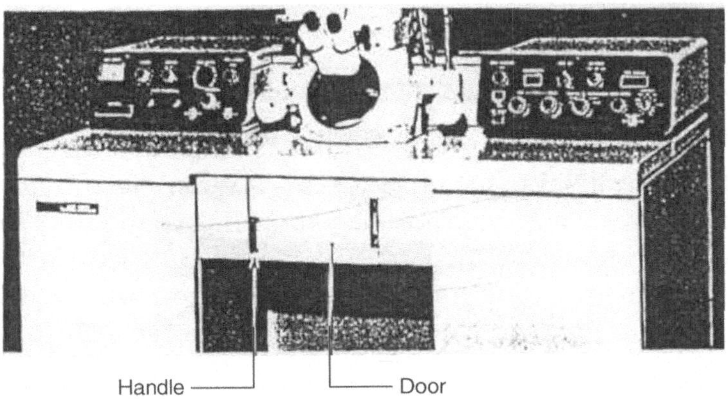

Figure 3.7 Viewing chamber and camera chamber. (*Source: TEM Instructions Manual* for JEM 100 SX JEOL Transmission Electron Microscope, JEOL Ltd., Tokyo, Japan. pp. 5–15, Fig. 5.11.)

Camera Chamber

The photographic chamber or the camera chamber is situated just below the screen. By lifting the screen, the image is allowed to fall on the photographic film directly. The best way to record images for permanent storage is through direct photography using high-resolution films. Since electrons require vacuum for their travel, this camera chamber is also a part of all electron microscopes, unlike the light microscopes where photography can be optional and the photographic attachment is external.

The photographic film in an electron microscope should record the image with a sufficiently short exposure time to yield adequate contrast without loss of details. Slow-speed and high-contrast emulsions are found to be most suitable for electron microscopy. Films are most subject to electrostatic charging when exposed to the beam leading to movement of the image and to small streaks on the image caused by discharges. The shutter to lift the screen should be carefully handled to avoid vibrations. It must be opened smoothly without shaking the apparatus and similarly closed quickly to minimize uneven exposure to the film. The camera chamber contains two magazines one for loading the films and the other for storing the exposed films till time of removal for further processing of the exposed films. Apart from this, in most microscopes, the provision for using roll films is also available.

To get the best photographic record, it is necessary to ensure proper focusing and exposure followed by correct processing of the negative. The exposed negative has to be developed, fixed and finally reproduced as a print for the recording process to complete. The actual processing does not differ essentially from ordinary photographic methods.

In recent microscopes, of course, a CCD camera attached to the screen facilitates digital display and storing of the images.

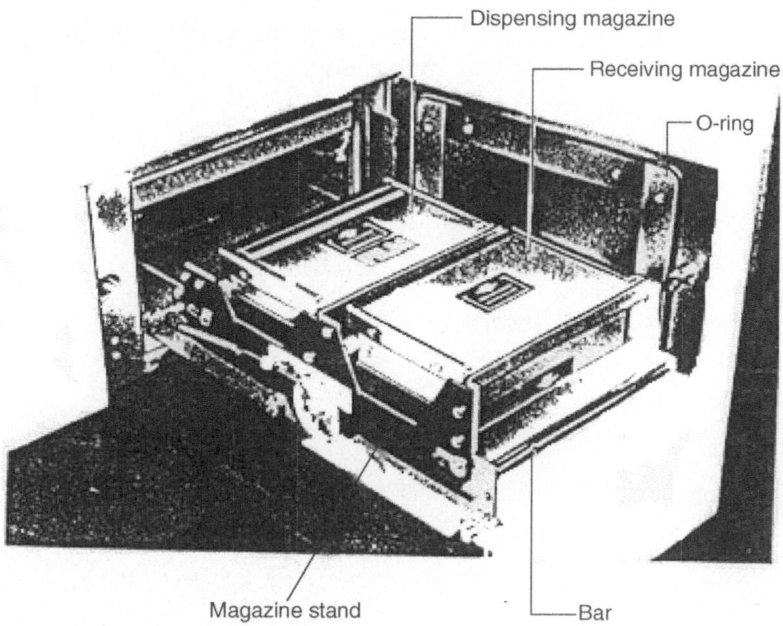

Figure 3.8 Camera chamber. (*Source: TEM Instructions Manual* for JEM 100 SX JEOL Transmission Electron Microscope, JEOL Ltd., Tokyo, Japan. pp. 5–15, Fig. 5.12.)

VACUUM SYSTEM

Electrons are greatly influenced by the medium through which they must pass. Accordingly, it is desirable that high vacuum be maintained in the column. A poor vacuum may cause high tension electrical discharge, specimen contamination, contrast reduction, oxidation of the electron source and damage to the specimen itself. To achieve the required vacuum, electron microscopes are usually evacuated with oil rotary pumps and oil diffusion pumps. Thermionic pumps are also in use in some microscopes.

COOLING WATER CIRCULATION

The oil diffusion needs circulation of cold water around the oil tank in order to prevent overheating and consequent cracking of the diffusion oil. Similarly, the electron lenses also require circulating cold water. The focal lengths of these lenses are varied by varying the current applied to them. A part of this applied current will be dissipated as heat thus raising the temperature of the lenses. Since the electrical and magnetic properties, in general, are highly sensitive to temperatures, the beam disappears when the temperature of the lenses exceeds the tolerance limits. To avoid such a situation, cold water is circulated around the lenses also.

Permanent magnet lenses are very stable, and do not need any cooling. (Cooling is required with electromagnetic lenses, because they are excited by high current. So heat produced at the lens must be removed by cooling.)

4

OPERATION AND MAINTENANCE OF TEM

The biological materials invariably consist of substances made up of light atoms having approximately the same density as the embedding media and supporting films and therefore, contrast is an intrinsic difficulty for the biologists. Thus, the investigator working with ultrathin sections is involved in a never-ending struggle to improve contrast. Any given specimen has a certain potential contrast which can only be partially realized by the most skillful microscopy and modern photography.

In an electron micrograph (the photograph taken in the electron microscope is known as electron micrograph), contrast depends, to a large extent, inversely upon the amount of "background fog" that the photographic film records. Fogging results from scattered electrons, one way or another reaching the photographic emulsion. These electrons travel at varying speeds, and therefore are not focusable. They do not particularly traverse the refined optical pathway but many of them spray the photographic film randomly after having bounced off the side walls of the microscope column. It is for this reason that steps are taken to minimize the number of scattered electrons which finally reach the viewing screen and photographic emulsion. If apertures of suitably small size are placed at critical levels along the microscope column, the useful beam can go through these tiny holes while scattered electrons, even though they may be nearby, will hit the aperture plate and be effectively removed. Of course the resolution of the photographic emulsion is better than the screen resolution. Therefore features are clearer in an electron micrograph than on the fluorescent screen.

Specimens with great intrinsic contrast like particulate materials and surface replicas which have been shadowed with heavy metals can be handled with unapertured microscopes. Unapertured microscopes are also used for the direct observation of dense particulate matter as in dealing with pigments, smokes, etc. Here there is no need to prepare ultrathin sections and thus processing and viewing becomes much easier.

Apertures are mainly used for objective and condenser lenses. Objective aperture is used mainly to increase contrast in bright field imaging. But astigmatism may arise when the apertures get contaminated in the column and thus astigmatism correction has to be periodically done using stigmator. Condenser lens apertures partly increase the contrast by removing the off-axis electron beam. "Space charge" effects occur if electrons are crowded too closely together near the electron emitter at the gun. This can happen due to generation of large number of thermions (electrons) near the surface of the electron emitter. There will be little or no emission of electrons due to this repulsive force. To overcome the space charge effect, a small DC potential is applied near the filament. This is analogous to the grid in a triode valve. By varying this potential (grid bias), the amount of electrons leaving the filament can be controlled.

The handling and maintenance of each type of microscope is given in the instruction manuals. The biologists working with tissue specimens have special problems which need some understanding and detailed consideration. Therefore, here we discuss commonplace problems of operation and maintenance without reference to particular types of instruments.

OPERATION STEPS

The basic operation steps are:

- Supply cooling water.
- Start the instrument by switching on the start up button.
- Wait for 20–25 minutes for high vacuum.
- Select high voltage/accelerating voltage
- Insert specimen.
- Increase beam current from '0' to filament saturation point.
- Select the field and screen the specimen with the image on the screen.
- Focus and take photographs (Use of binoculars, if attached to the EM, enables the image to be observed with greater clarity).

In modern microscopes, with autofocus and other auto options, operation might be simpler. But even then, the electron microscopist should have sufficient knowledge about the significance of the options so that he/she can take them into consideration while improving the quality of the micrographs as well as their interpretations.

HV SELECTION

The selection of HV depends on which is more important, image contrast or image resolution and brightness. A higher voltage will give better resolution and brightness at the sacrifice of contrast and vice versa in the case of a lower voltage. A low voltage is especially useful in examining specimens susceptible to damage by electron beam irradiation like biological materials and in generating contrast. But at very low voltages the beam intensity is too low to view the features. So for biological materials, 60–80 kV is generally used.

Contrast and High Voltage

The contrast is low in thin sections. As the quality of the objective lens improves, the contrast of the image decreases, since at least some of the contrast arises from lens defects. From thin specimens like biological materials which predominantly consist of carbon, the majority of electrons affected in passage through the object suffer inelastic as opposed to elastic scatter (i.e., they lose a few volts of energy by collision with orbital electrons in the atoms, generally with only a small deflection, rather than a relatively large deflection by collision with atomic nucleus). Thus a physical aperture in the objective lens will not stop more than a small proportion of the electrons affected by the object. Many of the beam electrons affected by passing through the object, and therefore carrying information about it, are thus allowed to reach the image diffusely spread by the chromatic aberrations of the lenses. Not only is the information lost, but also the contrast is decreased.

But, low voltage operation increases contrast because low accelerating voltage results in greater proportion of elastic scattering by a given object. Such scattering would enable more electrons to be stopped out by the objective aperture, and hence it increases the contrast of the image. But, the resolving power will be decreased. This is true in bright field imaging. In dark field imaging, the direct beam is stopped by the objective lens aperture and only the scattered beam is allowed to reach the screen.

Beam Current

Beam current is increased from '0'. As the beam current increases, the intensity of the image increases and at one position further increase in beam current does not alter the intensity. This is the saturation point/position. The life of the filament will be considerably decreased if the beam current is increased beyond the saturation point. This position is fixed while aligning the beam initially.

ALIGNMENT OF THE BEAM

There are several reasons for column alignment, some practical and some theoretical. An important theoretical reason is that both magnetic and electrostatic lenses are fundamentally spherical lenses. Inevitably such lenses distort. The distortion becomes increasingly severe as one moves away from the central axis, thus giving rise to spherical aberration. As we have already seen, this cannot easily be corrected as with optical lenses. But this aberration is particularly noticeable only at very low magnifications and hence are of concern only while making accurate measurements.

Electrical stability is also an important factor. A slight ripple of the accelerating voltage produces electrons moving at different speeds. This corresponds to using light of different wavelengths. It leads to the electron equivalent of chromatic aberration. The aberrant wavelengths focus at different levels and tend to blur the image, thus destroying resolution. Furthermore, electrons moving through an electromagnetic or static lens move in spiral paths, and not as light waves do through glass lenses. Thus if there is any change in the accelerating voltage, or if lens currents change, the image turns about the optical axis. Naturally such a movement, even though slight, would tend to destroy resolution. This effect increases rapidly as one moves away from the optical axis. If there is a reasonable degree of electrical stability in the microscope circuits, this effect is not important at low and moderate magnifications but is of serious concern while working at high magnifications. Typically the high voltage stability requirement is in the order of 1 ppm.

It is necessary to have the tip of the filament coinciding with the optic axis of the TEM column. Whenever a new filament is installed, the beam has to be made passing through the optic axis. All lenses are pre-aligned to the optic axis and hence the electron beam generated by the filament has to be coincided with this axis. Gun alignment coils are used to shift and tilt the electron beam in such a way that the beam will be at the optic axis. Alignment of practical importance is the alignment of the condenser aperture and the intermediate lens. In the aligned microscope one can change magnifications, vary illumination, change the spot size, make large focus changes, and so forth, all without having to readjust the source of illumination much or re-centre the image.

OBJECTIVE ALIGNMENT

The objective lens is aligned with the projector lens. Two types of alignment are actually possible. One is called the "current alignment" and the other "voltage alignment". Because electrons spiral through a magnetic lens, if the focusing current is varied, the image is seen to spiral about an optical axis. For current alignment the objective lens is moved about on top of the projector lens until this optical axis coincides with the centre of the viewing screen. This axis will not necessarily correspond well with the axis observed when electrons of differing speeds pass through the lens. This is called "voltage alignment". This is of great importance for high resolution. This is achieved by deliberately varying the acceleration voltage and finding the centre of image rotation.

After the objective and projector lenses have been aligned, the intermediate lens is added functionally to the system. In most microscopes, this lens is used as reducing lens to extend the lower ranges of magnification. This alignment is again done for operational convenience. The centre of rotation of the image is observed as the magnification control is turned to its lower ranges. This indicates the optical axis of the lens, and the lens is moved about until this coincides with the centre of the viewing screen (current alignment). If the intermediate lens is used for both low and high magnifications, one alignment will not do well for both and hence alignment is done for the magnifications mostly used.

ALIGNMENT OF THE ILLUMINATING SYSTEM

There are at least three elements to this system. The filament is the actual source of electrons. The electrons are given acceleration and direction by the applied voltage and also the shape and positioning of the hole in the anode cap. Then the condenser lens allows this beam to be modified convergently or divergently. When the beam leaves the condenser lens, its axis must nearly coincide with the optical axis of the objective lens.

Two basic steps are necessary to align the illuminating system. The first is the source alignment, which is concerned with the alignment of filament tip, anode and optical axis of the objective lens. The second step is the alignment of the condenser lens. In essence, the optical axis of the source, which is already parallel with that of the objective lens, is made to coincide with the latter. In the unaligned instrument, the illumination sweeps, perhaps greatly, as the intensity (condenser) control is varied. In the aligned instrument, the illumination expands symmetrically on either side of the cross-over image of the source. This is a condition of much practical convenience even though not of great importance in relation to resolution.

To achieve source alignment, the gun is tilted so that the beam moves in a direction diametrically opposite to the asymmetric cloud. When the image sweeps off the viewing screen, it is returned by gun translation. In practice, the source alignment should be done at low magnification, such that as the image of the source is deflected by tilting, its path can be followed as long a way as possible. Also the image of the entire source can be seen on the viewing screen. The source alignment is not very critical with a biased gun, and if it is obtained originally with a well-centred filament, subsequent replacement of the filament should not require much readjustment. Of course, the tilt once established must not be altered. Centring of the illumination between alignments should be done only by changing the gun translation.

FILAMENT REPLACEMENT AND ALIGNMENT

- The tip of the filament must be mechanically centred with reasonable accuracy. If centring is done very carefully each time, there will be little need for subsequent optical alignment.
- The distance of the filament tip from the anode is critical in a biased gun. If it is too close, excessive beam currents will be drawn which are apt to overload seriously the circuits of the microscope and destroy resolution. If the distance is too far, beam intensity will be weak.
- After the gun is reassembled, some translational adjustments of the beam may be necessary, so also the gun tilt adjustment.

Every time the gun is disassembled to change a filament, it is advisable to clean the anode cap also. As insulating contamination films build up around the aperture of this cap, bias relations are altered and beam current soon increases to dangerous levels. The insulating properties of these establish aberrant, asymmetrical, electrostatic fields which distort the beam. The insulation is apt to break down from time to time allowing discharge. This causes flicker, which is often rhythmical and can become severe. If the objective aperture is asymmetrically contaminated, alterations due to beam flicker will cause image movements. The critical region to be cleaned in the anode cap is the region around the aperture, particularly the side walls of the aperture. An unduly short filament life accompanied by an obviously rapid contamination of the anode cap suggests a small vacuum leak in the gun area.

ALIGNMENT OF THE CONDENSER LENS APERTURE

This is done by varying the condenser control through cross over. The illumination will sweep until alignment is achieved when it will expand symmetrically. Alignment is achieved by successive approximation alternating the two axes of corrective movement.

OBJECTIVE LENS COMPENSATION

Astigmatism resulting from aperture contamination can appear rapidly and deteriorate images seriously. It is probably better to check astigmatism everyday before starting to use the EM. The correction for the condenser and objective lens astigmatism can be carried out using corresponding stigmators.

Astigmatism can be seen most readily in diffraction images (Fresnel fringes) at edges. Small circular holes in fairly thick carbon films (Holey carbon films) is one of the best choices for objective astigmatism correction. These are supplied as standard calibration grids by various manufacturers. Compensation should be performed at the highest magnification that one expects to use and should be checked whether adequate for the whole range.

Figure 4.1 shows a through-focus series of micrographs of a small hole in a film at high magnification. (a) shows the diffraction pattern of an overfocused image with a characteristic apparent light band outlined by black; (b) is very close to true focus with just a suggestion of overfocusing; (c) also is very near true focus but shows a slight rim of white suggestive of underfocusing. The bright band is more conspicuous in (d) which is definitely underfocused. An edge such as this has much greater apparent contrast than that of (e) and is possible to be mistaken for true focus when this is not so. (f) the bright diffraction band of the decidedly underfocused edge is conspicuous. These diffraction bands are known as Fresnel fringes.

The overfocused diffraction image has a conspicuous light band outlined by a dark line which is fairly easy to see. It is symmetrical in all axes. The below focus or underfocus image looks different. The diffraction band is bright, and is not particularly visible since it does not contrast greatly with the illuminated background. But, it makes the apparent edge of the particle or the rim of the hole appear very sharply defined with great contrast. The in-focus edge in which the true edge of the hole is seen will not appear with as much contrast as the latter. The edge will be sharp, but relatively indistinct. When astigmatism exists, the focal plane is different for two axes. Thus if the poles of one axis are in focus, the poles of the other will show diffraction images corresponding to either the above or below focus pattern.

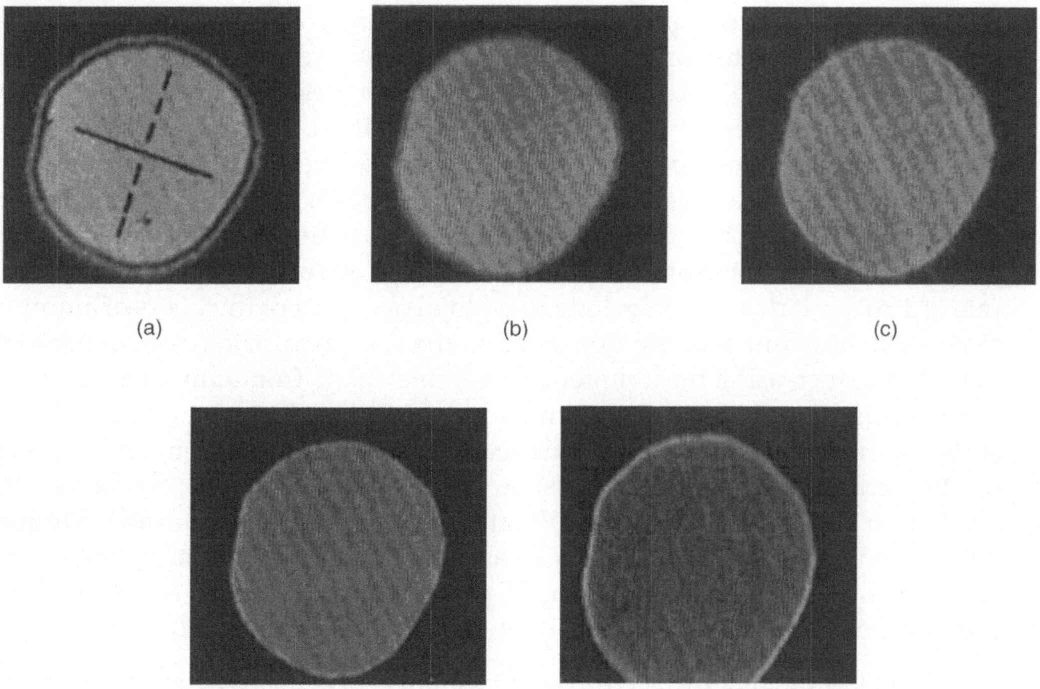

(a) (b) (c) (d) (e)

Figure 4.1 A through-focus series of micrographs of a small hole in a film at high magnification. [*Source:* Daniel, C. Pease. (1964). *Histological Techniques for Electron Microscopy,* 2nd edn. Academic Press, New York and London. p. 290, Fig. 34 (d, e, f, g, h).]

LIMITATIONS DUE TO SPECIMEN

1. ***Resolution*** Resolution is usually 1/10 of specimen thickness. So for specimen thickness of 500 Å, the resolution would be around 50 Å. Many other factors are also involved in limit of resolution. But the specimen has to be ultrathin and invariably such specimens lack contrast.
2. ***Artifacts*** Small granules or dots, especially on membranes may be due to the accretion of fixative, aggregated by the effect of an intense electron beam. Lead stain under prolonged electron beam exposure also shows distinct artifacts (*See* Chapter 18).

3. ***Damage to the specimen*** This may be due to heating and contamination.

   i. ***Heating*** At a magnification of 40,000×, electrons are spread out over an area 1,600,000,000× as large as that of the object. Such an electron beam can cause severe damage to the specimen by producing both heat and radiation. So, a small area of illumination would help in restricting damage to the specimen. Photographs can also be recorded with minimal intensity required for accurate focus.

   ii. ***Specimen contamination and anticontamination device*** As we have already seen in the previous chapter, organic molecules in EM like diffusion pump oil, vacuum grease specimen and even fingerprints act as sources of contamination. Carbon evaporated from these (by electron beam irradiation in vacuum) may reach the specimen and various apertures in the EM. Since these surfaces are exposed to the electron beam, a layer of carbon contamination forms on them. This may grow at an alarming rate, to a maximum of 8 Å/sec. But even if 1 Å/sec is the rate of contamination, long-time exposure definitely affects the specimen. Even at a rate of 1 Å/sec, it does not take long to build up a sufficient thickness of amorphous material to hide much of the fine detail in the specimen. Thus, contamination has the effect of increasing the apparent size of fine detail. Contamination layer on the specimen is formed rapidly at room temperature, but the effect is negligible if the surface is heated to 200°C, since the organic molecules do not remain long enough on the surface. But this is not possible for biological materials. Similarly, objects cooled to –80°C also fail to collect a layer of contamination. Hence an anticontamination device, which is an annular cylinder surrounding the specimen and cooled by liquid nitrogen, is used. But in such conditions, the throughput of EM may be less, as it takes considerable time to cool the anticontamination device.

Area of object illuminated by electron beam should also be restricted as far as possible, in order to,

- reduce amount of heat dissipated in it
- prevent radiation damage to unexamined parts of the object.

A single condenser lens projecting an image of the electron source on the object will illuminate an area of at least 50 μ in diameter. With use of double condenser lenses, this can be further reduced. Microscopes with double condenser lens have the advantage of brighter illumination.

INSTRUMENT STABILITY

- Drift may appear due to specimen instability rather than to trouble with the microscope. This is particularly with specimens which do not have much stability in the electron beam and when only plastic films are used as support films without carbon stabilization. In a micrograph, images will be fuzzy in one axis. This can be checked by examining rough contours of the specimen on the fluorescent screen itself. These should not show drift for more than a few seconds after the stage has been moved, which should be sufficient time for thermal equilibration.

- Most of the time, when there is image instability, the cause is a contaminating film or a particle of dust in one of the critical places in the microscope column. But it may be due to circuit instability also. So when image instability occurs, it is best to first clean up the column. Since objective aperture is so apt to be the cause of difficulty, it is usually better to examine image instability with this removed to rule out circuit instability.
- The specimen stage should enable the scanning of the specimen controllably, even when high magnifications are used. When the stage movement is stopped, the stage must not drift while a photo is recorded.
- Dirty switches and plug connections and dirty potentiometers may lead to beam instability.
- The flickering is mostly due to:

  i. Dirt or contaminating films on the objective lens aperture or pole piece, when the movement varies with the beam intensity. If it is the aperture that is involved, moving it slightly may give stability temporarily.

  ii. Dirt or lint near the specimen either on the supporting film or in the bore of the specimen holder or on the lip of the anode cap may be responsible. Then the instability will vary as the stage is moved.

  iii. Flickering illumination with a stable or nearly stable image indicates difficulty above the specimen level. The anode cap probably is dirty. This is mostly associated with unduly high beam currents which may overload the other circuits in which electronic instability is present as well. An unstable condenser lens circuit may produce flickering. Some flickering can be tolerated if it is ascertained that the image is stable in spite of it.

  iv. Contamination lower in the column, particularly in the projector pole piece, is not uncommon, and electronic instability due to dirty focusing potentiometers, poor contacts in switches and plugs may cause the difficulty.

PHOTOGRAPHY

Photography is not dealt with in detail here because the modern equipment are supplied with digital cameras and images projected on computer screens where the visual manipulations in digital images are possible, and these images are permanently stored in CDs. In older versions, photographic films or plates, which are high-contrast films especially used for electron microscopy have to be loaded in the loading magazine and after exposure they get collected in the receiving magazine, which is removed from the EM, taken to the dark room, the exposed films developed and printed.

New film boxes supplied by the manufacturer, stored in the refrigerator, can be opened only after keeping the boxes outside for 2 to 3 hours for the films to attain room temperature. Otherwise moisture condensation will take place on the emulsions. Even then, it will take at least several minutes to pump the camera chamber to the required degree of vacuum. This pumping time can be decreased considerably, however, by keeping the loaded films in a vacuum desiccator for 4 to 6 hours before loading them in the camera chamber of the EM. A vacuum desiccator is a part of many microscopes.

The photographs have to be recorded with minimum intensity required for accurate focus. But intensity of the beam and the exposure time have to be matched for good micrographs. If exposure time is prolonged, chances of drift is high. If the exposure time is too low, image formation may not be accurate. Hence generally 2 to 3 seconds exposure is given and the intensity adjusted accordingly.

The exposed films removed from the microscope have to be processed further in the dark room. The film is first kept in the developer (most commonly used is Kodak D19 developer) for 2 to 4 minutes, washed in water for 1 to 2 minutes, kept in the fixer solution for 8 to 10 minutes, washed in running water for 20 minutes or stored water for 40 minutes and finally air-dried. Developing time for more than 4 minutes may lead to fogging while too less a time may not be sufficient for features to develop on the negative. Washing is also very essential. These developed negatives have to be printed wherein positive original image of the specimen can be visualized on the photographic printing papers. Since the photographic resolution is better than the EM screen resolution, the image details are clearer in the electron micrograph than on the screen.

The operator today is relieved of all these time-consuming, laborious tasks because of the use of modern microscopes equipped with digital cameras.

X-RAY PROTECTION

EM produces X-rays from any surface which is struck by the beam. The parts receiving the heaviest load are the condenser and objective apertures and the object itself. But all these are surrounded by large masses of metal which shield the operator from exposure to X-rays. The anode and the viewing screen also receive considerable amount of X-rays. The anode is surrounded with lead which reduces the emitted X-rays below tolerance dosage. The viewing screen is usually thinly coated on a metal surface which is a good emitter of X-rays. The lower the magnification of the beam, the greater the current density and hence greater the X-ray production on any screen. So suitable thickness of lead glass is fitted to the viewing windows. On the final viewing screen, the current density is so low that in normal operation X-ray production is below tolerance dosage.

ERRORS OF INTERPRETATION

1. *Focusing* **Underfocus** is shown as a bright fringe outlining the hole in the holey carbon film. **Overfocus** is shown as a dark fringe around the hole. **Correct focus** is exhibited as almost structureless.

   Reversed contrast in the overfocused picture would be detected as false structure, but underfocusing also could be easily misinterpreted. The additional contrast at the film edge in the underfocused picture makes it appear sharper when observed in microscope than the properly focused edge.

2. *Astigmatism* Astigmatism of the objective lens is important and needs periodic correction. Astigmatism may give rise to a striated effect in the background whose direction corresponds to that of the astigmatism.

3. *Electron noise* Graininess in an enlarged photo may not be due to photographic grain but may be due to the random arrival of electrons at the photographic plate. Successive planes of the same area of the object will show a completely different arrangement of the electron noise at any given point in the image. Therefore, micrographs (photos) recorded to show any fine structure close to the limit of resolution of EM should be duplicated in order to demonstrate that the fine structural detail recorded is in fact a true feature and not an artifact.

4. *Shape discrimination* The resolution limit quoted for a microscope only expresses its capabilities in distinguishing two small objects separated by their diameter and all such objects at the limit of resolution appear spherical. As for square objects, as such, the smallest that can be identified with certainty may be about 3 times larger than the smallest object detected. For objects with a greater number of sides, the sizes increase roughly proportionally to the number of sides.

5. *Artifacts due to specimen processing* As discussed in a later chapter on artifacts, these may arise at any stage of processing and also during screening. (*See* Chapter 18.)

CALIBRATIONS

Magnification Calibration

Magnification exhibited in the control panel of EM is accurate enough for most purposes. But for quantitative measurements of sizes of particles, etc., when better accuracy (5%) is required, the instrument has to be calibrated, for which a deposit of latex particles or diatom along with the specimen is commonly used. When high precision is required, a direct comparison with optical measurements either on a selected diatom or on diffraction grating is carried out. But at very high magnification, rods of TMV (tobacco mosaic virus) is a reference standard. Ferritin can also be used for high magnification calibration.

Calibration standards as grids are commercially available, like grating, latex, etc.

i. ***Diatoms*** The finely structured siliceous shells of minute aquatic organisms can be used for calibration. Varieties of these are known with regular holes at spacings from several microns down to below the RP of LM. Suitable species is *Synedra fulgens* (Greville) Wm Smith which has a flat surface perforated with rows of holes. Spacing of transverse rows has an average of 0.66 μ, different specimens showing a variation of (±) 5% in their mean. The spacing is less than that of a 15,000 line grating, there being 7 to 8 rows in a 5 cm field at 10,000× magnification, so that it has a wider range of use than a grating. Owing to its fineness of structure and resistant nature, diatom is preferred to grating.

ii. ***Polystyrene latex and similar particles*** Particles of standard size which could be mixed with any specimen provide a direct measure of magnification in any field of view. For this, the particles must be

i. Less than 1 μ in size
ii. Have great uniformity
iii. Of some simple shape like sphere or cube

Dow Chemical company produced a batch of polystyrene latex, 0.25 μ in diameter, which have the following properties.

* chemically inert
* can be mixed with any preparation
* can be readily measured in final micrograph especially when metal-shadowed
* uniform in size (better than 5%)

Thus, this is an excellent magnification standard. A small drop of latex suspension with 1 in 2500 dilution can be placed on a grid and used. But, these particles can be used only for low to moderate magnifications, i.e., only up to 15,000 or so.

iii. ***TMV (tobacco mosaic virus)*** This is a standard for high magnification, particularly used for the examination of other viruses, bacterial flagella and other large molecules, etc. A droplet of the purified suspension in a concentration of about 0.1 mg/cc is applied to the grid, as in latex suspension.

TMV contains rods of variable length but of a regular width of 150 Å. Because of their low contrast, metal shadowing of the specimen is essential.

iv. ***Diffraction grating replicas: Parallel and crossed lines*** These are readily available with several EM consumable suppliers. A germanium-shadowed carbon replica of a ruled diffraction grating having specified number of lines per inch is used for the purpose of magnification calibration. Small number of lines is suitable for low magnifications while, for higher magnifications the number of grating lines should also be high, say for example, 15,240 lines/inch with actual grating spacing of 1667 nm is sufficient for low magnifications while, 543,864 lines per inch, with actual grating spacing of 463 nm is required for high magnification work. This is because, grating with a small number of lines presents too few lines for accurate measurement at high magnifications. A germanium-shadowed carbon replica of a ruled diffraction grating with two sets of lines ruled at right angles to each other is very useful for magnification calibration of TEM. Also used for detection and measurement of lens distortion.

v. ***Line grating replica with latex spheres*** The latex sphere is 0.261 microns and the grating replica is 2160 lines/mm. The calibration aid allows to double check the accuracy of magnification calibration.

vi. ***Catalase crystals*** Catalase crystals are mounted on a grid and negatively stained. They display very clear lattice plane spacings of approximately 8.75 nm and 6.85 nm and ideal for high magnification calibration.

Astigmatism Correction

- ***Holey carbon film*** This is used for correcting astigmatism and for evaluating the resolution of the transmission electron microscope, a thin film of carbon treated to obtain numerous round holes of various sizes on a 3.05 copper grid is used.
- ***Ferritin on holey carbon film*** This is used to correct the astigmatism of the TEM and to determine the resolution of the TEM. The ferritin molecules are close in proximity to the holey carbon film holes.

Quantitative Size Measurement

Particle-size standard Particle-size standards are certified for mean diameter and are traceable to the National Bureau of Standards (NBS). The line of standards in the submicrometre size range supplied by the manufacturer is called Nanosphere Size Standards and Monosized Polymer Microspheres. Nanosphere Size Standards are highly uniform polystyrene spheres calibrated in billionths of a metre (nanometre) with NBS traceable methodology. The 20–1000 nm range is suitable for checking the size of bacteria, viruses, ribosomes and sub-cellular components. Theses are packaged as aqueous suspensions in dropper-tipped bottles.

Monosized polymer microspheres are the most uniform spheres available. For the highest level of measurement accuracy in the 1 to 40 micron size, this series of size standards is recommended. Many other standards for SEM, diffraction, etc. are also available as calibration grids by several manufacturers.

OTHER INSTRUMENTAL PARAMETERS

There are certain other instrumental parameters which are of high value in biology particularly in diagnostic pathology. Most important ones are:

i. ***Specimen tilt*** This is an underutilized technique; most EMs are equipped with a goniometer stage, or it comes as an accessory, with a single tilt specimen holder. This method has been successfully used to identify Auer body precursors in promyelocytes.

   Specimen tilting can resolve problems such as whether a dense structure located in the intercellular border between two adjacent cells is a true desmosome or it is just a focal density of the adjacent plasma membranes.

ii. ***Selected area or microbeam diffraction*** This can decide rapidly whether an inclusion is a crystalline inclusion or an amorphous one.

iii. ***Backscattered electron imaging*** In this mode in scanning electron microscopy, thin sections, or 0.5 to 1 μ resin-embedded semi-thin sections can be analysed. Analytical electron microscopy (EDX) has been successfully applied in the study of environmental diseases.

iv. ***Scanning transmission imaging*** This technique has been used to study thick sections of unosmicated tissues from patients with interstitial lung disease. This method is an useful adjunct for analytical electron microscopy because with this method an area of approximately 20 nm can be analysed compared to the minimum of 50 nm obtainable in conventional TEM (CTEM).

Ultimately, it goes without saying that the high-resolution performance of an EM, however good the instrument and however perfect the specimen, depends on the expertise and experience of the operator/scientist.

5

SCANNING ELECTRON MICROSCOPY

INTRODUCTION

Scanning imaging method was introduced in the early days of electron microscopy but it gained familiarity in biological and medical fields only many years later. But, transmission electron microscopy was, by then, already established as an indispensable instrument for structural research because of its high resolving power. TEM extended the cellular level of the LM imaging of tissues and organs to subcellular level and was a major tool in the development of cell biology. However, LM and TEM were confined to thin specimens and could reveal only a limited view of bulk organic matter, restricted to one dimension (depth) to micrometres or nanometres respectively.

TEM–platinum–carbon replica method offered a limited 3D imaging of organic biological surfaces, even macromolecular details, but this method had the following disadvantages.

- it could be applied only on relatively smooth specimens
- it is restricted in depth to dimensions smaller than those of an average cell
- platinum–carbon does not evenly coat the surface and agglomerate into large aggregates on the surface

So attention was naturally drawn to the development of scanning electron microscope where bulk specimens could be viewed directly.

The scanning electron microscope was first invented by Manfred Von Argenne in 1930. Max Knoll in 1935 obtained the first SEM image of silicon steel showing electron channelling contrast. SEM was further developed by Prof. Sir Charles Oatley and his student Gary Stewart and was first marketed in 1965 by the Cambridge Instrument Company as the "Stereoscan".

SEM images the sample surface by scanning it with a high-energy beam of electrons in a raster scan pattern. The electrons interact with the atoms of the sample producing signals that contain information about the sample's surface topography, composition and other properties. Some of the most important features of SEM are easy changing over of magnification, large depth of field (depth of focus), and stereographic image display.

COMPARISON OF LM, TEM AND SEM

In the optical/light microscope (OM/LM), the light passing through a sufficiently thin specimen is magnified through glass lenses. In scanning electron microscope (SEM), an electron beam that is finely focused by electron lenses is scanned over the specimen and the brightness on the monitor (CRT) is modulated by the signals obtained. Figure 5.1 gives the comparative standard configurations of LM, TEM and SEM.

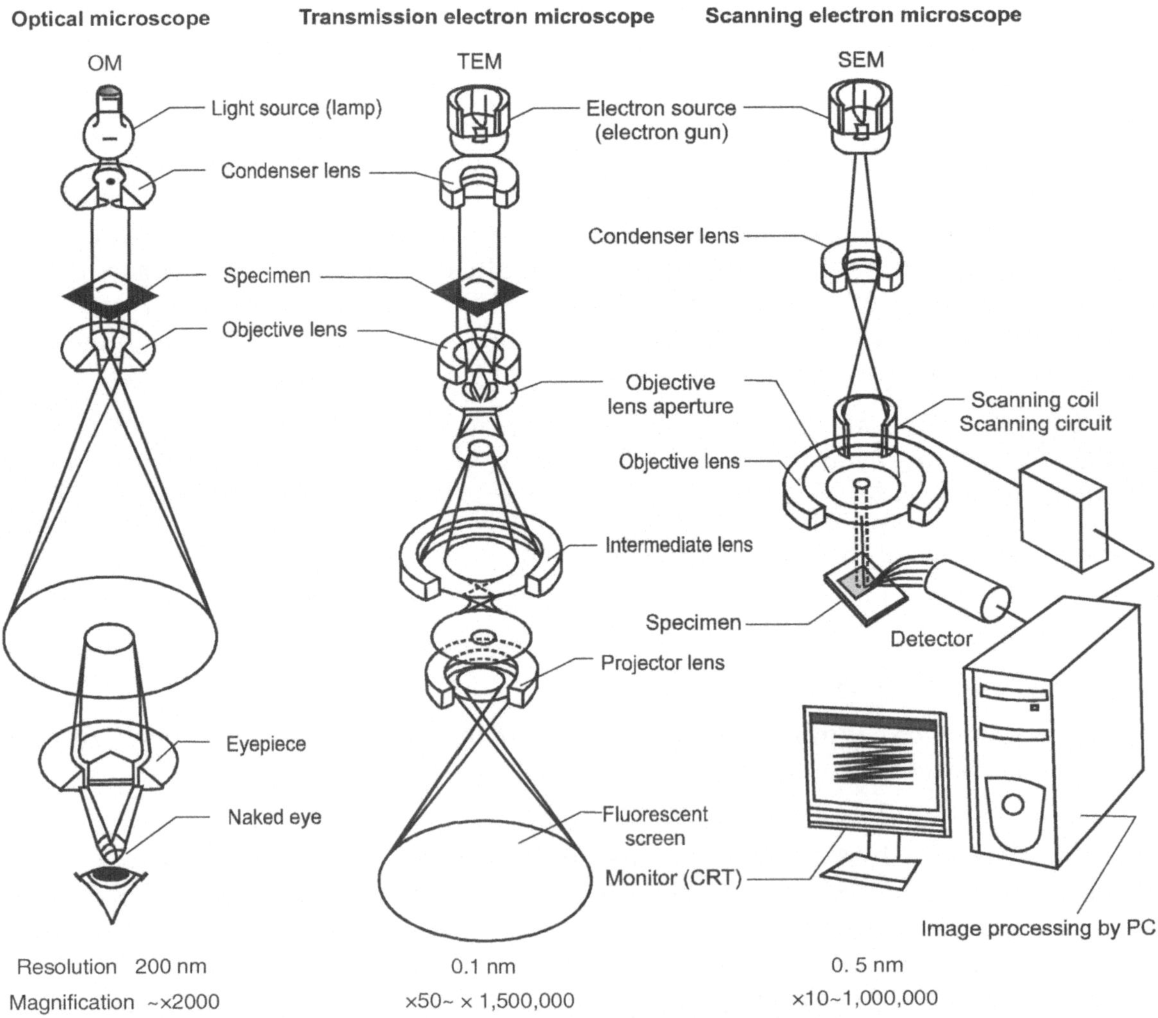

Figure 5.1 Standard configurations of LM, TEM and SEM. (*Source: JEOL SEM Manual.* p.40.)

	Scanning electron microscope	Transmission electron microscope
Ultimate resolution	0.5 nm (5 Å)	0.1 nm (1 Å)
Accelerating voltage	0.2~50 kV	20~1250 kV
Specimen shape	Bulk	Thin film
Obtainable image	Surface topography *Example:* Cone with a ball in it 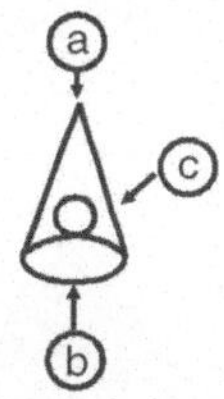 The image appears different with the direction from which it is seen. Image interpretation is simple. 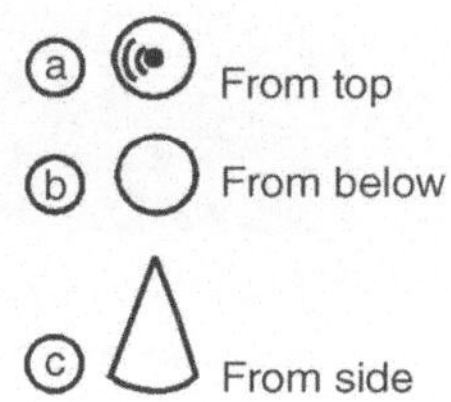 	Specimen inside *Example:* Cone with a ball in it 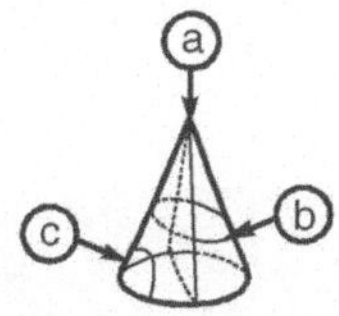 The thin section varies in shape with the position where it is formed 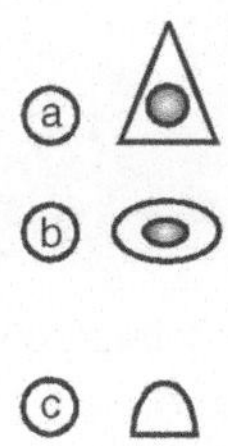
Features	Provides images appearing stereoscopic	Provides images having high resolution

Figure 5.2 Comparative features of SEM and TEM. (*Source: JEOL SEM Manual*. p. 43.)

In a transmission electron microscope (TEM), an electron beam is transmitted through an ultrathin specimen, and the transmitted electrons, along with the scattered electrons, form an image which is magnified by electron lenses.

Figure 5.2 narrates the major differences between the two types of EM, namely, TEM and SEM. Generally, SEM is used to observe a bulk specimen and the image obtained shows the morphology of its surface. With a rugged-surfaced specimen, therefore, its morphology may change depending on the direction from which it is viewed. With the TEM, it is necessary to make the specimen thin enough to transmit the electron beam, so thin sections cut from a bulk specimen are observed. Obtainable information differs with the direction in which the sections are cut out as shown in Figure 5.2.

Figure 5.3 gives a comparison of LM, TEM and SEM images of a rat glomerulus. (3D image in SEM and 2D image in TEM and LM).

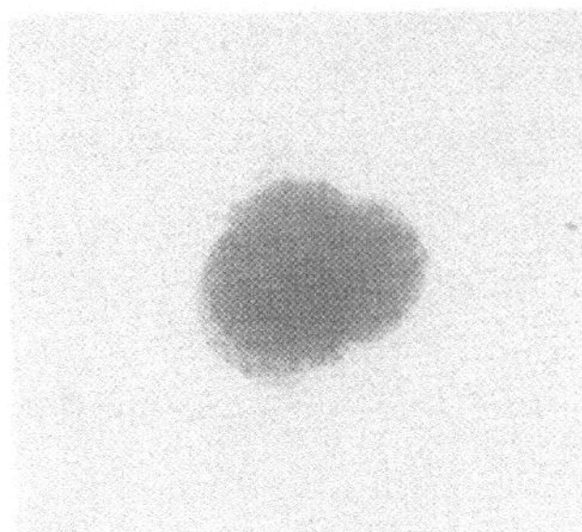

OM image: × 200
Specimen: Free cell (OsO_4-stained)

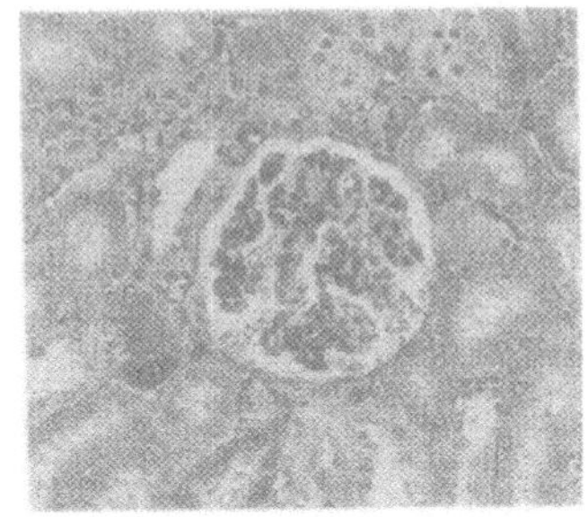

OM image: × 200
Specimen: Section (HE-stained)

(a)

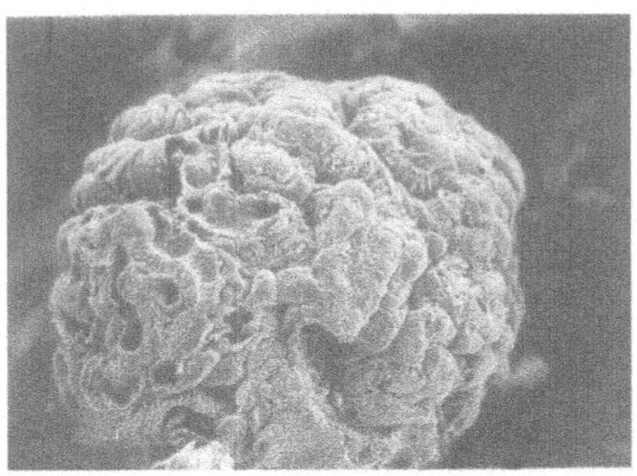

SEM image: ×750 (secondary electron image)
Specimen: Surface morphology of the bulk-state specimen

(b)

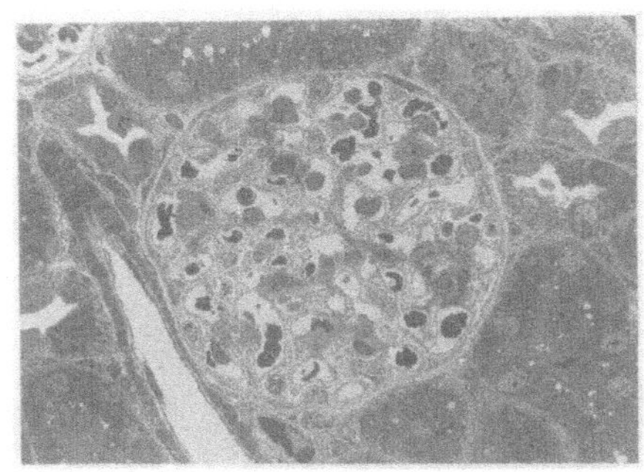

TEM image: ×750
Specimen: Internal structure of a thin section specimen

(c)

Figure 5.3 LM, TEM and SEM images of rat glomerulus. (*Source: JEOL SEM Manual.* p. 41.)

INTERACTION BETWEEN THE BEAM ELECTRONS AND THE SPECIMEN ATOMS

When a specimen is irradiated with an electron beam, the constituent atoms in the specimen produce various signals as shown in Figures 5.4 and 5.5.

Figure 5.4 illustrates the kind of signals obtained whereas Figure 5.5 indicates the depth within the specimen from which these signals are generated.

The terms mentioned in Figures 5.4 and 5.5 are briefly explained in the following section.

Electromotive force image This is an image which is obtained by detecting a voltage or current signal generated in a sample when irradiated by an electron beam. This is an important method for evaluating the integrity of a P–N junction or observing the crystal defect in a semiconductor device.

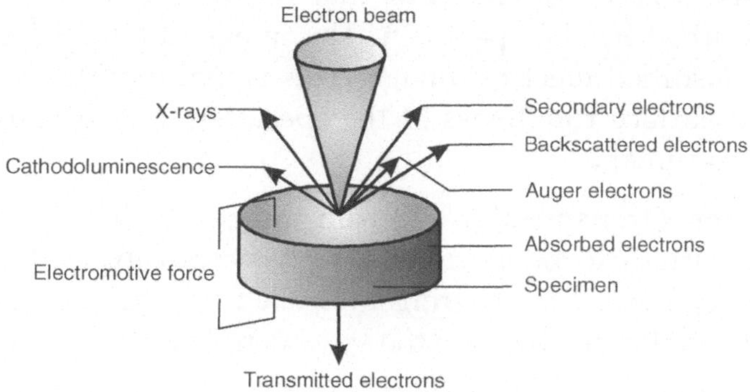

Figure 5.4 Interaction between electrons and specimen atoms. (*Source: SEM Manual*. p. 2.)

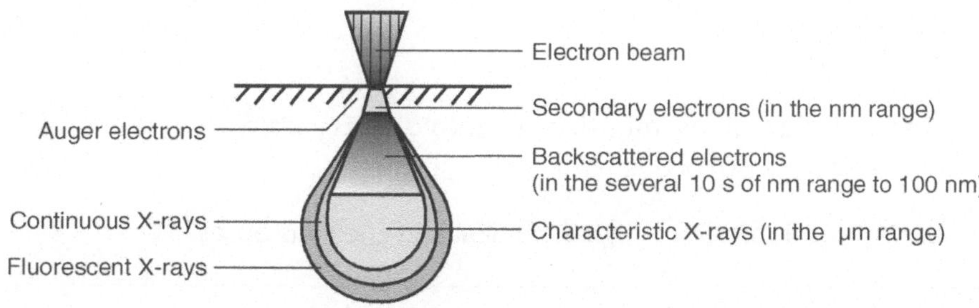

Figure 5.5 Depth of specimen generating various signals. (*Source: JEOL SEM Manual*. p. 2.)

Auger electron When the atoms of a specimen that are excited by electron beam irradiation return to the normal state, the released energy is used mainly to generate X-rays or light. But, a portion of the released energy is used to release electrons in the atoms of the specimen. This phenomenon is called the Auger effect and the electrons emitted are called Auger electrons.

Cathodoluminescence (CL) When a specimen is irradiated with an electron beam, the electrons in the valence electron bands of the atoms comprising the specimen are excited. Light emitted during the process in which the hole created in the valence electron bands recombine with the electrons is called "cathodoluminescence."

Characteristic X-rays When a substance is irradiated by X-rays or an electron beam, one of the shell electrons of the atom is released. Then, an electron in a higher energy shell drops into the vacancy. The energy difference, in this case, is released as an X-ray. This is called characteristic X-ray. The wavelength (or energy) of the characteristic X-rays is determined by the energy difference between the two shells, and the measurement of wavelength (or energy) identifies the type of the element. Each characteristic X-ray is named with the atomic symbol as follows K, L or M; α,β or γ; 1,2 or 3, etc. These symbols are defined by the wavelength and energy level.

Absorbed electrons These are electrons that irradiate a specimen, then lose their energy and are absorbed by the specimen. The image obtained from these absorbed electrons is called absorbed electron image. This signal is influenced by the average atomic number and surface roughness of the specimen, and is complementary to the backscattered electron image.

Transmission electrons/Transmission electron image If the specimen is sufficiently thin, the electrons that irradiate the specimen at high energy can pass through the specimen with almost no energy loss. Such electrons are called transmission electrons. These can be detected by a detector located beneath the specimen. The image formed by the detected signal is called a transmission electron image.

The backscattered electrons and secondary electrons are discussed further in this chapter in the following pages.

PURPOSES OF SEM

The various kinds of signals as mentioned above, carry different types of information and are used for different purposes as listed in Table 5.1.

Table 5.1 Signals in SEM and their purposes

Signal	Purpose of SEM (information carried)
Secondary electrons	Topographical observation of surface
Backscattered electrons	Compositional observation of surface
X-rays	Elemental analysis of specimen
Transmitted electrons	Internal structure observation
Cathodoluminescence	Internal characteristics observation
Electromotive force	Internal characteristics observation

Backscattered electron images are often used in analytical SEM along with the spectra made from the characteristic X-rays as clues to the elemental composition of the sample.

IMAGE FORMATION IN SEM

When the primary electron beam interacts with the sample, the electrons lose energy by repeated scattering and absorption within a teardrop-shaped volume of the specimen known as the **interaction volume** which extends from less than 100 nm to around 5 μ into the surface. The size of the interaction volume depends on the energy of the incident electrons, the atomic number of the specimen and the density of the specimen. The energy exchange between the electrons and the sample results in the reflection of high-energy electrons by elastic scattering, emission of secondary electrons by inelastic scattering and others as above.

When scanning a specimen surface with a finely focused electron beam, information will be emitted from each point of the scanning. The emitted information is converted into an electric signal, amplified and then fed into an observation CRT (cathode ray tube). On the CRT the information is used to control the brightness of the corresponding spot. The spot on the CRT is shown in real time with the electron beam scanning on the specimen surface. Thus the information emitted from the specimen surface is displayed on the CRT as an image. Therefore the image displayed is a distribution map of intensity of the signal emitted from the scanned area of the specimen. The type of information obtained can be changed by switching the signal. In this way, a specific desired characteristic of the specimen surface can be seen on the CRT in a magnified scale.

The image may be captured by photography from a high-resolution cathode ray tube, but in modern instruments, it is digitally captured and displayed on a computer monitor, as in TEM.

MAGNIFICATION IN SEM

The magnification of the displayed image, here, is defined as the ratio of the size of the image on the CRT to the size of the electron beam scanning on the specimen surface (Figure 5.6). Specific desired characteristics of the specimen surface can be seen on the CRT in a magnified scale. Thus the magnification results from the ratio of the dimensions of the raster on the specimen to the raster on the display device. Assuming that the display screen has a fixed size, higher magnification results from reducing the size of the raster on the specimen and vice versa. Magnification is therefore controlled by the current supplied by x, y scanning coils and not by objective lens power. Magnification can be controlled over a range of about 5 orders of magnitude from 25× to 2,50,000× or more. The magnification in modern instruments is up to 5,00,000× and therefore capable of imaging individual macromolecules on bulk surfaces. Since the image is not formed by an objective lens, the half angle can be very small resulting in a large depth of field. Therefore a SEM image focused at high magnification will be in focus at low magnifications also.

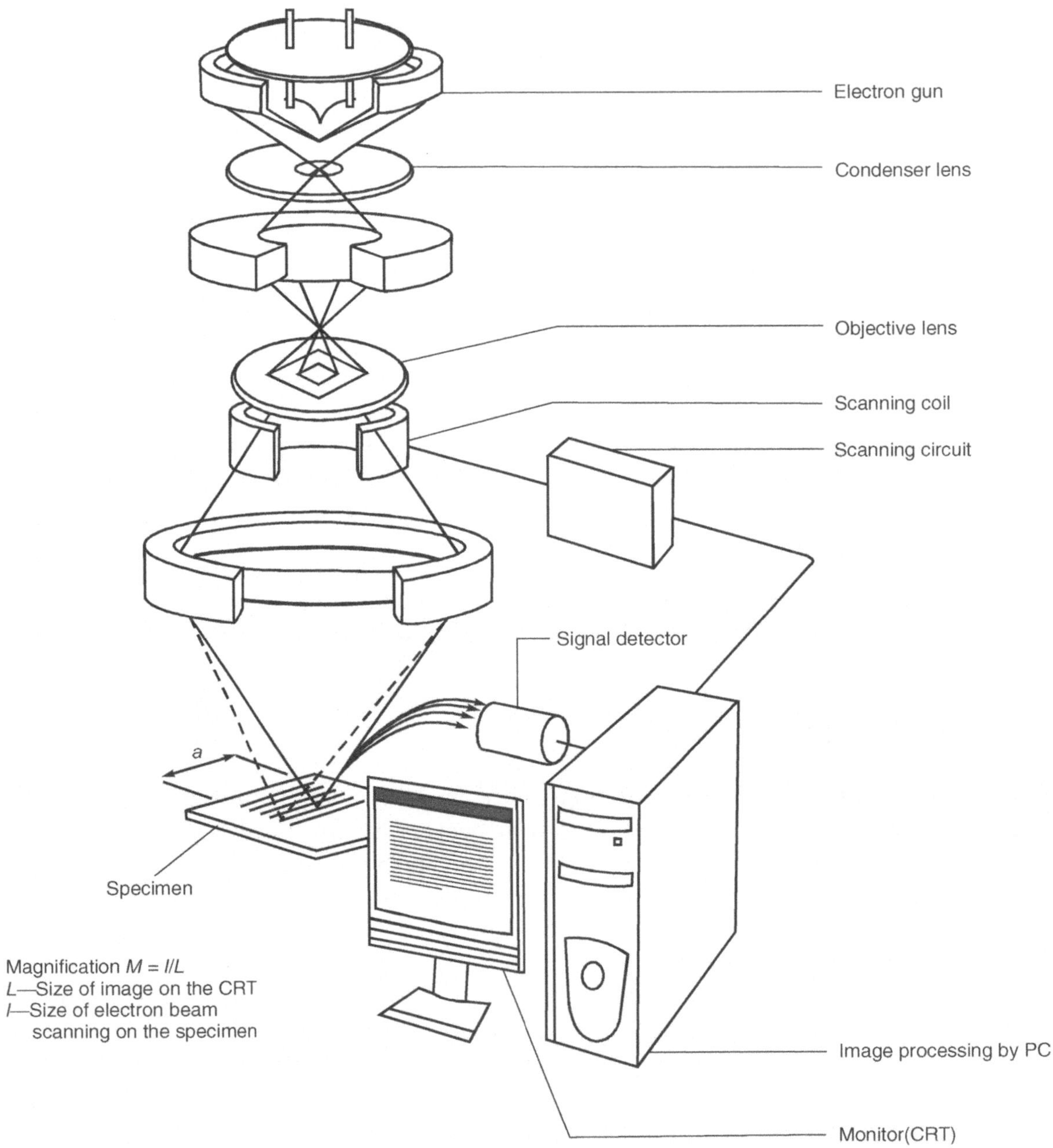

Figure 5.6 Principle of magnification in SEM. (*Source: JEOL SEM Manual*. p. 3.)

RESOLUTION OF THE SEM

The spatial resolution of the SEM depends on the size of the electron spot, which in turn depends on both the wavelength of the electrons and the magnetic electron-optical system which produces the scanning beam. The resolution is also limited by the size of the interaction volume, or the extent to which the material interacts with the electron beam. The spot size and the interaction volume both might be large compared to the distances between atoms, so the resolution of the SEM is not high enough to image individual atoms. But the SEM has compensating advantages, including,

- the ability to image a comparatively large area of the specimen
- the ability to image bulk materials (not just thin films or foils as in TEM)
- variety of analytical modes available for measuring the composition and the nature of the specimen.

The resolution in conventional SEM is restricted by both instrumental performance and the properties of the metals used for specimen coating like gold or platinum (Au or Pt). But now, in modern instruments, advanced technologies result in high resolution and high magnification in SEM. The ultimate resolution is determined by the signal generation area which depends on both the probe diameter and the range of signal electrons collected. (The electron probe is an electron beam that is focused finely on the surface of a specimen.) Depending on the instrument, the resolution can range between less than 1 nm and 20 nm. In general, SEM images are easier to interpret than TEM images.

DEPTH OF FOCUS

SEM stands out in performance from the other types of microscopes by its

- large depth of focus/ depth of field
- magnification range
- maximal size of specimens that can be examined.

SEM has a very high depth of focus which allows superior imaging of convoluted and topographically complex surfaces. The 3D aspect of the images is further increased by the fact that specimens can be physically tilted in the microscope from 0 degrees to 90 degrees using dynamic focusing. Since the image is generated in successive scan lines, the focus of the electron probe can be adjusted during scanning of the picture frame, thereby expanding the depth of field. Depth of field/depth of focus is the distance between two object planes where the in-focus images can be obtained. By selecting a small objective aperture and a long WD (working distance), a large depth of field is obtained.

WD (Working distance) The working distance (WD) is the distance from the lower face of the lower pole piece of the objective lens to the surface of the specimen. In order to observe a specimen whose surface is very irregular, necessitating an increase in the depth of field, the WD is increased. In this case, greater effectiveness can be obtained by selecting the optimum aperture diameter of the objective lens.

MAC (Magnification automatic control) In a SEM, if the working distance (WD) changes, the scanning range at the specimen surface changes, resulting in a magnification change. MAC controls the amplitude of the scanning current (magnification) according to the objective lens current (working distance) so as to maintain the amplitude constant and keep the magnification unchanged.

PARTS OF SEM

SEM consists of two main parts, as shown in Figure 5.7. They are

* The probing system
* The display system

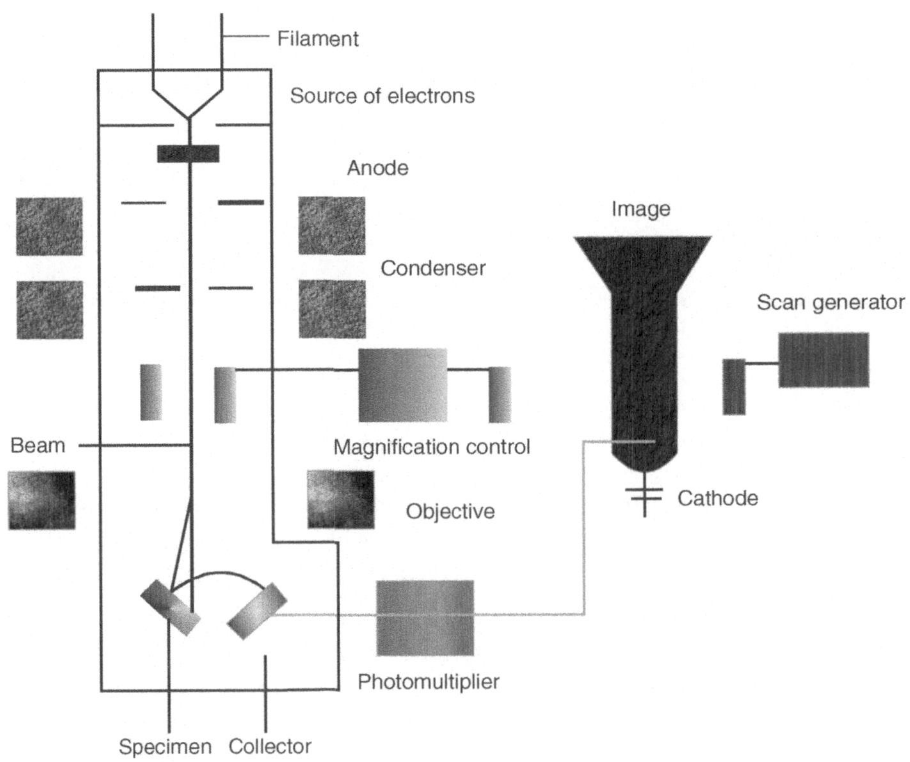

Figure 5.7 Parts of the scanning electron microscope

The Probing System

The probing system consists of the **electron gun and the lenses**. As mentioned in Chapter 3, both thermionic emission gun and field emission (FE) gun can be used as sources of electrons. FE-SEM is suited to high magnification morphological observation, whereas the SEM fitted with the thermionic emission gun is suited to applications that require a large current. The diameter of the initial focused spot of electrons, called the spot size, is 50 μm. To increase resolution at higher magnifications, electron source is demagnified, i.e., the spot size is decreased to 2 nm by using electromagnetic condenser lenses.

The electron beam which typically has an energy ranging from a few hundred eV to 40 keV is focused by one or two condenser lenses into a beam with a very fine focal spot sized 0.4 nm to 5 nm. The beam passes through pairs of scanning coils or pairs of deflector plates in the electron column, typically in the final lens, which deflect the beam horizontally and vertically so that it scans in a raster fashion over a rectangular area of the sample surface. SEM is used for diverse applications depending on the purpose and the information required. Its accelerating voltages range from several hundreds of volts to several tons of kV. The probe current also is variable from 10^{-6} to 10^{-12} A. Generally currents of 10^{-11} to 10^{-2} A are used for observation of secondary electron images, whereas 10^{-7} to 10^{-8} A are needed for elemental analysis with WDS, and observation of cathodoluminescence.

Generally, two condenser and one objective electromagnetic lenses act as demagnifiers to reduce the probe diameter. Their function is to focus the beam to a spot and not to image the specimen. The condenser lens controls the probe current and the objective lens focuses the electron probe on the specimen surface. (The probe is defined as the electron beam that is focused finely on the surface of a specimen.) The probe diameter depends on the accelerating voltage, probe current and working distance (WD) which is the distance between the underside of the objective lens to the specimen surface.

The Display System

This consists of the **detectors,** and **processing units**. When an electron beam strikes the specimen, many signals are generated which are collected and used for various purposes as follows:

- secondary electrons for information on surface morphology
- backscattered electrons for surface topography
- X-rays for elemental analysis

Separate detectors are used to collect each signal which is further amplified. The detectors are placed in the chamber surrounding the specimen at various positions depending on the signal to be collected as shown in Figure 5.8.

Figure 5.8 Specimen chamber in SEM

A number of different detectors can be incorporated into the chamber surrounding the specimen. Figure 5.8 shows the specimen chamber with different detectors. A detector placed in the column is called an "in-lens" detector. It produces a very different image compared to a conventionally located detector (outside the column, side-mounted).

Detectors

Detectors of secondary electrons The signal that is used most frequently in the SEM is the secondary electron signal. Due to their low energy, the electrons originate within a few nanometres from the sample surface. The most widely used type of secondary electron detector consists of a combination of a scintillator and a photomultiplier tube. The secondary electrons emitted are attracted by a grid at positive potential, located in front of the detector. The detector collects all the electrons coming towards it in all directions. These electrons are accelerated towards a scintillator by applying a strong positive voltage (10 to 12.5 kV). When the electrons hit the scintillator, they generate photons, which are guided by the light pipe to the photomultiplier tube where photoelectrons are produced. These photoelectrons are amplified and fed to the cathode ray tube (display screen), converting the electrical signal to an image signal.

The probe scans the specimen in a raster pattern synchronous with the electron beam of the CRT. A point-to-point correspondence between the specimen and the displayed image is thus achieved. Variations in secondary electron yield are then used to modulate the intensity of the electron beam in CRT, thus forming an image.

The depth from the specimen surface at which secondary electrons are released is only 10 nm; hence there is little scattering of the incident electrons at the specimen surface. This results in the best image resolution that can be obtained in a SEM. The contrast of a secondary electron image depends upon the surface roughness and profile of the sample, as well as upon differences in the composition of the sample.

The probe scans line by line on the specimen surface. The number of lines producing the image can be varied from 100 to 2000, thus giving different resolving power. To obtain a good photograph, slow scan rate is important so that the probe can stay at one spot for longer duration thus increasing the number of emitted secondary electrons. For viewing, fast TV scan rate can be used.

As the angle between probe and specimen surface varies, the signal strength and contrast of the image also vary. The detector is placed to one side of the specimen, which, preventing signals from other areas of specimen to reach the detector, enhances the specimen contrast. When the specimen is tilted, there is a difference in the image brightness depending on whether the tilted side of the specimen is directed to the secondary electron detector or the opposite side. With a long specimen, for example, the brightness differs between the side facing the detector and the opposite side. In such a case, directing the longitudinal axis of the specimen to the detector makes the brightness uniform.

The brightness of the signal depends on the number of secondary electrons reaching the detector. If the beam enters the sample perpendicular to the surface, then the activated region is uniform about the axis of the beam and a certain number of electrons escape from within the sample. As the angle of incidence increases, the 'escape' distance of one side of the beam will decrease and more secondary electrons will be emitted. Thus increasing the tilt angle increases the amount of secondary electrons. Therefore, steep surfaces and edges tend to be brighter than flat surfaces, which results in images with a well defined 3D appearance. Using this technique, image resolution less than 1 nm is possible.

The contrast in the secondary electron image corresponds to the number of secondary electrons emitted from the specimen surface and captured by the secondary electron detector. The number of emitted secondary electrons depends greatly on the following factors:

- incident angle of the electron beam to the specimen surface
- edge effect
- atomic number effect
- accelerating voltage effect
- effect of the magnetic field or electric field existing on the specimen surface.

Noise in secondary electron image originates from the secondary electrons that are generated as a result of backscattered electrons striking the metal parts of the microscope. The picture emerges as snowy without any resolvable details.

Detection of backscattered electrons Although secondary electron images are obtained most frequently with the SEM, backscattered electron images also provide important information.

After incident electrons are scattered within the specimen, some of them are backscattered while keeping a relatively high energy and emitted again from the specimen surface. These electrons are called "backscattered electrons" also called "reflected electrons". Backscattered electrons vary in their amount and direction with the composition, surface topography, crystallinity and magnetism of the specimen. Backscattered electrons consist of high-energy electrons originating in the electron beam, which are reflected or backscattered out of the specimen interaction volume. While the secondary electron image shows a contrast that depends mostly on the surface morphology, the backscattered electron image depends largely on the specimen composition, showing a contrast depending on the mean atomic numbers of the constituent elements of the specimen. Thus backscattered electrons may be used to detect contrast between areas with different chemical compositions, especially when the average atomic number of the various regions is different since the brightness of the backscattered image tends to increase with the atomic number. Since topographic contrast also is produced with a directional detector collecting these electrons from one side above the specimen, the contrast appears as illumination of the topography from that side. BSE can also be used to form an electron backscatter diffraction image (ESBD). This image can be used to determine the crystallographic structure of the specimen. The intensity of the backscattered electron signal depends on the specimen tilt angle.

The generation region of backscattered electrons is larger than that of secondary electrons, namely, several tens of nm. Therefore, backscattered electrons give poorer spatial resolution than secondary electrons. But because they have a larger energy than secondary electrons, they are less influenced by charge-up and specimen contamination.

The detector is a highly sensitive annular type semiconductor element which is located above the specimen. It detects electrons that are scattered backward from the specimen surface. The detector is divided into two or four detector elements, each of which produces a signal. These two or four signals are subjected to addition and subtraction processing in an operational amplifier, thus enabling the data pertaining to the effect of the specimen composition, surface topography of the specimen and stereoscopic feel, skillfully separated from each other and displayed. The signal resulting from addition is called a COMPO (composition) image, and the signal resulting from subtraction is called a TOPO (topographic) image. Thus the backscattered electron image depends upon the surface topography of the specimen and also the average atomic number of the elements contained in the specimen. The detector can be used for both low and high vacuum. These images are utilized as the most effective means of performing EPMA analysis.

(EPMA—The abbreviation of Electron Probe Microanalyser. This is an instrument which can analyse the composition of elements in a very small area on the solid specimen surface, without destroying the specimen (non-destructive analyser). An accelerated electron beam is used as a probe. It is also called an XMA.)

Thus three types of images can be obtained from BSEs.

1. *Composition image (COMPO)* This image is obtained by adding the signals obtained from two or four backscattered electron detectors located symmetrically about the electron beam. The signal represents the composition (mean atomic number) of the specimen, and is called a composition (COMPO) image. Thus, this represents the composition differences in the specimen. If the specimen surface is even, the backscattering coefficient of incident electrons is larger for larger atomic numbers thus resulting in a brighter image in the composition image mode.

2. *Topography image (TOPO)* This image is obtained by subtracting the signals obtained from two or four backscattered electron detectors located symmetrically about the electron beam. The signal represents the topography of the specimen surface, and is called a topographical image.

3. *Stereoscopic image or solid image (SHADOW)* This image is obtained by processing (combination of adding and subtracting) the signals that are obtained from two or four backscattered electron detectors located symmetrically about the electron beam. By mixing TOPO and COMPO signals, an enhanced 3-dimensional effect is obtained, and is called a solid (SHADOW) image. This emphasizes stereoscopic appearance by synthesizing composition and topography signals. Figure 5.9 illustrates the three types of BSE images.

(a) Composition image (COMPO) represents the composition differences (Z) in the specimen

(b) Topography image (TOPO) represents the topography of the specimen surface

Figure 5.9 Three types of images obtained by BSE (Specimen: Sand-containing rubber eraser; Magnification ×100. (*Source: JEOL SEM Manual*. p. 30.) (*Continues*)

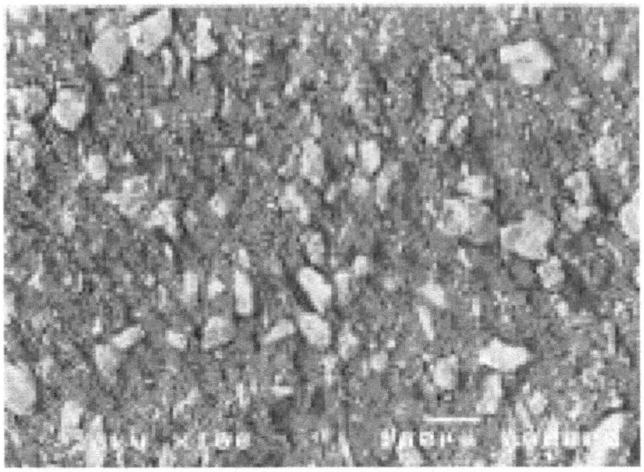

(c) Stereoscopic image (SHADOW) emphasizes stereoscopic appearance by synthesizing composition and topography signals

Figure 5.9 Three types of images obtained by BSE (Specimen: Sand-containing rubber eraser; Magnification ×100. (*Source: JEOL SEM Manual*. p. 30.)

Detection of X-rays: elemental analysis When an incident electron beam hits the atoms of a specimen, X-rays are emitted which are characteristic for the particular atom involved.

What are characteristic X-rays? When a specimen is irradiated with an electron beam, inner shell electrons of the specimen atoms are ejected, resulting in the formation of vacancies at their original positions. To fill these vacancies, outer shell electrons transfer to the vacancies and their excessive energy is emitted in the form of electromagnetic waves. Since the energies of individual shell electrons are determined by the type of elements, the X-rays are characteristic of the elements generated by electron transfer, thus the name, characteristic X-rays.

A special detector mounted in SEM collects these X-rays. By analysing the characteristic X-rays generated, we can identify the elements that constitute the specimen (qualitative analysis) which determines what elements are contained in a certain part of a specimen or make quantitative calculation of their weight concentrations (quantitative analysis) which determines the weight concentration of each composite element in a certain part of a specimen. If a number of elements are simultaneously irradiated by the electron beam, a spectrum of X-ray energy is emitted. Thus information on the types of elements, their concentration and distribution can be obtained by analysing such X-ray spectra.

There are two methods of analysing characteristic X-rays. They are:

1. using the wavelength-dispersive X-ray spectrometer (WDS)
2. using the energy-dispersive X-ray spectrometer (EDS)

At present, EDS is more widely used than WDS because of its operational ease and because it allows overall qualitative analysis to be carried out in a short period of time. Both have merits and demerits, making it necessary to selectively use them, depending on the purpose.

WDS X-rays generated from a specimen have different wavelengths depending on the elements. X-rays emitted from the specimen are dispersed in their wavelengths by the analysing crystal and the X-rays having the specified wavelength are detected by a proportional counter. After they are processed by an electric circuit such as an amplifier, X-ray intensities are counted and indicated. Different wavelengths are detected by using analysing crystals differing in position and lattice spacing. To enable all elements (except gas elements) to be analysed using a WDS, four analysing crystals are necessary. In a WDS where all four crystals are installed, all elements from boron to uranium can be analysed.

EDS This is an X-ray spectrometer which uses a semiconductor detector to detect the energies of X-rays emitted from the specimen surface and convert them into electric pulse heights. Characteristic X-rays emitted from the specimen have energies characteristic of individual elements. With the EDS, X-rays are detected by an Si(Li) semiconductor placed at the tip of the EDS detector. The height of current pulses generated by X-ray illumination is proportional to the energy of incident X-rays. By calibrating the multi-channel analyser in advance using a standard specimen, characteristic X-rays from an unknown specimen can be measured for element identification. Thus element distribution image (X-ray image), which is a two-dimensional representation of the distribution of specific elements that exist on the specimen surface, is obtained by elemental analysis. EDS can be attached to the TEM also.

Signal and Noise

SEM is a probe-forming (electron beam) and signal-detecting device. Signal is the result of the beam interacting with the specimen. Noise arises as a result of imperfections in the electronics of the detector and display system. The signal-to-noise ratio (S/N) can be increased by the following:

- placing the detector closer to the source of signal
- slowing down the scan (collect more signals per unit time)

Although the same amount of signal is produced throughout the specimen the topography of the surface will allow differing amounts of signal to reach a detector (Figure 5.10).

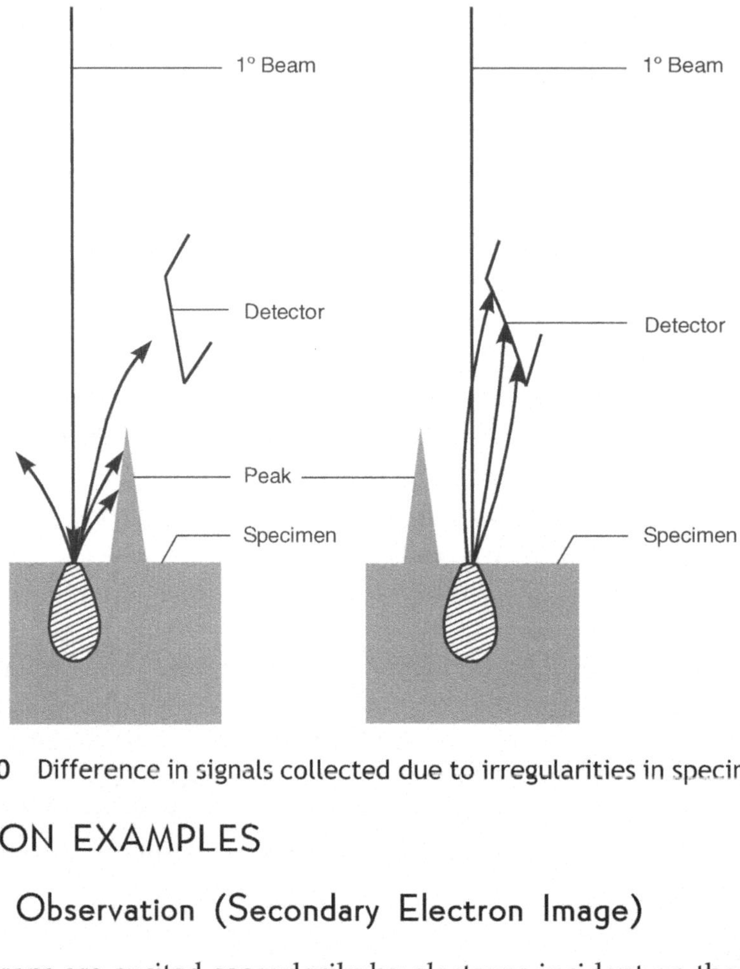

Figure 5.10 Difference in signals collected due to irregularities in specimen surface

OBSERVATION EXAMPLES

Topographical Observation (Secondary Electron Image)

Secondary electrons are excited secondarily by electrons incident on the specimen. Since their generation region is as shallow as approximately 10 nm, the diffusion of incident electrons within the specimen has little influence on the image, thus allowing the best resolution to be obtained. The contrast of secondary electron image depends on the tilt angle and topography of the specimen surface.

The number of secondary electrons emitted from a specimen depends greatly on the incident angle of the electron beam to the specimen surface. In other words, the secondary electron signal depends on the surface undulation of the specimen. Moreover, since the secondary electron energy is very low, they are emitted only from a thin layer on the specimen surface. Thus the secondary electron signal is considered to be the most suitable signal for observing the surface topography of a specimen.

Secondary electron image is applicable to observe large objects such as insects and animal tissues, as well as small objects such as bacteria and viruses. It is a powerful tool in entomological taxonomy as well as taxonomy of spores.

Figure 5.11 illustrates the top view and side view of a tick taken using the SEM at two different magnifications.

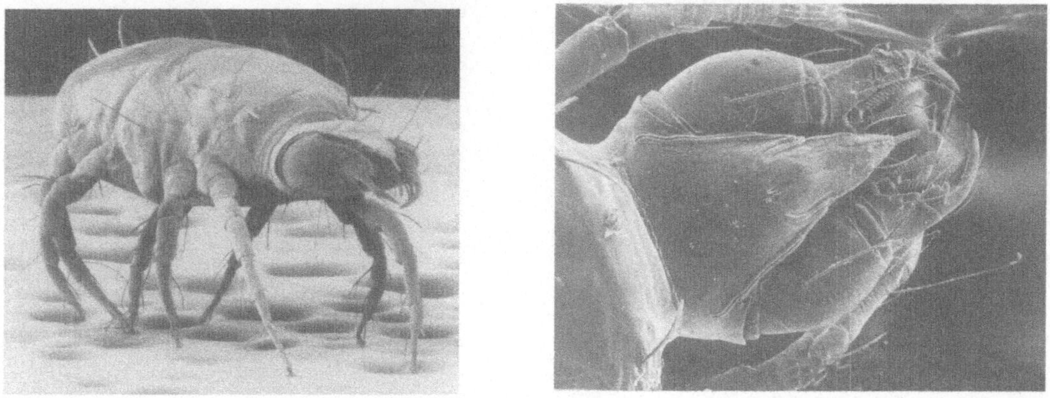

×100 ×360

Figure 5.11 Entomological specimens (Specimen: Tick) (*Source: JEOL SEM Manual.* p. 5.)

Compositional Observation (Backscattered Electron Image—BSI)

Since energy of BSE is higher than that of secondary electrons it carries information from deeper layers of specimen surface. The number and scattering direction of the BSE are determined by the average atomic number of the substances composing the specimen surface and the incident angle of the electron beam to the specimen surface.

A comparison between secondary electron image and backscattered electron image of two specimens is illustrated in Figure 5.12.

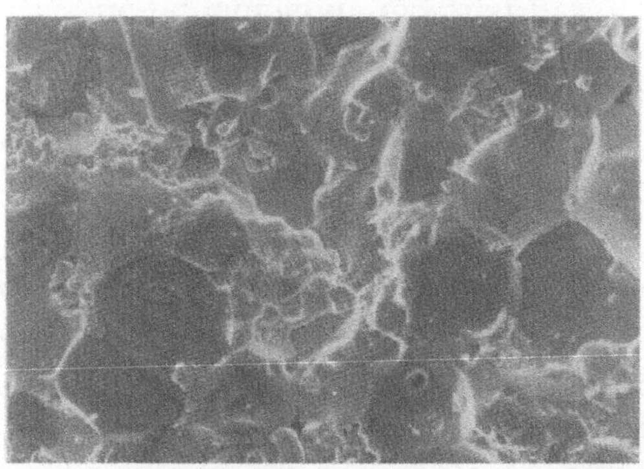

× 2000

Secondary electron image
(Accelerating voltage 15 kV)

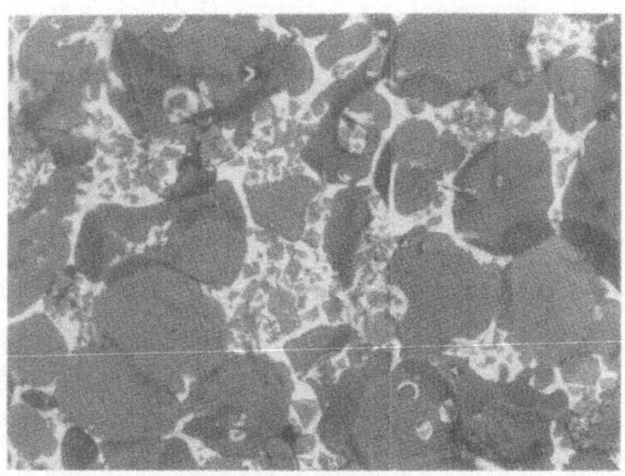

× 2000

Backscattered electron image (Composition image)
(Accelerating voltage 15 kV)

(a)

Figure 5.12 Comparison between SEI and BEI of (a) Varistor and (b) Superposed grids (*Source: JEOL SEM Manual.* p. 7 and p. 28.) (*Continues*)

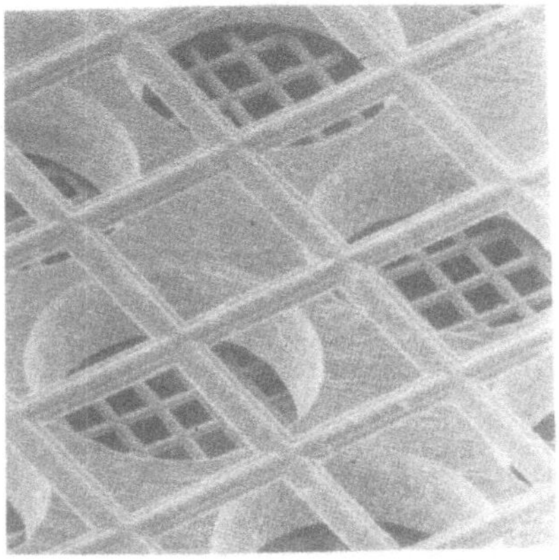

Secondary electron image

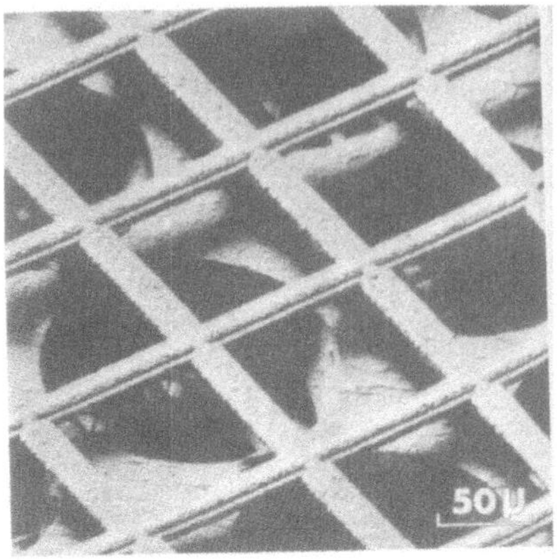

Backscattered electron image

(b)

Figure 5.12 Comparison between SEI and BEI of a) Varistor and b) Superposed grids (*Source: JEOL SEM Manual*. p. 7 and p. 28.)

As in Figure 5.12, the surface topography of a specimen can be seen with the SEI and the composition difference (grain and grain boundary) can be clearly distinguished with the BSI.

Elemental Analysis

By detecting and analysing the characteristic X-rays the elements contained in the specimen (qualitative analysis) can be identified and the weight concentration of the contained elements (quantitative analysis) can be determined. The electron beams are very finely focused. Therefore, by using the spot mode, one can perform an elemental analysis of a very small area on a specimen surface. Also by scanning the electron beam over a specified area, an averaged element concentration can be obtained.

Inner Structure Observation (TEM Image)

If a specimen is thin enough for the irradiated electron beam to penetrate it, electrons will be scattered during penetration of the specimen. The degree of scattering depends on the product of the density of the specimen substance δ and the specimen thickness (t). Namely, electrons penetrating through a not dense area (small "$\delta \times t$" value) will not be as largely scattered as the electrons penetrating through a dense area (large "$\delta \times t$" value). Thus, a larger signal is detected through the not dense area. When a thin section is cut from a bulk specimen and observed with transmitted electrons, an information about the inner structure of the thin section can be obtained. Here, the detector is placed below the specimen. The TEM image of the tail of a mouse sperm is given in Figure 5.13.

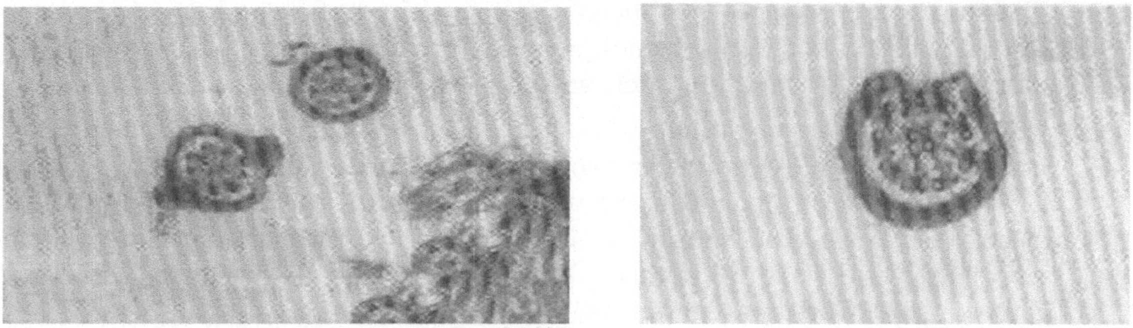

×50,000 ×100,000

Figure 5.13 TEM image of a tail of a mouse sperm in a SEM (Accelerating voltage 35 kV) (*Source: JEOL SEM Manual*. p. 9.)

Light Signal Observation (Cathodoluminescence Image)

When a specimen is irradiated with an electron beam, the electrons in the valence electron bands of the atoms comprising the specimen are excited. Light emitted during the process in which the hole created in the valence electron bands recombine with the electrons is called cathodoluminescence. This phenomenon is exhibited especially by certain kinds of specimens (sulphide, oxide, mineral, semiconductor, etc.) and is used for detection of impurities or structural defects in a specimen, and for research using antibodies in biology. Cathodoluminescence detection using a SEM is especially useful compared with other methods. This is because, the character of a cathodoluminescence image can be studied in a micro-area and compared with a secondary electron image, backscattered electron image and other information.

Apart from the above, electron beam induced current (EBIC) image, crystal structure analysis and magnetic domain observation are all possible with the SEM. These modes are mainly used for non-biological samples and hence not further discussed here.

SPECIMEN PREPARATION AND OBSERVATION TECHNIQUE

Because of the wide variety of SEM applications, many kinds of specimens can be observed in SEM. Therefore, many different specimen preparation techniques have been developed depending on the object and the purpose of the study. In every case though, there are three major points that must be taken into consideration during specimen preparation. They are:

1. *The specimen must be clean* SEM observes the surface layer of a specimen. Therefore, it is essential that the specimen surface be clean.
2. *The original morphologic construction must be maintained.* If a specimen containing water or gas is brought into a vacuum, the specimen may shrink or deform.

Therefore, for biological specimens, fixation, dehydration and drying are done prior to observation. Cryo-techniques where the specimen is frozen to fix the morphological structure are often used for emulsion specimens.

3. *The specimen must not acquire an electrostatic charge.* When a specimen is irradiated with an electron beam, some electrons are emitted from the specimen as secondary electrons and backscattered electrons. The rest of the irradiated electrons may be absorbed in the specimen. However, if the specimen has no electric conductivity, the absorbed electrons can charge the specimen. This charge causes many errors in observations.

Some methods to minimize/prevent a specimen from acquiring an electrostatic charge are:

- Metal coating of the specimen surface
- Observations with low accelerating voltage
- Observations under low vacuum

Biological specimens normally require fixation to preserve their structures. This is usually performed by using buffered chemical fixatives as discussed in Chapter 6. Then they are dehydrated using organic solvents which are replaced with a transitional fluid such as liquid carbon dioxide in critical point drying. If the SEM is equipped with a cold stage for cryomicroscopy, cryofixation may be used and low-temperature scanning electron microscopy performed on the cryogenically fixed specimens. Cryo-fixed specimens may be cryofractured under vacuum in a special apparatus to reveal internal structure, sputter coated and transferred onto the SEM cryostage while still frozen.

Let us discuss the individual steps briefly:

The processing protocol for biological specimens involves the following steps:

- Surface cleaning
- Primary fixation
- Washing
- Post-fixation (optional)
- Washing
- Dehydration
- Drying
- Critical point drying (CPD)
- Mounting
- Surface coating

Obtaining a Clean Surface

In all SEM works, the goal is to observe the specimen surface as it actually exists. Because of this, if a specimen surface has been contaminated or covered by unwanted materials, the contaminants must be removed. Or if the object to be observed or analysed exists inside a specimen, the desired object must be disclosed on the specimen surface. Although elemental analysis work by a SEM can be done with a rough-surfaced (natural surface) specimen, a flat-surfaced specimen gives more accurate analysis.

The surface of an object to be studied must have the following characteristics:

- Must be free from foreign particles or cell debris
- Must be stable when placed under high vacuum
- Should remain stable after exposure to electron beam
- Must emit sufficient number of secondary electrons
- Should develop as few charges as possible

As extracellular products such as mucus, blood and other body fluids and also tissue fragments may obscure the surface to be examined, they must be removed by gentle washing with normal saline (especially to remove mucus), or working buffer or with distilled water. The specimen to be washed is placed in the wash solution and the surface is cleaned with '0' (zero) number brush or with finger gently.

Tissue culture cells, cell suspensions or microorganisms can be centrifuged and further processed in suspension or attached to a suitable base (coverslip) before carrying out the preparative procedures.

Fixation

This is done to fix the protein and lipid substances in their living morphological shape, so as to prevent the specimen from deformation when in a vacuum. Apart from hard objects such as teeth, bone and wood, all other biological materials must be stabilized in order to prevent structural changes.

Fixation can be done by chemical or physical methods: chemical—using primary fixative (glutaraldehyde) and secondary fixative (Osmium tetroxide) and physical—using cryotechniques (*See* Chapters 6 and 16).

The procedure followed for primary and secondary fixation is identical to what is practised for TEM.

The excess fixative has to be removed by washing in buffer, especially after secondary fixation with osmium because un-reacted osmium tetroxide may be reduced be dehydrating agents causing precipitates to form on the specimen thus obscuring the surface morphology.

Dehydration

This is carried out using graded series of acetone or alcohol. The fixation, washing and dehydration methods are similar to what is practised for TEM as given in the chapters that follow.

Drying

Different methods are employed for drying:

Air-drying This is the simplest method and is usually done for blood or serum by preparing a smear on a slide that is air-dried. But this method has many disadvantages:

- creates maximum distortion
- causes collapse and shrinkage of specimen
- results in flattening of specimen surface because of compressive forces of surface tension of liquid–gas interface
- causes cracking of surface

Chemical drying Some chemicals like, HDMS (1,1,1,1,3,3,3-hexamethyl disilazane) can be used for drying the specimen.

Critical point drying This is the most commonly used technique to dry biological samples. The greatest problem encountered during drying a specimen is the surface tension of the liquid. This causes shrinkage and wrinkles. A method of drying a specimen by adjusting the pressure and temperature so that the distinction between liquid and gas phases disappears is called critical point drying method. Generally, carbon dioxide, which is non-toxic, inexpensive and can be obtained readily, is used for this method. The apparatus is shown in Figure 5.14.

In this method the sample is transferred from an organic dehydration medium (acetone) to a drying medium (liquid CO_2 or Freon 13) in a chamber that is cooled and put under pressure. When the dehydrating agent (acetone) has been completely removed and replaced with liquid CO_2, the chamber is warmed to its critical temperature where the density for the drying medium is same in both liquid and gaseous phase. Phase boundary disappears and CO_2 can be released gradually to avoid condensation. When the temperature is raised slightly above critical temperature and the entire CO_2 is vented out, the sample gets dried completely. The specimen should not be exposed to air (to avoid air-drying) while transferring from one medium to another. After drying, the sample has to be protected from excessive fluctuations in relative humidity by keeping the specimen in a dry and dust-free environment. A desiccator can be used to store the dried samples.

Figure 5.14 Critical point dryer

The most commonly used media for critical point drying are

- Liquid CO_2—critical point 31.5°C at 1100 p.s.i. (pressure) (p.s.i.—pounds per square inch)
- Freon 13—critical point 28.8°C at 560 p.s.i. (pressure)

Freeze-drying method This is a method of drying a specimen containing water by freezing it, using, for example, liquid nitrogen, then sublimating the water in a vacuum. It is used for biological specimens however, it is necessary to be careful of water crystals that grow when the specimen is frozen.

Mounting

The dried material is mounted on metallic stubs with the help of conducting paints or conductive double-sided tapes. The mounting medium should be heat-resistant. While mounting, the following points should be taken care of:

- conductive paint should not come over the surface to be scanned
- paint should not lose its adhesive properties and should be able to withstand tilting
- it should make good electrical contact with both specimen and stub.

Large specimens can be stuck directly on to the stub with the help of conducting paint (silver or copper paint). For smaller specimens, the stub is first covered with a piece of double-sided adhesive tape and the specimen is stuck down onto this with the help of a magnifier or binocular microscope. Specimen should be oriented in such a way that the area of interest is optimally exposed to the electron beam.

Surface Coating

Biological specimens are poor electrical conductors. When a non-conductive specimen is directly illuminated with an electron beam, its electrons with a negative charge collect locally (specimen charge-up), thus preventing normal emission of secondary electrons. This charge-up causes some unusual phenomena such as abnormal contrast, image deformation and shift.

If the sample acquires electrostatic charge during observation of a SEM image, the contrast of the image will become abnormal and unstable. Also, it will be impossible to obtain a sharp image. In order to prevent this, the sample is provided with a metal coating, or the image is observed under a low accelerating voltage. The surface conductivity can be increased if the specimen is coated with a thin layer of metal. The specimen is coated with a heavy metal such as gold, gold–palladium alloy or platinum, while for X-ray analysis, the specimen is coated with a light substance such as carbon or aluminum, using a vacuum-deposition device or ion-sputtering device. Rough-surfaced specimens must be evenly coated from every direction.

Metal coating also increases the following.

- Mechanical strength of fragile biological materials
- Number of secondary electrons emitted from specimen surface, which leads to improvement in image formation.

The metal to be coated must be selected based on the purpose of observation and magnification. When the coating is too thick, its particles become visible while at the same time the structures of interest are sometimes buried.

The thickness of the coating should be optimum. Usually 20 nm thickness (15–30 nm) is sufficient. It is also necessary to uniformly coat the specimen from all directions by rotating and tilting it during evaporation. Gold is the most commonly used metal for coating as it is inert (does not form oxides) and forms uniform and stable coating for repetitive examinations. Other metals used for coating are palladium, platinum, silver, gold–palladium, thin alloys, etc. Au and Au–Pd are generally used as they are easy to get and they generate secondary electrons well. Recently however, high-melting metals such as Pt and W have been used for high magnification observation, because of their good granularity. The sputtering device is not used for carbon coating.

Coating can be done in two ways:

- by thermal evaporation method
- by sputter coating method

Thermal evaporation method The principle behind vacuum coating method is that many metals, when heated in vacuum, evaporate rapidly into a monoatomic state. This coating is carried out using a vacuum deposition device. This device has a glass bell jar that has about 250 mm inner diameter and about 260 mm height in which a high vacuum of 10^{-3} to 10^{-4} Pa is obtained. The coating material, a piece of metal, is wound around a tungsten wire, or put into a tungsten basket inside the bell jar which is subsequently evacuated. When the tungsten is heated by an electric current in vacuum, the metal evaporates and the evaporated atoms land on the specimen forming a uniform film on the surface. Generally, the specimen is rotated and tilted during the evaporation to obtain an uniformly coated film. For carbon coating this is the method available. Carbon rods are used as the source of carbon. This device has multiple functions:

- carbon coating for X-ray analysis
- coating the specimen surface for SEM
- carbon support film preparation
- preparation of replicas
- shadow casting
- cleaning the apertures
- when the charge-up on the specimen surface cannot be prevented simply by coating with Au because of surface roughness, Au coating may be done after carbon evaporation.

For SEM studies with this device, the coating may be relatively uneven and the chances of surface charge accumulation are higher. Hence sputter-coating is the method of choice.

Sputter coating method Sputter-coating allows uniform coating on the specimen even on the parts that are not directly facing the metal to be evaporated. Hence it is the most acceptable technique. But the sputtering device is not suitable for carbon coating.

There are two types of sputtering devices:

- Magnetron sputtering device (Low vacuum—several Pa* order)
- Ion beam sputtering device (High vacuum—10^{-3} order)

Note

(*Pa is abbreviation for Pascal. This symbol represents a unit of pressure.

1 Pa = 10 μ (micro) bar, 133 Pa = 1 torr = 1 mm Hg)

In the magnetron sputtering device, the cathode and anode are coaxially arranged, and a coaxial cylinder magnet is placed behind the cathode. The advantage of this construction is that specimen damage due to electron or ion bombardment is reduced.

In the ion beam sputtering device, an inert gas ion beam is irradiated on the metal used for coating the specimen. The sputtered metal is deposited on the specimen. Very finely coated films can be obtained with this device. Specimen damage is also very low, because the ions do not hit the specimen surface.

The ion sputtering device is used for preparation of SEM specimens (Figure 5.15). It creates a glow discharge in a low vacuum of about 10 to 1 Pa. The specimen is placed on a cooled base plate which functions as an anode. The cathode, which is the metal to be evaporated (for example, gold) is positioned above the specimen. The chamber in which the cathode and anode are placed is evacuated. Inert gas such as argon is flooded into the chamber and is again evacuated. A negative high voltage is applied to the cathode. This results in glow discharge and argon gas atoms get ionized. The positively charged ions move towards the cathode and strike a negative potential cathode target at high velocity, sputter the metal of the cathode and deposit the sputtered metal on the surface of the specimen. The dislodged metal atoms due to numerous collisions with gas molecules get deflected several times and hit the specimen at different angles. Thus a uniform thin layer is deposited on the specimen.

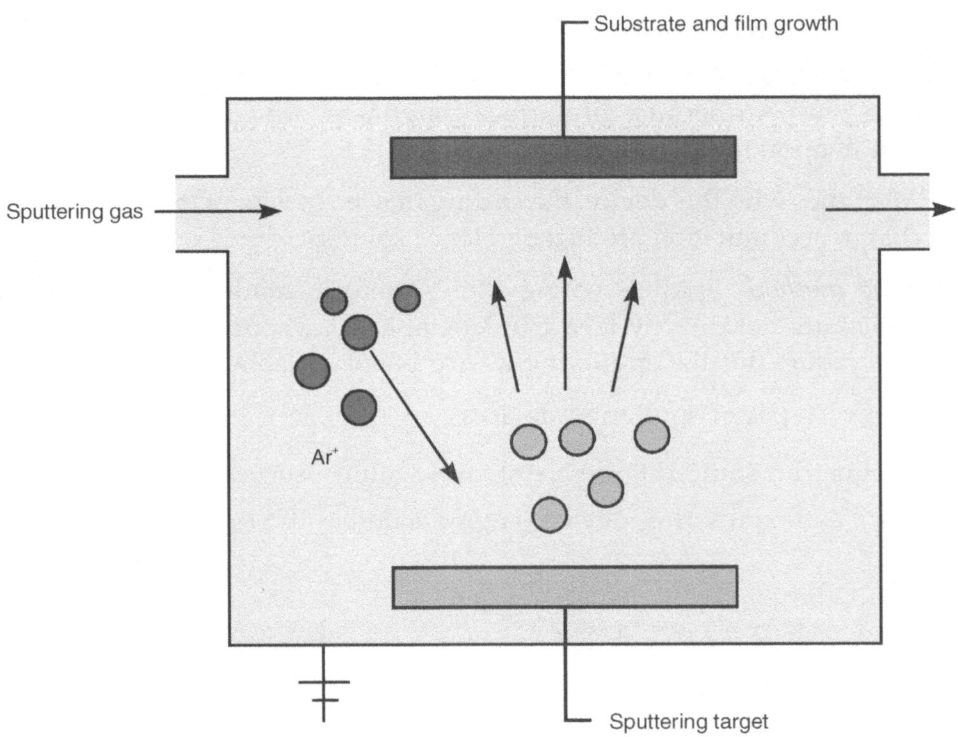

Figure 5.15 Principle of sputter (low vacuum) coating

Alternatively, this device can be used to shoot ions at the surface of a specimen instead of the cathode target, so as to remove (etch) an unwanted film on the specimen surface.

The thickness of coating depends on:

- Current applied
- Anode-to-cathode distance
- Sputtering time

After coating, the specimen can be introduced into the specimen chamber and viewed in the SEM.

Disadvantages of coating But the coating process may distort internal structure of the specimen and at times may disturb the surface features also. Particularly, delicate and fragile specimens may undergo distortions by coating. The coating, however thin, has a finite thickness which can obscure the underlying fine details of the specimen for very high magnification images. There is also a loss of natural contrast of the specimen surface due to coating. Moreover, X-ray analysis is difficult with a coating of a heavy metal, so carbon coatings are routinely used in conventional SEMs for this purpose.

Prevention of Charge-up by the Sample

Biological, cloth and powder specimens cannot often be photographed clearly, with some portions looking too bright and some too dark. This is because these specimens are partly charged-up.

To prevent this, it is necessary to give the specimen surface uniform conductivity as follows:

1. When fixing the specimen on a specimen stub, apply conductive paint (carbon paint or the like) to specimen portions which are hard to coat (Figure 5.16).
2. In the case of powder, if the particles are piled up on each other, charge-up easily takes place, causing them to move during observation. To prevent this, after the adhesive for fixing the powder is dried, blow the piled-up particles using a hand blower (Figure 5.17). Different adhesives need to be used depending on the size of particles. Mainly, double-sided adhesive tape, manicure liquid, and aluminium foil are used. When using double-sided tape, applying conductive paint (carbon paint) to the four corners would be effective.

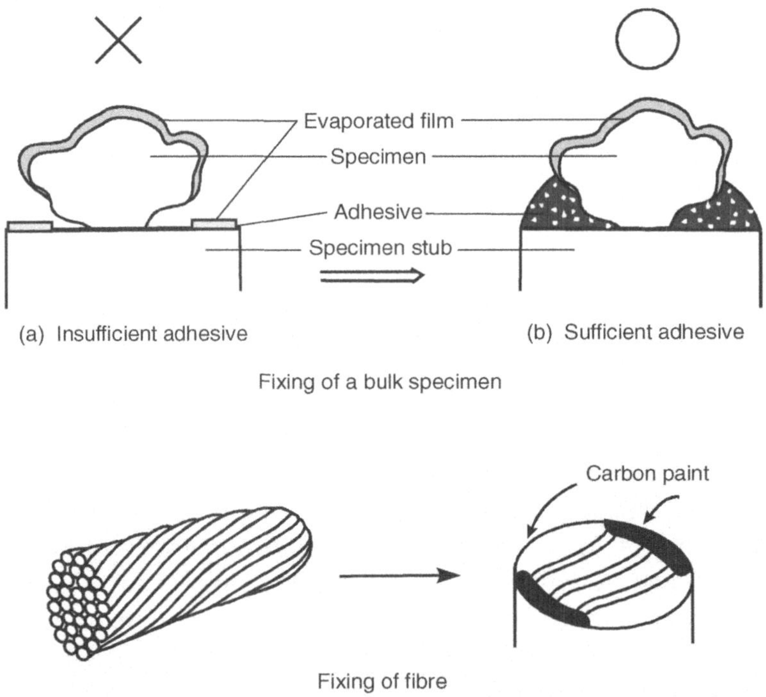

Figure 5.16 Fixing the specimen (*Source: JEOL SEM Obsn. Manual.* p.15, Fig. 23.)

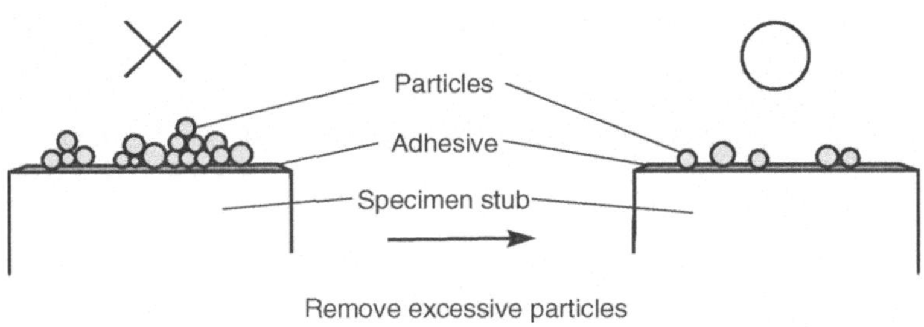

Figure 5.17 Preparation of powder specimen (*Source: JEOL SEM Obsn. Manual.* p.15, Fig. 24.)

LOW VACUUM (LV) OBSERVATION

One method of observing an electrically non-conductive specimen without metal coating is low vacuum observation.

The low vacuum SEM (LV-SEM) is a SEM in which an additional pumping system has been added so that the specimen chamber can be maintained in a low vacuum state while the column is in a high vacuum state. In such a condition, gas molecules surrounding the specimen are ionized by the incident electron beam and the electrons emitted from the specimen. This results in neutralization of the electrostatic charge on the specimen surface. Thus, a non-conductive specimen can be observed or elemental analysis can be carried out without metal coating. A hydrous specimen or a specimen containing oils can also be observed with an LV-SEM. The loss of natural contrast in the specimen due to surface coating can be avoided using this technique.

Substances such as foodstuff, cosmetics, paints and emulsive specimens possess much water or oil, but cannot be treated with chemicals and drying processes. In such cases, the so-called cryomethods can be used. These liquid specimens are frozen by liquid nitrogen to fix the structure. But for specimens which do not include so much water and oil as emulsive foodstuff and cosmetics, a low vacuum SEM may be suitable for observations.

ANALYTICAL SCANNING ELECTRON MICROSCOPE (ANALYTICAL SEM)

A SEM was originally used to observe the surface morphology of an object and cannot perform elemental analysis with its pure SEM function only, but if X-ray detectors, which detect the characteristic X-rays emitted from the specimen, are installed in the SEM, it becomes possible to identify and quantitatively analyse the chemical composition in micro-areas on the specimen while observing the SEM image. A scanning microscope which is provided with this function is called analytical SEM.

ESEM (ENVIRONMENTAL SEM)

Non-conducting specimens may be imaged uncoated using specialized SEM instrumentations such as the environmental SEM (ESEM) or field emission gun SEMs (FEGSEM) operated at low voltages. In ESEM the specimen is placed in an internal chamber at higher pressure than the vacuum in the electron optical column. Positively charged ions generated by beam interactions with the gas help to neutralize the negative charges on the specimen surface and provide amplification of the secondary electron signal. The pressure of gas in the chamber can be controlled and the type of gas used can be varied according to need. Coating is thus unnecessary and X-ray analysis unhindered.

The initial commercial development of the SEM in the late 1980s allowed samples to be observed in low-pressure gaseous environments (e.g., 1–50 Torr) and high relative humidity (up to 100%). This was made possible by the development of a secondary electron detector capable of operating in the presence of water vapour and by the use of pressure-limiting apertures with differential pumping in the path of the electron beam to separate the vacuum regions around the gun and lenses from the sample chamber. The first commercial ESEMs were produced by the ElectroScan Corporation in USA in 1998. ESEM is especially useful for non-metallic and biological materials because coating with carbon or metal is not required. Uncoated plastics and elastomers and other biological samples which are sensitive to the process of applying coating can be easily analysed. Wet biological specimens can be viewed directly in this. Apart from these, the ESEM allows dynamic experiments to be performed and photographed like for example, watching the process of paint drying.

The different types of SEM include the following.

- Conventional SEM (high vacuum)
- Low pressure SEM (coating can be avoided)
- STEM (combined SEM and TEM)
- ESEM (environmental SEM)
- FESEM (field emission SEM)
- FIBSEM (focused ion beam SEM)
- SEMSCA (structural chemical analyzer) (combined Raman spectroscopy and SEM)
- Combined SEM/FIB/AFM

HOW SEM OPERATES

The electron beam produced by the electron gun is finely focused on the specimen and it irradiates the specimen surface. The specimen surface is scanned point-to-point with this small (diameter) electron probe in both X and Y directions. Secondary and backscattered electron signals generated by the electron beam irradiation are collected by the detector. The detected signals are amplified and used to modulate the brightness on the CRT brightness. Thus, secondary and backscattered electron images are produced and displayed on the image display unit. Forming a small electron probe and detecting the signals efficiently make it possible to obtain high resolution.

The total amount of current to be irradiated on the specimen is called the probe current. It is controlled between approximately 10^{-12} A to 10^{-6} A. The control is done by varying the excitation of the condenser lens of the SEM.

IMAGE CHANGES CAUSED BY INTERACTIONS BETWEEN ELECTRON PROBE AND SPECIMEN

Influence of Accelerating Voltage and Probe Diameter on Image Quality

In SEM, finer surface structure images can generally be obtained with lower accelerating voltages. At higher accelerating voltages, the beam penetration and diffusion area become larger, resulting in unnecessary signals (e.g., backscattered electrons) being generated from within the specimen. And these signals reduce the image contrast and thus the fine structural details.

Theoretically, considering the electron probe diameter alone, the higher the accelerating voltage, the smaller is the electron probe. However, there are certain demerits in increasing the accelerating voltage. They are:

- Lack of detailed structures of specimen surfaces
- Remarkable edge effect
- Higher possibility of charge-up
- Higher possibility of specimen damage

So, it is desirable to use low accelerating voltages for observation of low-concentration substances. At low accelerating voltage, while surface microstructures can be observed, it is difficult to obtain sharp micrographs at high magnification. In such a case, clear images can be obtained by shortening the WD (working distance)or reducing the electron probe diameter. Probe diameter depends on the accelerating voltage, probe current and working distance. (WD is the distance from the underside of the objective lens to the specimen surface and probe current is the total amount of current to be irradiated on the specimen.)

When a high accelerating voltage is used, more scattered electrons are produced from the constituent substances within the specimen. This not only eliminates the contrast of surface microstructures, but produces a different contrast due to backscattered electrons from the substances within the specimen.

In SEM, the smaller the electron probe diameter on the specimen, the higher the magnification and resolution. It is therefore necessary to select a probe current suited for the magnification and observation conditions (accelerating voltage, tilt, etc.) and the specimen.

Influence of Edge Effect on Image Quality

Among the contrast factors for secondary electrons, the tilt effect and edge effect are both due to the specimen surface morphology. Secondary electron emission largely depends on the incident angle of the probe on the specimen surface, and the higher the angle, larger is the emission caused. The objects generally have uneven surfaces. There are many edges all over them which contribute most to the contrast of secondary electron images. On the other hand, large quantities of secondary electrons are generated from the protrusions and the circumferences of objects on the specimen surface, causing them to appear brighter than the even portions. This is known as "edge effect" which is illustrated in Figure 5.18.

The degree of edge effect depends on the accelerating voltage. The lower the accelerating voltage, the smaller the penetration depth of incident electrons into the specimen. This reduces bright edge portions, thus resulting in the microstructures present in them being seen more clearly. Higher the accelerating voltage, greater is the edge effect, making the edges brighter.

Normally, secondary electron images contain some backscattered electron signals. Therefore, if the tilt direction of the specimen surface and the position of the secondary electron detector are geometrically in agreement with each other, more backscattered electrons from the tilted portions are mixed, causing them to be seen more brightly due to synergism.

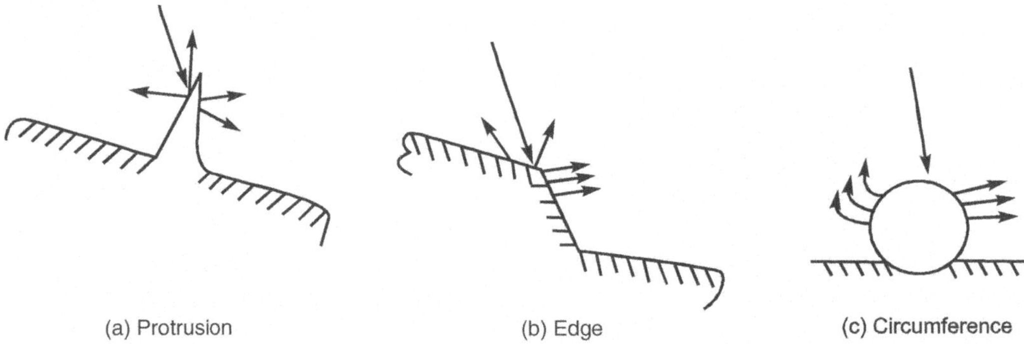

Figure 5.18 Edge effect (secondary electron emission differing with surface condition) (*Source: JEOL SEM Obsn. Manual.* p.9, Fig. 11.)

Influence of Specimen Tilt

Specimen tilt is aimed at:

- improving the image quality of secondary electron images
- observing topographic features and observing the sides of the specimen
- obtaining stereomicrographs

Figure 5.19 shows a specimen image taken at tilt angle of 0 and 45°. It is evident from this figure that the image with the 45° tilt angle is of more smooth quality and is stereoscopic as compared to that with the 0° angle. However, when the specimen is tilted, lengths observed are different from their true values. When measuring pattern widths, etc., therefore, it is necessary to measure without specimen tilting or to correct the values obtained from a tilted state.

With SEM images it is sometimes difficult to correctly judge their topographical features. In such a case, observation of stereo-SEM images makes it easy to understand the structure of the specimen. Besides, stereo-observation allows unexpected information to be obtained even from specimens of simple structure.

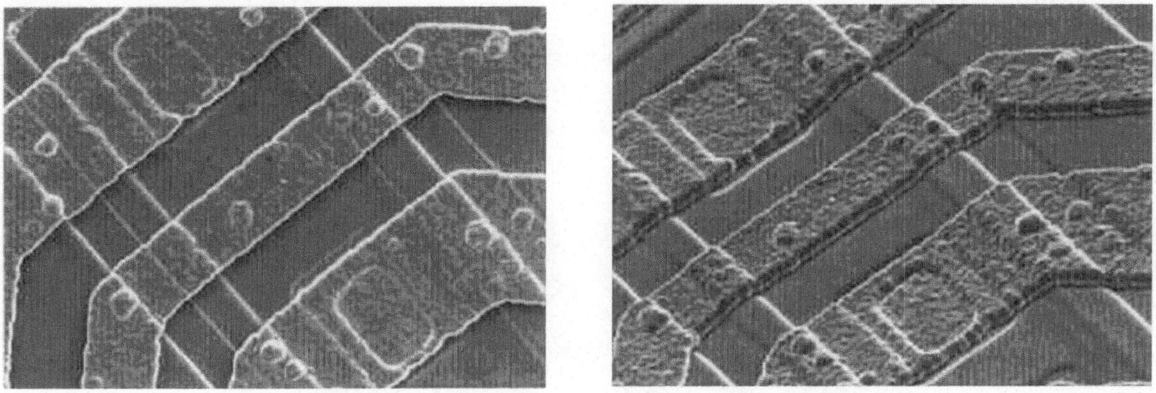

(a) Tilt angle 0° (b) Tilt angle 45°

Figure 5.19 Effect of tilt angle on image quality (*Source: JEOL SEM Obsn. Manual.* p.10, Fig. 13.)

Specimen Damage by Electron Beam

The loss of electron beam energy in the specimen occurs mostly in the form of heat generation at the irradiated point. The temperature increase at an irradiated point is dependent on:

* the electron beam accelerating voltage and dosage
* scanning area
* scanning time
* heat conductivity of the specimen

Polymer materials and biological specimens, which are generally not resistant to heat, are easily damaged by the electron beam, because of their low heat conductivity. To avoid this damage, the following should be taken into consideration.

* Using low accelerating voltage.
* Decreasing electron beam intensity.
* Shortening exposure time, even though this reduces photograph smoothness slightly.
* Photographing large scanning areas with low magnifications.
* Controlling the thickness of coating metal on the specimen surface.

It is also advisable to adjust beforehand the astigmatism and brightness using another field of view and then photograph the actual field as quickly as possible.

Contamination

When the electron probe is irradiated on a part of the specimen for a long time, its image may lose sharpness and become dark as seen in Figure 5.20. This is caused by the residual gas in the vicinity of the specimen being struck by the electron probe. This phenomenon is called specimen contamination.

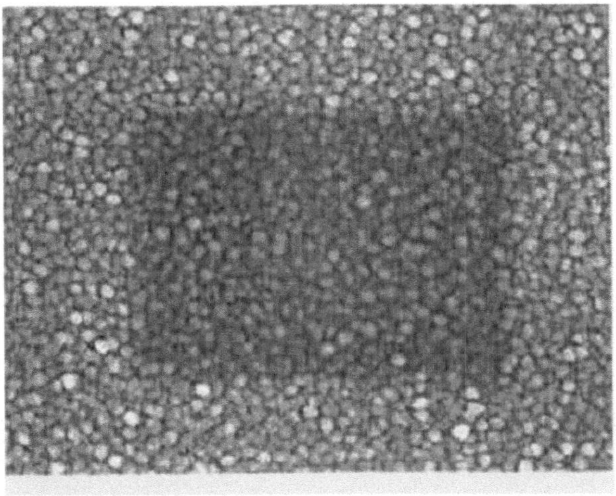

Figure 5.20 Specimen contamination after prolonged scanning (*Source: JEOL SEM Obsn. Manual.* p.17, Fig. 27.)

The residual gases in the specimen chamber, which cause contamination may arise due to the following.

1. Gas may arise from the instrument itself.
2. Gas may be brought into the instrument by the specimen and the specimen itself gives off gas.

To prevent specimen contamination, special attention must be paid to the following matters:

i. Use the minimum amount of double-sided adhesive tape or conductive paint, and completely dry it before putting the specimen in the instrument.
ii. In some cases, contamination can also be reduced by drying the adhesive with a drier or the like.
iii. Reduce the specimen size to the minimum possible.
iv. Since some embedding agents and resins give off much gas, they need to be selected carefully. Also since organic gas is given off when the resin surface is irradiated with an electron probe, irradiate the smallest possible surface area or coat the surface with a conductive material.

INFLUENCE OF OPERATIONAL TECHNIQUES AND EXTERNAL DISTURBANCE ON IMAGES

Influence of Working Distance, OL Aperture and Astigmatism

The smaller the OL (objective lens) aperture diameter and the longer the WD the greater is the depth of field obtained.

The aberration caused by the inaccuracy of the machining and material of the pole piece is called astigmatism. An image is judged as astigmatism-free if it has no unidirectional defocusing when the objective lens is changed to under- or over-focus at a little high magnification.

Influence of External Disturbances on Images

External disturbances like stray magnetic field, mechanical vibrations, etc., may influence the image and cause such phenomena as image distortion and jagged edge lines. The disturbances in SEM images are caused mostly by insufficient installation conditions such as

- ❋ Leakage of magnetic field from the distribution board, high-tension cable, etc. located near the instrument.
- ❋ Low floor strength
- ❋ Improper grounding

Therefore these have to be checked and avoided carefully before installation. Figure 5.21 shows the effect of external field on image quality.

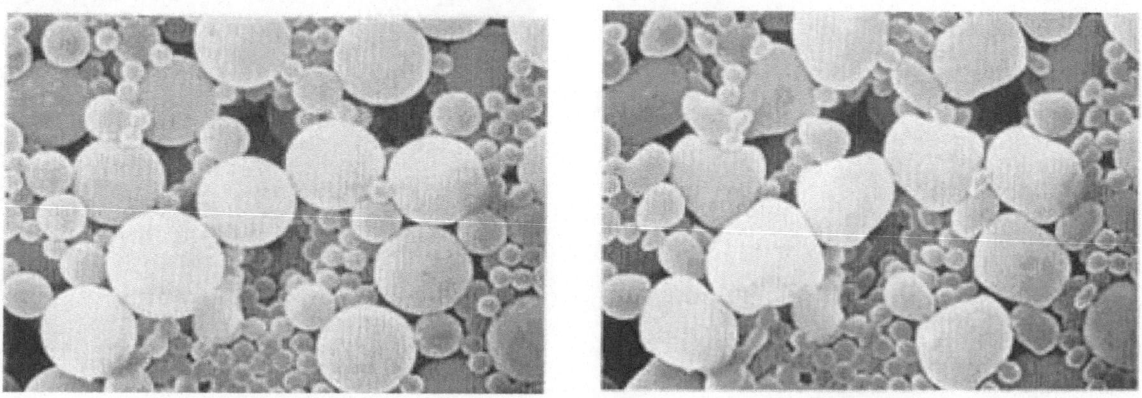

(a) Not influenced by external field (normal)   (b) Influenced by external field

Figure 5.21 Influence of external magnetic field (*Source: JEOL SEM Obsn. Manual*. p.27, Fig. 40.)

Influence of Specimen Processing Techniques

Biological specimens are apt to become defective during specimen preparation, because they require processes such as fixing, washing, dehydration, etc. Their defects can be classified as

- ✳ specimen deformation
- ✳ impurity precipitation
- ✳ impurity coating

To avoid such deformations as much as possible, the critical point drying method is used instead of air-drying. The difference in image quality between the two methods of drying is evident from Figure 5.22.

Impurity precipitation is a phenomenon where the crystal material contained in the fixing, washing and dehydrating agents precipitates during the drying process. Thus sufficient care must be taken when handling the above-mentioned chemicals.

Regarding surface coating for biological materials, the metal has to be selected depending on the purpose of observation and magnification. When the coating is too thick, its particles become visible while at the same time the structures of interest are buried. It is necessary to select a coating suitable for the magnification at which observation has to be made. In the case of specimens with rough surface, the charge-up effect cannot be prevented simply by coating with Au. In such cases, Au coating can be done after carbon evaporation. But it is absolutely necessary to uniformly coat the specimen from all directions by rotating and tilting the specimen during coating.

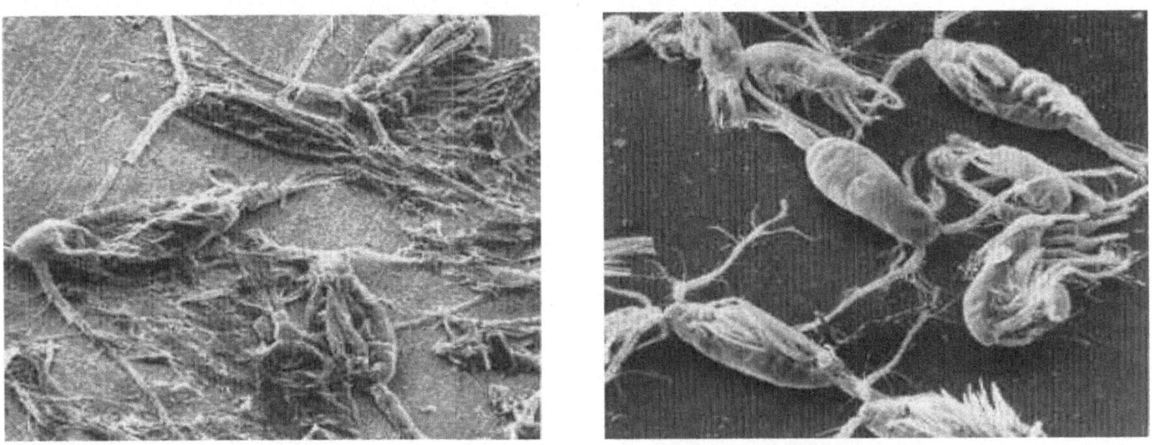

a) Air-drying (b) Critical point drying

Figure 5.22 Comparison between air-drying and critical point drying. (*Source: JEOL SEM Obsn. Manual.* p.25, Fig. 37.)

DISTURBANCES CAUSED BY INSTRUMENTAL DEFECTS

Insufficient Filament Heating

When the filament tip is not at a high-enough temperature due to insufficient filament heating, a proper cross over point may not be obtained, making it impossible to obtain a sharp image. Also, if the filament is heated too much, the filament is evaporated excessively, which results in the generation of whiskers, instability of the electron probe, or in a shorter filament life. It is thus important to maintain the temperature optimally.

Incorrect Alignment and Centring of Objective Aperture

The beam has to be aligned periodically, particularly after

- Changing the filament
- Replacing the objective aperture after cleaning
- When the column is disassembled for cleaning

The OL (objective lens) aperture should be set at the centre of the pole piece. When the aperture shifts from this position, however, the astigmatism of the image becomes extremely large, making it impossible to obtain high-resolution images. Incorrect alignment may lead to distortion of the image.

Influence of Discharge of Detector and CRT

The secondary electron detector and CRT are supplied with a high voltage. The poor connection of cables, exfoliation of the fluorescent paint, presence of dust on the metal ring and fluorescent plane of the detector all can cause discharges. The effect of these on image quality is similar in appearance to those of unusual accelerating voltage, unsatisfactory gun emission and specimen charge-up.

After a long use, the surface of CRT becomes dusty; therefore it should be cleaned from time to time.

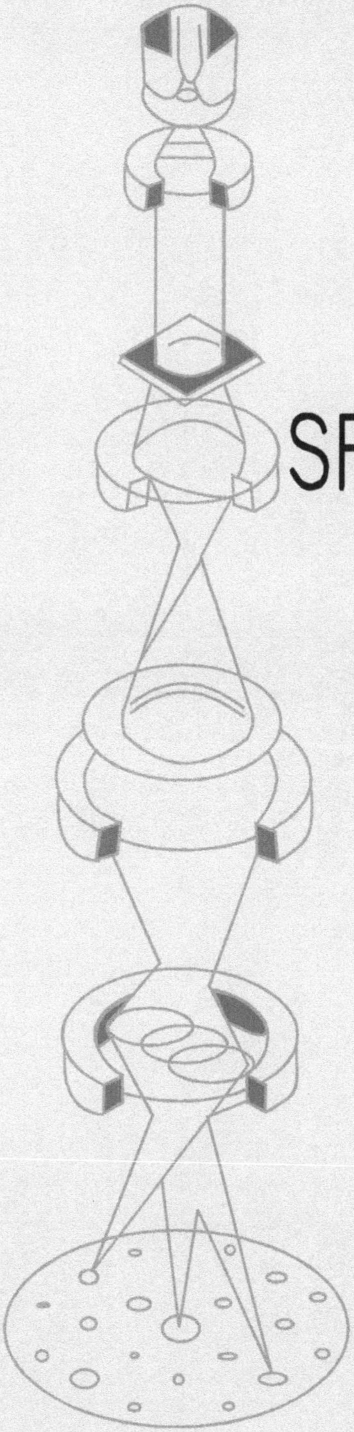

PART II

SPECIMEN PREPARATION TECHNIQUES

Electron microscopy makes an invaluable contribution not only in Anatomy, i.e., structural aspects, but also in Physiology and Pathology as well. The structure of the cell and its constituents as well as its structural relationships with neighbouring cells and adjacent tissues are evident by looking at their morphology directly at high magnifications and high resolution offered by the TEM. Normal ultrastructure has, indeed, helped to explain the functions of many cells. In fact, the existence of previously unsuspected functions can be revealed using the ultrastructure of the cells. For example, presence of secretory granules may indicate the endocrine potentialities of the cell. The consequences of an experimental procedure, either alone or in combination with cytochemical and immunocytochemical methods can be observed using the sophisticated tool, TEM. But to achieve the fruits of TEM technology, the specimen has to be suitably prepared for observation in the microscope. Actually, for biological specimens, the processing is very critical. Since the thickness of the specimen to be screened in TEM is limited, around 100 nm, ulltrathin sections have to be prepared for TEM.

The tissue specimen has to be preserved to prevent cellular damage and the consequent morphological alterations, and finally cut into ultrathin sections. Thus, in short, the specimen has to be preserved first as it is removed from its source, the water in the tissue has to be removed and simultaneously replaced with a material which can make the tissue hard enough for further sectioning and staining and finally viewing in the EM.

The outline of the protocol is as follows:

The tissue specimens for EM are collected in the primary fixative itself and cut into very small bits of 1mm × 1mm immediately. This cutting is conveniently done by placing the tissue on a sheet of dental wax or glass slide adding enough fixative to keep the tissue moist, and then cutting it into tiny pieces using a fresh single-edged razor blade or dissecting knife. The pieces are then transferred immediately to pre-labelled glass vials containing the fixative. The specimen bits are kept in the same vial during further processing steps, fixation, washing, dehydration and infiltration, i.e., till they are transferred to the embedding moulds.

Most important point to remember here is that the fixation should be initiated immediately on collecting the specimen. Delay in fixation or preserving the tissue in saline many lead to necrosis or other types of artifacts. Figure II.1 reveals the features of routine fixation while Figure II.2 shows the effects of preservation of cells in saline for 1 hour before routine fixation. The artifacts are clearly visible in Figure II.2.

The steps involved in the preparation of specimen for EM are shown in the flowchart that follows.

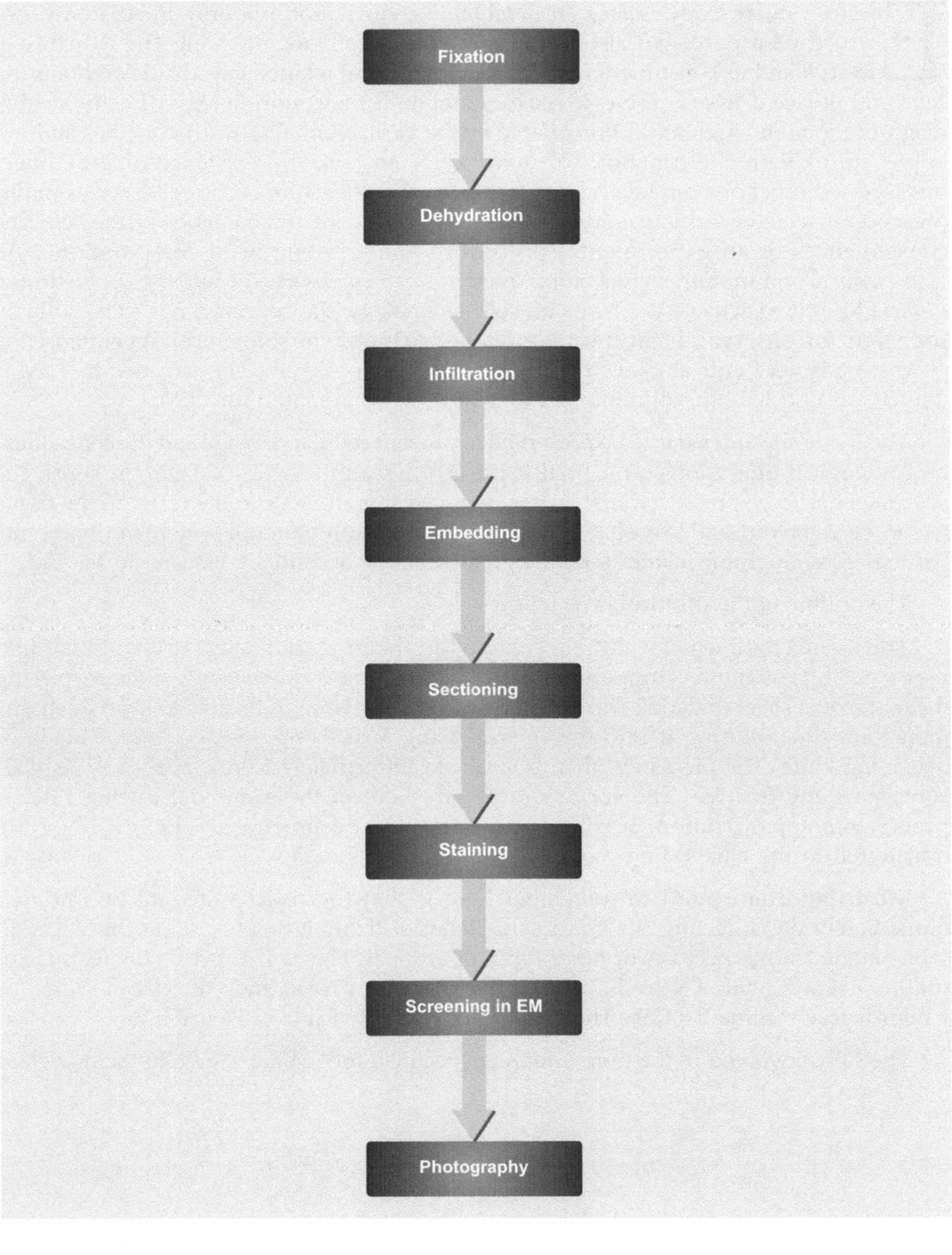
Fixation
Dehydration
Infiltration
Embedding
Sectioning
Staining
Screening in EM
Photography

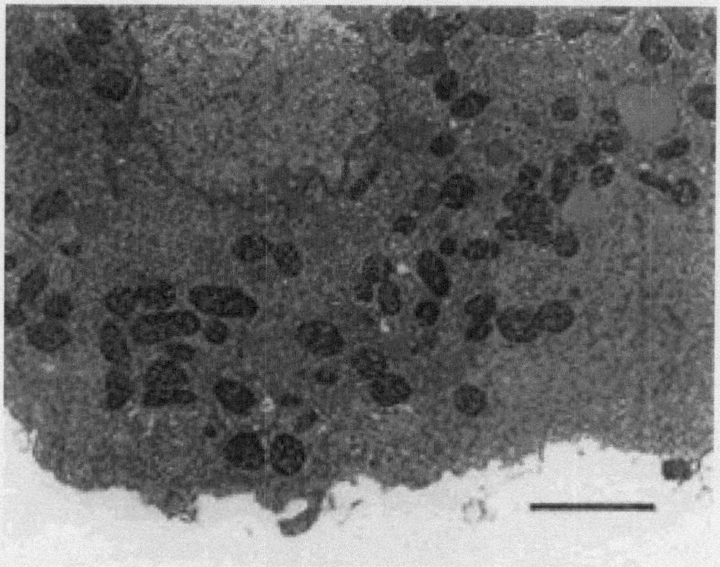

Figure II.1 Cell fixed by the routine chemical method

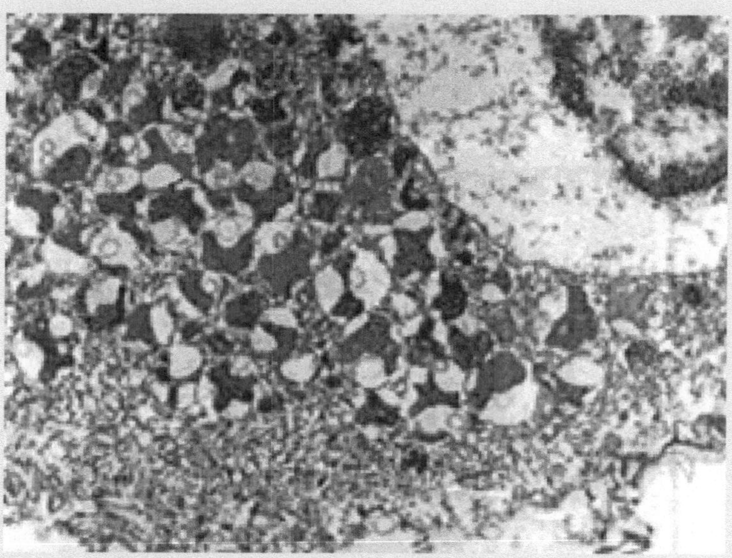

Figure II.2 Cell stored in physiological/unbuffered saline for 1 hour prior to routine aldehyde fixation

This emphasizes the need for strictly following the processing schedule *in toto* to get reliable and reproducible photographs in the EM.

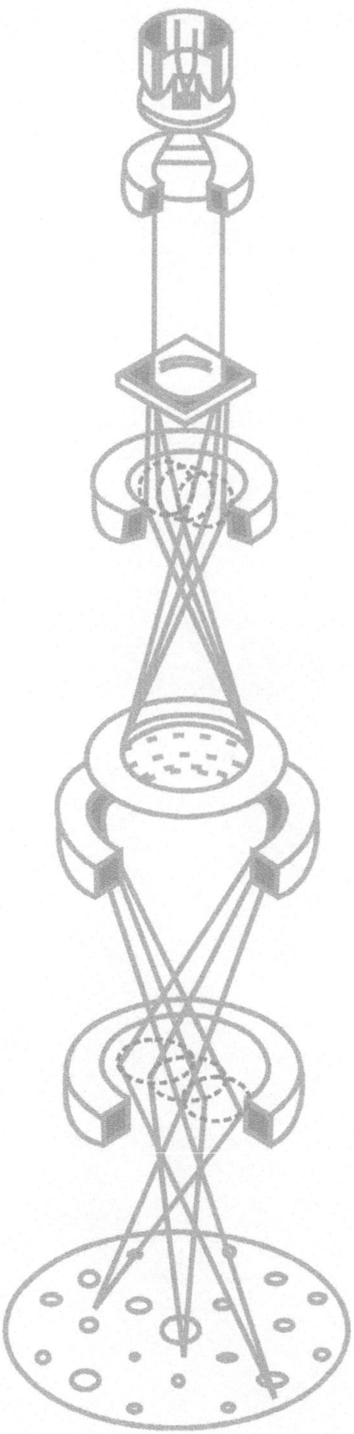

SECTION I

PROCESSING OF TISSUE SPECIMENS

- Fixation
- Washing and Dehydration
- Infiltration and Embedding
- Sectioning—Semithin and Ultrathin
- Staining

6

FIXATION

Living specimens have to be immobilized and all biochemical processes including enzyme reactions must be halted such that there are minimal changes in their ultrastructure. This initial step of morphological preservation is called "Fixation". Any failure at this fixation stage renders the whole project useless.

Fixation actually kills the cells and arrests all enzyme reactions preventing autolysis. Thus it preserves the morphology of the tissue. This fixation should be strong enough to prevent as far as possible any adverse effects of the subsequent dehydration and embedding procedures as well as irradiation with electron beam.

AIM OF FIXATION

The aim of the initial step, fixation, is to preserve the ultrastructure with as little alteration as possible when compared with living conditions. It is also to stabilize cellular organization to such an extent that the ultrastructural relations are preserved despite the subsequent rather drastic treatments of dehydration, embedding and exposure to the electron beam. However this stabilization modifies certain structures such as the contraction or relaxation of muscle fibres or the triggering of secretions by solutions containing calcium.

Two processes of good preservation of ultrastructure are by:

- formation of a stable protein gel by the major proteins present in the tissue. Protein gels are formed during fixation by formation of stable cross bridges between adjacent protein molecules.
- stabilization of membranes, both, external membranes of cells and internal membranes of cellular organelles.

CRITERIA FOR GOOD PRESERVATION OF ULTRASTRUCTURE

1. Final appearance of specimen should not conflict with evidence gained from examination of living specimen viewed with light microscope, including advanced techniques like confocal microscope, polarizing microscope, etc.
2. Final appearance should not depend on the method of fixation, except in ways that can be explained by known chemical and physical principles.

   Particular structures and their interrelationships are not unique for one method of fixation which offers strong evidence against them being completely artifacts. For example, rapidly frozen material which is then dehydrated by freeze-substitution before embedding, provides valuable confirmation of many of the results obtained by standard chemical fixation and dehydration methods. In other words, cryofixation and chemical fixation should show similar features.
3. Final appearance should be consistent with those biochemical and physiological data which have structural implications. For example, if two membranes of mitochondria show imperfect appearance, it indicates the result of damage during processing.

METHODS OF FIXATION

There are basically two methods of fixation—physical methods and chemical methods. **Cryoimmobilization fixation** of non-chemically treated tissue is a viable alternative. **Initial irradiation with microwaves** to produce physical fixation has also been recently advocated. But the most popular method is chemical fixation as it is easier to carry out and handle.

But, whatever be the method of fixation employed, it should be started as early as possible such that post-mortem changes are kept minimum.

Physical Methods

This method involves two types of fixation:

i. Cryo-fixation (cryotechniques) (*See* chapter 16—Cryotechniques in electron microscopy)

ii. Microwave fixation (using microwave energy) (*See* chapter 15—Microwave techniques in electron microscopy)

Chemical Methods of Fixation

Chemical methods of fixation are more widely used, where certain chemicals called '**fixatives**' are used for fixing the tissues. The choice of fixatives and other procedural parameters vary depending on the nature of the specimen and the study requirement. Since there is no one perfect fixative to suit all types of specimens, it is necessary to use different fixatives for different structural elements within a tissue. But the general aim is to preserve the tissue or organ as a whole. Subtle differences in fixation can also occur at the cellular level, and the best preservative of mitochondria may not be compatible, for example, with the best preservation of rough endoplasmic reticulum.

Choice of fixative The choice of the fixative depends on the nature of work.

1. *For accurate preservation of ultrastructure* For this purpose, 2.5% glutaraldehyde followed by 1% osmium tetroxide and then enbloc staining with uranyl acetate.
2. *For some cytochemical studies* For this, lower concentration of glutaraldehyde is preferable and uranyl acetate enbloc staining is omitted.
3. *For enzyme cytochemistry* Primary aldehyde fixation has to be carried out at 4°C and tailored to particular enzyme under study and cytochemical techniques applied before post-fixation with osmium.
4. *For immunocytochemistry* Primary fixation is even more critical here. Low concentration of aldehyde (0.05 to 0.2%) is required without post-fixation to preserve epitopes under study with inevitable loss of some ultrastructural details. Aldehyde fixation needs to be followed by dehydration at low temperatures. This is particularly for pre-embedding techniques. (*See* chapter 17 for details).

Concentration, duration and temperature are variables which may produce improvements, but of course, have to be selected after exhaustive study. Use of additives is also recommended for preservation of specific features.

The most popularly known fixative is formalin which is routinely used in light microscopy. It is used even to preserve cadavers (dead bodies) in mortuaries. Formalin has a fast penetrating power, penetrating 2 mm in 2.4 hours but has a slow fixing action which is reversible. Moreover, commercial formalin is impure and contains other impurities including traces of formic acid and considerable amounts of methanol, which may destroy the ultrastructure. So, formalin as such is unsuitable for electron microscopy. But, methanol-free formalin, which is freshly prepared by depolymerization of paraformaldehyde, is used in combination with glutaraldehyde in electron microscopy especially in immunoelectron microscopy where rapid initial fixation by using fast penetrating fixative is mandatory for preserving the ultrastructure.

In EM for ultrastructural studies, usually another aldehyde, glutaraldehyde is used which gives excellent preservation. It is a slow-penetrating fixative unlike formaldehyde, penetrating 2 mm in 24 hours but its fixing action is fast and irreversible. Hence it is considered as the most suitable fixative for ultrastructural studies. Because of its slow-penetrating power, the specimen is cut into very small bits, 1 × 1 mm size, before the actual fixation starts. For good ultrastructural preservation, samples for fixation should be no thicker than about 250 micron in one dimension. Fixation time of 30 minutes to one hour is usually sufficient for thin slices of tissue but for cells and tissue cultures even shorter duration is enough for stabilization of intracellular organelles. But more prolonged fixation may improve the overall preservation of ultrastructure. Plant tissues usually require longer fixation times than animal tissues.

Glutaraldehyde and formaldehyde (freshly prepared from depolymerization of paraformaldehyde) combination is also used where higher concentration of glutaraldehyde is used for ultrastructural studies and higher concentration of formaldehyde for cytochemical and immunocytochemical studies. Formaldehyde alone is not good for ultrastructural studies. But it is less disruptive for functional proteins like enzymes and antigens than glutaraldehyde and is hence used for cytochemical and immunological studies. Acrolein, rarely used routinely, penetrates cell membranes rapidly and is useful for immunofixation of larger specimens.

Chemistry of fixation Table 6.1 gives the reactions between some common fixatives and the various cellular constituents.

Table 6.1 Fixatives and constituents of cells

Fixatives	Cellular constituents				
	Prot.	N.A	Lipids	Polysac.	Pho.lipids
Aldehydes	++	–	–	–	–
OsO_4	+*	–	++**	–	++
$KMnO_4$	++	Destroyed	+	+(Glycogen)	++

Prot. = Proteins; N.A. = Nucleic acids; Polysac. = Polysaccharides; Pho.lipids = Phospholipids

\+ Moderate; ++ Strong; –Weak or no reaction

* Fixation for longer duration leaches out proteins.

** Mainly unsaturated lipids

Aldehydes Aldehydes react primarily with protein molecules. Protein gel is formed by formation of cross bridges between adjacent protein molecules. Most of the reactions are with free amino groups of proteins, principally those on lysine residues. Reactions with individual protein may render them insoluble and therefore less likely to move, but more important reaction is between adjacent protein molecules with the formation of covalent bridges which firmly anchor the protein molecules in place.

The properties of bridges vary with the aldehyde used and conditions of formation. Bridges formed by formaldehyde are less stable than glutaraldehyde and these bridges are more stable with higher concentration of fixatives.

Glutaraldehyde Glutaraldehyde, the most popular fixative for ultrastructural studies, is a pentacarbon dialdehyde having two potentially highly reactive —CHO (aldehydic) groups and hence cross links between adjacent amines could occur readily. It reacts rapidly and irreversibly with amino acids particularly, terminal amino groups of lysine residues, *N*-terminal amino groups of proteins, particularly lysine, and SH groups of cysteine.

Reactions with amines and cyclic imines result in the formation of pyridine products which involves uptake of oxygen molecules and so anoxia can have an inhibitory effect on this aspect of fixation. Uptake of oxygen further makes these reactions irreversible and hence glutaraldehyde produces a very stable fixation than formaldehyde. Uptake of oxygen moreover may lead to anoxia for large specimens and therefore specimens have to be cut into thin slices for glutaraldehyde fixation.

Reactions between —CHO (aldehydic group) in the fixative and amino groups of the proteins in the tissue under study, lead to net production of acid, and therefore adequate buffering to maintain the pH is essential during glutaraldehyde fixation. Phosphate, cacodylate and PIPES buffers have pH values below 7 and hence are able to take up adequate amounts of acid without the pH falling too much. All three buffers carry a net negative charge at physiological pH and hence diffuse through cell membranes more slowly than the uncharged molecule of fixative. Therefore, cell interiors may become acidic during fixation. To minimize this, the fixing solution should contain high concentrations of buffer and maintain osmolarity also. Anyway the pH of the fixative should be compatible with cellular pH (7.0 to 7.4) during fixation.

Phosphate buffer has the advantage of being cheaper and nontoxic but gets easily contaminated with microorganisms. So it has to be freshly prepared and cannot be stored for long. Cacodylate is compatible with calcium and all heavy metals used in cytochemistry. It strongly inhibits growth of microorganisms, inhibits tissue respiration also and therefore, does not get contaminated with microorganisms easily. Thus the working solution prepared can be stored for a longer time in the refrigerator. But, the disadvantage is, that it contains arsenic which is toxic and is also very costly. Recently, PIPES or MOPS buffer which are nontoxic and cheap are also being tried.

The fixative solution should match tissue environment with respect to pH, osmolarity and ionic constitution. Osmolarity should be such that the cells neither swell nor shrink. The ionic concentration of fixative is such that materials are neither extracted nor precipitated during fixation. Aldehyde fixatives do not destroy all semipermeability properties of lipid membranes whether of cell or its organelles. So osmolarity is important for all solutions for primary aldehyde fixatives. But the secondary osmium fixative destroys all permeability barrier. So osmolarity is not important during post-fixation with osmium.

0.1 M cacodylate buffer is hypo-osmotic slightly but overall, the fixative is hyperosmotic. Thus osmolarity is adjusted by altering the concentration of buffer or by addition of inert osmotically active substances as below:

- NaCl (sodium chloride) is the best constituent to make up necessary osmolarity.
- Glucose and sucrose should not be added with primary aldehyde fixative because they can cause artifacts such as swelling of mitochondria.
- High concentration of potassium is also not used because it can cause cytoplasmic changes before fixation is complete.
- So only low concentration of divalent ions like 2 to 10 mM magnesium, or 1 to 3 mM calcium can be incorporated into the fixative along with 0.1 M phosphate, 0.1 M cacodylate, or 0.1 M PIPES buffer.

Thus glutaraldehyde with rapid crosslinking forms a 3D network throughout the cell cytoplasm within a few minutes even at low concentrations.

Reactions of glutaraldehyde with other cell constituents

1. DNA is well-fixed because aldehydes react avidly with histones. But direct reaction with DNA or RNA does not occur at room temperature.
2. A major proportion of glycogen is retained because glycogen gets enmasked in the crosslinked matrix of proteins.
3. As for lipids, it reacts with some lipids like phosphotidyl ethanolamine which carry a free amino or imino group. It also reacts with protein components of lipoproteins in cell membranes but this can cause release of lipid components. Glutaraldehyde fixation alone does not protect most lipids from extraction during dehydration and embedding. So to preserve lipids, secondary fixation with osmium is necessary.

25% glutaraldehyde is the usual stock solution which is generally stored at low temperatures. If the solution turns yellow or the pH becomes less than 4, then it should be discarded. A fixative consisting of 1 to 2.5% glutaraldehyde in 0.1 M phosphate, or cacodylate or PIPES buffer gives satisfactory primary fixation.

Formaldehyde It is prepared freshly from paraformaldehyde to avoid methanol and formic acid present in commercial formalin preparations. Formaldehyde is a gentler fixative, but penetrates more rapidly than glutaraldehyde. Many of its reactions with tissues are at least partially reversible. Formaldehyde penetrates rapidly and hence may modify semipermeability of cell membrane more than glutaraldehyde. So usually, a combination of glutaraldehyde and formaldehyde is used as primary fixative for cytochemical techniques.

Reactions with cell constituents

- Formaldehyde forms methylene cross bridges with amino groups particularly terminal groups of lysine residues.
- It also reacts with DNA and histones.
- It does not react strongly with lipids and carbohydrates.

4% Formaldehyde + 5% Glutaraldehyde + 0.05% $CaCl_2$ (Calcium chloride) in 0.08 M cacodylate buffer at pH 7.2 is known as Karnovsky's fixative which is used for immunoelectron microscopy.

Osmium tetroxide The primary aldehyde fixative does not fix lipids and membranes and hence secondary post-fixation with osmium tetroxide is essential for ultrastructural studies. Osmium is the only fixative that preserves delicate cytoplasmic processes and causes no alteration in mitochondria and fat droplets, when checked with dark-field microscopy. But osmium as a primary fixative has a destructive effect on proteins. So it is used only as a secondary fixative with an advantage of being a stain cum fixative. Osmium fixation has to be carried out at low temperatures and for shorter duration, say, 1 to 2 hours at 1% concentration. Use of buffer is optional with OsO_4 unlike glutaraldehyde where it is essential.

Osmium reacts with lipids, carbohydrates and proteins. It reacts avidly with olefinic double bonds of unsaturated lipids. It is reduced by free SH groups of cysteine and thus osmium gets linked to S atoms of proteins. Osmium which is bound to lipid or carbohydrate readily binds to nitrogen atoms in proteins (as in histidine, tryptophan, arginine, lysine, etc.) thereby forming cross links between different tissue components which help to stabilize cytoplasmic structures and surface membranes. Many of these compounds probably react with alcohols during dehydration leading to the deposition of black osmium dioxide (OsO_2) which is resistant to further dehydration and embedding. Thus the specimen turns black on exposure to osmium.

Osmium has high affinity for nitrogenous compounds in general, and amines in particular. A mixture of osmium tetroxide and imidazole gives better staining of lipids and lipoproteins than osmium alone.

Glycogen structure is enhanced by post-fixation with osmium tetroxide in the presence of ferrocyanide/ferricyanide. A mixture containing 1.0% OsO_4 + 1.5% potassium ferricyanide is a popular fixative for ultrastructural studies.

Aqueous osmium tetroxide buffered by chromate–dichromate mixture is also used. The chromate contributes to the fixation and staining of lipids and carbohydrates with improved preservation of cytoplasmic material and of myelin. Hence it is used as a secondary fixative for studies of the nervous system.

Primary fixation with mixtures of osmium tetroxide and certain metallic iodides, particularly those of zinc and cadmium, produce rather special staining effects. These mixtures are referred to as "champy–maillet" fixatives, which stain lysosomes, Golgi regions and some other structures known to contain lipids. They also produce intense but not reproducible staining of synaptic vesicles both in monoaminergic nerve endings and in some cholinergic endings.

Uranyl acetate Though uranyl acetate is primarily used as a stain for ultrathin sections, it gives a very definite improvement in ultrastructural preservation of many cell components when used as a third fixative after primary aldehyde fixation and post-fixation with osmium tetroxide. It is a third fixative for ultrastructural studies because it gives good preservation and staining of membrane structures. Like osmium, it serves a dual role of fixing and staining specimens.

Uranyl acetate is particularly effective in fixation of membranes as a consequence of stabilization of phospholipids. It stains membranes and also cellular organelles like rough endoplasmic reticulum (RER), smooth endoplasmic reticulum (SER) and extracellular proteins such as collagen. This type of staining with uranyl acetate as a third fixative is generally referred to as "en bloc staining." This en bloc staining preserves cytological details and reduces risk of introducing contamination during staining.

Reactions with cell constituents Uranyl acetate in aqueous solution contains both positive and negative ionic complexes and therefore fixes both acidic and basic groups on cell constituents.

i. Uranyl acetate reacts avidly with nucleic acids probably combining with phosphate groups and firmly fixes and stains both DNA and RNA. It binds also strongly to phosphate groups of phospholipids of membranes and reduces the extraction of their lipid content during dehydration.
ii. Proteins react with both their acidic and basic groups.
iii. But, carbohydrates react only if they possess acidic groups.

The routine procedure adopted for enbloc staining is to immerse samples in the dark (because, uranyl acetate is light-sensitive) in 0.5 to 1.0% aqueous uranyl acetate for 1–2 hours at room temperature after initial aldehyde fixation and post-fixation with osmium tetroxide.

Other fixatives Apart from the above-mentioned fixatives—glutaradehyde, formaldehdye, osmium tetroxide, and uranyl acetate—no other fixative has general acceptance for routine electron microscopy. But some fixatives with limited purpose are listed below:

Acrolein This fixes tissues more rapidly and is therefore of value for fixation of large blocks of tissues. But a much safer alternative is to use glutaraldehyde–formaldehyde mixture. Acrolein is most useful for fixation of plant tissues and microorganisms with dense cell wall and for specimens covered in impermeable layers such as chitin or wax.

Tannic acid This is a slow-penetrating fixative. So it is mainly used in conjunction with glutaraldehyde or osmium tetroxide. Thus it acts more as an additive. It is used as a mordant during staining as it increases uptake of heavy metals such as lead and uranium salts by extracellular components. Tannic acid is applicable for a range of substances.

It increases contrast of outer lamina of cell membranes and also of extracellular structures like elastic fibres and cell junctions (e.g., desmosomes). It also stabilizes and stains microtubules and other proteinaceous structures including actin filaments.

Potassium dichromate It has two applications when used as a fixative.

i. It acts as a buffer for osmium tetroxide fixative because it has a strong buffering action at pH 7.6. It improves fixation and staining of lipids and carbohydrates. So it is used as a secondary fixative after primary aldehyde fixation especially for studies of nervous system because of its excellent demonstration of myelin.

ii. It is used for demonstration of biogenic amines such as noradrenaline and dopamine because of combination of chromium with products formed by the action of aldehyde fixatives on biogenic amines.

It has been shown that,

- only 5 HT (5-hydroxytryptamine) is demonstrated if formaldehyde alone is used.
- noradrenaline and dopamines are also demonstrated if glutaraldehyde is added to formaldehyde.
- adrenaline is demonstrated only if acrolein is present in the primary fixative.

Potassium permanganate This is not at all used as an EM fixative now. It does not fix proteins and RNA and therefore, is unable to preserve many cellular components. It causes ribosomes to disappear and membranes to rupture. But it preserves DNA and glycogen. It has the ability to penetrate thick cell walls and waxy surface coats and hence is favoured only for botanical and insect samples.

METHODS OF APPLICATION OF FIXATIVES

There are primarily four methods of application of fixatives which again depend on the nature of the specimen. They are:

1. *Fixation in situ* for small specimens such as, tissue cultures, cell suspensions and pellets of isolated granules.
2. *Fixation by simple immersion* for biopsy specimens
3. *Fixation by injection* for invertebrates, especially insects.
4. *Fixation by perfusion method* (whenever possible) for vertebrates. Fixing method depends on time taken for fixation to reach all cells.

Whatever be the method employed, fixation should be started as early as possible especially for birds and mammals which have high body temperature. The choice of method depends on percentage of extracellular space and its degree of interconnection. When these are high as in tissues with lots of connective tissue, articular cartilage, etc., immersion fixation is the best. In highly packed tissues with little extracellular space as in avian brain, endocrine tissue with epithelium, etc., perfusion fixation is most suitable. In rapidly metabolizing tissues like skeletal muscle, cardiac muscle, etc., perfusion fixation offers greatest benefits.

Surface Application of Fixatives (*in vivo Fixation*)

Fine structure of many animal tissues depends on a continuous supply of blood and therefore fixation has to be started while the animal is still alive.

This method is carried out by dripping the fixative onto the exposed tissue in an anaesthetized animal. It is useful only for superficial layers of the tissue, especially when there is no capsule around the tissue. The capsule if present, can be slit open, but this may damage underlying tissue. This method is most useful for study of:

- underside of skin
- skeletal muscle
- tendons
- connective tissue

Preliminary dissection is done to avoid damaging the tissue while preserving nerve and blood supplies intact. The tissue is then flooded with ample supply of fixative cooled to 0 to 4°C, to reduce to a minimum the continuation of any metabolic activity in the tissue. Blood supply should be stopped as soon as the fixative is applied. Higher concentration of fixative is applied.

4% glutaraldehyde + 4% formaldehyde in 0.1 M cacodylate buffer at pH 7.4 is the fixative of choice here. Formaldehyde is important because of the necessity for rapid penetration. The tissue is kept continuously bathed with fixative for 20 minutes by adding drops of fixative every few seconds either manually or through a gravity feed. As excess fixative accumulates, it is withdrawn using a pipette or by slightly tilting the animal to drain off the excess fixative. At the end of twenty minutes, the tissue is dissected out and cut into pieces so as to preserve the orientation of the well fixed superficial layer. The tissue pieces are further fixed by immersion forl hour at room temperature or 1 to 3 hours at 4°C in the same fixative.

Fixation by Injection (*in vivo* Fixation)

The internal structure of an organ or tissue can be fixed by injection of fixative through a hypodermic needle. Usually, glutaraldehyde–formaldehyde mixture is used which is injected using a syringe over a period of 5 to 20 minutes. The region of tissue around the injection site is then carefully dissected out so as to include only well-fixed tissue.

Thin slices are cut and fixed in the same fixative for 1 hour at room temperature or for 1 to 3 hours at 0 to 4°C and processed further.

This method is mainly suitable for hollow structures such as, stomach, bladder, length of intestine, etc. Injection is in progress for 2 to 3 minutes after which a small hole is cut well away from the injection site to allow escape of excess fluid. The inner surfaces of hollow structures appear well-fixed by this technique.

Fixation by Vascular Perfusion (*in vivo Fixation*)

This method offers greatest benefits. It is most suited for tissues in which cells are tightly packed with little extracellular space. When organs are too large or heterogeneous in density or when dissection of a particular tissue type is time-consuming, fixation is carried out by perfusing the intact organs with fixative. With perfusion, the blood is first displaced by a suitable saline solution and then immediately by the fixative. Hence fixation starts before any autolytic or other degradative changes have a chance to occur. Also, this method can be carried out without handling the tissue, i.e., without any need to cut fresh tissue which may lead to excess pressure on the fragile tissue and distortion of its components. In this method, the fixative is perfused through the blood supply in an anaesthetized animal. Moreover in this method several tissues can be fixed simultaneously.

The main disadvantage of this method is the fact that large quantities of fixative are required for whole body perfusion. For an adult rat, 300 to 400 ml of the fixative may be required. Added to that, this method cannot be applied for human samples and is possible only for animals which can be sacrificed.

For many animal studies, fixation by perfusion, whether of the whole animal or of isolated regions of tissue such as kidney, heart or lungs is much to be preferred.

This fixation can be carried out by two methods:

1. constant pressure method
2. constant volume method

In constant pressure method, the reservoir of fixative is kept at a set height above the animal which produces a non-pulsatile supply of fixative at a well-controlled pressure. In the latter method, a roller pump produces a pulsatile delivery of fixative to the vascular system but at a pressure which may be difficult to control.

In both methods, the animal is quickly but fully anaesthetized and minimum amount of dissection is carried out to insert a cannula in an appropriate blood vessel. The solutions are introduced into the left ventricle of the heart or into a major vessel of an appropriate anaesthetized animal. That is, the fixative is introduced into the arterial system and allowed to egress from a cut made in the venous system. Outlet for blood and perfusion solutions is made in the right atrium/ventricle. For example, a needle is inserted into the left ventricle and the solution is allowed to flow from the left ventricle to the aorta when it is distributed to most of the capillary beds of the body. It returns to the heart via the venous system and exits the heart through a cut made in the right atrium or ventricle. Initially, physiological saline (for mammals) or Ringer's solution (for non-mammalian vertebrates) sometimes with an anticoagulant is perfused at room temperature to wash out the blood. Organs will blanche when cleared of blood. This is followed by fixative which is perfused for ten to fifteen minutes. Required tissues are dissected out and are sometimes fixed further by immersion.

Steps in perfusion fixation procedure

1. Animals are anaesthetized with Nembutal (sodium pentobarbitone) 30 mg/ kg body weight.
2. The heart is exposed.
3. The ascending aorta is cannulated with a catheter of 1 mm diameter through an incision in the left ventricle.
4. The right auricle is incised for the outflow of fixative.
5. Perfusion with normal saline is done to wash out blood.
6. Fixative is allowed to flow rapidly for a minute and then the flow is reduced to about 6 ml/minute for 10 to 15 minutes.
7. Proper fixation is indicated by the loss of trembling movements, hardening of organs and stiffening of limbs.
8. Following perfusion, the organs are dissected out and tissues are cut into smaller size 1 to 2 mm^3 and further fixed with the same fixative for 2 to 12 hours at 4°C.

Fixation by Immersion (*in vitro* Fixation)

This is the only method that is applicable for human tissue samples. Most suitable for biopsies and surgical specimens. Organs and tissues such as skin which can tolerate a temporary interruption to their blood supply and still retain their function and structure are adequately fixed by immersion. The tissue resected out is cut into small bits and completely immersed in the fixative which is usually 2 to 3% buffered glutaraldehyde. The fixation is carried out at 4°C or at room temperature for 4 to 8 hours or even longer depending on the nature of the specimen. The specimen should not be allowed to dry at any stage during processing starting from collection of the tissue and initial fixation. So, the usual practice is to keep the fixative in the operation theatre itself such that the duration between resection of the tissue and its immersion in the fixative is minimized. The specimen is immediately brought to the EM laboratory where it is further cut into bits of 1 × 1 mm size because, glutaraldehyde, the primary fixative, has a very slow penetrating power. While cutting into small bits also, care must be taken that the specimen is not dried. A drop or two of the fixative is placed on a clean glass slide and the tissue specimen received is placed on this drop. By using a sharp, clean razor blade or scalpel, the specimen is cut in the centre into two halves to equalize the stress on both sides. Drops of fixative may be added periodically if needed. By repeating this process several times, bits of 1 × 1 mm may be obtained. These are collected on the blade using the forceps and dipped into the fixative. It is necessary to ensure that all the bits are completely immersed in the fixative. For this, sufficient volume of fixative should be added. The fixative should be usually several times the volume of the specimen. The fixation is done preferably at room temperature for 30 minutes to 1 hour. If it is desired to be carried out in the refrigerator, the fixative has to be precooled. The fixation is done for 1 to 4 hours at 0 to 4°C. Instead of a glass slide, dental wax can also be used while cutting the specimen into small bits.

The main disadvantages of the immersion fixation method are:

- Diffusion is slow, post-mortem changes can occur in deeper layers of thick pieces of tissue.
- Buffer and other constituents may penetrate tissues more rapidly than fixative itself and cause morphological changes.
- The initial effect of the fixative, which is more in glutaraldehyde, is to reduce inner diffusion of fixative into deeper layers, i.e., rate of diffusion differs with depth. So deeper layers may become anoxic. Anoxia and autolysis can alter appearance of certain organelles like mitochondria. Therefore it is absolutely essential to use small specimens as mentioned above.

The choice of fixation method depends on the type of specimen and the nature of the study.

FIXATION TEMPERATURE

The optimum temperature for initial aldehyde fixation also depends on the type of specimen:

- For mammalian cells in culture or in suspension, which show rapid surface changes when exposed to a sudden drop in temperature, the initial contact between tissue and fixative should be at environmental temperature of the tissue.
- For biopsies, by immediate immersion, some fixatives penetrate before the temperature has fallen significantly. So for fixing tissue of warm-blooded animals, the fixative is used at room temperature, placing tissues in the refrigerator later if fixation at 4°C is required.
- For monolayer cultures, fixation is carried out at the temperature of incubation.

- ✳ For botanical specimens and tissues from cold-blooded animals, the ambient temperature is used.
- ✳ In fixation by perfusion, the body temperature is probably maintained, at least for a brief period, which should be sufficient to prevent changes from occurring while the tissue later cools.

Primary fixation at lower temperatures reduces extraction of lipids and other cell constituents but longer period of fixation is required. For purely ultrastructural studies, room temperature may be used. Secondary osmium tetroxide fixation is carried out at low temperatures, probably at 4°C. Tertiary uranyl acetate fixation is done at room temperature but in dark.

FIXATION TIME

For cell suspensions and monolayer cultures, 10 to 30 minutes is enough. For biopsies and other specimens fixed by immersion method, fixation should be carried out for longer periods especially when fixation is carried out at low temperatures, 1 to 4 hours at room temperature or longer periods, say, 4 to 6 hours or overnight at low temperatures. If need be, the specimen can be stored at low temperatures for longer periods using cacodylate buffer. For post-fixation with osmium, 1 to 2 hours fixation at low temperatures is sufficient.

RAPID FIXATION METHOD (*See* Chapter 13 also)

The standard schedule for processing the specimens for electron microscopy from fixation to the final block takes 4 days or more, but the procedure can be made much more rapid when necessary. Particularly, when the electron microscope is used to aid in making a clinical diagnosis, speed in processing the specimen is very essential. For this purpose the fixation schedule is also altered as below.

1. Fix by immersion, tissue bits of size less than 1 mm at least in one direction for 15 to 20 minutes in 2.5% glutaraldehyde in 0.1 M cacodylate or PIPES buffer, pH 7.2 to 7.4, containing 2 to 3 mM calcium chloride.
2. Wash for 5 minutes each in 3 changes of the same buffer.
3. Post-fix in unbuffered osmium tetroxide, wash briefly in distilled water, treat for 20 minutes in 1% aqueous uranyl acetate, and wash briefly in distilled water again.

Microwave irradiation can also be used for rapid processing, if the proper facility is available.

ADDITIVES FOR FIXATIVES

Various additives to the routine aldehyde fixatives are also advocated to improve some aspects of appearance of specimen. Some examples are as follows:

- digitonin red for preservation of lipids, particularly free cholesterol
- ruthenium red for enhancing staining of cell surfaces and various extracellular substances. It is useful as a fixative cum stain. It is useful mainly for connective tissue, as it stains the collagen fibres and reveals an otherwise invisible network of fibrils and proteoglycans in the extracellular matrix. It is particularly used to demonstrate glycocalyx and other surface-associated carbohydrate components.
- ruthenium hexamine trichloride for preservation of proteoglycans in cartilage
- dichromate for biogenic amines
- potassium ferricyanide added to osmium (1% OsO_4 + 1.5% potassium ferricyanide) for obtaining strong staining of lipid membranes.
- malachite green for some cellular lipids
- lead citrate added during fixation enhances the contrast of certain carbohydrates like dextran
- picric acid and other trinitro compounds added to glutaraldehyde fixative are useful in the following ways.
  - Membrane system is well-fixed.
  - Myelin figures are rare.
  - Excellent preservation of SER is obtained in steroid-secreting cells.
  - Useful particularly in immunocytochemical studies.

Trinitro derivatives are advantageous in the following ways.

- They destroy membranes and allow rapid penetration of fixative.
- They act as protein precipitants also, thus enhancing preservation.

FIXATION PROCEDURES

Fixation of Animal Tissues

Primary fixation is with 2–4% buffered glutaraldehyde at room temperature for 4 to 8 hours or more. The tissue is washed with same buffer twice or thrice for 10 to 15 minutes each time.

Secondary fixation is performed with 1% buffered osmium tetroxide at 4°C for 1–2 hours. Again it is washed with buffer twice or thrice for 10–15 minutes each time.

Fixation of Isolated Cells (*See* Chapter 12)

Isolated cells include cells cultured in suspensions, protozoa, bacteria, blood cells, etc.

With these types of isolated cells, there is no problem with penetration of fixative. But the main difficulty is only in handling the cells, especially when they are in small quantities.

Primary fixative glutaraldehyde 0.5 to 4% in 0.1 M cacodylate buffer containing 2 to 3 mM calcium. The temperature of fixation should be the same as that of isolated cells because change in temperature can rapidly affect the external morphology of mammalian cells. So cells cultured in suspension are fixed at incubation temperatures while cells obtained by an isolation procedure are fixed at the final temperature of that procedure which is usually room temperature or 4°C.

Isolated cells or cell fractions are fixed in suspension and then centrifuged or centrifuged to a pellet and then fixed.

Fixation of cells in suspension in medium The fixative has to be added while cells are still in suspension in medium. 2.5% glutaraldehyde in 0.1 M cacodylate buffer at pH 7.2 containing 3 mM $CaCl_2$ is the preferred fixative. Fix for 30 minutes at room temperature, centrifuge and post-fix in osmium tetroxide.

Prepare some of the suspending medium and add to it double the amount of aldehyde to give the final required concentration. Then add equal volume of this double-strength fixative and cell suspension to a centrifuge tube, mix at room temperature. Fix for 30 minutes at room temperature and centrifuge. Many pellets fixed in glutaraldehyde are firm and show no tendency to disintegrate in the buffer and can be removed from the centrifuge tube at this stage. Since the pellet has to be removed now, it is preferable to use eppendorf tubes and use spatula to remove the pellet or cut the plastic tube to release the pellet. Place the pellet on dental wax in a drop of buffer, cut into small pieces, if required, and put in a vial containing osmium tetroxide (1%) at room temperature for 1 hour and 1% aqueous uranyl acetate (UA) for 1 hour at room temperature. (UA is strongly recommended.)

Fixation of cells and organelles in a pellet Some types of isolated cells, such as bacteria, RBCs, cell organelles such as mitochondria, can be fixed after centrifugation into a pellet. The supernatant is removed without disturbing the pellet and then aldehyde fixative is added. The centrifugation speed is kept minimum. If pellets are centrifuged at too high speed the pellet becomes very compact. Then penetration of the fixative is slow. In that case, fixation time can be extended to 4 hours.

The fixative should be at least 20 times larger in volume than the pellet. It is run slowly down the side of the centrifuge tube keeping the pellet intact. Fix for 10 minutes to 1 hour. After primary fixation the pellet is centrifuged again before washing and secondary fixation.

Encapsulating methods for pellets Certain pellets are not cohesive enough to remain intact after initial fixation and disintegrate when buffer is added. Such pellets are encapsulated by surrounding them with a soft gel of agarose or calcium alginate.

Agarose Agar solidifies at 55°C. This high temperature may damage cells. Hence agarose is used. This has low melting point and remains fluid at 32–37°C.

Calcium alginate Pellets can be encapsulated at any temperature between 0 and 25°C.

This method is used mainly for:

- mammalian oocytes
- plant and animal cells
- protozoa
- yeasts
- bacteria
- cell organelles such as mitochondria

Pellets can be encapsulated before dehydration at any stage, but it is preferable to do it early to avoid centrifugation at each stage.

Fixation of Cells and Viruses on Millipore Filters

The pore size for filters used for fixation of various cells and viruses are as follows:

0.45 μm for mammalian cells and some bacteria

0.22 μm for smaller bacteria

50 nm for viruses

Cells are collected by filtering the suspension through the Millipore membrane or a sterile filter placed in a small culture dish and the concentrated suspension is placed gently in the centre. The filter is laid on several layers of absorbent (blotting) paper to soak up and draw through the medium by surface tension. The filter is incubated at 37°C for 2–3 hours to allow cells to attach firmly to the filter. The cells on filter are fixed in small vials and dehydrated. Then the filter is cut into pieces before infiltration. Propylene oxide will dissolve the filter and so xylene or toluene are used as intermediate solvents.

Fixation of Organ and Monolayer Cell Cultures (*See* **Chapter 12**)

These specimens are very thin, 2 to 4% glutaraldehyde with formaldehyde is used as the primary fixative. 2 to 3 mM $CaCl_2$ should be added to the fixative.

1. Organ cultures are easy to fix especially if cultured on Millipore filter.
   - Lift them carefully from medium.

- ✳ Rinse in balanced salt solution to remove excess medium.
- ✳ Place in primary fixative in a small vial for fixation by immersion.

2. Colonies of cells (tissue culture cells, fungi, bacteria, etc.) growing on agar or other solid medium are fixed by the same method.
3. When cells are cultured on a substrate rich in proteins (such as gelatin or albumin): include sufficiently high concentration of buffer in fixative to counteract production of acid during reaction between proteins of substrate and glutaraldehyde. After primary fixation, the solid medium with cells attached, is peeled off the culture dish and reprocessed in vial.
4. Monolayers of cells grown on slides or coverslips or other substrates are very sensitive to changes in environment. So the substrate is placed in a small culture dish containing growth medium such that the cells can be fixed *in situ* at the temperature at which they have been incubated. The external morphology changes dramatically following drop in temperature, but intracellular organelles are not so seriously affected.

FIXATION METHODS FOR SPECIALIZED TECHNIQUES, (CYTOCHEMISTRY, IMMUNOCYTOCHEMISTRY)

These methods require only milder fixation initially than for routine ultrastructural studies, because, in these techniques it is essential to demonstrate localization of particular chemical entity which takes precedence over preservation of ultrastructure. Especially in case of cytochemical techniques for enzymes where some diffusion of reaction product is inevitable, high magnification micrograph is not required.

There are three categories.

Fixation for General Cytochemical Techniques

For most cytochemical techniques, aldehyde fixative is used at ambient temperature, and precise composition of the fixative is maintained for best preservation of ultrastructure. Mixed aldehydes are used, i.e., mixture of glutaraldehyde and formaldehyde. Only exception to use of mixed aldehydes is where cytochemical technique depends on the detection of aldehyde groups, e.g., some methods for DNA and methods involving use of periodic acid.

With cytochemical techniques, the only staining present in the final ultrathin sections should be that due to the specific chemical constituent that is to be demonstrated. So osmium tetroxide post-fixation is also generally omitted. Ultrathin sections are examined unstained although further adjacent sections can be stained with lead citrate and/or uranyl acetate to reveal finer details of ultrastructure if required.

In a few techniques, it is necessary to cut thin slices of aldehyde-fixed material on a vibrotome and to stain or treat these slices before proceeding further. This is required in cases where treatment with enzymes is necessary either as a control or for ultimate staining or as a means of distinguishing between different tissue components which would all produce the same type of final staining. Some cytochemical stains do not penetrate ultrathin resin sections. So they have to be applied to thin slices after aldehyde fixation.

Fixation for Enzyme Cytochemical Techniques

Many enzymes may not survive any standard form of embedding procedure. So staining should be applied to thin slices after aldehyde fixation. For the same reason, enzyme cytochemistry is not widely used at the ultrastructural level.

A few enzymes like alkaline phosphatase survive fixation at room temperature. Most enzymes are reduced less by formaldehyde than by glutaraldehyde, but a small amount of glutaraldehyde to improve preservation of ultrastructure has to be included. So fixation using 4% formaldehyde (from paraformaldehyde) + 1% glutaraldehyde in 0.1 M cacodylate or PIPES buffer at pH 7.2 to 7.4 for 1–2 hours at 4°C is routinely followed. Fixation by perfusion is also possible if necessary. Some enzymes will tolerate only formaldehyde and some only short fixation. The following steps are applied:

- Fix in aldehyde.
- Wash in buffer.
- Cut on vibrotome to 50–250 micron thick sections.
- Incubate at 0–4°C. (sometimes short pre-incubation and post-incubation may be necessary).
- Wash in buffer and post-fix with unbuffered osmium tetroxide without addition of ferri- or ferrocyanide.
- If Os obscures enzyme staining, use 4% glutaraldehdye at room temperature for 1–2 hours. Treatment with uranyl acetate as third fixative is not done.

Fixation for Immunocytochemical Techniques

The method depends on the antigen to be studied. Few antigens can withstand routine fixation and embedding procedures. So gentle fixation is used.

The best way to choose optimum fixation conditions is as follows:

Cut a series of cryostat sections at one particular thickness from an unfixed block of tissue and subject these to a graded set of fixation schedules to assess the strength of antigenic response obtainable at LM level. Fixation should be at 4°C in an isotonic fixative buffered to pH 7–7.4 with 0.1 M cacodylate containing 2 to 3 mM $CaCl_2$. Osmium or uranyl post-fixation is also not used. Low-temperature embedding may be necessary for retaining adequate level of antigenic reactivity. (*See* Chapter 17).

Fixation of Botanical Specimens

Ultrastructural changes occur rapidly in plants also mainly due to water loss. So immediate fixation is essential.

Problems unique in fixation of plants arise from:

- impermeability of cell wall
- presence of vacuole containing materials
- presence of intercellular air spaces

Growing, meristematic tissues such as root tips are easy to handle because their cell wall is more permeable and vacuoles smaller, so routine fixation can be followed. For older, especially woody tissues, the standard procedure must be modified. The biggest problem is that such plant specimens float in the fixative and hence for fixation excess air must be removed by vacuum treatment. Vacuum should be applied gently initially and increased gradually over a couple of minutes and held at vacuum not exceeding half an atmosphere, i.e., 380 mm Hg for a few more minutes. Single treatment should be enough. If the bits still float, the treatment has to be repeated two more times over a period of half an hour. Any specimen which still does not sink should be rejected. For these more woody tissues, fixation time needs to be increased. The time for each of the individual steps in the whole fixation procedure is to be increased by a factor of up to 4. Plant tissues must be cut into thin slices to allow penetration of fixative.

Choice of fixative for the botanical specimens depends on the nature of specimen and aim of investigation as in the case of animal tissues. Variations are generally in osmolarity and pH of the fixative. In general, preferred mixtures are not very different from those for animal tissues. Therefore, the fixative of choice is glutaraldehyde or a glutaraldehyde–formaldehyde mixture in a buffered solution. Phosphate buffer is preferred to cacodylate, and PIPES is not used. With cacodylate buffer, 2–3 mM $CaCl_2$ can be added but not with phosphate buffer. Typically osmium is also buffered. Most plant materials are fixed best at pH 6.8 rather than at 7.4, but higher pH is usually preferred at initial stages of fixation. Fixation should be mostly at room temperature and can be extended to several hours. Buffered osmium tetroxide is used for one hour or more for post-fixation. Tertiary en bloc fixation with uranyl acetate is also used.

Potassium permanganate fixative is used where very waxy materials are involved, but massive extraction of intracellular material, especially nucleoproteins occur with this fixative.

Microwave-assisted Specimen Preparation

The advantages of using microwave techniques in specimen preparation are as follows.

- decrease in fixation times from hours to minutes or seconds
- enhanced structural preservation
- better retention of lipids and soluble proteins
- better preservation of antigenic reactivity

Fixation time is only 10–60 seconds for primary glutaraldehyde fixation and maximum 60 seconds for post-fixation with OsO_4.

(*See* Chapter 15 for further details.)

Cryofixation—Ultrarapid Freezing

Superior ultrastructural preservation can be achieved by processing the samples at low temperatures only. Plunge-freezing is the most commonly used technique. (*See* Chapter 16 for further details.)

HANDLING AND SAFETY PRECAUTIONS FOR FIXATIVES

Exposure to chemicals is the greatest cause for concern. There are certain risks of handling chemicals and hazards peculiar to fixation procedures. Fixatives have a hardening effect on skin even in dilute solutions. Prolonged contact may cause cracking of skin, dermatitis, etc.

(*See* Appendix 1 for further details on this topic.)

7

WASHING AND DEHYDRATION

WASHING AND STORAGE

After glutaraldehyde fixation, the specimen bits have to be washed thoroughly in the buffer, the same as that used for the preparation of the fixative. Washing is done to remove traces of unreacted excess fixative. The osmolarity and the pH of the buffer solution used for washing should also be the same as that used for the primary fixative. Ideally, the washing solution should be made up with calcium ions in the same buffer and with the same osmolarity as the primary fixative in order to reduce the changes in the environment of the specimen to a minimum. If traces of unreacted aldehyde fixative is left in the specimen, they may react with later fixation steps. So they have to be completely removed before proceeding further.

Usually two or three changes of the washing solution over a period of 1 to 2 hours is considered sufficient for thin slices of tissue, the general pattern being washing thrice for ten minutes duration each time. At this stage, the specimen can be stored in the buffer itself for short durations. But if phosphate buffer is used only overnight storing in the buffer is possible. This is because, divalent calcium cannot be added to the buffer and moreover, phosphate buffer allows growth of microorganisms. If cacodylate buffer is used, the specimen can be stored for a longer time in the buffer itself after washing. With cacodylate, calcium can be added and it does not allow growth of microorganisms as it contains arsenic. Moreover, cacodylate will counteract any acid that may be released by the specimen during storage. It also depresses any autolysis which might occur. Calcium is added to the fixative to a final concentration of 2 to 3 mM. With buffers other than cacodylate, a crystal of thymol can be added to prevent growth of microbes.

But if long-time storage is required, it is preferable to add fixative to the buffer for the following reasons.

1. to preserve the ultrastructure.

2. to prevent reversal of cross linking which might occur particularly with formaldehyde.

Of course, for cytochemical and immunocytochemical studies no long-time storage is recommended.

After post-fixation with osmium tetroxide also, it is mandatory to wash the specimen, to remove excess fixative, with the same buffer for similar duration as mentioned for primary aldehyde fixation. If unreacted osmium is not removed by washing, it will react with the dehydrating agent forming again precipitates of reduced osmium. If the third fixative, uranyl acetate, is also used, washing is still carried out after fixation to remove the excess.

In all these washings, in fact, in any change of chemicals it should be remembered that solutions should not be drained completely and at least a drop or two of the previous solution should be retained such that the specimen is never allowed to dry during processing.

The washing period after primary fixation with aldehyde, is the ideal time to dissect out selected areas of a tissue to form pieces for eventual embedding. After treatment with OsO_4, most tissues are too brittle to dissect and individual structures are too difficult to identify because of the general blackening.

DEHYDRATION PRINCIPLE AND METHODS

After preserving the morphology of the specimen through fixation methods, the specimen has to be cut into ultrathin sections to be viewed in the electron microscope. Since the biological tissues, plant or animal, contain water as the major cellular constituent, it is impossible to cut thin sections, leave alone ultrathin sections at this stage. With high water content the specimens cannot be loaded in the high-vacuum column of the electron microscope. So, the intra- and intercellular water has to be removed and the specimen made hard-enough for further sectioning. If the water is removed as such the cells will shrink leading to disruption of the morphology. Therefore the water is not removed totally, though this step is referred to as dehydration step, it is only replaced with a material which can be hardened enabling the specimen to be cut subsequently. This hardening material added is called the embedding medium. For light microscopy, i.e., for routine histopathology, "paraffin wax" is used to replace the water. But for ultrathin sectioning paraffin cannot be used because paraffin is hard enough to give sections with thickness only in microns and not less as is required for electron microscopy. Here chemicals called resins are used as infiltration and embedding media. These embedding media, both paraffin and resins (most of the routinely used ones), are immiscible with water. So replacement of water by embedding medium is carried out as a two-step process in which initially water is replaced by the dehydrating agent which is in turn replaced by the embedding medium. Alcohol or acetone is employed as the dehydrating agent since both are freely miscible with both water and the embedding medium (paraffin and resins). If water-miscible resins, such as some acrylic resins, are used, then this dehydration step can be eliminated.

Ethanol and acetone are the two dehydrating agents most commonly used in ultrastructural studies. The tissue after washing is treated with a series of ascending concentrations (graded series) of the dehydrating agent usually starting from 30% alcohol or acetone to 100% through 50%, 70%, 80%, and 90%. These steps are carried out at room temperature for a duration of 10 minutes at each step. At 100% concentration the specimen is kept twice for ten minutes each. Some workers use 30%, 50%, 70%, 90%, 95% acetone for 30 minutes at each step at 4°C and then in dry acetone for 30 minutes at room temperature twice. Dry acetone is prepared by adding sufficient amount of anhydrous copper sulphate to absolute acetone, shaking well and leaving aside for sometime. The solution has to be filtered before use.

The usual dehydrating agents, acetone and ethanol, give similar results with the majority of specimens. Acetone has a lower viscosity than ethanol and the dehydration process is more rapid. It is also a less aggressive solvent for most lipids, but it is very hygroscopic and incomplete dehydration may result unless the acetone is carefully dried and kept dry in a sealed bottle. In addition acetone is a radical scavenger and can inhibit the polymerization of acrylic resins. There is still some difference of opinion about the relative merits of ethanol and acetone as the main dehydrating agent, but ethanol is preferred by many as it is easier to handle, it is not very volatile, is relatively non-toxic and is readily miscible with water. But acetone has an advantage over ethanol as it is more readily miscible with clearing agents like xylene, toluene, propylene oxide, etc.

Polar acrylic resins and some epoxy resins are miscible with water. So they themselves act as dehydrating agents or they can be used in partial dehydration schedules in which dehydration is completed in mixtures of the dehydrating agent, the resin and water.

Dehydration of cryofixed specimens is done by freeze-substitution.

For preservation of antigens, dehydration is carried out at low temperatures by progressive lowering of temperature or by freeze-substitution.

For X-ray microanalysis, dehydration by freeze-drying is used to minimize redistribution of soluble ions.

Dehydration time, i.e., the length of each dehydration step should be standardized to avoid leaching out of soluble components and at the same time allow removal of water. The length of each dehydration step is dependent on the tissue. Connective tissue, plant tissue and senescent cells are dehydrated for long time and soft tissues like testes are dehydrated for short time.

CHEMICAL AND MORPHOLOGICAL EFFECTS OF DEHYDRATION

- ✳ Major chemical effect of dehydration is extraction of lipids resulting in shrinkage of cells and their components.
- ✳ Physical state of proteins, mucopolysaccharides and nucleic acids are also affected by dehydration but extraction losses are lower, occurring mainly in dilute solvent solutions.

Dimensional Changes During Dehydration

The overall effect is the shrinkage of specimen as a consequence of osmotic effects and as water and lipids are extracted.

The dimensional changes can be divided into two stages:

1. Significant swelling of 15 to 20% by volume in 30% and 50% ethanol which is followed by shrinkage.
2. Initial swelling is nearly reversed in 70% ethanol and then there is massive shrinkage in 80% ethanol, the overall reduction in volume reaching 30% in absolute alcohol.

Dehydration with more gradual steps has little effect on these volume changes; the most rapid shrinkage still occurs at 80% ethanol and overall shrinkage is only reduced to 25% by volume. If specimen is passed directly from water to 70% there is only a transient period of swelling, but subsequent shrinkage is not affected.

The extent of swelling in 30% and 50% ethanol, which may well cause irreversible structural damage to the specimen, depends on the fixation procedure. It is less in tissues fixed with glutaraldehyde followed by osmium. It can be completely prevented by the addition of divalent cations (for example, 1 mM to 0.1 M calcium ions) at any stage before or during dehydration, preferably with the fixative. The overall shrinkage of specimens during dehydration can be as high as 30 to 40% by volume for some specimens, such as isolated cells, and the possibility of differential shrinkage must always be remembered. For some specimens shrinkage can be reduced by continuous dehydration but the most important factor is the fixation procedure. The shrinkage can be minimized by tertiary fixation with uranyl acetate. Dilute ethanolic solutions extract proteins and other tissue constituents and induce swelling, so directly to 70% is preferred by some.

Retention of Lipids

As much as 95% of the total lipids may be lost during the dehydration of an aldehyde-fixed specimen, while secondary fixation with osmium tetroxide reduces this loss considerably and further fixation with uranyl acetate still more. Aqueous solutions of ethanol of 70% or less dissolve relatively little lipid from specimens fixed with glutaraldehyde and osmium tetroxide, the majority of the lipid loss occurring in the higher ethanol concentrations. Thus the partial dehydration schedule in which the specimen is passed from 70% ethanol to a mixture of 70% ethanol and an embedding medium reduces loss of lipids and therefore preferred by some workers. Lipid extraction can be reduced even further by using shorter times for the steps in the dehydration procedure and by reducing the temperature to 0°C. Acetone is a more gentle lipid solvent than ethanol.

Extraction of lipids and proteins during dehydration is accompanied by shrinkage of tissues, cells and subcellular components. This extraction has been shown to lead to a decrease in the periodicity of nerve myelin and the shrinkage of chromosomes with an accompanying production of abrupt bands in spindle microtubules. Some striated muscle filaments shorten by as much as 10% during dehydration in ethanol, though there is little change in dimensions during dehydration in acetone.

In general, optimum preservation of lipids is obtained with a primary aldehyde fixative containing calcium, an osmium tetroxide fixative and using uranyl acetate as a tertiary fixative. The use of these three fixatives in sequence gives the best preservation of lipid membranes and consequently the best ultrastructural results.

Thus in conclusion, the following methods can be expected to reduce the loss of lipids during dehydration.

- Add calcium to the primary fixative and washing solutions.
- Fix the specimens with osmium tetroxide before dehydration.
- Include potassium ferricyanide with the osmium tetroxide.
- Fix the specimens with uranyl acetate before dehydration.
- Select a dehydrating agent with minimal solubility for lipids.
- Dehydrate specimens as rapidly as possible.
- Use low temperature for the dehydration sequence.
- Use a partial dehydration schedule, if possible.
- Since propylene oxide is a very 'energetic' solvent for lipids, toluene or xylene can be used as intermediate solvents.

DEHYDRATION SCHEDULES

In the dehydration procedure, the fixed specimen is passed through a graded series of solutions of increasing concentration of the dehydrating agent in water, ending with pure (absolute) dehydrating agent. Stock solutions of 100% dehydrating agents should be kept dry in tightly sealed bottles and these should be opened for as short a time as possible, particularly in humid conditions. It is often convenient to keep the specimen in the vial or tube or any small container in which it was fixed throughout the dehydration procedure. If plastic vessels are used, they must be resistant to solvents, such as propylene oxide. During dehydration, one solution is carefully removed with a fine Pasteur pipette or decanted carefully and the next solution added. This process is carried out as rapidly as possible to prevent the sample from drying out and the vials are then capped firmly to prevent absorption of water from the air. It is essential to ensure that there is a large excess of each solution, at least 20 times the volume of the specimen. As with fixation, better dehydration is obtained with small and thin specimens.

Dehydration is normally carried out at room temperature or in the cold. If dehydration is started in the cold, the temperature is raised to RT (room temperature) when the specimens are in 90% or 95% solutions of the dehydrating agent in water. The steps of the dehydration schedule are not critical and excessive dehydration should be avoided since it can result in greater extraction of lipids and degradation of ultrastructure. When specimens have to be stored during processing, as already mentioned, it can be done in the buffer wash following primary fixation whenever possible. Specimens which have to be stored overnight during dehydration should be kept in 70% ethanol or acetone in the cold to minimize the loss of lipids.

Longer dehydration times are necessary for cells with dense walls, such as plants and fungi. More delicate specimens, such as suspensions of cells and tissue cultures, may need to be dehydrated through a more closely spaced series of concentrations of the dehydrating agent than the standard schedule but with shorter times for each step. Continuous dehydration methods using sophisticated exchange apparatus are required for specimens that are particularly sensitive to osmotic effects.

Rapid Dehydration

When required, such as for urgent pathological diagnosis, the dehydration can be made considerably more rapid. This is as a continuation of rapid fixation as already discussed in chapter 6. Specimens must be 0.1 mm or less in at least one dimension and the specimens should be agitated throughout dehydration. Dehydration time can also be decreased by continuous dehydration technique. The schedule for rapid dehydration is as follows:

70% ethanol	3 min.
80% ethanol	3 min.
90% ethanol	3 min.

100% ethanol	2 × 5 min.
Propylene oxide	2 × 5 min.

Partial Dehydration

Some epoxy and acrylic resins are miscible with aqueous solutions of ethanol and enable the higher concentrations of ethanol and propylene oxide to be omitted from the dehydration schedule, thus avoiding steps during which majority of lipid extraction occurs. Epon 812 is miscible with 70% ethanol in water and in the schedule proposed by Idelman (1964) the specimen is passed from 70% ethanol to a 1 : 1 mixture of 70% ethanol and Epon embedding medium. It is infiltrated with this mixture for 30 min. at room temperature and then for 30 min. at 37°C before infiltration with the embedding medium in the usual way. Stein and Stein (1971) found that the loss of some lipids is reduced if the specimen is first infiltrated with the Epon resin alone, since the acid anhydride hardeners and the tertiary amine accelerator appear to increase the solvent action of the complete embedding medium. Lipid extraction is reduced even further by using shorter steps in the dehydration procedure and by decreasing the temperature. The partial dehydration at 0°C before embedding in Epon is as follows:

70% ethanol in water	2 × 5 min.
95% ethanol in water	2 × 5 min.
Epon 812	3 × 1 h

The specimens are then infiltrated with complete Epon embedding medium, overnight at 4°C, before embedding in the usual way.

DEHYDRATION BY FREEZE-SUBSTITUTION

See Chapter 16—Cryoelectron microscopy.

CLEARING

Clearing is a process of replacement of the dehydrating agent with a suitable solvent that facilitates penetration of resin into tissues. Although most epoxy resins are miscible with ethanol and acetone, the most common dehydrating agents used, they mix more readily with some other organic solvents like, propylene oxide and these intermediate solvents are routinely used following dehydration of well-fixed specimens. Xylene and toluene are also suitable as intermediate solvents for epoxy resins. But propylene oxide is preferred since it is a low-viscosity epoxy diluent which reacts with the anhydride hardener in the embedding medium and becomes an integral part of the cross-linked polymer. Consequently no problem arises even if a small amount of propylene oxide remains in the tissues. Propylene oxide mixes in all proportions with ethanol, acetone and epoxy resins. It quickly replaces the dehydrating agent and then diffuses out of the tissue allowing easy infiltration of the epoxy resin. But being highly toxic, it must be handled with full safety precautions. Clearing is generally carried out twice for a duration similar to that used for dehydration (15 minutes each or 30 minutes each).

8

INFILTRATION AND EMBEDDING

The purpose of the infiltration and embedding steps is to replace the dehydrating agent or the intermediate solvent with a resin monomer which can be cured or polymerized to produce a solid block with good sectioning properties. The specimen is passed through an ascending series of a mixture of the liquid embedding medium and the dehydrating agent, until it is the pure embedding medium and this step is referred to as **infiltration** as the liquid embedding medium is allowed to infiltrate the specimen by replacing the dehydrating agent. Then the specimen is transferred to moulds for curing or polymerization which is called **embedding.** The final outcome of this step is a hard block of the polymerized resin with the specimen embedded at one end such that the block can be held in the microtome with the specimen at the free end from which sections can be cut.

For optical microscopy, paraffin wax blocks are prepared which are sectioned in the microtome using metallic knives. But paraffin blocks can give only sections of micron thickness, which is unsuitable for EM studies. For EM, the section thickness has to be of the order of angstroms in order to facilitate the penetration of electrons as well as to achieve a good resolution. Hence, the specimen blocks have to be still harder than the paraffin blocks. To prepare hard blocks for ultrathin sectioning "**resins**" are used in place of paraffin wax. These resins are highly viscous liquids in the monomeric forms. When they are allowed to polymerize, a hard block is obtained.

INFILTRATION

AIM

The aim of infiltration is to obtain uniform penetration of the embedding medium throughout the specimen, which is achieved by gradually replacing the dehydrating agent or the intermediate solvent (clearing agent) with the embedding medium.

The specimen bits well fixed with the two fixatives, primary aldehyde and the secondary osmium, are dehydrated with acetone or alcohol. Now the bits have to be hardened for further sectioning. This is achieved by infiltrating the specimen with the liquid embedding medium initially and polymerizing it to make it hard. Thus, infiltration actually means, replacing the dehydrating agent with the embedding medium in steps, by increasing the concentration of the resin at each step.

For light microscopic routine histopathology, the embedding medium used is paraffin wax. In the molten state it is allowed to infiltrate through the specimen replacing the organic solvent and then embedded in moulds. The wax on cooling solidifies thus forming a hard paraffin block with the specimen at one end such that the block can be mounted in the microtome with the specimen at the free end from which sections can be cut. Four to five micron thick sections are cut from this block using metallic knives. But, the embedding medium used in electron microscopy is a **resin** because ultrathin sections cannot be cut with paraffin blocks.

After the treatment with the clearing agent, infiltration of the specimen with the resin is carried out. This is done by gradually decreasing the concentration of the clearing agent and increasing the concentration of the embedding medium proportionately. Thus, the specimen, after treatment with dehydrating agent and clearing agent, is passed through a series of mixtures of the clearing agent and the liquid embedding medium, until finally, passing it into the pure embedding medium. If no clearing agent is used, then a mixture of acetone and the resin mixture is used for infiltration. Direct transfer of absolute ethanol to the resin (epoxide monomer, the most commonly used resin) mixture, as a result of slow exchange leads to poor blocks. Remedy for this is use of propylene oxide as an intermediate solvent. On the other hand, transfer of absolute acetone directly to the epoxide monomer is possible. The specimens are then transferred to capsules or embedding moulds for curing or polymerization of the resin, as the case may be.

The timing and number of steps in the process of infiltration depend on the type of embedding resin and on the size and nature of the specimen. For example, longer infiltration times are required for large specimens and dense tissues, and also when thick cell walls are present, as in plants. In the case of hard tissues, slow diffusion of the embedding medium is a problem. This can be overcome by carrying out the steps of infiltration under vacuum. For plant tissues vacuum treatment is routinely employed. This treatment also eliminates the air bubbles around the tissue.

To speed up the process of infiltration when viscous embedding media are used, the specimen vials can be placed on a slow rotary shaker. As an alternative to using a shaker, a continuous stream of gas bubbles can be used to keep the embedding medium circulating around the specimens, for which purpose, usually, dehydrated air is used for epoxy resins and nitrogen gas for methacrylate resins. Delicate tissue specimens may be adversely affected by sudden changes in viscosity and it is therefore advisable to increase the concentration of infiltrating/embedding medium in a continuous process as detailed in the following sections.

The entire processing including infiltration can be carried out in an automatic processor commercially available. These units are expensive and more suitable when large number of samples are processed simultaneously. In most of the laboratories it is carried out manually by changing the solutions after each treatment. Infiltration of the specimen is done using liquid resins. Though different kinds of resins are available, the choice depends on the study requirement and nature of the specimen.

COMPONENTS OF THE INFILTRATION/EMBEDDING MEDIUM

The resin mixture usually consists of the following components:

- Resin
- Hardener
- Accelerator

A plasticizer or a flexibilizer is also used. Sometimes more than one epoxy resin is also used in some laboratories, the accelerator is not included in the infiltration medium but only in the embedding medium while in a few laboratories the accelerator is added in the infiltration medium itself.

KINDS OF RESINS/TYPES OF EMBEDDING MEDIA

Chemically the resins can be classified into three main types:

- Epoxy, resins which are most widely used worldwide.
- Polyester resins, with limited use because less easily available and some of the components are not stable during storage.
- Methacrylate, not routinely used now because it polymerizes unevenly and is not stable under electron bombardment. But water-soluble methacrylates are used in cytochemistry.

Some of the commonly used resins are listed below:

i. **Epoxy resins**
   - Epon 812 (aliphatic resin)
   - Araldite CY 212, 502, 506, 6005, etc. (aromatic resins)
   - Maraglas 655

ii. Polyester resins

- Vestopal W
- Nanoplast
- Selectron
- Rhodester
- Polyethylene glycols

iii. Methacrylate resins

- *n*-butyl and methyl methacrylates
- Glycol methacrylates
- Hydroxypropyl methacrylates

The major difference between the different types of resins is in their viscosities. Longer infiltration times are required for more viscous media. Addition of plasticizers lessens the viscosity of the embedding medium.

Resins can also be classified broadly into two major types:

1. Water-miscible resins, e.g., acrylic resins
2. Water-immiscible resins, e.g., epoxy resins

Examples of water-miscible acrylic resins include:

- Glycol methacrylate (GMA)
- Hydroxypropyl methacrylate (HPMA)
- LR gold and LR white resins
- Lowicryl resins

As these acrylic resins are most used in immunoelectron microscopy, they are discussed in the respective chapter (Chapter 17).

Examples of water-miscible epoxy resins include:

- Aquon
- Durcupan
- Aqembed

Examples of water-immiscible epoxy resins include:

- Araldite CY212
- Araldite GY502, 506 6005
- Araldite MY 753

- Epon 812
- Spurr
- Maraglas
- ERL 4206

Hardeners The following are the hardeners used.

- DDSA (Dodecenyl succinic anhydride)
- HHPA (Hexahydrophthalate)
- MNA or NMA (Nadic methyl anhydride)
- NSA (Nonenyl succinic anhydride)

Accelerators

- DMP-30 [2,4,6 Tris(dimethyl aminomethyl)phenol]
- BDMA (benzyl dimethyl amine)
- DMAE (dimethyl amino ethanol)

Additives

1. *Plasticizer*
   - DBP (dibutyl phthalate)
2. *Flexibilizers*
   - Thiokol LP-8
   - DER-732

Epon 812 resin is the most widely used worldwide.

Epoxy resins, the most commonly used resins, are polyaryl ethers of glycerol bearing terminal epoxy groups. Epoxy resins like Araldite and Epon 812 are highly viscous but show good uniformity of infiltration, good uniformity of hardening, and only slight volume shrinkage during embedding. Hence, epoxy resins are widely used for ultrastructural studies. On the other hand, Spurr which is a low-viscosity epoxide resin shows problems with uniformity in infiltration as well as in hardening. Though the sectioning properties of the blocks are good, the staining is not very good. Spurr is used mainly for plant specimens. Some acrylic resins can be polymerized at low temperatures and therefore useful for immunoelectron microscopy.

A large variety of epoxy resins and their accelerators are available which are suitable for electron microscopy. In practice, in the majority of laboratories a standard embedding medium and procedure is adopted and then used for many years without modification for a great range of specimens. Most of the epoxy resin embedding media for infiltration have the same basic components consisting of an epoxy resin, a hardener and an accelerator, sometimes with the addition of a plasticizer.

1. ***Epoxy resin*** It is a synthetic thermosetting resin containing epoxy groups.
2. ***Anhydride hardener*** It reacts with the epoxide and OH groups of the resin molecules to give three-dimensional cross-linked structures. With epoxy resins, it is not polymerization but only addition reaction (referred to as "curing") which accounts for the small accompanying shrinkage.

   The common hardeners used are:

   i. *DDSA (Dodecenyl succinic anhydride)* It is an anhydride prone to hydrolysis and hence to be stored at room temperature. It is a low-temperature hardener and is a long-chain anhydride and so confers plasticity on the cured resin. Epon 812 is an aliphatic epoxy resin. It does not give sufficient hardness with DDSA alone and so a second hardener NMA or MNA is also used.

   ii. *NMA or MNA (Nadic methyl anhydride)* Nadic is a registered trademark of the Allied Chemical Corporation of USA. The unsaturated "Nadic" ring of MNA enhances beam stability of the cured resin.
3. ***Accelerator*** These are added at the time of curing or polymerization to reduce the temperature during embedding procedure so that blocks can be kept at a lower temperature of 60°C at which the structural details in the specimen are not damaged. They promote "epoxy to epoxy" and "epoxy to OH" reactions but do not themselves serve as direct cross linking agents and thus act as **catalysts.** The most commonly used ones are DMP-30 [2,4,6 Tris(dimethyl aminomethyl) phenol] and BDMA (benzyldimethylamine). DMP-30 has a higher viscosity than BDMA and leads to difficulty in mixing up evenly with other components and moreover it diffuses slowly. Hence BDMA is preferred to DMP-30.

Plasticizers and flexibilizers are used as additives for epoxy resin embedding media.

4. ***Plasticizers*** These decrease the viscosity of the embedding medium and improve sectioning properties of the final blocks. The hardness of the final block decreases with the addition of the plasticizer, DBP, the only plasticizer compatible with epoxy resins. DBP is added to prevent the block from becoming brittle and difficult to section.
5. ***Flexibilizer*** Instead of a plasticizer, a flexibilizer can also be added to improve the sectioning properties of the block. In fact, flexibilizer is preferred, because it reacts with the epoxy resin and becomes a part of the cross-linked structure and thus is less likely to be lost under electron bombardment.

INFILTRATION SCHEDULE

Preparation of the Infiltrating Medium

- Resin (Araldite CY212 or Epon 812)
- DDSA (Dodecenyl succinic anhydride)—a hardener
- MNA (Methyl nadic anhydride)—another hardener
- A plasticizer, dibutyl phthalate, can also be included in the mixture.

The above ingredients are added in the same order with vigorous stirring after each addition. The air bubbles are allowed to settle before use. The composition of the resin mixture may vary depending on the nature of the specimen. By varying the proportions of MNA and DDSA, the block consistency can be altered. For example, for hard tissues, hard blocks can be prepared by increasing the concentration of NMA in the mixture. Similarly, by increasing the concentration of DDSA relatively, softer blocks to suit soft specimens can be prepared. Thus, the facility of varying the hardness of the embedding medium is obtained by using the double hardeners, DDSA and NMA.

Schedule of Infiltration

The specimen bits after treatment with ethanol/acetone followed by clearing agent is treated with the infiltration medium as per the schedule outlined below.

1 part resin mixture + 3 parts clearing agent (25% resin)—2 hours at RT

2 parts resin mixture + 2 parts clearing agent (50% resin)—2 hours at RT

3 parts resin mixture + 1 part clearing agent (75% resin)—2 hours at RT

100% resin—2 hours at RT—Twice

Infiltration with only 50% resin at RT overnight and then 100% resin twice for 2 hours duration each time is also practised in some laboratories.

UNIFORMITY OF INFILTRATION

Epoxy resins like Araldite and Epon are highly viscous, though Araldite is less viscous than Epon. Spurr has low viscosity, but is very highly toxic. Hence in the case of high-viscosity epoxy resins, for uniform infiltration of the resin, intermediate solvents such as propylene oxide are required. Moreover embedding media with high viscosity can exert mechanical and osmotic stresses and can damage some very delicate specimens during infiltration. So the resin is warmed briefly to between 55 and 60°C just before use as viscosity decreases when temperature increases.

Viscosity of the complete embedding medium is influenced by the viscosity of the added hardeners also and it can be reduced by the addition of a plasticizer. For uniform infiltration, there should be no difference in the molecular sizes of the various components of the resin mixture. Though differences are seen in Spurr resin, Epon or Araldite resin mixtures do not show any such differences.

INTERACTIONS WITH SPECIMENS

Changes in the specimen during embedding and infiltration is less and in fact minimal when compared to those that occur during fixation and dehydration. Some of the changes which occur during infiltration and embedding are:

- Epoxy resins (Epon and Araldite) can cross-link with proteins and react with nucleic acids.
- Most obvious effect of interaction with specimens is extraction of lipids but this can be minimized with good fixation.
- Most epoxy resins including Araldite, Epon and Spurr have relatively little extractive effect on lipids during standard infiltration/embedding procedures.
- But, acrylic resins are powerful lipid solvents and can easily affect the appearance of membranes in specimens not fixed with osmium tetroxide.

SPECIAL METHODS OF INFILTRATION

Continuous Infiltration

Some delicate tissues and cells are susceptible to damage during infiltration. Infiltration with viscous epoxy resins, and procedures involving the continuous increase in the concentration of the embedding medium are then preferable to the standard stepwise method. No sudden changes in viscosity then occur and the specimen is not subjected to mechanical or osmotic stress. Overall shrinkage of the specimen is also reduced.

A simple and well-tried method is to use a specimen rotator. After dehydration in ethanol and propylene oxide, tissue pieces are placed in 2 to 4 ml of propylene oxide in vials on a rotator in an efficient fume cupboard. The liquid should completely cover the bottoms of the vials when they are in an inclined position. The device is set to rotate slowly and then a similar quantity of complete epoxy resin embedding medium is added continuously in drops. The uncovered vials are kept rotating overnight (minimum 8 hours) allowing the propylene oxide to evaporate almost completely. A change of complete resin mixture can then be made before curing, but this is not essential, since small residual amounts of propylene oxide do not affect the curing of epoxy resins.

Vacuum Infiltration

The application of a slight vacuum sometimes aids the infiltration of the embedding medium into large or dense specimens, or into those that contain large air spaces, such as botanical or freeze-dried tissues, although this is not necessary for the majority of specimens. It is only necessary to apply the vacuum at the initial steps of the procedure to accelerate the replacement of the dehydrating agent by the resin, otherwise there is a danger that volatile components of the embedding medium, such as accelerators, will be removed, resulting in significant changes in the resin formulation and unsatisfactory blocks. Vacuum embedding chambers are available commercially but the simplest method of vacuum infiltration is to place the specimen in the infiltrating solutions in uncovered containers in a vacuum desiccator. The pressure is reduced to about half an atmosphere but not less, by pumping with a rotary pump and this partial vacuum is maintained during infiltration with the embedding medium. The vacuum should be released gently over several minutes at the end of each stage of infiltration.

Rapid Infiltration

The infiltration of the embedding medium into the specimen and the subsequent curing of the resin are the most time-consuming steps in the processing of specimens for electron microscopy, but the total infiltration time can be reduced from more than 10 hours to less than 1 hour using propylene oxide as the intermediate solvent in the rapid technique schedule as given below.

Propylene oxide (2 × 5 min.)

Propylene oxide : epoxy embedding medium 1 : 1 (5–15 min.)

Propylene oxide : epoxy embedding medium 1 : 3 (15–20 min.)

Epoxy embedding medium (10 min.)

Rapid infiltration of epoxy resins using microwave irradiation has also been proposed but little time is saved, as compared with the schedule given above. Moreover, it is not advisable to use flammable solvents in a microwave oven.

EMBEDDING

ATTRIBUTES OF EMBEDDING MEDIA

Ideally an embedding medium should have the following properties:

* easy availability
* uniformity from one batch to another

* solubility in dehydrating agents
* low viscosity as a monomer
* polymerize or cure uniformly
* little volume change on polymerization or curing
* ease of sectioning
* section staining
* stability in the electron beam

Uniformity of Hardening

The transition from the liquid monomers to a hardened block should be a uniform process with very little volume change, resulting in blocks with good sectioning properties. The hardening process is different for acrylic and epoxy resins.

Acrylic monomers are linked or polymerized by radical chain reactions in a process which is controlled by free-radical-producing initiators and accelerators. Polymerization is accompanied by considerable shrinkage and is affected by the presence of impurities. The reaction is exothermic, so that heat is generated at sites within the hardening block. This leads to uneven polymerization and can also result in structural damage to the specimen which is known as **polymerization damage** (*See* Chapter 18, Figure 18.14). More uniform blocks can be obtained by using cross linking agents to reduce shrinkage and carrying out polymerization at low temperatures. On the other hand, the hardening of epoxy resins is by addition of polyfunctional agents such as anhydrides which react with epoxide and OH groups of epoxy resin monomer. This is not polymerization strictly but only an addition reaction which is called **curing**.

This is

* accompanied by very little volume shrinkage (about 2% volume)
* not exothermic
* not affected by impurities

Therefore, uniform hardening is easily obtained with well-infiltrated specimen as long as the temperature is kept below 60°C. With higher temperatures the addition reaction becomes exothermic with adverse effects on the properties of the final block. Therefore the temperature should not be increased above 60°C.

Ease of Sectioning

The final block should have good sectioning properties. Properly prepared epoxy and acrylic resin blocks have the following characteristics thus enabling thin sections to be cut easily with little compression.

- uniform consistency
- sufficient hardness

The sectioning properties of the epoxy resins, Araldite and Epon, and the non-polar acrylic resins such as Lowicryl HM20 are somewhat better than those of the low-viscosity epoxy resins such as Spurr and the polar acrylic resins such as Lowicryl K_4M and the methacrylates. Sectioning of the epoxy resins is easier when the plasticizer **dibutyl phthalate** is present in the resin mixture.

The sectioning properties of the final block depends greatly on the hardness of the block and also on the ratio of anhydride to epoxide which actually implies the extent and nature of cross links formed during polymerization. Sectioning properties improve when A/E ratio is lowered from 1 to 0.6, i.e., reduction in anhydride favours shorter polymers with fewer cross links.

Section Staining

The staining of sections either to increase contrast or for cytochemical studies should be uniform. The polar acrylic resins are readily infiltrated by aqueous stains and consequently good contrast is obtained in electron microscopy. Aqueous stains do not infiltrate as readily into non-polar embedding media, such as most epoxy resins, but adequate contrast is obtained with a combination of pre-embedding staining with osmium tetroxide and uranyl acetate (en bloc staining) and post-embedding section staining with lead. Problems arise, however, if the infiltration of the embedding medium is not uniform. Aqueous stains penetrate more readily into structures which are only poorly infiltrated by non-polar embedding media. Therefore caution must always be exercised in the interpretation of the distribution of stains.

Stability in the Electron Beam

The damage induced in thin resin sections in the EM should be minimized to ensure the preservation of the ultrastructure of well-fixed and well-embedded specimens. The initial and immediate effect of irradiation with electrons is the fragmentation and disruption of molecules and the removal of components of the embedding medium, with a resultant loss of mass and shrinkage of the section. It has been found that the resistance to this damage has been highest for the epoxy resins, particularly for the aromatic Araldites, followed by Spurr resin and then the aliphatic Epon. Epoxy resins and anhydride hardeners form three-dimensional polymers which are very stable and maintain their structure during irradiation with electrons, even when there is some loss of mass. This loss is detectable in Epon sections as a slight decrease in density, or **clearing** of the section in the area irradiated.

The acrylic resins are the most sensitive, since they consist of long chains of linear polymers which are readily broken. In consequence, they can lose as much as 70% of their mass in the EM under normal operating conditions. The resistance of the sections to electron irradiation can be increased by the incorporation of aromatic acrylic resins into the formulation and by the use of divinyl benzene as a cross-linking agent. In general, however, contrast is more variable and membranes are less well defined in acrylic than in epoxy resin sections.

The only way to reduce radiation damage significantly is to examine the sections at very low temperatures of –120°C or less, using a cold stage in the electron microscope, which is not always possible. Therefore, it is preferable to choose a resin with good resistance to beam damage like Araldite or Epon.

As examination of the sections continues, more breakdown occurs and gross changes are observed, such as

- increased shrinkage,
- formation of holes, and,
- movement of the whole section.

The absence of these effects is frequently cited as evidence of 'stability' of the sections, but it does not provide evidence that the initial structural damage has been avoided. At very high electron doses the sections become very hot and all embedding resins are effectively carbonized in much the same way.

EPON EMBEDDING

Epon embedding has the following advantages.

- It polymerizes uniformly, because polymerization is an addition reaction.
- It shows very little change of volume of the specimen, i.e., 2% shrinkage only. This is because, as already mentioned, the polymerization is an addition process and also, the resin is highly associated in the cured state.
- It is stable in the electron beam.
- Polymerization is not dependent on an initiator and hence impurities do not affect polymerization unlike methacrylates, in which components of the specimen itself may initiate polymerization.

But the major disadvantage is its high viscosity which necessitates a lengthy infiltration process.

Epoxy resins may be polymerized by a variety of bifunctional setting agents which add across the epoxy groups of the resin molecules to give three-dimensional structures.

These setting agents are commonly aliphatic polyamines or aromatic anhydrides. The reaction between epoxy resin and DDSA is slow and amine accelerator (catalyst) is added to hasten polymerization.

Epoxy resins are easy to handle and are capable of preserving ultrastructural detail down to the macromolecular level, and therefore are the most suitable embedding media for electron microscopy.

STORAGE OF RESINS AND PREPARATION OF RESIN MIXTURES

Aromatic and aliphatic epoxy resins and the anhydride hardeners, DDSA and MNA, can be stored indefinitely (for many years) if they are kept in air-tight, securely sealed containers at room temperature. They should not be stored in a refrigerator or deep-freeze cabinet, since this increases the risk of condensation of water vapour inside containers which are opened before they reach room temperature. Accelerator amines are readily inactivated by water. Therefore, these bottles should be kept open for as short a time as possible and should be replaced with a fresh bottle frequently. Especially, DMP-30 should be replaced more frequently since it is particularly sensitive to inactivation by water vapour.

It is essential to mix the components of the embedding medium very thoroughly to obtain uniform polymerization. Inadequate mixing is one of the main causes of sectioning problems with epoxy resins. At room temperature no amount of stirring or shaking gives complete mixing even though this is the practice in many laboratories. The resins and hardeners may appear to be fully mixed, but this will not be so. The simplest way to obtain good mixing is to warm the resin, the hardener and also the containers required to 60°C. The warm solutions are either measured by weight or by volume and poured into a warm conical flask containing the required quantity of the accelerator. In general, slight variations in the relative amounts of the various components are not important, except for the accelerator, because addition of too much accelerator makes the block brittle and difficult to section. The warm embedding medium is mixed by swirling or stirring with a warm glass rod. A few bubbles may form during stirring but these will go if the mixture is allowed to stand at 60°C for a short time. If the components of the embedding medium are not warmed, then a much longer mixing time is required and should be stirred well and allowed to stand for sometime till the bubbles disappear. Araldite and Epon are directly soluble in ethanol and acetone, but the diffusion process is slow, so removal of every last trace of the dehydrating agent is difficult, and any remaining ethanol or acetone has an adverse effect on the properties of the final block. This problem is overcome by using the intermediate solvent, propylene oxide, which reacts with the anhydride hardener in the embedding medium, so that small amounts remaining in the tissue do no harm. The accelerator used is DMP-30. After addition of DMP-30, the resin mixture is stirred well and left aside for the air bubbles to disappear.

Low-viscosity epoxy resins like Spurr's resin can be used. But these low-viscosity embedding media can cause structural changes in tissues which are not seen with Araldite and Epon.

Mixtures of aliphatic 'Epon' epoxy resins and anhydride hardeners take a long time to harden at room temperature but they gradually begin to increase in viscosity as soon as they are prepared and this increase is even faster after the accelerator has been added. Storage of mixtures of epoxy resins and hardeners should therefore be avoided. Therefore, it is advisable to prepare the embedding medium just before use when the specimens are in the final stages of dehydration. These increases in viscosity occur even in the cold and contamination with water vapour is a danger for mixtures stored in a refrigerator or deep-freezer. In particular, complete embedding media containing the accelerator should not be stored. It should be freshly prepared. But, if necessary the embedding medium can be stored at 4°C in a bottle with a well-fitting cap for several weeks, or for months at –20°C. During infiltration the medium is kept at room temperature and warmed to 60°C and stirred well before use.

WATER-MISCIBLE EPOXY RESINS

Embedding resins that are miscible with water can act as dehydrating agents, thus avoiding the use of ethanol, acetone and propylene oxide and so preserving some components of the specimen which might otherwise be extracted. Water-miscible acrylic resins are more popular than the water-miscible epoxy resins. These acrylic resins are mainly used in immunoelectron microscopy and discussed further in Chapter 17.

ANHYDRIDE–EPOXIDE RATIO

The sectioning properties of the final block depend not only on its hardness but also on the extent and nature of the cross links formed during polymerization. It is reported that a linear, shorter polymer with few cross links is easiest to section. And that the ratio of anhydride to epoxy is an important factor. 1 : 1 formulation was initially suggested for the addition reaction to go to completion. But a fully cross linked polymer would be too brittle to section. Hence, the A : E ratio is limited to 0.7 to 1.0 to limit the cross-linking, particularly for Araldite and Epon. The actual value of this ratio is not very critical as long as the curing temperature of 60°C is not exceeded. At higher temperatures the formation of cross links progresses much faster and a lower ratio is required to obtain satisfactory blocks.

CURING TEMPERATURE

At 60°C, which is the recommended curing temperature for epoxy resins, the formation of cross links reaches a maximum in 16 to 24 hours. A linear polymer with few cross links is favoured at lower temperatures, and hence a three-stage process of curing using incubation at 35, 45 and then finally at 60°C is preferred by some. But Gluert and Lewis suggest that when epoxy resins are cured in one stage at 60°C similar blocks are obtained and so the more lengthy three-stage process is unnecessary, although it may have some value when embedding temperature-labile specimens. Higher temperatures are avoided. But when rapid processing is required, higher temperatures must be used. Then, it is advisable to start the cure at 70°C and to raise the temperature to 95°C only after 45 min. This avoids any exothermic reaction and minimizes the possible solvent action of the embedding medium at high temperatures, producing blocks with better sectioning properties for both light and electron microscopy.

This standard temperature of 60°C is too high for some immunocytochemical studies, but very long incubation time of many days is required to harden epoxy resins at lower temperatures. The effect of heat on the specimen can be reduced by starting at the lower temperature of 35°C and then increasing the temperature in steps of 45 and 60°C to complete the cure. With this schedule the majority of the epoxy resin will have immobilized (but not hardened) before the temperature rises and so the adverse effects of heat on antigens will be minimized.

Blocks of Araldite and Epon can be sectioned after curing overnight (about 16 hours) but the sectioning of some embedding media improves if a longer time of 24 to 48 hours is used. Temperature is more important than time.

But there is some further slow curing even at room temperature during the following days so that sectioning properties often seem to improve simply with prolonged standing after curing/polymerization.

The most serious difficulty of Epon sections result probably from surface displacements while sections are being cut. The end effect is a "smudging" or "mottling" of the sections (*See* Figure 18.15 in Chapter 18). But this effect is rarely, if ever, seen when a diamond knife is employed, and yet may be unavoidable when a glass knife is used. This would suggest that adhesive forces exist during sectioning between some component of the block and glass. Because of this it can be imagined that surface particles are dragged along as the knife passes the block, or the section glides over the knife edge. This effect is associated with blocks that do not seem to be properly elastic, and which are prone to produce "chattered" sections. This problem does not seem to occur when a good lot of Epon is used.

EMBEDDING MOULDS

These provide blocks of a convenient size for thin sectioning. Ideally, specimens in moulds must be oriented properly and the resulting blocks must be polymerized evenly. Also, it should be easy to remove these blocks from the mould capsule.

Moulds made of different kinds of materials are available. They are,

- Gelatin capsules (no. 00, 0, 1, 2 sizes are usually used).
- Polyethylene capsules, commercially available BEEM* or TAAB capsules, which have pyramid-shaped ends and yield preformed blocks which require minimum of trimming.
- Polypropylene capsules used for polymerization at high temperatures.

* Registered trademark of Better Equipment for Electron Microscopy, Inc.

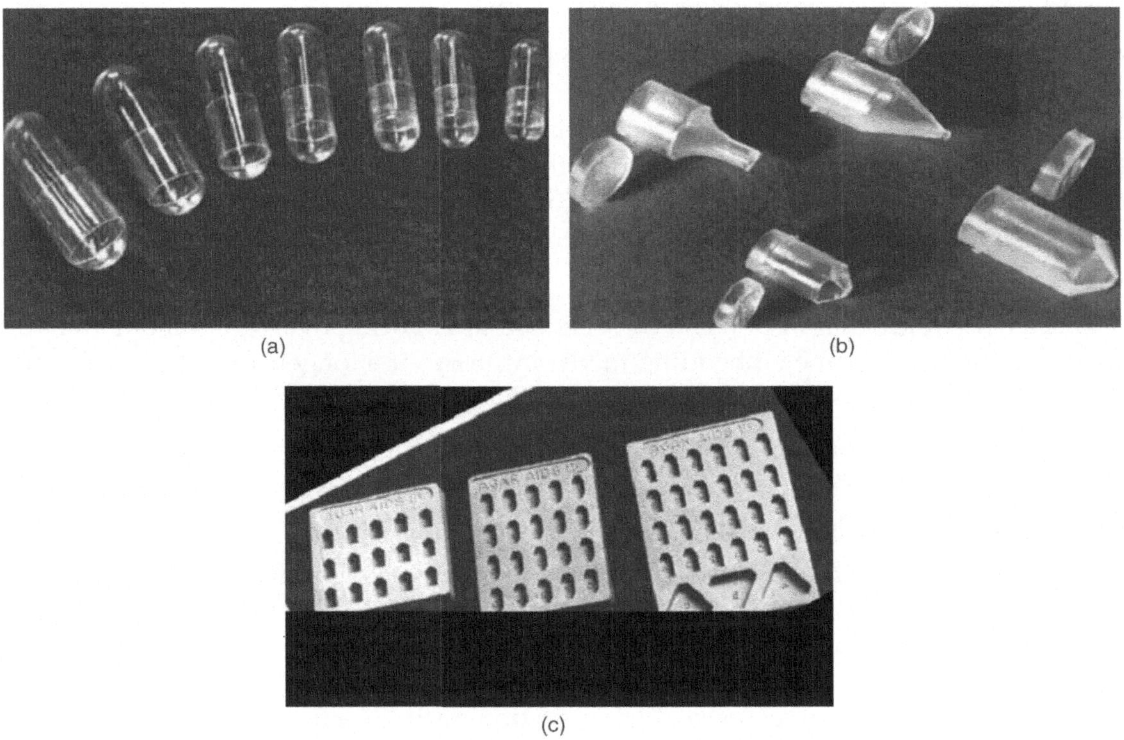

Figure 8.1 Embedding moulds. (a) Gelatin capsules giving cylindrical blocks with dome-shaped ends, (b) Polyethylene BEEM capsules of different shapes, (c) Shaped silicon rubber flat moulds

These moulds are available in the following shapes as seen in Figure 8.1.

- Flat embedding moulds
- Cylindrical embedding moulds like gelatin capsules, BEEM capsules, Easymolds, etc.

Specimens infiltrated with the resin are embedded in capsules or flat/cylindrical moulds to produce blocks of a convenient size and shape for ultramicrotomy.

Flat moulds are used when specific orientation of the specimen during polymerization is essential.

Some of the commonly used types of moulds are briefly described.

Capsules

Gelatin capsules are suited for resins including acrylic resins. Polypropylene capsules are suited for embedding at higher temperatures. Polyethylene capsules are specially designed capsules (e.g., BEEM capsules) that produce blocks with ends in various forms such as truncated pyramid or conical tip which would

* minimize the amount of trimming required before sectioning, and
* limit movement of the specimen within the capsule.

Flat Moulds

Flat embedding moulds are useful in the following ways.

* They are suitable for long specimen and for thin tissue slices.
* They allow more accurate orientation of the specimen in the block.
* They are re-usable.

Disadvantages of flat moulds The tissue may sink to the bottom of the mould during polymerization or curing and thus cause difficulty while trimming. One solution to this problem is to half-fill the mould with resin and cure it. The tissue is then transferred to this hard base and the mould is filled up with fresh resin, such that the specimen is embedded in the middle of a flat block. Specimens sometimes move as hardening of the embedding medium commences and it is then necessary to adjust the position and orientation of the specimen in the mould when the surrounding medium has just started to become viscous.

The caps of BEEM polyethylene capsules make convenient flat embedding moulds for small specimens, as do the polyethylene caps of some glass vials. Transparent versions of flat moulds are also available which are more easily visible and suitable for polymerization of resins by UV irradiation.

Flat embedding moulds with double-tapered ends which enable specimens to be embedded at both ends of a casting, and moulds with triple-tapered ends are available now.

Unlike flat moulds, gelatin capsules require a suitable support, such as the lid of a cardboard box punched with an array of holes or special trays which are available commercially. The capsules and the embedding moulds must be completely dry.

Easymoulds

The capsules employed for embedding require a support to hold them in position during the entire process of polymerization or curing as the case may be. Easymoulds are the new, convenient, tray-and-mould combinations designed for better embeddings (Figure 8.2). The tip of each mould is only 0.3 mm thick and is designed to be transparent.

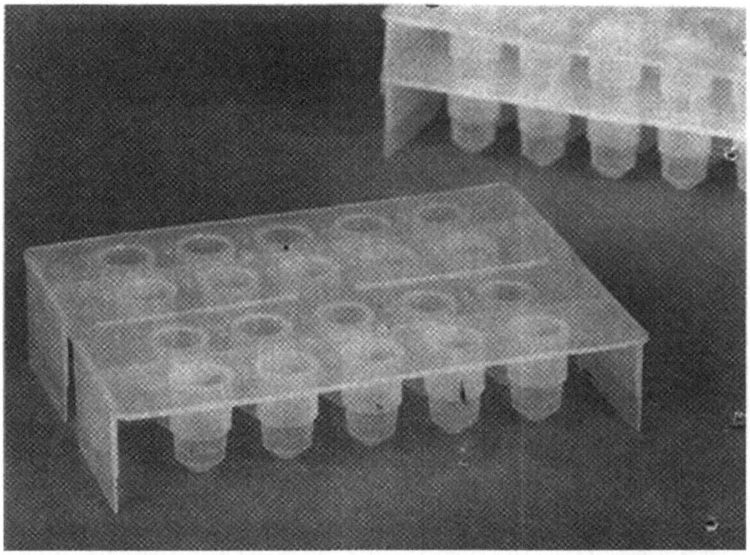

Figure 8.2 Easymoulds (*Source: Leica Manual*)

That is, one can look through the capsule to orient the specimen before polymerization. The conical tip of the mould, together with careful alignment, will reduce trimming and realignment and thereby save time. The easymould tip forms a block face that is wide enough to provide a large sectioning area. This wide surface helps to prevent air bubbles being trapped around the specimen during polymerization. Easymould trays raise the moulds away from the floor of the polymerization oven, allowing air of a constant temperature to circulate freely around the specimen tip during the hardening process. The thinness of the mould tip minimizes interference with heat exchange to give ideal conditions for even polymerization around the specimen. Simply by compressing the tip the blocks will just pop out and thus can be removed easily. Moreover the mould spaces are already numbered for ready reference. Used Easymoulds can be used for the storage of blocks which have already been sectioned.

CURING AND POLYMERIZATION

- Following infiltration with the embedding medium, each specimen bit is removed from the vial with fine tweezers and placed in the bottom of the capsule or mould.

- The capsule is filled to about two-thirds with fresh embedding medium with added catalyst and the specimen is guided back to the exact bottom of the capsule if necessary.
- Viscous embedding media should be warmed to about 60°C before use.
- A label indicating the experiment and/or the sample number (any identification mark) is placed in each capsule such that the number can be visible from outside (write with pencil).
- The capsules are filled up with the same resin embedding medium prepared with the catalyst.
- Capsules need not be capped if Epon and Araldite resins are used but it is advisable to cover Spurr's medium since some of the components of the medium are volatile.
- Specimens sometimes move within the capsule/mould as hardening of the embedding medium commences and then the position and orientation of specimen in the mould/ capsule can be adjusted.

The capsules or moulds are placed in an oven at the selected temperature (usually 60°C) for the required time, say, about 24 to 48 hours. Generally, higher temperature should be avoided since this will modify the hardening process. More cross links may be formed and the blocks will become brittle and less easy to section. But for rapid processing, the temperature is increased. Here, the specimen should be as small as possible and epoxy resin must be used because higher temperatures enhance the uneven polymerization of acrylic resins and poor preservation of ultrastructure. Vacuum ovens can also be used during polymerization but due to the inherent dangers from hot resin vapours this is not recommended routinely.

While using acrylic resins the capsules containing acrylic resins should be placed in holes drilled in an aluminium block so that the heat evolved during polymerization is conducted away as efficiently as possible and all the resin is kept at a uniform temperature.

SPECIAL METHODS OF EMBEDDING

Ultraviolet irradiation Embedding media can also be cured or polymerized by UV irradiation instead of by heat. A wavelength of 350 to 360 nm at a low intensity is required for curing and polymerization. Uniform irradiation of the specimens is essential for consistent results, which is more easily obtained in a special chamber, suitably designed. The only advantage of UV irradiation is that it enables resins to be hardened at lower temperatures. In the case of epoxy resins, this is not of much value because temperatures cannot be reduced below room temperature and also the procedure is lengthy. But some acrylic resins can be polymerized at very low temperatures by UV irradiation.

Microwave irradiation Resins can also be cured or polymerized using electromagnetic waves in a microwave oven of the type used for fixation. The advantage of using microwave irradiation for embedding is that the procedure is very rapid. But the commercial units have to be modified and designed for laboratory use. Since the electromagnetic field is not uniform, the oven has to be calibrated before each embedding. Embedding media containing the hardener, MNA, discolour during curing by microwave irradiation and so it is advisable to use an Araldite resin, which only requires the hardener DDSA (*See* Chapter 15).

Rapid embedding Araldite and Epon resins can be cured rapidly only by using high temperatures. The preferred procedure is to place the capsules or embedding moulds containing the specimens in the embedding medium at 70°C for 45 min. and then at 95°C for a further 45 min. The embedding oven should be placed in an efficient fume cupboard when these high temperatures are used. Specimens should not be placed directly in an oven at a high temperature of 95 to 100°C since this procedure is more likely to result in uneven and brittle blocks. Rapid curing of epoxy resins can also be obtained in a microwave oven.

Embedding after dehydration by freeze-substitution Specimens dehydrated at –80 to –90°C by freeze-substitution are embedded in an epoxy resin at room temperature when good preservation of ultrastructure is the main aim.

After embedding, the blocks can be removed from the moulds and brought to room temperature and used for sectioning using the ultramicrotome.

9

SECTIONING (SEMITHIN AND ULTRATHIN SECTIONS)

The resolution of the electron micrograph is inversely proportional to the thickness of the sections. The limited penetration of the electron beam also requires thin sections for screening. In an electron microscope operating up to 100 kV accelerating voltage, the sections should be in the range of 10 to 100 nm thick and these sections are termed "ultrathin" sections. This is the practical upper limit of the specimen thickness. So the whole cell of about 20 μm thick should be sliced into at least 200 sections for a good electron micrograph. To cut such sections of a cell an ultramicrotome is used.

HISTORY OF ULTRAMICROTOMY

Pease and Baker in 1948 showed that ultrathin sectioning at reasonable cutting speeds was possible if a suitable embedding medium was used. They were able to modify a conventional microtome for ultrathin sectioning. Newman *et al.* in 1948 introduced butyl methacrylate which became the embedding medium of choice those days. They also first employed a thermal expansion principle to drive a microtome. Microtomy was aided greatly when sections were floated away from the knife-edge on fluid surfaces, an idea first employed by Claude in 1948. In 1950, Latta and Hartmann discovered that fracture edges of glass could be used as knife-edge. During 1952–1953 first micrographs of ultrathin sections were produced. In 1953, the commercial version of Porter–Blum microtome was made available. Technological innovations have led to the modern ultramicrotomes which facilitate easy and smooth trimming, ultrathin sectioning as well as identification of section thickness.

The quality of the sections depends on the

- knife angle
- clearance angle (the angle between knife-edge and block face)
- speed of the cut

- ⁂ temperature
- ⁂ sharpness of the cutting edge

As sections are produced, they will adhere to one another along one edge forming a "ribbon" which can be picked up on specimen grids for further screening in EM.

Requirements for Ultramicrotomy

1. *Specimen* Well-fixed, dehydrated, infiltrated, resin-embedded and well-polymerized tissue blocks are essential prerequisites for ultrathin sectioning. Consistency of the block should not be soft and elastic but hard; the free plastic surface should not be tacky and dull but smooth and shiny.
2. *Knife* Either diamond or glass knives are used in ultramicrotomy. Glass knives are common because of their easy availability and low cost. Glass strips are cut into squares, these squares are cut diagonally at an angle of 40° to 50°. The sharp edge obtained serves as a cutting edge, which is checked with binocular microscopes under transmitted light for sharpness and cleanliness.
3. *Knife maker* A good knife maker is essential for preparing good glass knives for use in ultramicrotome.
4. *Troughs* To avoid adherence of sections with the knife due to heat generation, troughs are used. Troughs are made of insulating adhesive tapes, metals or plastic. Troughs are attached to knife-edges with sealing wax and filled with distilled water or with mixture of water and alcohol on which the cut sections float.
5 *Ultramicrotome* A good ultramicrotome is capable of cutting up to 20 nm thick sections. Both semi-thin and ultrathin sections can be cut using this instrument. Trimming of the block can also be carried out using a special stage.
6. *Pair of forceps*
7. *Eye-lash brush* A small hair such as an eyelash or bit of horse tail is fixed to one end of a small wooden piece of stick using a drop of wax or glue. This is employed to manipulate the floating sections, removing the unwanted ones and to clean the knife-edge without causing damage to it.
8. *Grids* Grids are metallic discs, about 3 mm in diameter on which the ultrathin sections are collected. Only these grids can be inserted into the specimen stage of the EM and not glass slides as the practice in LM. Copper grids are commonly used though nickel, gold and platinum grids are also available.

TRIMMING THE BLOCK

This is the first step in the sectioning process. During embedding, a drop of the resin mixture with added catalyst is placed initially in the mould and then only the processed specimen bit is placed and centred such that the hardened block will have a layer of the resin covering the specimen. Probably this layer can protect the specimen from contamination and damage during removal of the block from the mould. It essentially means that the specimen is not exposed in the prepared block. Therefore, to expose the specimen for sectioning as well as to set the desired alignment of the sectioning plane and to bring the block face to a suitable size and shape, the extra resin has to be removed. This process is called **trimming**.

The cutting plane is trimmed to expose the specimen as well as to acquire the geometric shape of **trapezoid (pyramidal)** so that the sections in series would form a ribbon. Trimming can be done either manually using a razor blade or by mechanically using a trimmer.

While trimming, the following points have to be considered:

- Trim as small as possible. Thick semithin sections can be cut from 5 mm^2 or larger areas but ultrathin sections can be cut from as small a block face as possible (not more than 1 mm^2).
- Do not trim the sides of the pyramid too steeply as this reduces the stability.
- If possible trim away all free embedding medium (Figure 9.1).

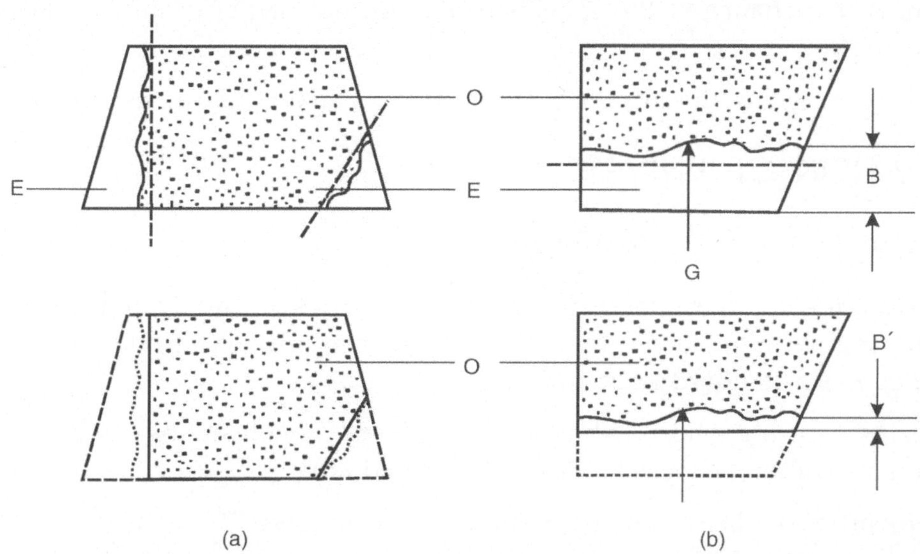

Figure 9.1 Trimming the specimen in the block face (a) Free embedding medium E without specimen O is generally to be removed. (b) If the free interface G is to be examined, the width B of the adjacent specimen-free zone E should be shortened to the minimum B´. (*Source: Leica Manual).*

- Pyramid should be flat and free from plastic debris.
- The trimmed block should have enough specimen details to be examined.
- It is advisable to have the material in the entire cutting face.
- Good contact between the sections, providing a stable ribbon formation, is obtained, if the width of the block face is approximately $1^1/_2$ times its height.
- For serial sections, two opposite sides of the cutting face (trapezoid or pyramidal) must be accurately parallel to each other and should have smooth sloping sides. These two opposite parallel sides should be aligned parallel to the knife-edge while sectioning (Figure 9.15).
- The trapezium-shaped block face enables easy cutting of serial sections.

All cavities like blood vessels, alveoli, intercellular spaces and edges and portions which contain predominantly or exclusively free embedding medium, should be completely removed. Similarly hard materials and areas containing inclusions which cannot be cut easily like fat cells, collagen, etc. should also be removed as much as possible.

Trimming may be simplified by the correct choice of embedding mould. Many plastic moulds (like Easymoulds) produce a pre-formed shape that requires a minimum of trimming. When a specific orientation is required, flat embedding moulds can be used, since there is less chance of the specimen moving out of the desired position during polymerization.

For sectioning using the ultramicrotome, glass or diamond knives are used.

PREPARATION OF KNIVES

Principles of Glass Knife Making

Glass knives are made instantaneously using glass strips manufactured for the purpose of knife making. A knife maker is used for preparation of knives from these glass strips. Figure 9.2 shows a standard knife maker.

It contains a scoring wheel with which a line score is made on the glass strip to be cut. By applying pressure, the glass breaks at the scored area.

The properties of glass for making knives are as follows:

- Should have a hardness of seven or more
- Should be completely free of air bubbles
- Glass surfaces should be flat and perfectly parallel to each other

Figure 9.2 Knife maker. (*Courtesy:* N. Murali, Lab India Instruments, Chennai.)

Glass strips for ultramicrotomy are produced from selected glass made by a special manufacturing process and are commercially available. They are usually 400 mm long and 25 mm wide. In thickness they range from 6.4, to 8 and 10 mm. When handling the strips, the edges should not be contacted because these ultimately form the knife-edge. The strips and the squares cut from them can be held by the wide upper and lower surfaces. But the corners of a partly broken strip or square should never be touched, since these will form the knife-edge.

Each glass strip is washed exclusively with cool tap water which has a low concentration of a mild laboratory detergent added to it and not with organic solvents including acetone. A sharp brush or cloth can be used for cleaning but the sharp edges of the strips should never be touched. The glass is completely rinsed of all detergent and then flushed thoroughly with distilled water. It is then dried gently with a clean, soft, lint-free cloth which will not scratch the edges of the strip. The knife maker also has to be cleaned repeatedly from glass splitters with a brush during knife making.

In the **balanced break method** of cutting glass strips (Figure 9.3), a standard strip of glass (manufactured for the purpose of knife making) usually 400 mm in length is scored and broken into 2 equal halves, each 200 mm long. With an equal mass of glass on each side of the score, the break is balanced and the freshly fractured surfaces are plane. By continuing to divide each piece produced into 2 equal halves, up to 16 squares can be made. All squares produced have straight sides and precise right-angled corners.

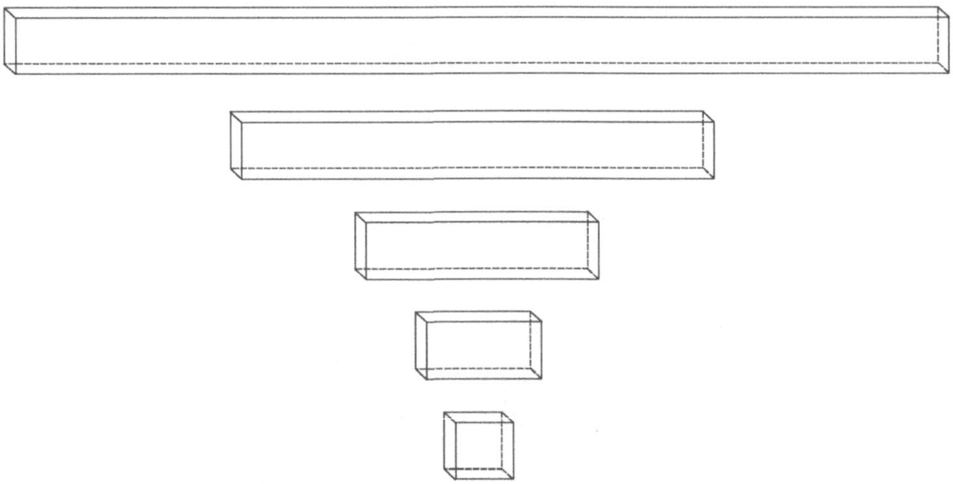

Figure 9.3 Balanced break method of knife making. (*Source: Leica Manual.*)

Both end pieces (squares) of a glass strip should not be used, as they are geometrically not perfect as shown in Figure 9.4.

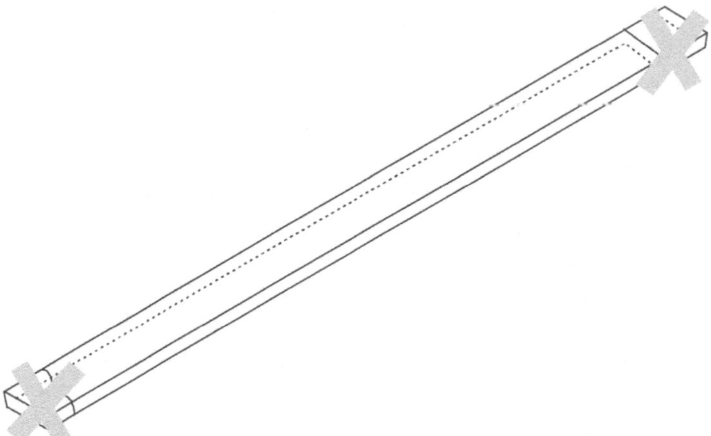

Figure 9.4 Knife strip with ends marked for discordance. (*Source: Leica Manual.*)

When a diagonal cut is made on the square, two right-angled triangles are produced each with a sharp edge at one end and a small counterpiece at the other. The sharp edge is used as the knife-edge (Figure 9.5).

The knife angle is an important criterion for good sectioning. The angle is adjustable to a few pre-set values in some knife makers. In some, it is predetermined and fixed and hence not alterable. The angle is usually 45° for routine sectioning and less for cryosectioning.

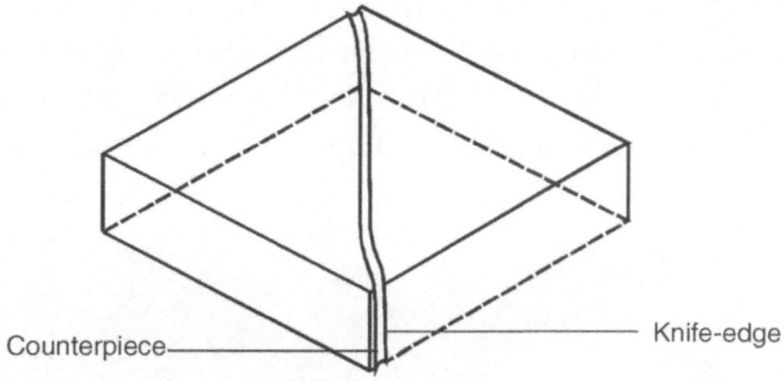

Figure 9.5 Two glass knives cut from a square piece. (At both ends one sharp knife-edge and a counterpiece can be seen.) (*Source: Leica Manual.)*

Each knife-edge is examined under dark-field illumination using a stereomicroscope, which is normally fixed in the ultramicrotome itself. When viewed in dark-field illumination, the edge will appear white against a black background.

The right side of the edge has visible stress marks which reduce the quality of knife and the left corner is also unsuitable for sectioning because it is disturbed by the stress line, which starts from the left-hand corner of the cutting edge. The stress line is also known as **Wallner line.** The stress line indicates the side where the break started, and the irregularities increase in magnitude towards the opposite side. Thus from the above, it is clear that the central part of the edge is most useful for ultrathin sectioning, as shown in Figure 9.6. The length of the useful edge is found to be longer in all knives produced with small counterpieces.

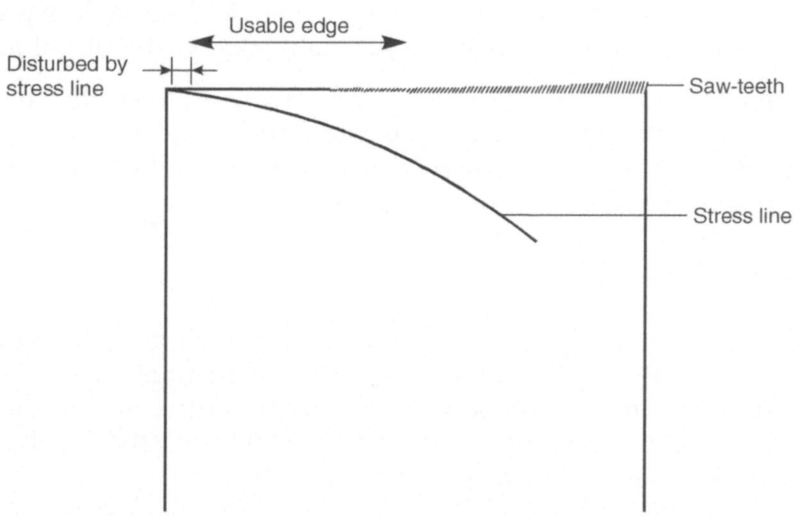

Figure 9.6 Diagrammatic representation of the edge of a glass knife. (*Source:* Audrey, M. Glauert. (Ed.). (1974). *Practical Methods in Electron Microscopy.* Vol. 3. American Elsevier, North Holland. p. 259, Fig. 3.10.*)*

The storability of glass knives is practically unlimited if the knives are kept clean, protected against dust and preserved in separate knife storage boxes. Figure 9.7 shows a few boxes with sliding covers along with knives stored in them. While placing or removing the knives from the boxes, care should be taken not to damage or touch the knife-edge.

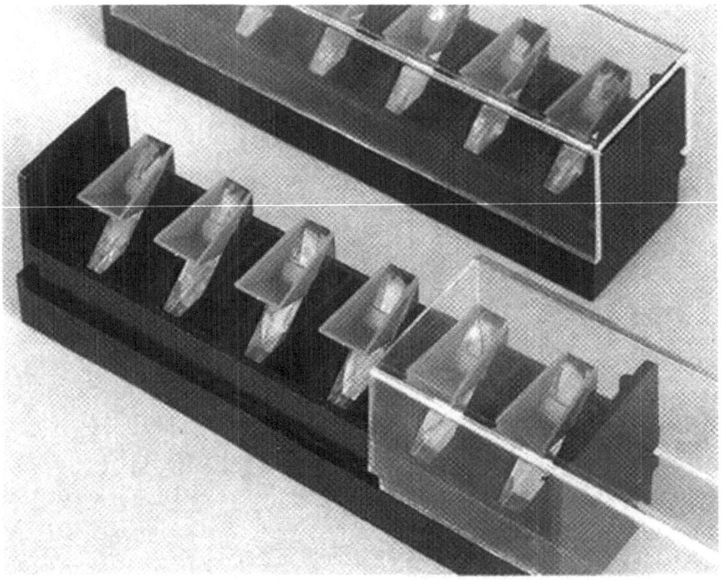

Figure 9.7 Knife storage box. *(Source: Leica Multiplate Manual.)*

But the "service life" of the glass knife is very small and hence it is preferable to cut initially a few 0.5 μm sections and then immediately take off the ultrathin sections.

Glass cutting edge should not be contacted in any case. Even the slightest pressure destroys it irreversibly and any contact leaves traces and therefore every knife is checked before use with the dark-field illumination and all knives which do not appear to be completely clean and free of defects are discarded.

Mounting of the Knife boat

To complete the knife for sectioning, a knife boat or trough is mounted onto the knife back. This boat will be later on filled with the floatation medium which is water or very dilute acetone, on which the cut sections will float. These troughs are usually made of pre-shaped metal or plastic or an adhesive tape. A metal or plastic trough clips round the glass knife and is then sealed with wax to make it water-tight.

Tape troughs have the advantage of being disposable but will normally provide a smaller, although adequate, liquid surface. Molten wax or nail lacquer can be used as a sealant. The simplest way to prepare a tape trough is to wrap an adhesive tape from one side to the other and in order to make the boat water-tight, seal the lower part of the boat with the knife, using the sealant. Care should be taken to avoid contaminating the cutting edge or the front face of the knife. The trough should not be too large because sections can be handled most easily if they are in a rather confined space. Therefore narrow adhesive tape should be used and sealed tightly to hold the floating medium. Water-proof adhesive tapes are more suitable for this purpose. Figure 9.8 shows the tape around the knife-edge and the position of the blade to cut the excess.

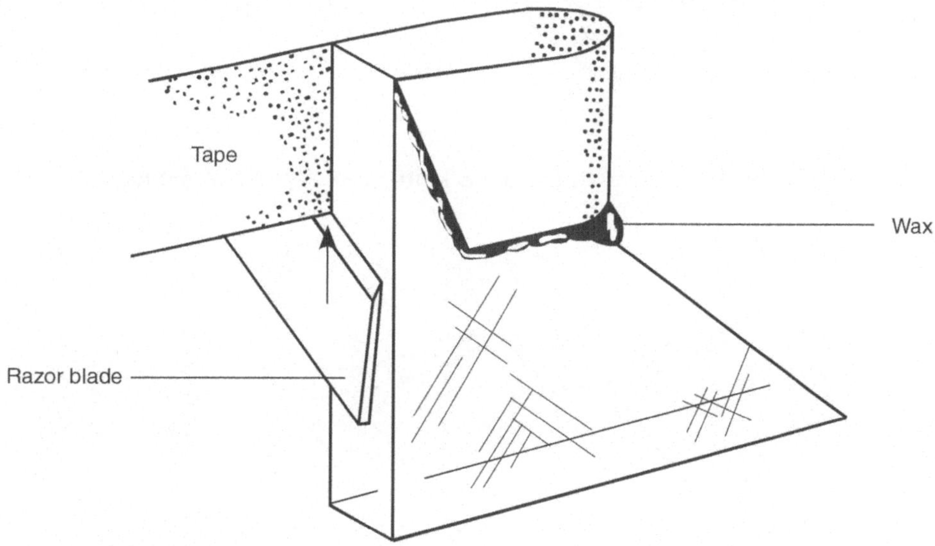

Figure 9.8 Tape fixed to the glass knife with wax. (*Source:* Audrey, M. Glauert. (Ed.). (1974). *Practical Methods in Electron Microscopy.* Vol. 3. American Elsevier, North Holland. p. 264, Fig. 3.14.)

Making tape troughs can be a slow and untidy procedure. The size and shape of tape troughs is not reproducible; therefore the water level must be varied in order to get a good reflection.

Instead of tapes, metal boats, made of bronze or stainless steel, were used previously. But now, plastic boats are available which are easier to handle and fix. Precleaned plastic boats called **trufs** are available commercially. The boat is fixed to the knife back by means of wax. With the truf, fixing the boat has become quick and easy, providing good results every time. Moreover, since the size and shape is always the same, the water level will also be the same and this will give the same perfect reflection. The size of these boats provide large areas for the floating sections and so long ribbons without the sections "crashing" or sticking to the walls of the trufs can be easily produced. Figure 9.9 shows a glass knife and a truf while Figure 9.10 shows the truf fixed to the knife-edge.

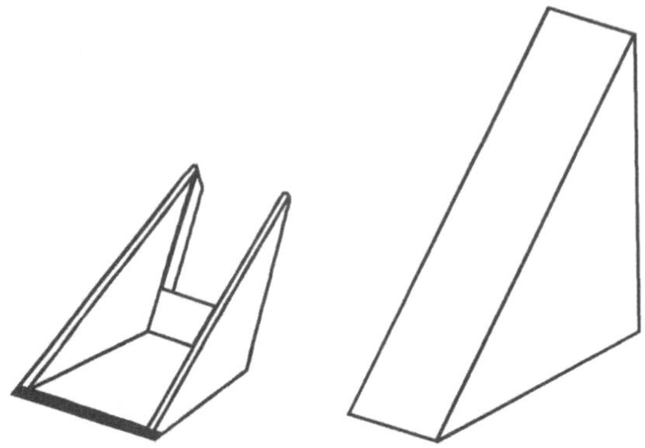

(a) (b)

Figure 9.9 (a) Trough (b) Glass knife. (*Source: Leica Manual.)*

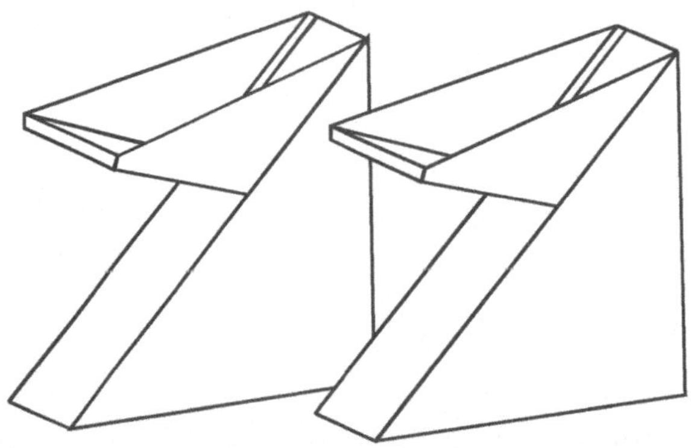

Figure 9.10 Trufs fixed to the glass knives using wax. (*Source: Leica Manual.)*

For the purpose of fixing the truf to the knife, a **Multiplate heater** is extremely useful. The "multiplate" is a small hotplate. This hotplate has a wax bath, and two additional plates with different temperature levels, one, for the drying and staining of sections and the other for prewarming of the glass knives before mounting the truf as shown in Figure 9.11.

The wax melting deck is kept at approximately 95°C to provide wax for mounting the trufs. By placing the trufs on this deck, the wax is distributed evenly on the edges so that they can be attached tightly on the knives. The drying deck has a temperature of approximately 80°C, ideal for drying and attaching semithin sections to slides. The deck is large enough to dry 6 slides at a time. This deck is also used for prewarming the glass knives without touching the knife-edge as shown in the figure. By mounting a raised plate over half of the drying deck, the heating deck is formed. Here the temperature is approximately 60°C and just right for heating during staining (i.e., toluidine blue) for semithin sections.

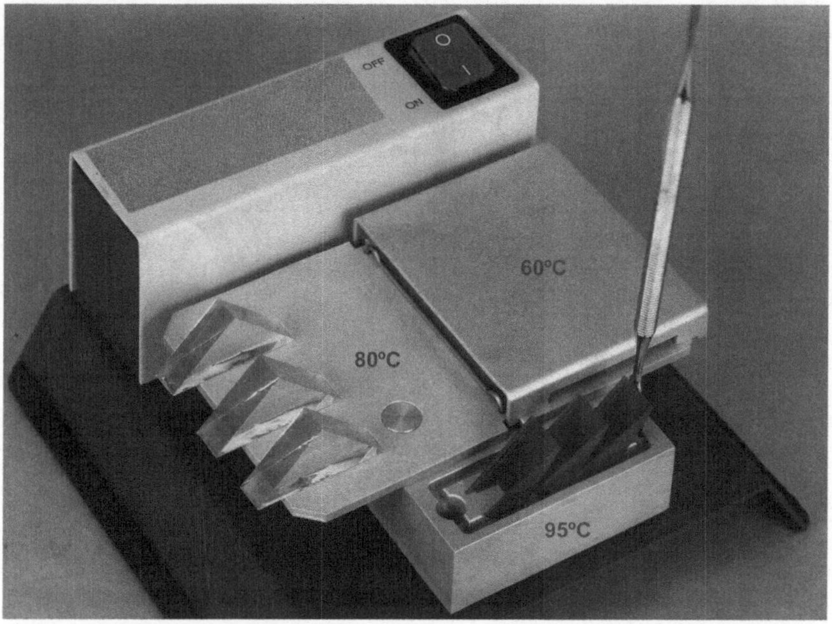

Figure 9.11 Multiplate heater. (*Source: Leica Manual.*)

Diamond Knife

Diamond knives were first introduced by Fernández-Morán in 1956 who demonstrated that industrial diamonds could be ground so that they have satisfactory edges for ultrathin sectioning. An obvious advantage of diamond knives is their durability. The edge is obtained by the cleavage of certain industrial diamonds along suitable natural planes, followed by grinding and polishing with ultrafine abrasives. The diamond is sintered into a metal shaft for handling purposes and this shaft is then fixed in a diamond knife holder incorporating a trough.

Diamond knives are relatively hydrophobic and impurities increase this property. So wetting of the knife-edge becomes a problem. Because of this hydrophobic characteristic of the knife-edge, the sections may not properly glide away from the knife-edge, and compression effects will result. In such cases, the diamond knives are best placed in water overnight and then coatings of foreign materials including debris removed with a small pith rod dipped in alcohol. Before sectioning also, when the trough is filled with water or dilute acetone, the edge has to be wetted with a sliver of wood or pith provided by the manufacturers. The knife with the fixed boat is shown in Figure 9.12.

Figure 9.12 Diamond knives with fixed boats

Diamond knives have sharp, scorefree, durable cutting edge and so they are particularly advantageous for certain types of "difficult" tissues. They are mainly useful for:

- ✳ preparation of thinner and qualitatively better sections
- ✳ cutting of nonhomogeneous specimens like hard inclusions
- ✳ cutting of hard specimens like calcified tissue
- ✳ cutting of difficult tissues like dense connective tissues containing heavy collagenous fibres.

Drawbacks

- ✳ Fragility. The edge, being thin, is reasonably flexible but can be chipped off due to carelessness.
- ✳ High cost.
- ✳ Though hard and durable, they are very brittle and can be easily damaged by misuse.
- ✳ Should not be used for cutting sections thicker than 1 μ.
- ✳ Should never be used for trimming.

Advantages

- ✳ They last indefinitely. Cutting edge remains sharp for a considerable amount of time.
- ✳ They can be re-sharpened.

- They are permanently mounted in a holder which incorporates a trough.
- They can cut specimens which are too hard to be cut with glass knives.
- They can cut thinner sections for high-resolution work.

Cleaning of the diamond knife-edge

- The diamond edges are cleaned by rubbing gently and carefully with a thin sliver of wood or pith strips supplied by the manufacturer.
- Strong acids and organic solvents may remove adhesives used for fixing the boat to the knife and hence should be avoided for cleaning.
- While cleaning, the knife-edge is not damaged if, all horizontal forces perpendicular to the edge are avoided, because diamonds are extremely sensitive only in this direction. Most important factor is to avoid rubbing it in a transverse direction.

Sections should always be removed from the knife-edge so that they cannot dry onto the sectioning facet of the knife as the trough liquid evaporates, because then it is very difficult to remove them. During sectioning, small particles of the embedding material may adhere to the knife-edge and will cause scratches on the sections if they are allowed to remain.

The edge should be cleaned with a stick, soft wood, balsa or pith which has been dipped into acetone or alcohol. One end of the cleaning tool is cut flat and then drawn across the whole knife-edge with a gentle but steady pressure so that the knife-edge just cuts into the cleaning stick. This wiping movement must only be made parallel to the edge and never obliquely or at right angles to it. The edge should never be touched by anything other than the specimen or the cleaning tool. After use, the unwanted sections should be removed immediately while the knife is still wet. The knife and the trough are then carefully rinsed with distilled water and allowed to dry. Acid should never be used to clean a diamond knife. The clean knife should be left in its holder and stored in its box.

Diamond knives can also be washed using magnetic stirring action with special stirring bar which gives a fast, clean and safe method to remove particles that are left on the knife-edge after sectioning.

Diamond knives are usually supplied with two angles 45° and 35°, the smaller angle for biological materials and the larger one for metals and other hard materials.

Knives with 45° knife angle are used for:

- routine sectioning of biological samples
- hard, brittle industrial samples such as ceramics
- electron spectroscopic imaging, 3D reconstruction

- ✳ alternating sectioning semithin/ultrathin; HVEM
- ✳ surfacing samples for STM, AFM, SEM, etc.
- ✳ sectioning of Lowicryl blocks
- ✳ sectioning of inhomogeneous samples such as undecalcified bone, dental materials, etc.

Knives with 35° knife angle is used for:

- ✳ Low-temperature situations
- ✳ Frozen biological samples such as cryo-protected specimens, frozen hydrated specimens, etc.
- ✳ Industrial samples such as polymers and rubber

There are certain advantages of using cryo-diamond knives against glass knives. They are as follows:

- ✳ Cryo-diamond knives produce thinner sections.
- ✳ They produce ultrathin and semithin sections with the same knife.
- ✳ They produce long ribbons of sections easily.
- ✳ They show less compression and greater uniformity of sections.

Recently, we even have Ultrasonic Oscillating Diamond Knife for room temperature sectioning which produces ultrathin sections free from compression. This is a boon because, compression, which is the shortening of the section compared to the sample block height is one major obstacle in microtomy, and compression in room temperature sections can be as much as 20% which readily limits the quality of the sections.

ULTRAMICROTOME

An instrument used for sectioning is called a microtome and the one that is employed for ultrathin sectioning in the EM laboratory is referred to as an **ultramicrotome.**

Ultramicrotome Design

This instrument consists of the following main parts (Figure 9.13).

1. specimen holder that moves in an angular segment arc
2. specimen-holding arm with ultrafine capability for advancement and retraction on return stroke
3. knife-holding stage with adjustments for lateral swings and tilt ability

4. a stereomicroscope
5. control panel for cutting speed control, automatic feed advancement, multi-illumination system, etc.

Modern instruments have 3 light sources:

1. *Incident light or top light* The pre-aligned LED illuminator built into the microscope carrier is so aligned to the stereomicroscope that an exactly plane and horizontal water surface in the boat of the knife reflects the light from it into the objective of the stereomicroscope.
2. *Backlighting* The second light source is in the base of the ultramicrotome which illuminates the knife and specimen from below.
3. *Transillumination* The third light source can be mounted into the segment arc and illuminates the specimen from behind.

These three light sources are useful for:

- adjustment of the water level
- measuring the size of the block face
- observation of the cutting process
- judgement of the section thickness
- alignment of the knives
- knife inspection for contamination and flaws (nicks) before sectioning
- transillumination for locating and identifying interesting structures or small samples within the specimen block when trimming

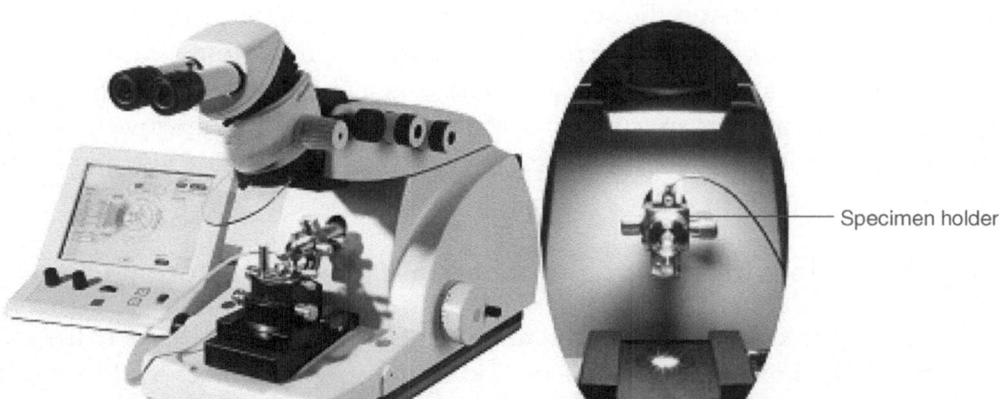

Figure 9.13 Ultramicrotome. (*Courtesy:* N. Murali, Lab India Instruments, Chennai.)

In the ultramicrotome the knife is kept stationary to avoid movement of the liquid in the boat fixed to it, whilst the specimen block is moved past it in the vertical plane. A section is detached during the downward stroke, but if the specimen block is returned upward along the same path, the section is immediately picked up by the specimen block as it passes the knife-edge. To avoid this problem, the specimen block is made to pass the knife-edge only once, i.e., on the downward cutting stroke, returning via another path. An arm carrying the specimen block is moved downwards past a sharp cutting edge, the knife, and a very small movement of the specimen arm towards the knife between each cut results in a thin slice or section being detached from the face of the specimen block.

When the specimen block is moved past a cutting edge, forces are transmitted from that cutting edge to the specimen causing a thin layer of the material to be detached. Considerable friction is encountered as the section moves down the sectioning facet of the knife but this becomes almost negligible if a suitable liquid such as water is present at the knife-edge. The section then floats easily onto the liquid surface.

Any microtome that is used for ultrathin sectioning is usually fitted with a binocular "dissecting" microscope to view and pick up the floating sections. These microtomes are equipped with a focused light source, hence it is possible to perform the examination of the knife-edge with the knife mounted on the instrument. A beam of light is focused onto the knife-edge so that it appears as a narrow band of light showing against a dark background. Irregularities will be shown quite clearly since they scatter the light. Only areas of the knife-edge free from imperfections when viewed at an angle of 30°, should be used for ultrathin sectioning. A prominent stress line will be seen originating from the start of the break and prevents the extreme corner being used for sectioning. The amount of usable edge will depend on the angle of the knife and the method of breaking. Areas showing small defects will produce some scratches on the sections but may be used for semithin sections or for trimming the blocks. The binocular microscope fitted with a microtome should be arranged with a decided tilt. This is not only a satisfactory arrangement for visualizing the interference colours of the sections as they are cut, but it is also convenient for viewing the trough and the block and for manipulations in this area. The best light source for viewing sections by reflected light is a fluorescent lamp.

Mounting the Knife Fitted with the Boat in the Ultramicrotome

To avoid damaging the knife, it is better to attach the specimen holder, or head, to the specimen arm before fixing the knife, although orientation of the specimen block will be made later. Most instruments have a reference mark for setting the knife height. The knife should be clamped tightly. Knife holder in the microtome holds the front face of the knife. This face fits smugly against a metal plate or plastic which then truly parallels the knife face.

Proper knife tilt is the most important thing for successful sectioning and care must be taken in determining it. A slight tilt is necessary and a tilt of about 3° to 6° seems best. This is known as "clearance angle" which is actually the angle between knife-edge and block face. Thus, the knife is inclined at an angle to the block face and not parallel to the face. The block should not touch the back of the knife. Too much tilt encourages "chattering". It promotes chatter because the knife-edge tends to dig into the block and then to bounce back when elastic forces become great enough. When the knife is leaning slightly backwards, the block coming down misses the edge and rubs against the back of the knife. This compresses the block a little and the same thing happens again on the next cycle. Thus no sections at all are cut for a number of cycles of the microtome, but then suddenly a thick section is cut when the block recoil catches up with the additive effects of compression and the knife digs in.

Orientation of the Block in the Microtome

The trimmed block is fitted in the block holder of the ultramicrotome. The block should be firmly clamped in its holder to avoid any movement during sectioning. When inserting the trimmed block in the specimen holder of the ultramicrotome, the end of the specimen block should not protrude more than a few mm from the holder to ensure adequate rigidity. Blocks that are cut too pointed or those which protrude too far out of the holder tend to vibrate and are hence to be avoided. (Figure 9.14). The primary concern here is to minimize the forces which tend to deform the knife-edge. The narrower the block, the less chances there will be for "chatter" and "digging".

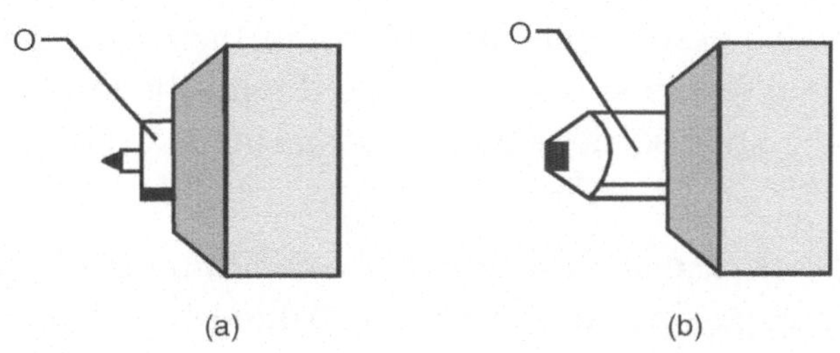

Figure 9.14 Clamping of the specimen block "O". (a) Block cut too pointed (b) Block protruding too far out of the holder. (*Source: Leica Manual.*)

As we have already seen, while trimming the block, the trapezoid should have the upper and lower edges exactly parallel to one another. To make it possible for the sections to adhere together forming a straight ribbon, the upper and lower edges of the block face must be parallel to the knife-edge. The block face must be parallel to both the knife-edge and to the cutting plane if the whole of the surface is to be cut.

Figure 9.15 shows the proper orientation of the two edges of the block face and the knife-edge. The upper and lower edges of the block face are exactly parallel to one another as well to the knife-edge.

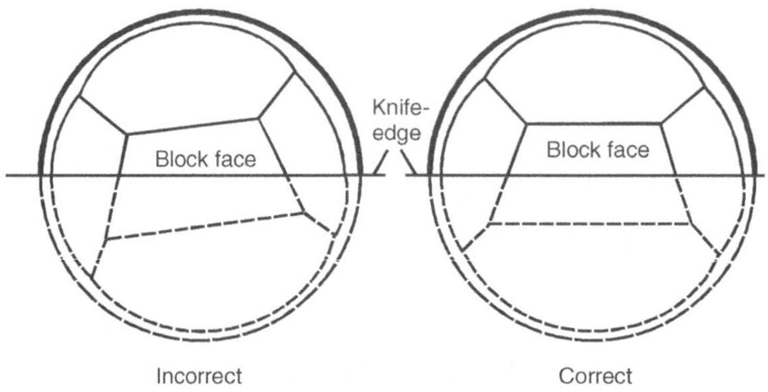

Figure 9.15 Alignment of the block face and the knife-edge. (The broken lines indicate the part of the face that is behind the knife when viewed from the front). (*Source:* Audrey, M. Glauert. (Ed.). (1974). *Practical Methods in Electron Miocroscopy.* Vol. 3. American Elsevier, North Holland. p. 290, Fig. 6.1.)

Dense connective tissue with its quantities of tough collagenous fibres, if present, is particularly difficult to cut smoothly. If this involves only a part of a block, it should be placed on the trailing side. Thus, to cut skin, the epidermis should reach the knife first, followed by the dermis. In cutting a portion of a large artery, the endothelium should be first and the adventia last, etc. In this way, coarse chatter effects and tears are at least confined to the connective tissue zone and the rest of the tissue spared. Very narrow but long blocks (longitudinal sections) can be mounted with the long axis somewhat oblique to the knife-edge. As the knife sections them, only a small part of its edge is being used at any given moment.

The trapezoidal block face should first strike the knife blade with the narrow edge and not with the long edge since the shock on striking the blade is proportional to the width of the edge. (Figure 9.16). Continuous ribbons of sections are obtained with sufficient contact between the successive sections and hence it is necessary to align the upper and lower block edges parallel to one another and both edges exactly parallel to the knife-edge.

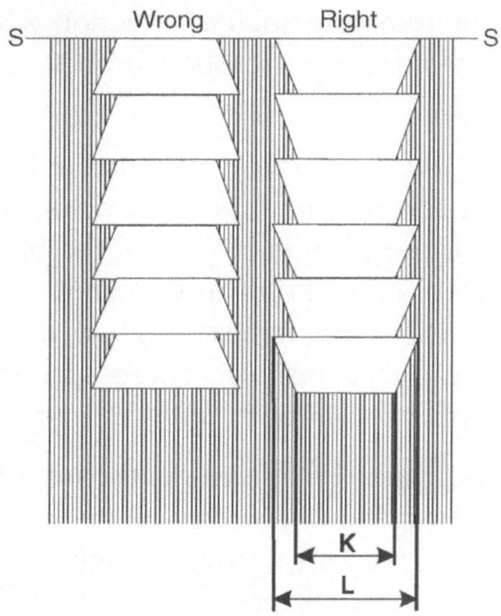

Figure 9.16 Contact between the knife-edge and the block face. The trapezoidal shapes or irregular rectangles should always first contact the knife SS with the shorter edge K and not with the longer edge L. (*Source: Leica Manual.*)

Filling the Trough

The fluid (water or dilute acetone) level is another important adjustment to be made after arranging the block face and the knife-edge as mentioned above.

Clean distilled water is perfectly adequate. Its physical properties make it almost ideal for this purpose; it has a relatively high surface tension, that allows the sections to float and a low viscosity that allows the sections to move freely on its surface. The low evaporation rate enables its level to be maintained without the need for frequent adjustments. In the early days, when methacrylate embeddings were prepared, a dilute solution of ethanol or acetone was used to fill the boat as this helped to stretch out wrinkles in the sections. But these have little stretching effect on epoxy and polyester resin sections. Solutions or liquids other than water may be employed only when the sectioning substances are affected by water or when sectioning at low temperatures.

Using a syringe or a pipette, the liquid is added at the rear of the trough to avoid touching the knife-edge, until the knife-edge is completely wetted, and the liquid surface is approximately horizontal. When difficulty is encountered in wetting the knife-edge, as is often found with a new diamond knife, more liquid than necessary should be added and then slowly removed until the desired level is reached. The fluorescent light is positioned to obtain a silvery white image over as large an area of the liquid surface as possible. It may not be possible to obtain such a reflection when liquids other than water are used. Some modern microtomes are provided with a "**Peristaltic water pump**" for filling the boat.

The fluid level should wet the knife-edge but not the block face. The water level in the trough or rather its meniscus, is also an important factor for good and serial sectioning. If the level is high, i.e., even if one drop is in excess, the block face will be wet and the fluid will be carried over the edge of the knife and down its front face, thus contaminating the knife-edge. On the other hand, if the meniscus is low, the sections will not glide away from the knife-edge in an unimpeded fashion. There will be some back pressure and a tendency for the sections to rumple and not flatten out properly.

It is very important that the water, or otherwise, the liquid used, be clean and without contamination. Demineralized or distilled water alone should be used to fill the boat. If acetone solution is necessary, reagent-grade acetone diluted with demineralized or distilled water should be used. Tap water may precipitate mineral salts and so should not be used.

Cutting Methodology

Prerequisites

- A clean knife with proper knife angle
- Proper tilt (clearance angle)
- Liquid meniscus at the proper level
- Sufficiently small block
- Quiet room without any vibrations
- The operator should have no contact with the microtome table; must not have his elbow on the table; nor his foot be hooked over a table rung.
- While clamping the specimen and knife, very stable seating of knife and specimen is essential for satisfactory sectioning.
- The specimen should not be pointed to a needle shape nor protrude too far out of the holder.
- Low-frequency vibrations can be eliminated when the blocks protrude from the holder by about the amount their diameter.
- Tightening of the clamping just before sectioning is necessary because the block is always loosened after the trimming; clamping surfaces should be clean, free from plastic shavings or glass splinters for tight clamping.

- The upper and lower edges of the block face should be parallel to each other and also to the knife-edge.
- Cutting speed is set to about 2 to 5 mm/sec for glass knives and 0.5 to 2 mm/sec for diamond knives.
- For semithin sections and for initial cutting, the section thickness should be less than 1 μm, since fissures result if the sections are too thick. Such fissures are seen on the block surface in top view as black slits or sharply delineated black areas and lead to the carrying along of water from the collecting boat and to the formation of water bridges between the cutting surface of the block and the front area of the knife. Since they usually run deeply into the interior of the block, these blocks are discarded.

THE PRODUCTION OF SECTIONS

The first contact between the specimen and the knife is most important as both can be damaged if care is not taken. The specimen arm is operated either by manual or automatic control and the knife is advanced in very small increments approximately 1 μm or < 1 μm between each stroke. If the advance is too great, there is a danger of cutting a very thick section which will blunt the cutting edge. The choice of the cutting speed will vary for different specimens but the initial speed should not be too slow, or, the liquid will be dragged over the back of the knife before the first section is cut. An average speed of 2–3 mm/sec. can be used as a starting point and then the speed may be varied according to the properties of the material being sectioned. When cutting soft specimen blocks, a faster speed than is desirable may be required to detach the first section and once sectioning has commenced the speed can be reduced.

As the advance takes pace automatically between each stroke, it is important that a section is produced at every stroke once the contact has been made, otherwise the specimen block will rub against the clearance facet of the knife thus quickly damaging the knife-edge. Moreover, if a section is not produced in one stroke, the following section will be approximately twice as thick as the advance setting.

The sections should be allowed to float freely on the liquid surface. If the knife-edge is dry, or the leading edge of the ribbon is impeded, the sections will be creased. Sectioning is continued until sufficient sections have been obtained or until sectioning becomes erratic. The arm movement and advance are stopped and the sections collected. Wrinkles which appear as the section is cut but slowly disappear indicate a softness, while sections from a firm block will always be perfectly smooth.

SECTION THICKNESS

The smallest thickness obtainable depends, mainly, on

- size of the block face
- quality of the embedding
- sharpness of the cutting edge

When sections floating on the surface of water are viewed through the binocular microscopes, interference colours are seen. Light is reflected from both the upper surface of the section and from the section–water interface and these light waves moving out of phase produce interference colours. These colours are related to the thickness of the section and the refractive index of the sectioned material.

It has become easier to estimate section thickness after the discovery that interference colours could be used in estimating section thickness. This discovery was first formalized by Porter and Blum in 1953. Peachey in 1958 presented the most detailed correlation of section thickness and interference colours. He produced a colour scale which is applicable to all embedding materials. The colours form a continuous spectrum and not distinct steps. So these values can only be taken as a guide to the actual thickness.

Table 9.1 gives the correlation between the interference colour and section thickness and Figure 9.17 shows the same as a colour chart which is provided along with each ultramicrotome by the manufacturers.

Table 9.1 Correlation between the interference colours and section thickness

Colour	Thickness (in nm)
Grey	60
Silver	60–90
Gold	90–150
Purple	150–190
Blue	190–240
Green	240–280
Yellow	280–320

Most sections within the silver to pale gold range (60–90 nm) are suitable for normal EM work; the thinner ones will be required for high resolution work. The thin sections lack contrast and hence have to be stained or metal-shadowed as necessary before screening in the EM.

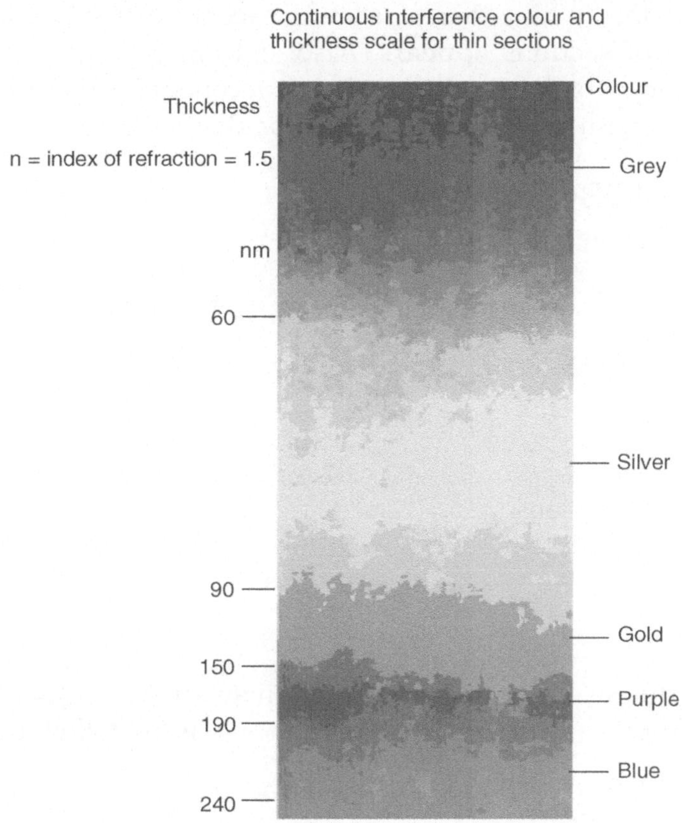

Figure 9.17 Colour chart for assessing thickness of floating sections. (*Source:* Audrey, M. Glauert. (Ed.). (1974). *Practical Methods in Electron Microscopy.* Vol. 3. American Elsevier, North Holland. p. 297, Table 6.1.)

Limiting Section Thickness

The limiting section thickness is greater than or equal to 30 nm for glass knives and greater than or equal to 15 nm for diamond knives under ideal conditions. There ideal conditions include

- best specimen composition
- small, homogeneous section area
- correct knife and specimen adjustments
- proper trimming
- optimally clean and sharp knife-edge
- absolutely quiet workplace, without vibrations

If it goes beyond the limiting thickness, every second section is skipped or absolutely irregular sequence of sections appears. Also, extremely thin sections are strongly compressed and show very small contrast. So it is recommended to mount these sections on unfilmed narrow mesh grids, and then stain or shadow-cast.

SECTIONING PROBLEMS AND COMMON SECTIONING ARTIFACTS

The most common problems encountered during sectioning are listed below:

- uneven section thickness
- vibrations
- knife marks
- wrinkles in the sections
- section contamination

Irregular Sectioning

If sections are produced, but only, at each alternate stroke or vary in thickness from section to section, it probably means that there is one of the following problems.

- The cutting edge is not sharp, which may be due to blunt knife-edge; so knife has to be changed.
- The quality of embedding is not good, e.g., a very soft block. Then the trimmed block can be left at 60°C overnight for hardening.

Poor Ribbon Formation

This could be due to one or more of the following reasons.

- upper and lower edges of the block face are not straight; the ribbon will curve towards one side
- liquid level too high
- upper edge is not parallel to knife-edge

Sections Dragging over the Knife-Edge

This occurs with soft blocks because the section is not cleanly detached from the upper edge of the block face, but is torn away leaving a ragged edge.

Wetting of Block Face

When the liquid level is too high, it will be drawn over onto the back of the knife by the specimen as it passes. Once this happens, the block face must be blotted dry and the excess fluid removed. When sectioning large areas, more than 2 mm^2, it is often necessary to reduce the liquid level as far as possible without allowing it to recede from the knife-edge. With smaller areas, the level is not so critical.

Holes in the Sections

These may occur due to poor embeddings, especially, when there are areas of hard tissue in the specimen, e.g., calcified structures, cuticle layers in botanical specimens, etc.

Scratches and Knife Marks in the Sections

Any straight mark running through a series of sections perpendicular to the knife-edge will have been produced by a defect in the edge. A portion of the glass knife-edge always displays saw teeth, as we have already seen. But, these areas are usually avoided for sectioning. And if present, the profile of these saw teeth is visible only in the first or second section of a ribbon and the following sections are entirely free of faults. These, with normal section thicknesses around 70 nm, usually produce no kind of disturbance. But, the real knife defects result through contact of the knife-edge with a hard object (forceps, glass needle, fingers) or by dust grains, which are pressed by the block against the cutting edge. Similar defects result from hard inclusions in the specimen block. They produce notches which split the sections along the entire length of the specimen block. They reduce the section quality and often lead to separation of the sections. The knife marks appear singly or at irregular intervals and split the section, while the saw teeth produce fine streaks at approximately regular intervals but which extend over broad regions of the section.

In Figure 9.18 the right-hand side of the ribbon of sections shows deep scratches resulting in tearing and creasing of the sections. As the scratch on the left does not appear in the leading part of the first section the damage has been caused by a hard particle within the specimen.

If scratches are observed, a fresh part of the edge or a new knife should immediately be tried. The first section or two may still contain the score marks, even after changing the knife, because, the fault in the knife may have damaged the block face. If the scratch continues to persist in exactly the same place on the sections, the fault will be with the specimen. A hard region within the specimen, can damage the knife-edge at its first contact which therefore produces a scratch mark. The damage, if caused by a foreign body, can be trimmed away. Or if it is inherent in the specimen, e.g., asbestos particles in lung tissue, the problem may only be solved by resorting to a diamond knife. A similar result may be seen where particles of the material are torn out from areas that have not been well impregnated with the embedding medium. These particles often adhere to the knife-edge causing scratch marks on subsequent sections until they are dislodged.

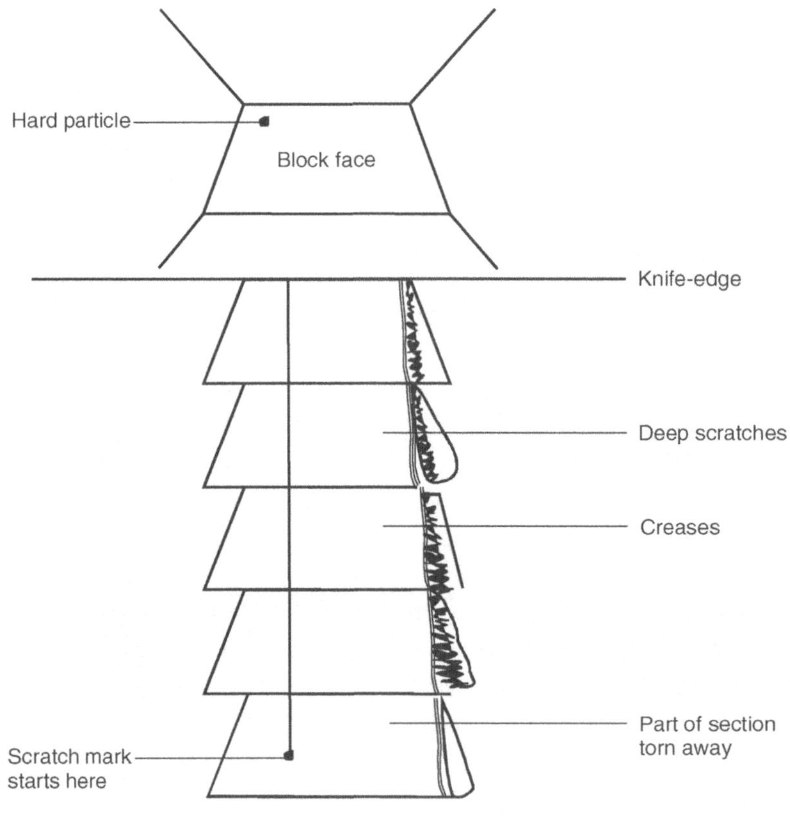

Figure 9.18 Scratches and creases in floating sections. (*Source:* Audrey, M. Glauert. (Ed.). (1974). *Practical Methods in Electron Microscopy.* Vol. 3. American Elsevier, North Holland. p. 303, Fig. 7.2.)

In Figure 9.19, the first part of each of the section is free from scratches. An area containing calcium has not been well-impregnated with the embedding medium and has been therefore torn out. Small particles have adhered to the knife-edge causing scratch marks only in the latter part of the sections. As each fresh section is cut, the particles are dislodged from the edge so that the scratches only re-appear after the calcium area has been cut.

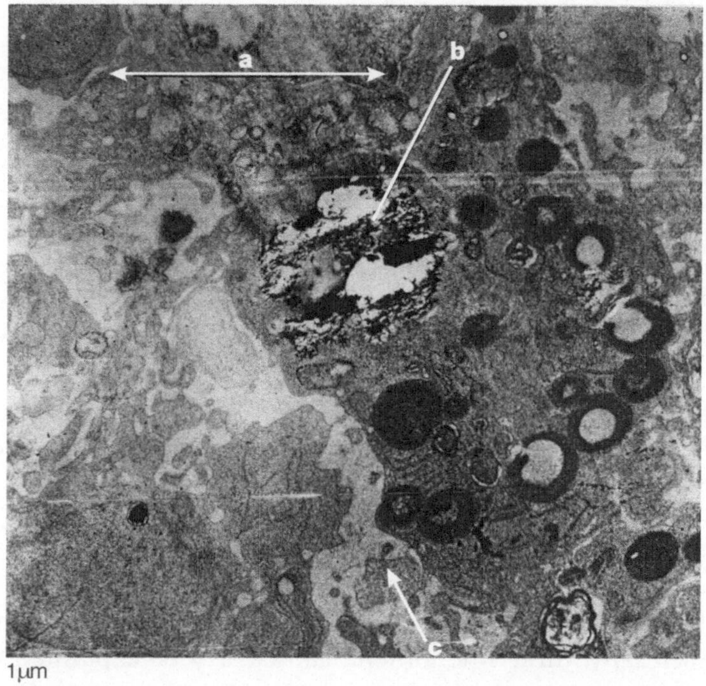

Figure 9.19 Scratches in the sections as seen in EM. (a) Region of scratches; (b) Material torn out; (c) Direction of cut. (*Source:* Audrey, M. Glauert. (Ed.). (1974). *Practical Methods in Electron Microscopy.* Vol. 3. American Elsevier, North Holland. p. 305, Fig. 7.3.)

Occasionally, the sections appear covered by fine scratches when viewed at about 5000× in EM. They are caused by poor quality of knife-edge. These scratches may be confused with chatter, another sectioning artifact; but scratches will be perpendicular to the knife-edge and chatter parallel to it.

Compression

There is normally no difference between the width of the cut section and the block face, but the length of the section is usually shorter; this shortening is referred to as "compression". Wrinkles and folds of the sections are associated with compression. These are seen as dark wavy lines on the section approximately parallel to the knife-edge but do not necessarily extend the full width of the section. In EM badly compressed sections will exhibit an altered structural orientation in the direction of the cutting.

Organic vapours can be used for stretching the floating sections. A brush or paper dipped in chloroform or ethylene dichloride may be held close to the sections without touching them for this purpose. Sections from soft blocks have a greater tendency to compress than those from hard blocks. Epon resin sections are more difficult to expand.

Variations in Thickness within a Section

These may arise due to the following three causes:

- If the demarcation between the two thicknesses is perpendicular to the knife-edge, it indicates that the cutting edge is not equally sharp in both regions. It is possible that one side of the section will be cut only at every second stroke. Then in that case the knife has to be changed.
- If the demarcation runs irregularly through the sections, the fault will be caused by the uneven consistency of the block, then, the block has to be trimmed or changed.
- If the thickness variations are in the form of bands running parallel to the knife-edge, they are usually caused by vibrations between the knife and specimen block; the fault may arise from the instrument or from external sources.

Irregular Section Thicknesses

Irregular section thicknesses can have very different causes and can appear in very different manifestations. Essentially, three phenomena can clearly be separated from one another as given below:

1. Different thicknesses from section to section with uniform thicknesses within the individual sections is attributable to manipulations and external influences on the ultramicrotome.
2. Zones of deviating section thickness parallel to the knife-edge is attributable to individual impulses during the taking of the section concerned. Here, the same phenomenon takes place in the subsequent section in the "negative", i.e., a zone of increased thickness is followed in the subsequent section by a zone of reduced thickness.
3. Zones of deviating section thickness which do not run parallel to the knife-edge but follow a specimen structure, i.e., zones of variable thickness running irregularly within a section. This is caused by zones of variable cutting consistency in the specimen. For example if a larger blood vessel is located in the cutting area or if there is a zone with free embedding medium at the object edge, the corresponding parts are often missing in a few sections and appear with double thickness in the subsequent section. Free embedding medium mostly displays a completely different cutting consistency than an enclosed object. Glass knives cannot be used for cutting such specimens having inhomogeneity in the specimen. Diamond knives cut the inhomogeneous preparation smoothly and without problems. This type of irregularity is never due to faults of the ultramicrotome. Figure 9.20 shows the types of variations.

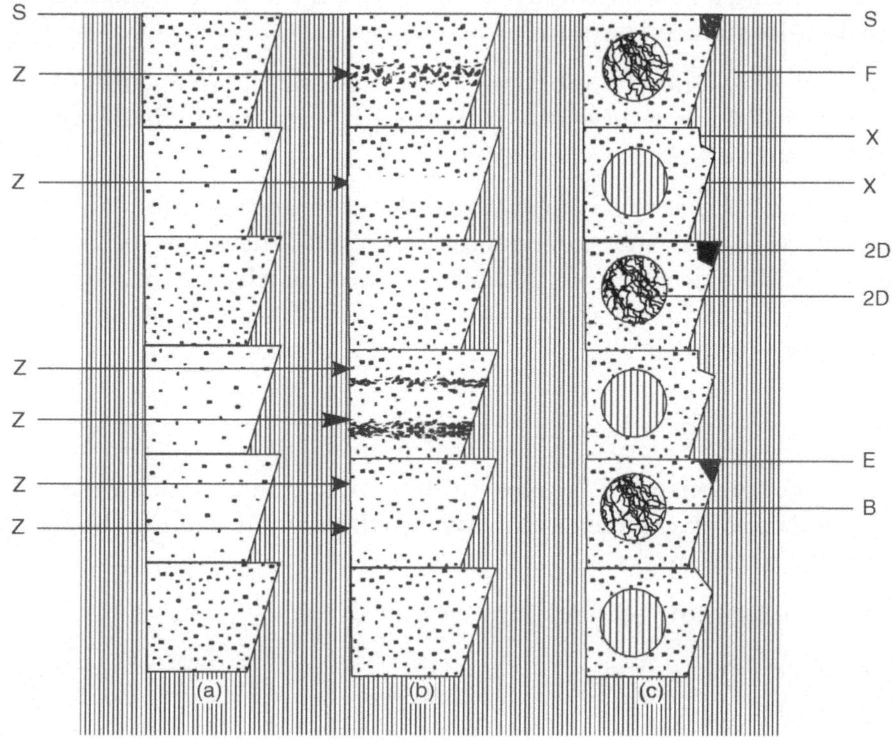

Figure 9.20 Irregularities of the section thicknesses: The ribbons of sections on the liquid surface F of the trough are illustrated. (*Source: Leica Manual.*)

i. Variable section sequence with uniform thickness within the individual sections. (Figure 9.20a).
ii. Zones Z of variable thickness in the individual sections, which run parallel to the knife-edge SS (Figure 9.20b).
iii. Regions of different thickness corresponding to structures with different cutting consistency, for example, blood vessel B or free embedding medium E. Here the corresponding parts often are missing (X) and then appear with double thickness (2D) in the subsequent section (Figure 9.20c).

In types (i) and (ii), faulty functioning of the ultramicrotome mechanics, is far rarer than operating errors and external influences. (i) to (iii) can overlap.

Chatter

It refers usually to the effect of high or low frequency vibrations set up between the knife and the specimen. The result is closely spaced variations in section thickness parallel to the knife-edge which can be seen only in EM (*See* Figure 18.21). Chatter, can be high-frequency or low-frequency chatter.

High-frequency chatter The chatter marks run perpendicular to knife marks and are visible only in the EM. They arise due to methodological errors like, poor embedding, erroneously selected cutting parameters, etc.

Low-frequency chatter These arise due to apparatus settings including building vibrations. These marks can be seen in the stereomicroscope itself. They, due to the large distances between the individual wave peaks, display no noteworthy disturbances, since under magnifications greater than 5000×, an impression of a completely uniform section thickness is already obtained.

Vibrations

Vibrations are synonymous with "waves, undulations, chatter, chatter marks" which are defined as, periodic wavy changes in thickness, whereby alternating thicker and thinner zones run parallel to the marks.

Vibrations (waves in the sections) are produced by

- incorrect embedding
- faulty trimming
- erroneously adjusted knife
- insufficient clamping of the knife or the specimen
- external influences, like, building vibration

Contamination

Contamination arises mainly from the solutions used, most important, from water. Hence freshly prepared distilled water should be used; all vessels and instruments, including the forceps, pipettes, etc., should be thoroughly cleaned. Grids are preferably washed in acetone in an ultrasonic cleaner.

Problems under Special Conditions

- With Epon blocks, the most serious problem while sectioning with glass knives is the one that results from surface displacements while sections are being cut. The end effect is a "smudging" or "mottling" of the sections (*See* Figure 18.15). This effect is rarely, if ever, seen when a diamond knife is employed for sectioning. This could suggest that adhesive forces exist during sectioning between some component of the block and glass. Because of this, it can be imagined that surface particles are dragged along as the knife passes the block or sections glide over the knife-edge.

* When the sections float away from the knife-edge singly and do not adhere to one another and the cohesion between sections in the ribbon is absent, it indicates that the upper border of cutting surface is not parallel to the knife-edge and also that the upper and lower borders of the cutting surface are not smooth and linear.

Thus, the prerequisites for getting satisfactory sections in the thickness range of 70 nm, i.e., silver colour range are:

* satisfactorily, embedded, easily cuttable specimen
* correct shape (trimming)
* satisfactory knife-edge
* appropriate cutting method

Good sections in the collecting boat do not necessarily assure good images in EM, because good image requires the following prerequisites to be fulfilled.

* elimination of section compression through spreading out
* appropriate collection of sections on the grids
* careful staining of the sections
* absolute cleanliness including all instruments and solutions

Two additional circumstances leading to faulty sections are:

* specimens difficult to cut, like, inhomogeneous specimens with hard inclusions
* faulty functioning or erroneously set up ultramicrotome

Specimens difficult to cut It becomes difficult to cut specimens because of their inherent nature or because the specimen is poorly fixed or embedded.

Some specimens are hard and tough naturally or contain inclusions difficult to cut, e.g., plant or insect tissues, silicose lungs, cartilage, collagenous connective tissue, fatty tissues, etc., or on the whole, hard like bones. For such specimens, diamond knives can be used for sectioning. But, diamond cutting edge is very rapidly and irreversibly destroyed by larger hard inclusions, e.g., silicates in diameter above 0.5 μm or by massive hard objects. Thus the cutting of individual extremely thin asbestos fibres or hydroxyapatite crystals normally presents no difficulty. On the contrary, thicker glass fibres destroy the knife-edge immediately. Objects difficult to cut in certain area dimensions are often arranged perpendicular to the knife cutting edge because the cutting force is substantially reduced in this manner. To exclude the shock, which results from the first contact of a block edge oriented parallel to the knife cutting edge, which is thus manifested as waves at the border of the section, the block face can be cut to form a triangle and this triangular block may be placed with the point of the triangle against the cutting edge. This way, the formation of vibrations can be practically eliminated. If only glass knives are available, the cutting of specimens with hard inclusions should generally be avoided, when the glass knife is already destroyed in the first section. By reducing the cutting area, it may be possible to obtain a few usable sections. Increasing the cutting speed (more than or equal to 10 mm/sec) may also help.

Blocks containing specimens that are poorly processed (fixed/embedded) are discarded. If the specimen is very rare, and therefore too precious to discard, the initial cutting area is reduced to the minimum and cut with glass or diamond knives comparatively at very reduced cutting speeds.

Erroneous setting up/Faulty functioning of microtome The ultramicrotome has to be placed in a quiet room far from large machines, including AC units. An **antivibration base plate** is provided by the manufacturers which may be placed on any stable and even table. The ultramicrotome stationed on this, is protected from vibrations. The erroneous setting up of the instrument results in the following.

- irregular section sequences
- non-uniform thickness in the sections
- low-frequency chatter

This usually happens when the microtome is housed,

- in a large room in which several persons are working.
- in a room subjected to irregular air movements, i.e., in the neighbourhood of windows, doors, heaters, AC units, etc. (AC units should be shut off during sectioning, particularly EM sectioning.)
- near large machines like workshops, elevators, voltage stabilizers, transformers, etc.
- when shock absorbers are not functioning or are worn out.

SECTION COMPRESSION AND SPREADING/FLATTENING OF SECTIONS

The floating sections may display compression artifacts. With methacrylate sections, they have to be softened and spread with the vapour fumes of a volatile solvent. But with Epon sections this is not absolutely essential. Nevertheless, the sections are spread by holding a filter paper saturated with an organic solvent like chloroform, xylene, trichloroethylene, etc. as close as possible above the sections on the water surface but without touching the sections. A hot pen called "wax pen" is also available commercially which helps to flatten and separate tissue sections and also reduces compression in ultrathin sections.

Advantages of spreading

* the section extends so that its thickness decreases by about one shade of colour of the interference scale, i.e., pale gold to silver.
* any deformations caused by compressions are eliminated (i.e., ellipsoid features again become circular).

Although no section is free from compression, it is better to limit compression by the correct choice of cutting conditions rather than to try to remove it later.

600 Å (60 nm) sections are optimal for high resolution work. Silver sections (60 to 90 nm) are ideal for most practical purposes. Gold sections (90–150 nm) are useful for low magnification and autoradiography. In fact gold sections can be used for TEM work also, provided higher accelerating voltage, say 120 kV, is employed. Purple, blue, green and yellow ranging from 150–320 nm are thick sections and not suitable for TEM but can be used for LM.

After spreading, the semithin sections are picked up on the glass slide, while only the ultrathin sections are picked up on the grids for use in EM.

GRIDS

The ultrathin sections floating on water have to be picked up on **grids** because only these can be inserted into the specimen chamber of the EM. The grids are thin mesh of the size 3 mm diameter. Grids made of copper are commonly used. Grids made of nickel, gold, platinum are also available which are used for specific purposes, particularly when chemical treatment of the sections may affect the grids, for example, nickel grids are used for immunoelectron microscopy. Copper grids with palladium plating are newly introduced grids where the plating offers better grid strength and avoids tarnishing. Similarly copper or nickel grids with gold plating (gilded) are particularly suited for immunogold labelling techniques and also autoradiography techniques. The gold-gilded grids decrease the risk of chemical reaction in many specimen treatment procedures and are an alternative to the more expensive pure gold grids. Copper grids with one surface coated with inert Rhodium are also available. This coating will eliminate tarnishing, give side identification and reduce the bar thickness. Beryllium grids are superior to all other materials for *in situ* analysis in TEM, because for practical purposes they give off no detectable background radiation which could interfere with the analysis. New technology utilizes diamond thin films to create TEM grids with some very unique properties and functions. Diamond displays thermal conductivity roughly four times that of copper and thus heat is carried away more effectively. For high-resolution TEM work, longer exposure times are possible as the good thermal conductivity of diamond means less specimen heating and, therefore, less specimen drift; NUM grids (numbered grids in random order), DOT grids (2 dots marked on one side of the grid for identification), NOTCH grids (a mark stamped on one side of the grid to facilitate the handling of the grids), FINDER grids (have one straight and round cut out from the rim which assists in the orientation of the grid), COORDINATOR grids (grids with handles); NUMERIC INDEX grids (increase the support value of the grids and also facilitate easy identification), are all commercially available now.

Copper grids being the cheapest, are most commonly and routinely used. The grids are available as meshed or with slots. Grids of 100 to 400 mesh size are most commonly used in TEM. If the grids are of large mesh size or if grids with many slots are used, a coating of support film is essential to render support to the sections.

The surfaces of the grids generally have a different appearance, one side matt and the other shiny. The matt surface is slightly rougher and is often selected for the mounting of the support film and sections. The sections are lifted onto the matted surface of the grid, which affords sticking of the sections firmly.

Many different kinds of grids are now available. Some of them are shown in Figure 9.21.

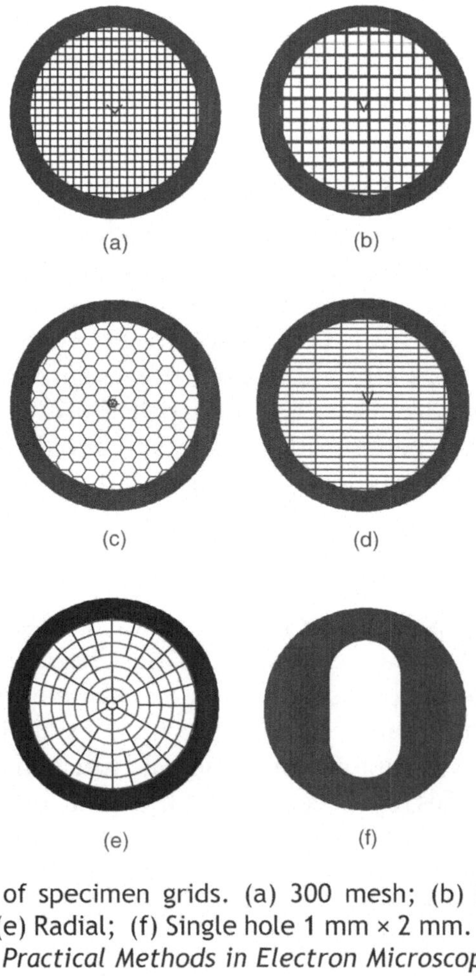

Figure 9.21 A selection of specimen grids. (a) 300 mesh; (b) 200 Special; (c) Hexagonal; (d) 75 × 300; (e) Radial; (f) Single hole 1 mm × 2 mm. (*Source:* Audrey, M. Glauert. (Ed.). (1974). *Practical Methods in Electron Microscopy.* Vol. 3. American Elsevier, North Holland. p. 314, Fig. 8.2.)

The mesh size refers to the number of openings to the linear inch and for most routine work the 200 or 300 mesh size will be adequate. Smaller mesh values may be needed when the sections are very small and would not receive adequate support from the grid bars. The hexagonal type gives more support to very thin sections and the single hole grids are used to give an uninterrupted view of a ribbon of sections. Identification marks on the bars enable areas to be quickly located in the electron microscope.

Grids are very fragile and hence have to be handled with utmost care. They have to be lifted with fine forceps only at the edge. Otherwise, the forceps may damage the sections collected, because the sections may be spread over the grid. During handling the grid should not bend, get folded or in any way get altered in shape. If it falls down on handling, it is difficult to spot it on the floor, and even if successfully, spotted it cannot be picked up, and even if picked up, it will be contaminated and not worth screening in the EM. Hence it will have to be discarded.

If the sections are large enough, to cover several of the grid bars, no additional support is usually needed, though ideally grids with carbon support films are suggested for picking up sections. But with small sections or with single-hole grids and also with specimens in suspensions, a supporting film is required. The grids should give adequate support to the sections without obscuring too great an area. Although the support film gives support to the sections it also reduces the contrast and if not well-made it may disintegrate in the electron beam, causing the sections to be lost and also contaminating the EM. The carbon support film provides a conducting layer which helps dissipate the heat from the electron beam.

The grids can be coated either with carbon or plastic polymer for supporting the sections.

Film coating is done on the matted surface of the grid.

2–3% Formvar (polyvinyl formaldehyde) in ethylene dichloride or chloroform, or collodion in chloroform is used for coating the grids. Carbon coating is done by a process of evaporation under vacuum on a grid already coated with formvar. After double coating, the formvar layer is dissolved either with ethylene dichloride or chloroform leaving behind the carbon film. Coated grids are useful for supporting particulate specimens like viruses and bacteria and other suspensions, including nanoparticles. Coated grids with carbon alone or carbon with formvar are commercially available.

Used grids can be cleaned but it is not worth the effort unless they are an expensive type, because invariably some debris remains on the grids, which interferes with future viewing.

The grid box is the most widely used method for storing and transporting grids in the electron microscopy laboratory. It contains indexed compartments for 50 or 100 grids and has a clear sliding cover. Each box is individually numbered for easy reference. The grids with sections can be stored in covered Petri dishes also with proper labelling. A grid box to hold 100 grids are shown in Figure 9.22.

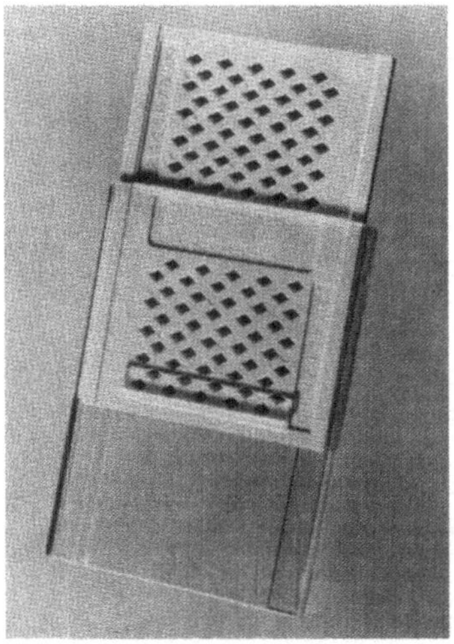

Figure 9.22 Grid box to store 100 grids. (*Source: Leica Multiplate Manual.*)

PICKING UP SECTIONS ON THE GRID/MOUNTING ON GRIDS

Collection of cut sections, especially ultra-thin sections calls for steady hands and some practice. Unwanted sections floating in the boat must be removed first by firmly prodding them one at a time, with a single hair to which they will stick. The hair must be wiped carefully between folded lens tissue to remove debris before touching the good sections. The ribbon of sections can be detached from the knife-edge by touching the top of the last section with a clean hair and pulling it gently away from the knife-edge. The sections should only be touched on the upper surface; if attempt is made to move the sections by touching their outer edges, they are likely to adhere to the wet hair.

Figure 9.23 shows the method of mounting of sections, especially of ribbons of sections S on grid N. The grid N held with the forceps P, is introduced either from above (a) or from below (b) on the section ribbon. After the mounting, the water W must immediately be sucked off with clean filter paper F from both sides of the grid, as well as between the arms of the forceps (c).

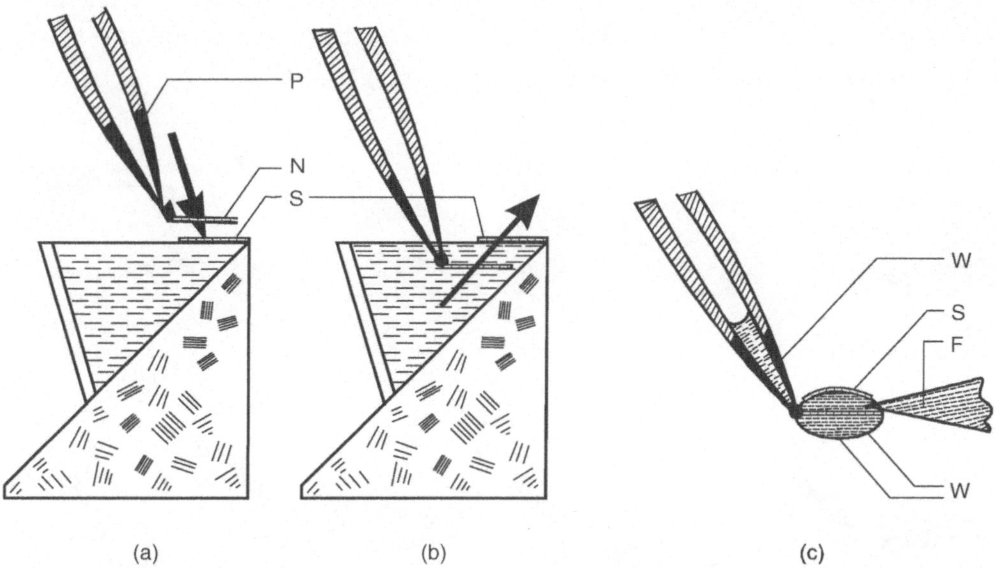

Figure 9.23 Mounting of sections. (*Source: Leica Manual.*)

METHODS OF PICKING UP OF SECTIONS

Picking up above method The sections are transferred to the grid as shown in Figure 9.23a, in that the grid is placed from above on the ribbon of sections with the sticky side in contact with the sections to be picked up and slowly pulling it away with the sections. It is very difficult to place the grid absolutely flat onto the sections with the result that they are often creased when using this technique. This procedure also leads regularly to rejections and folds.

Fishing from below method Therefore, "fishing from below" method is used to pick up sections as in Figure 9.23b. The grid with the sticky side up is dipped in floating liquid first and placed horizontally underneath the sections and then slowly raised lifting out the sections on a drop of water. The grid should be raised slowly through the liquid, or the sections will float away from it. Then the grid is dried as mentioned above.

Loop method A small loop of thin platinum or other wire can be used to pick up sections. The "perfect loop" can be used to pick up sections consistently without causing any damage to the sections. The outside diameter of the loop is the same as the grid and the inside diameter is slightly larger than the observation area of the electron microscope. The thickness is about 40 microns. Due to the fact that the loop and the grid are of the same diameter they are attracted to one another when in water and attach together through the surface tension of water. Even if the section touches the inside of the grid during blotting, the touching area is minor and therefore the section is not damaged. When the grid is removed from the loop, the sections remain in place on the grid. The area equals the observation field (about 2 mm diameter) of the electron microscope and thus sections can be fully observed. Figure 9.24 illustrates these steps clearly.

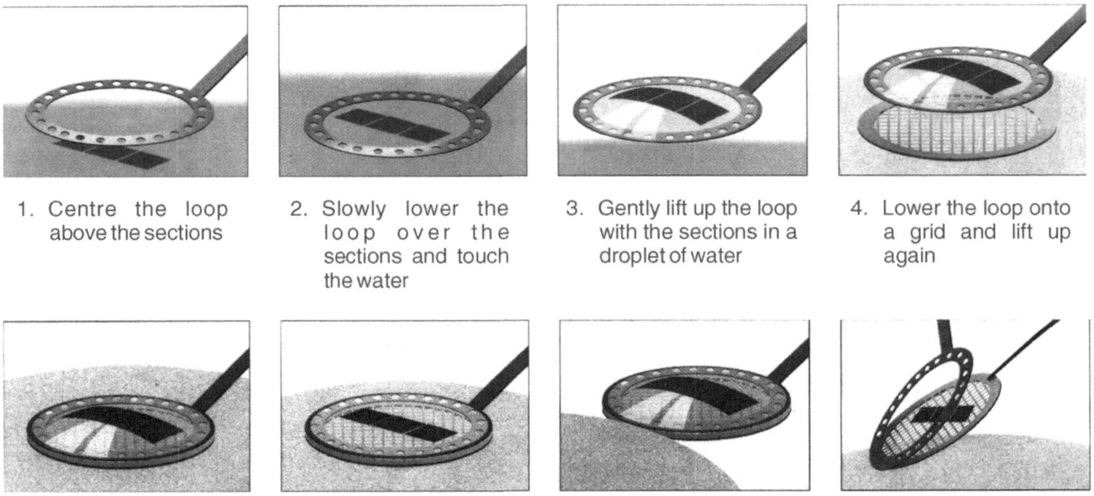

1. Centre the loop above the sections
2. Slowly lower the loop over the sections and touch the water
3. Gently lift up the loop with the sections in a droplet of water
4. Lower the loop onto a grid and lift up again
5. The grid holds to the loop by surface tension
6. Lower the loop to the filter paper to remove water
7. For coated grids touch with filter paper to remove water
8. Separate the grid from the loop with an eyelash

Figure 9.24 Loop method to pick up floating ultrathin sections on grids. (*Source: Leica Loop Manual.*)

It is important to carefully remove all adhering water with a clean filter paper strip immediately after taking up the sections. Water must be immediately sucked off with clean filter paper from both sides of the grid as well as between the arms of the forceps. Otherwise, the water between the arms of the forceps, due to surface tension effects may draw the grid up while the grid is being released from the forceps, thus ruining the sections collected on the grid.

SEMITHIN/SURVEY SECTIONS

The one micron sections cut initially with the ultramicrotome for LM are referred to as "semithin" sections as they are thicker than EM sections but thinner than the conventional LM sections prepared for histopathology. These sections are also known as "survey sections" because they enable the electron microscopist to determine the following:

- ✳ the usefulness of the block from which sections have been cut as to whether they contain the area of interest
- ✳ the state of the embedded specimen
- ✳ the quality of fixation

- selection of area for ultrathin sectioning
- orientation may be checked
- uninteresting or poorly processed material may be discarded
- the size and position of the cutting face for final trimming, i.e., when searching for specific region, locate areas of interest in survey sections and then trim the specimen block to include only the required areas
- LM is an intermediate stage and will help to give a better understanding of the material as a whole.

These resin-embedded LM sections reveal more details than the conventional HPE sections because of their reduced thickness. More details can be seen in such sections with unparalleled resolution particularly with phase-contrast microscopy. But because of the fact that routine H&E (haematoxylin and eosin) stains cannot be applied for these sections, pathologists do not favour them. The thickness of survey sections is 0.5 μ to 2 μ. The section area may be large to include more specimen.

Survey sections can be picked up on a suitable support such as a coverslip, or a fine paint brush, or a glass rod with round edge, which can be rolled on the sections and by rolling it in reverse direction on a drop of liquid (usually water) on the slide. When using a brush, the brush should be placed under the sections and carefully raised. The section will wrap itself round the brush, but will unfold as soon as the section touches the water on a slide. The section can also be transferred by carrying it on a drop of liquid supported by a fine wire loop. A "perfect loop" can also be used for picking up these sections. Alternatively, a small square coverslip's pointed end can be inserted slowly into the water in the boat as far as it can go without disturbing the sections. The floating sections can be slowly directed towards the coverslip end with the help of an eyelash brush and brought just above the coverslip. On lifting the coverslip up slowly the sections will be caught on it. The coverslip is inverted and mounted on a standard glass slide. The semithin sections are then stained as given below.

The glass slide with the sections is gently warmed. This helps to stretch out any wrinkles that have been formed in the sections and as the water evaporates, the sections become firmly attached to the slide. The heat should not be too intense or air bubbles will be formed under the sections and may become trapped as the sections dry. The multiplate heater is highly suitable for this purpose. The sections are then stained with toluidine blue or methylene blue or azure II for 0.5 to 2.0 min. depending on the section thickness. Then the sections are washed thoroughly in running water, dried and mounted with DPX.

The staining of grids are discussed in detail in Chapter 10.

10

STAINING

A. STAINING OF ULTRATHIN SECTIONS FOR ELECTRON MICROSCOPY

In order to view the details of the features in the sample clearly there should be contrast between the various features as well as between the specific areas in the specimen and their environment in the magnified and resolved image. The contrast is introduced by staining the specimen. The stains used in light microscopy are basically coloured dyes. They have affinity towards certain regions rather certain chemical entities in the specimen and thus get bound to those regions thereby imparting their colour to them. In this way contrast is introduced into the specimen. There is no universal stain and depending on the nature of the feature to be identified, different stains are applied. Staining is the last step in processing. The slides with the sections are immersed in the stain solution and after allowing sufficient time for the reaction, the excess stain is completely washed off.

In the case of EM, since the electron wavelength is very low, no colour is possible. So coloured dyes are of no use. Instead special stains called "electron stains" are used for staining ultrathin sections for EM. These electron stains are basically heavy metals. These heavy metals by blocking or scattering the electrons improve image contrast. Thus the main purpose of staining is to increase electron opacity thereby increasing contrast and thus leading to better resolution. Only those heavy metals with affinity for selective regions of the material under study are chosen for staining. Thus staining for EM is addition of heavy atoms to the specimen. These scatter electrons more than light atoms and so enhance the final contrast of the picture. Therefore, heavy metals and their salts are chief electron stains. Phosphotungstic acid was the stain originally used for methacrylate embedding. But for electron-resistant Epon/Araldite embedding, alkaline lead citrate and uranyl acetate are most suitable.

This staining is sometimes referred to as "positive staining" as the stains penetrate the specimen and react with the various components in them. This is in contrast to the "negative staining" which is dealt with in detail in Chapter 14.

The most common electron stains for ultrastructural studies are uranium and lead. The osmium which is used as the secondary fixative, being a heavy metal, acts as a stain also. It fixes as well as stains the membranes due to its high affinity for lipids, particularly the unsaturated lipids. The uranyl acetate, when used as the third fixative after osmium tetroxide and buffer wash, also imparts its staining property. Uranium has affinity towards structures containing nucleic acids and its staining action at this stage is referred to as "en bloc staining". Uranium can also be used as a regular stain for ultrathin sections.

Generally, the stains should have the following properties:

1. Highly soluble in aqueous medium
2. Highly electron-dense
3. High melting point

STAINS USED IN ELECTRON MICROSCOPY

Some of the stains used in electron microscopy are listed below:

1. ***Uranyl acetate*** 2% aqueous solution is used either for en bloc staining or for staining ultrathin sections. The pH of the solution is 3 to 4. Longer duration staining for 1 to 2 hours is the recommended time. However, 1% solution in ethanol stains in 15–20 min. As this stain is light-sensitive, it has to be stored in brown bottles. Uranyl acetate dissolves slowly and has to be prepared 24 hours before use.
2. ***Lead citrate*** Soluble at alkaline pH (around 11). 0.2% aqueous solution gives impressive results with 5 min. staining time. This stain is regarded as a "useful all-purpose stain".
3. ***Phosphotungstic acid*** 2% aqueous solution staining for 5 min. is recommended. This is the best stain for staining connective tissue.
4. ***Potassium permanganate*** This is a fixative cum stain. Sections are stained for less than 5 min. in 1% $KMnO_4$ in water. 5% citric acid is used for washing after the staining.
5. ***Other stains*** Vanadium salts and chromyl chlorate have also been used as stains but the routine stains are only uranyl acetate followed by lead citrate. It is the standard "double staining" method for ultrastructural studies.

Some of the structures obtained by different stains are as follows:

Bismuth	Basic proteins, phosphates, lipids, polysaccharides
Iron	Acid mucopolysaccharides
Lanthanum	Glycoproteins
Lead	Wide range of structures including glycogen, mitotic spindles, smooth endoplasmic reticulum vesicles, etc.

Osmium	Lipids
Uranium	Nucleic acid, collagen
Silver	Golgi apparatus
Ruthenium red	Extracellular material

METHODS OF STAINING

1. *En bloc staining for uranyl acetate* After fixing and washing, the tissue is treated with uranyl acetate and then processed further. Osmium fixing of lipids is also a method of en bloc staining. Though en bloc staining with uranium has been shown to preserve ultrastructure better, staining before embedding may make tissue hard and result in poor sectioning.
2. Staining ultrathin sections using any of the three methods given below have the following advantages.
   - Extraction and displacement of structures are minimal.
   - Less chances of distortion.
   - More contrast enhancement.
   - More uniform staining.

   i. *Floating of grids* The grid with the ultrathin sections can be placed on a surface of a drop of stain in such a way that sections remain in contact with the stain.

   ii. *Immersion of grids* Grids are immersed in drops of stain with the sections in contact with the stain. Though in this method, the problem of deposition of contaminants on the specimen is high, picking up and handling of grids is relatively easier.

   iii. *Staining kits (automatic stainer)* It is a grid staining device that can stain up to 40 grids at one time. The unit has a flexible plastic plate that has precision cut slits in it which firmly hold the grid rim in position without damaging the sections. The flexible plate is mounted on the plate holder to open the slits to enable the insertion of specimen grids. The loaded plate with specimen grids is then removed from the holder, inverted and placed on the staining dish. After staining is complete, the plate is put in the drying position. The grids are dried and screened in the EM. In this method there are certain disadvantages like precipitation, need of large amount of stain, problem of inadequate washing, etc.
3. *Microwave-assisted staining* Use of microwaves for staining has been shown to improve contrast as well as reduce staining time.

A piece of parafilm or dental wax can be used generally for keeping the stain drops during the staining process. Instead, a "stain pad/ grid mat" can also be employed

(Figures 10.1 and 10.2). Grid mats are silicone rubber mats, with numbered compartments. They are good for organizing grids as the grids will not slide or jump between compartments. They are ideal for staining grids. One drop of the stain solution can be placed in each compartment, the numbers of which guide in easy identification of the specimen sections on the grids. It is also easy to pick up grids from the grid mat without damaging forceps tips. In addition, silicone is resistant to all chemicals and solvents. Mats are available for square and round Petri dishes. After staining, the grids are carefully removed, washed and air-dried before screening in the EM. The "staining pad" is made from white silicone, a non-reactive material. The pad has several cells. Each cell is half-sphere shaped with an opening. A few drops of staining solution is added to the wells, and the grids are then immersed and retrieved as per staining procedure.

Figure 10.1 Grid mats in Petri dishes and lids

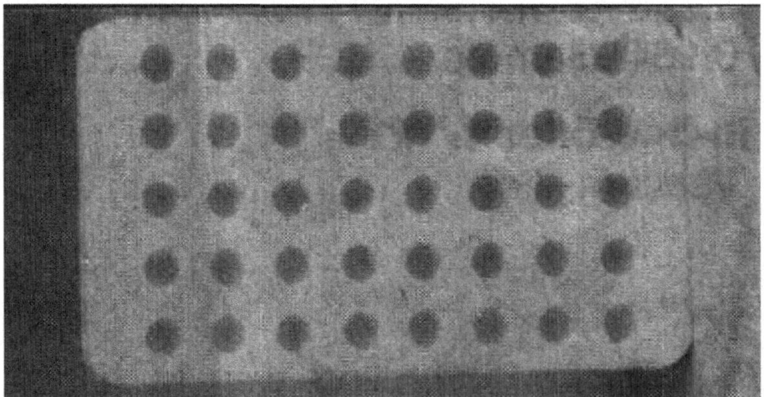

Figure 10.2 Stain pad

Some authors emphasize that after staining has been completed (required minimum time), the sections are washed only briefly in distilled water, by drawing the grid once rapidly through the water and then drying immediately with filter paper. According to them, a long or especially intense washing (jet stream from a plastic bottle) is more likely to wash the staining medium out of the section, since the precipitates are not very soluble.

But cleanliness during preparation, storing and using the stain solutions is more important. If precipitates are found on the edge of the liquid level or at the bottom of the supply bottle, a uranyl acetate solution must be filtered. If this happens in a lead citrate solution, it should be discarded and fresh solution made up again.

It is preferable to complete the staining of sections on the day they are cut and mounted. Otherwise, weak staining may result perhaps because of slow alterations in the embedment which affect the penetration of stains. Also, unsupported sections seem less likely to survive the hazards of staining after prolonged standing than freshly made specimens.

The characteristics of some of the most common stains are discussed in the following sections.

URANYL STAINS

Watson in 1958 was the first to use uranyl acetate (UA) as an electron stain. UA stains proteins fairly intensely. But cytomembranes are not stained vividly. It binds to nucleoproteins specifically.

The solubility of UA is quite temperature-dependent. In dilute alcoholic solution, its solubility increases, so does its tendency to decompose. So mostly freshly prepared alcoholic solution is used as a stain. The most convenient way of preparing a fresh solution is to add an appropriate quantity of the salt to the solvent in a small covered container and put this in a shaker for 20 to 30 min. to dissolve the crystals. Alternatively, 2% aqueous solution or alcoholic solution of uranyl acetate is prepared preferably a day earlier. The solution is filtered and used for staining. In some laboratories, this is centrifuged and stored in the refrigerator. Drops of this solution are placed on a parafilm or stain mat. UA is light-sensitive and hence precautions should be taken for this also. The staining time is usually from roughly 30 min. to several hours depending on the embedment. For EPON, 20 to 30 minutes is sufficient. After the staining, each grid is washed in dilute alcohol used for the preparation of the stain, then in distilled water and finally air-dried before screening in the EM.

Uranyl nitrate or uranyl formate has also been tried in place of the acetate, but it is not as effective a stain as the acetate.

Uranyl acetate is used as the primary stain. Lead staining after uranium is more effective and requires less staining time than either of them used alone. Uranium acts as a mordant for lead staining.

LEAD STAINING

In 1950 Watson introduced lead hydroxide staining for EM and it is now recognized as the most valuable general-purpose stain. It provides necessary contrast to visualize most cytological features. It should be noted that tissue heavily stained with lead is quite sensitive to "beam damage" if excessive illumination is used.

Lead binds to negatively charged components such as —OH groups and areas that react with osmium, such as membranes. There are three mechanisms of attachment of the lead to different reacting groups in the specimen.

1. The most important reaction results in the staining of cytomembranes. This reaction is regarded as dependent upon the presence of negatively charged reduced osmium on the polar groups of the phosphatides.
2. Staining of substances such as glycogen may be based on the chelation of lead by —OH (hydroxyl) groups of carbohydrates.
3. Finally, nucleoprotein staining has a preference for RNA-containing proteins while DNA-containing proteins are preferentially stained by UA.

The major problem with the lead stain is the reaction between lead and atmospheric carbon dioxide to form black lead carbonate precipitates in the specimen. The water used for preparation of lead stain should be boiled well to remove dissolved gases. Moreover, keeping hygroscopic sodium hydroxide pellets surrounding the sections during the staining process is effective in preventing formation of lead carbonate precipitates. Also, avoidance of breathing on sections during lead staining in order to maintain a CO_2 atmosphere is essential.

REYNOLD'S LEAD CITRATE

This stain uses citrate as the chelating agent in sufficient excess to sequester all of the lead present and thus quite effectively prevents it from combining with carbonate to form lead carbonate precipitate in the specimen. Tissue binding sites have greater affinity for the lead cation than citrate and thus are well-stained.

1.33 g of lead nitrate + 1.76 g of sodium citrate are added to 30 ml of distilled water, shaken vigorously and allowed to stand with intermittent shaking for 30 min. Lead nitrate is converted to lead citrate. Then 8 ml of 1N (1 Normal) sodium hydroxide is added. This solution should be freshly prepared to make it carbonate-free using sufficiently boiled distilled water or double-distilled water. Then this solution is diluted to 50 ml and mixed well by inversion. Lead citrate dissolves. The pH should be around 11 and the solution should be clear. This solution is ready for use as the stain. This can be stored in glass or polyethylene bottles and if fairly turbid can be centrifuged before use. If turbidity increases, the solution has to be discarded. Lead citrate can also be prepared directly from the compound in dilute sodium hydroxide solution for the preparation of which water free carbon dioxide is preferably used.

A parafilm or a piece of dental wax or stain mat/pad can be used for staining. A moist filter paper is placed in a Petri dish on which parafilm or stain pad is kept. Sodium hydroxide pellets are placed in the surrounding without touching the stain drops. Grids with sections are floated upside down on single drops of stain solution or immersed with the sections up. The Petri dish is closed with the lid which prevents the evaporation of the stain. The sodium hydroxide, being highly hygroscopic, absorbs moisture from the atmosphere thereby preventing the lead from combining with atmospheric carbon dioxide. Breathing heavily over the drops also may introduce carbon dioxide promoting lead carbonate precipitation. The duration of lead staining again depends on the embedment used for block making. It is shortest for methacrylate and longest for Araldite and in between for Epon. Normally, 5 to 10 min. staining time is sufficient for Epon-embedded sections, which have already been stained with uranyl acetate. After staining, each grid is washed in 0.02N NaOH, then in distilled water and allowed to dry.

If lead staining is not adequate, uranium staining can be started again followed by repeat lead staining. Similarly, in specimens with inherently low contrast such as skin sections, contrast can be improved by staining first with lead for one minute, followed by uranyl acetate for 5 to 10 min. and then finally again with lead for one minute.

PHOSPHOTUNGSTIC ACID (PTA) STAINING

This stain was introduced originally by Hall *et al.* in 1945, during their studies of collagen. It is somewhat a selective stain, demonstrating particularly extracellular connective tissue components including collagen, elastin and the mucopolysaccharides of basement membranes and other structures. PTA can be dissolved in either aqueous or alcoholic solutions. It is most preferred for "negative staining" as explained in Chapter 11.

COPPER COMPLEX STAINING

A new single step staining method using Copper (II) Uranyl acetate (Cu–UA) complex in place of the routine double staining with uranyl acetate followed by lead citrate has been reported by Lokanadham *et al.* (2000). Ultrathin sections on copper or nickel grids are stained with this copper complex for 10 seconds and excess stain washed in alcohol: water mixture (1 : 9). The stain has been successfully applied to ultrathin sections of several tissue samples. In Figure 10.3 ultrathin section from a lymph node specimen stained with this copper complex is shown, which can be compared with those stained with uranyl acetate followed by lead citrate in Figure 10.4. While staining time for uranyl acetate is 15 to 20 minutes with additional 5 to 10 minutes for lead staining, copper complex staining requires less than 30 seconds. Thus the copper complex staining method is

- rapid
- reliable
- reproducible
- shows enchaced image quality.

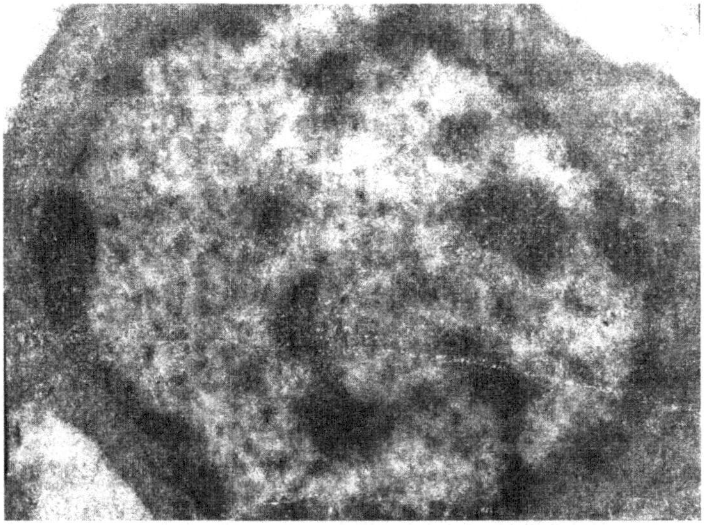

Figure 10.3 Copper complex staining (10,000×)

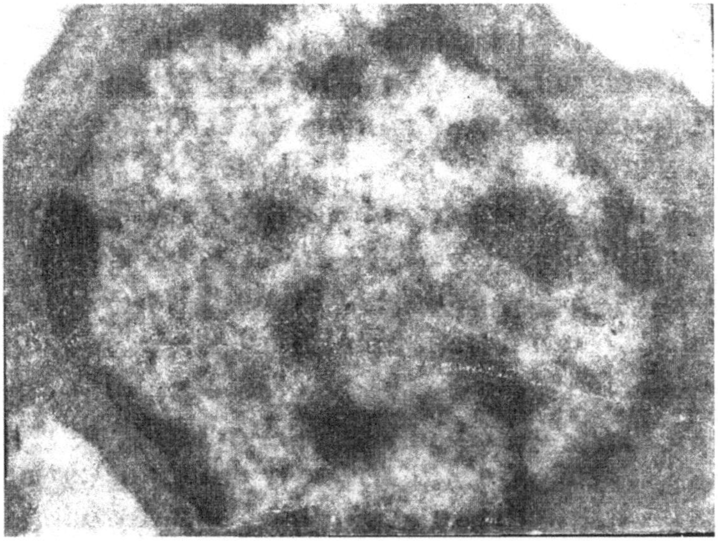

Figure 10.4 Uranyl acetate staining (10,000×)

LIMITATIONS DURING STAINING

1. When old stains are used, the reaction between stain and the specimen may not be strong.

2. Prolonged staining times (longer than 10–15 min. for uranyl acetate and longer than 3–5 min. for lead) can result in stain precipitation due to
   - drying at the surface of the drops of stain
   - increased evaporation of stain solvents

   and these precipitates are not easily removed by washing.
3. Thicker sections stain faster probably because of availability of more sites for reaction with stains.
4. *Contamination* Sometimes, serious contamination of grids caused by "dirt" and /or stain precipitates renders them useless. Micrographs with such contaminations are unacceptable, particularly, for publication.

Dirt Dirt adheres well only to a dry section and dirt on most sections come from surfaces of stain and wash fluids. Therefore, the grids have to be kept wet throughout the staining process and should be dried only after the staining is finished by touching filter paper points to the edge of the grid on the forceps side of the grid. When many dirt particles are seen on the sections, such sections have to be discarded, fresh ones cut and stained.

Bacteria and other organisms Another source of contamination is bacteria which can grow on filters or wash bottles that store distilled or deionized water for use in the boat of the knife during sectioning or for use in preparation of stain solutions as well as in final wash after staining. By using freshly boiled water, this contamination can be removed.

Stain precipitates The uranyl acetate precipitates appear as small, needlelike objects on the sections and can result from not filtering the stain properly and immediately before use or even from evaporation during prolonged staining. If alcoholic uranyl acetate is used as the stain, the sections should be washed with the same concentration of alcohol used for preparation of the stain. This should be followed by washing in decreasing concentrations of alcohol down to water in order to prevent rapid drying and possible precipitate formation.

In the case of lead, the precipitates appear large and spherical (*See* Figure 18.19). A common cause of lead precipitates and poor staining intensity with lead citrate is incorrect pH. A pH lower than 12, causes dense particles of various sizes and shapes. When the pH is lowered to 8, heavy precipitation occurs. As pH approaches 14, staining is severely reduced and there may be loss of contrast also. So it is best to use CO_2-free water for preparation of lead stain and IN NaOH to correct the pH. The most common cause of faulty pH is the use of NaOH pellets instead of solution. Since it is not possible to weigh pellets accurately due to their size and since the pellets may absorb water from the atmosphere, the actual pH of the final lead solution may be unknown.

Therefore, it is advisable to use commercially available, CO_2-free NaOH solution for the preparation of lead stains.

Preparation and storage of lead stain are done preferably in plastic containers rather than glass containers. Since lead stains produce precipitates of $PbCO_3$, it is difficult to store them for long. A new stable lead solution which is free from precipitates when kept at RT (room temperature) for over one year has been developed by Takamasa Hanaichi *et al.* (1986).

The stock lead solution, according to them, is prepared as follows:

Calcinic lead citrate*	0.2 g
Lead nitrate	0.15 g
Lead acetate	0.15 g
Sodium citrate	1.0 g
Distilled water	41 ml

* Calcinic lead citrate is obtained by heating lead citrate crystals for several hours at 200 to 300°C until the colour changes to a light brownish yellow. Overheated lead has a dark brownish or black colour which cannot be used.

The above reagents are placed in a volumetric flask and mixed well to produce a yellowish milky solution. Then 9 ml of lN NaOH is added and mixed well until the solution becomes clear with a light yellowish colour. The solution is transferred to an amber glass bottle and stored. This solution can be stored at RT or in the refrigerator for over one year.

EXCESS STAIN REMOVAL

In case of over-staining the excess stain can be removed by salvage methods.

1. To remove excess uranyl acetate (either for en bloc staining or grid staining), a chelator oxalic acid can be used.
2. To destain sections with both uranyl acetate and lead citrate, the sections are kept in aqueous 10% glacial acetic acid for 30 secs. to 1 min. followed by several rinses in distilled water. This removes most of the lead stain precipitates and also the original stain from the sections.
3. The stain precipitates can also be eliminated by the following method:

   The grid with epoxide-embedded sections is placed on a drop of 0.5 NHCl for 1 to 2 min. Then it is rinsed in deionized water and blotted dry on filter paper.

   These destained sections can then be restained with uranyl acetate and /or lead citrate because the acids have a tendency to bleach the sections.

But these salvage techniques are not always successful especially after extensive exposure to the electron beam. Therefore, it is better to remove grids from electron beam as soon as possible, destain, wash and restain as quickly as possible. Effect of electron bombardment on excess lead stain is shown in Figure 18.20.

PEPPERY PRECIPITATE

Sometimes a fine, electron-dense "Peppery" precipitate appears in sections after post-fixation cum staining with osmium. Specimens which exhibit this fine 'osmium' related precipitate should be recut and picked up on nickel grids. The sections can then be oxidized with 1 to 2% freshly prepared periodic acid for 5 to 10 min. at room temperature, washed with deionized water and stained in the conventional manner with uranyl acetate and lead citrate.

B. STAINING OF THICK SECTIONS FOR CONVENTIONAL MICROSCOPY: SEMITHIN OR SURVEY SECTIONS

As we have already seen, these semithin or survey sections are collected on glass slides for screening under the conventional light microscope. Routine haematoxylin and eosin cannot be used for staining these sections due to the resin embedment which neither dissolves easily nor allows the stain to penetrate. Hence either the embedment has to be removed by some treatment or other stains compatible with resin embedment have to be employed for staining these sections.

It has been indicated that epoxy resins formulated with DDSA (dodecenyl succinic anhydride) can be dissolved in a mixture of sodium methoxide, benzene and methyl alcohol. Alcoholic KOH (potassium hydroxide) at 65 to 75°C may also be tried. But these drastic treatments may destroy the ultrastructure.

Kindblom (1983) has suggested the following procedure. The 1 μ sections are treated with 1% KOH (potassium hydroxide) in 99.5% ethanol for 5–15 min. at 70–80°C, then washed in 99.5% ethanol with acetic acid, 50% ethanol and finally with distilled water. The sections can then be stained with routine H & E (haematoxylin and eosin) or any other special stains also.

Alternatively, many staining reactions can be carried out even without removing the plastic embedments. Alkaline stains with a pH of about 11 usually can be expected to be much more effective than when used near neutrality or in acid solution. The easiest way to achieve alkalinity is to dissolve the stain in a dilute borax solution rather than water. Alcoholic stains can be expected to penetrate cross-linked plastics better than aqueous ones and should be used when possible.

The basic dyes which can be handled most easily and which prove to be most generally effective with any embedment are the Nissel stains, particularly "toluidine blue". Alkaline toluidine blue to stain Epon sections is prepared by dissolving 0.1% toluidine blue in 2.5% sodium carbonate solution at pH 11. The sections are stained for a few to thirty minutes, followed by a wash in water and then very brief dips in 90 to 100% alcohol which serves to differentiate the stain. Otherwise, 1% solution dissolved in 1% borax is used at 85°C where staining occurs in Epon/Araldite sections in 1 to 2 minutes. A convenient and usually effective way of staining with alkaline toluidine blue is to place a slide with sections on a hotplate and flood the slide with stain. When a little steam or a few bubbles are seen, it can be assumed that staining is complete. It is indicated that strongly alkaline toluidine blue solutions are not stable and thus should be made up freshly before use.

Monica. A. Abrew *et al.* (1993) report a toluidine blue/basic fuchsin stain for these resin-embedded sections. The stain solution is prepared thus:

Solution A Toluidine blue O and sodium borate, 1% each in distilled water.

Solution B 1% basic fuchsin in 50% ethanol. The working solution is prepared by adding 30 ml to 60 ml distilled water.

Both solutions are stable in amber glass-stoppered bottles for 6 months when stored at room temperature. The sections are covered with a few drops of solution A for 3 to 5 minutes on a hotplate at 55–60°C. Excess stain is washed off in running tap water. After drying on the hotplate, the sections are then stained by covering them with a few drops of solution B for 1 minute at room temperature. They are washed well in running tap water, dried on a hotplate, and mounted with mounting medium. The colours that are obtained with this staining technique are as follows:

Nuclei	Dark blue
Cytoplasm and connective tissue	Various shades and intensities of red

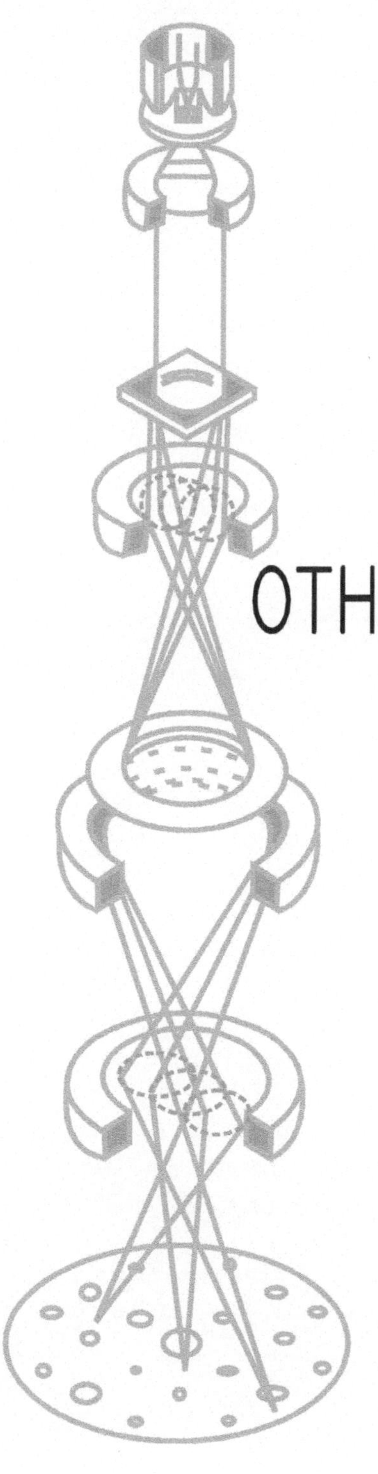

SECTION II

PROCESSING OF OTHER TYPES OF SPECIMENS

- Paraffin Block Sections
- Cell Cultures and Fine Needle Aspirates

11

PARAFFIN BLOCK SECTIONS

Quite often in a diagnostic set-up, it may not be possible to get fresh specimens for EM. Because, the realization for the need for ultrastructural study may arise only after the histopathological diagnosis, in which case fresh biopsies would not be available. In such situations, the available paraffin block sections can be used for EM work. Apart from this, sometimes, it is desirable to examine material initially fixed in formalin in the EM. So formalin-fixed paraffin-embedded sections prepared for routine histopathology are also important source material for EM.

Specimens fixed in formalin and immediately processed for routine histopathology, i.e., paraffin-embedded, are far superior for EM work than specimens preserved in formalin alone. Probably immediate further processing may eliminate overfixation artifacts.

Rarely, the specimen as soon as it is removed from the patient, is put in formalin immediately and then small bits cut from this specimen put in EM fixative and sent to EM laboratory. The formaldehyde fixation artifacts can be made out while screening in the EM.

PROCESSING OF PARAFFIN BLOCKS

Thicker sections from prepared paraffin blocks, 20 to 40 μ thick or even more, depending on the specimen availability, can be cut and processed for EM after removal of wax.

There are two methods for removal of wax. They are:

1. using xylene
2. using inhibisol

Using Xylene

Method 1

1. De-wax the section by heating at 50°C to melt the wax, and then by immersion in xylene at room temperature, for 2 changes of 15 min. each.

2. Rehydrate in descending concentrations (100, 95, 75, 50 and 15%) of ethanol, allowing 10 min. for each stage, followed by a wash in the buffer to be used for fixation.
3. Fix, dehydrate and embed using the standard schedule.

Method 2

1. De-wax the sections by heating at 50°C to melt the wax, followed by immersion in xylene at room temperature for 2 changes for 30 min. each.
2. Treat the sections with acetone for 30 min. followed by acetone : alcohol 1 : 1 mixture for 30 min.
3. This is followed by treatment with 1 : 1 alcohol: water mixture for 30 min. and then 30% and 20% alcohol for 15 min. each followed by distilled water for 15 min.
4. Then treat with buffer to be used for fixation for 15 min.
5. Fix, dehydrate and embed using the standard schedule.

Using Inhibisol (1,1,1-trichloroethane)

1. Dewax the section by immersion in inhibisol at 50°C for 2 changes of 15 min. each.
2. Rehydrate in 100, 90 and 70% ethanol at 50°C, allowing 15 min. for each stage.
3. Wash in distilled water at room temperature, and then with buffer used for fixation.
4. Fix, dehydrate and embed using standard schedule.

Although ultrathin sections from the reprocessed specimen offer less than optimal cellular preservation, they are still sufficient for identifying diagnostic characteristics. The diagnostic features are clearer, if the initial processing for wax embedding is carried out properly. Some shrinkage that occurred during wax embedding is seen in these reprocessed sections while screening in the EM. Many times, prolonged search has to be made even to identify a single diagnostic characteristic feature. On the other hand, if the initial fixation in formalin is not controlled but prolonged or the specimen is preserved in formalin for a longer duration, say, for days, before further processing, then the ultrastructural features could not be delineated. In those conditions only nuclear membrane and granular cytoplasmic material could be made out. Even cell membrane could not be clearly identified.

SOME EXAMPLES

Mesothelioma

The three most important diagnostic characteristics of mesothelioma—1) long, complex microvilli, 2) keratin filaments and 3) microvilli–collagen interaction can be seen in the formalin-fixed wax-embedded sections shown in Figure 20.22.

Histiocytosis X

The diagnostic Birbeck granules can be clearly visualized in Figure 20.28 and 20.29, where only fine needle aspirated specimen is processed for EM.

On the contrary, a similar specimen, a sample of histiocytosis X, with histological and immunological diagnosis, does not show any cytoplasmic details leave alone the characteristic Birbeck granules when only paraffin block sections were made available for EM as seen in the following figures. This sample is from a patient with lesions in the vulva, oral cavity and axilla. After the histological diagnosis the available paraffin block sections were processed for EM. Only the grooved nuclei can be made out (Figure 11.1).

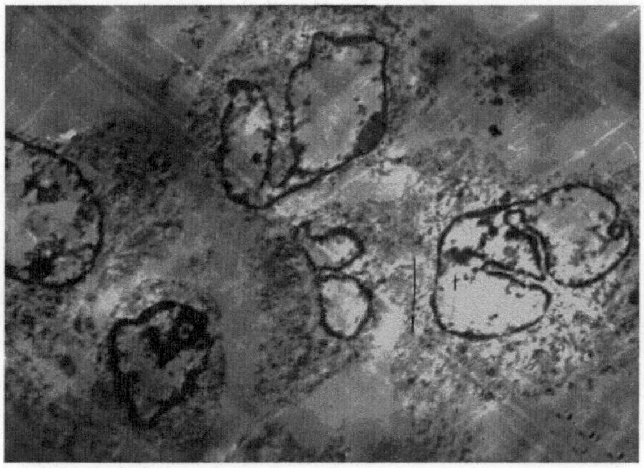

Figure 11.1 Characteristic grooved nuclei in histiocytosis X (6000×)

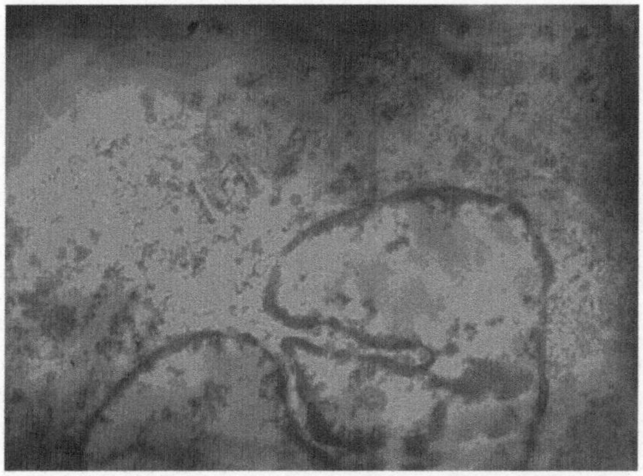

Figure 11.2 A part of the cell in the same sample (15000×)

In Figure 11.2, nuclear grooves are clearly seen but the cytoplasmic features could not be made out including the pathognomonic Birbeck granules.

Follicular Dendritic Cell Sarcoma

The ultrastructural features of this tumour include cells with oval or elongated nuclei with or without indentations; small and usually single central nucleolus; sparse organelles mostly free ribosomes; dense network of long and intertwining cytoplasmic processes and desmosomes between processes. But the paraffin block sections processed for EM and screened show only the nuclear outline and dense nuclear granular material as in Figure 11.3 while in Figure 11.4 the intertwining processes can be made out.

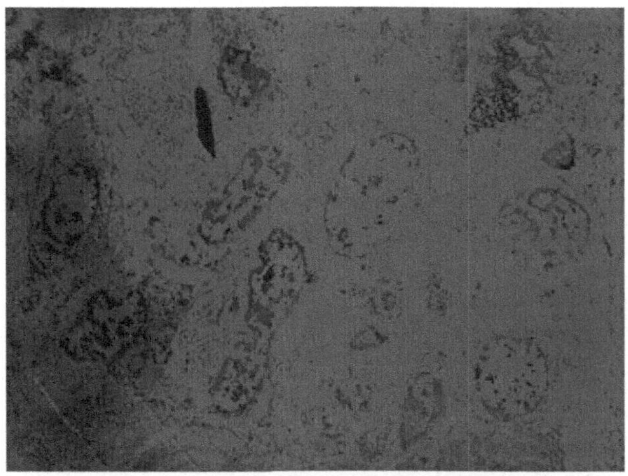

Figure 11.3 A few pleomorphic cells (2000×)

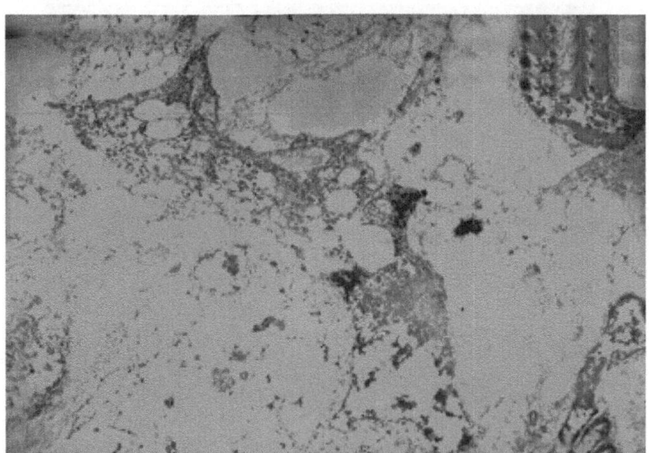

Figure 11.4 Intertwining processes (4000×)

Hairy Cell Leukaemia

The cell show only nuclear outline (Figure 11.5). The cytoplasmic details and also the characteristic hairlike membrane projections could not be made out in the ultrathin sections prepared for EM from paraffin block sections.

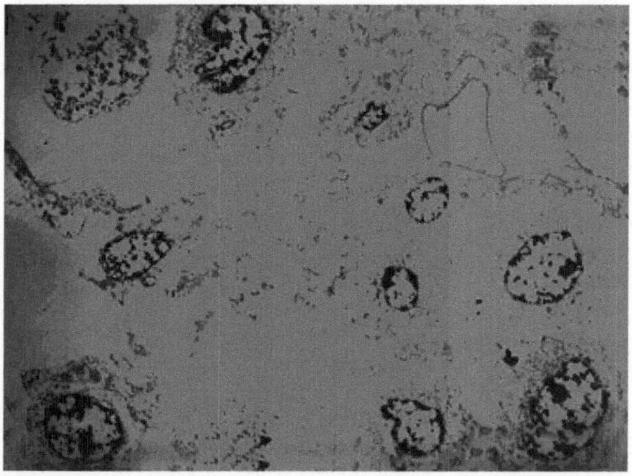

Figure 11.5 A few cells of hairy cell leukaemia (2500×)

Many specimens which were received as paraffin block sections and processed for ultrastructural study and screened in the EM have been included in the later chapters.

Quite a few examples have been cited particularly in Chapter 23 including samples of AOT (adenomatoid odontogenic tumour) (Figures 23.7 to 23.10), SOT (squamous odontogenic tumour) (Figures 23.21 to 23.23) and granular cell ameloblastoma (Figures 23.16 to 23.18) which were received as paraffin block sections only.

Thus, it can be concluded that the usefulness of paraffin block sections for EM depends on the initial processing done. If the biopsied or surgical specimen is kept air-dried or even preserved for a prolonged period in formalin, it may produce deleterious effects, making the whole efforts of processing and screening in EM less worthy.

12

CELL CULTURES AND FINE NEEDLE ASPIRATES

The cultured cells have to be removed from the culture flask or dish or any other container for further processing for electron microscopy. The cells can be removed from the container either by treatment with enzymes like collagenase or trypsin or they can be scraped out gently using a rubber "policeman" or a suitable spatula. Figures 12.1 and 12.2 show a cell removed from the culture flask by both these methods. The cells scraped out or trypsinized are immediately fixed in glutaraldehyde, pelletized, the pellet considered as a piece of tissue and processed further for EM. The ultrathin sections can be screened in EM finally.

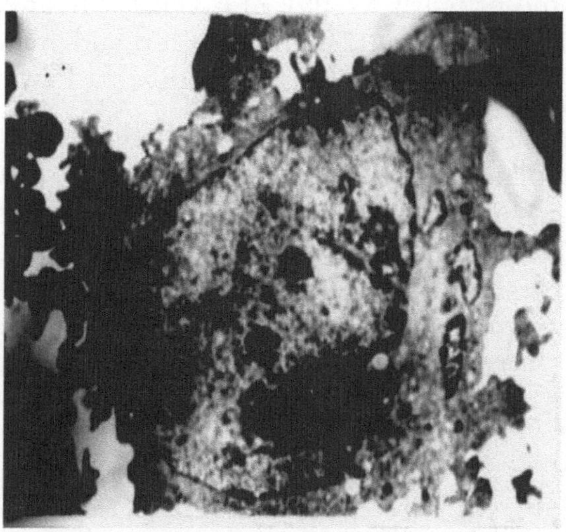

Figure 12.1 One cell from cultured Ewings sarcoma cell line, trypsinized (5000×)

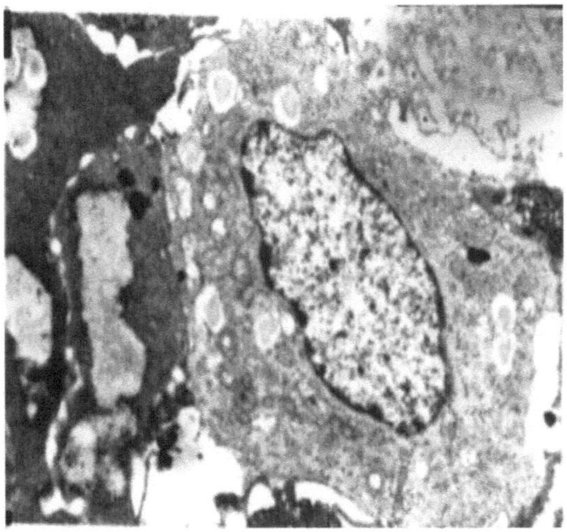

Figure 12.2 One cell from ES cell line, non-trypsinized; scraped out (6000×)

FIXATION

Fixation of Organ and Monolayer Cell Cultures

Organ cultures are very easy to fix, particularly if they have been cultured on a Millipore filter, since it is not difficult to lift them carefully from the medium in which they have been grown. They are then rinsed in a balanced salt solution to remove excess medium and placed directly in the primary fixative in a vial for fixation by immersion. These specimens are very thin. 2–4% glutaraldehyde with formaldehyde is used as the primary fixative. 2–3 mM $CaCl_2$ should be added to the fixative.

Colonies of cells (tissue culture cells, bacteria, fungi, etc.), growing on agar or other solid media in a culture plate are fixed by the same method. When cells are cultured on a substrate rich in proteins (such as gelatin or albumin), a sufficiently high concentration of buffer should be added to the fixative, because the reaction between protein and glutaraldehyde produces acid. After the primary fixation, the solid medium with the cells attached, is peeled off the culture plate and processed in a small vial.

Monolayers of cells growing on slides or coverslips or other substrates are very sensitive to changes in their environment. Therefore, the substrate is placed in a small culture plate containing the growth medium so that the cells can be fixed at the temperature at which they are incubated. Although external cell morphology often changes following change in temperature, the intracellular organelles are not seriously affected.

Fixation of Isolated Cells

With fixation of isolated cells (cells cultured in suspension, protozoa, bacteria, blood cells, etc.), the difficulty is mainly with handling the cells, especially when they are in small quantities, because penetration of the fixative is not a problem. The primary fixative is usually 0.5 to 4% glutaraldehyde in 0.1 M cacodylate buffer containing 2–3 mM calcium chloride. The temperature during fixation should be the same as that of the isolated cells since changes in temperature can rapidly affect the external morphology of mammalian cells. Consequently cells cultured in suspension are fixed at the incubation temperature, while cells which have been obtained by an isolation procedure are fixed at the final temperature of that procedure. This is mostly either room temperature or 4°C. Samples of isolated cells are handled in any one of the two ways mentioned below.

1. They are fixed in suspension and then centrifuged down to form a pellet.
2. They are centrifuged gently first and the pellet is fixed by immersion.

Cell fractions are also handled similarly.

Fixation of Cells in Suspension in Medium

For most isolated cells good fixation is obtained by adding the fixative while the cells are still in suspension in medium. The simplest method is to prepare some of the suspending medium and add to it double the amount of aldehyde fixative to give the final required concentration. Then add equal volumes of this double-strength fixative and the cell suspension to a centrifuge tube and mix them gently. Because diffusion is not a problem, the concentration of the fixative can be lower than that used for tissue specimens and the fixation time is also shorter. But, however, the concentration should not be lower than 0.5% and the fixation time not shorter than 20 min.

After the fixation, the suspension is centrifuged to allow the specimen to form a pellet. The speed of centrifugation should be kept to a minimum, just needed for pellet formation. Many pellets which have been fixed in glutaraldehyde are firm and show no tendency to disintegrate in the buffer during washing.

Fixation of Cells and Organelles in a Pellet

Some types of isolated cells, such as bacterial cells, erythrocytes and cell organelles such as mitochondria can be fixed adequately after they have been centrifuged to a pellet. The supernatant is removed without disturbing the pellet and then the aldehyde fixative is added. The fixative should be several times larger in volume than the pellet.

Encapsulating methods for pellets Some pellets tend to disintegrate when the buffer is added for washing and material is easily lost. In such cases, the pellet is encapsulated by surrounding it with a soft gel of agarose or calcium alginate. Though the encapsulation can be done at any stage before dehydration, it is preferable to do it as early as possible. This is applicable, particularly for plant and animal cells, mammalian oocytes, protozoa, yeasts, bacteria and cell organelles such as mitochondria.

EMBEDDING

Embedding Methods for Cell Monolayers

The thin layers of cells are easily disturbed and so it is preferable not to remove them from the substrate before embedding. Consequently, the substrate is so chosen that the hardened block can be separated from it or the cells are grown on a substrate which can be sectioned. The most suited substrate which can be separated easily is a plastic surface. Most types of culture dishes and coverslips are suitable. Alternatively the cells can be cultured in the snap-on cap of a BEEM capsule.

The monolayer of cells is fixed, dehydrated and infiltrated with the embedding medium by standard methods. Gelatin or BEEM capsules are then filled with the embedding medium and are inverted over selected areas of the monolayer. The embedding medium is then cured in the usual way and the capsule is removed from the substrate, leaving the monolayer embedded in the surface of the block. It is easy to remove capsules containing Epon if this is done after about 12 hours at 60°C before the resin has completely hardened. After removal, the blocks are returned to the embedding oven for final curing. Otherwise, fully hardened blocks are separated from the substrate immediately after removal from the embedding oven while they are still warm. In this **inverted capsule** technique the layer of cells is embedded in the surface of the block and sections can only be cut parallel to the plane of the monolayer. Sectioning is easier if a flat embedding mould is used and the surface of the block is covered with a thin layer of resin.

Tissue culture monolayer embedding moulds are readily available now, which can be used for embedding cells in monolayer culture for both light and electron microscopy. These moulds have several advantages. They eliminate the need for traditional BEEM capsules. The improved moulds can be employed to transfer the majority of cells from the entire surface of a coverslip to avoid losses of large amounts of cellular information. In addition, other problems such as surface blemishes caused by air entrapment associated with BEEM capsules are eliminated. Moreover, the mould uses only a minimum amount of resin.

Monolayer embedded and sectioned on the substrate When the substrate itself can be sectioned, then it does not have to be removed from the embedded layer of cells. This way, damage to the cells is avoided and it is also possible to study the relationships between the cells and their substrate in detail. The substrate must be suitable for the good attachment and growth of cells and must section easily without separating from the monolayer.

Slides made out of epoxy resins, such as Araldite, Epon or Spurr's also make very good substrates provided the toxic factors in them are removed by treatment with salt solutions before the cells are added. It is easy to make a slide by curing a thin (2 to 4 mm) layer of the epoxy resin in a container with a flat level bottom, such as one of the larger flat silicone rubber embedding moulds. The epoxy resin slides are sterilized by irradiation with UV light and transferred to a polystyrene tissue culture dish. The cells are added and incubated for the required time and are then fixed, dehydrated and infiltrated with the same epoxy resin *in situ*. The monolayer of cells is then embedded by covering it with a thin, 0.1 to 0.5 mm, layer of embedding medium, or by inverting the slide over a flat mould containing the embedding medium. The resulting sandwich is sectioned in the same way as in the flat embedding method. The same Epon mixture is used for embedding as is used for the preparation of the slides to ensure that the final 'sandwich' has a uniform hardness and consistency.

Alternatively, cells are cultured on permeable substrates, such as Millipore Filters or dialysis membranes. These allow better diffusion and infiltration of culture media, fixatives and embedding media.

Keiji Kawamoto *et al.* (1980) have reported simplified *in situ* preparation of cultured cell monolayers for electron microscopy. Though embedding *in situ* is the method of choice for ultrastructural studies of cultured cell monolayers, the separation of the embedded cells from the growth surface is difficult, sometimes requiring use of liquid nitrogen or pliers which may damage the ultrastructure. To overcome this difficulty, these authors have simplified the preparative procedure by using Micro Test Plates and by dissolving the plastic growth surface of *in situ* fixed, dehydrated and Epon-embedded cells with xylene. Dissolution of the plastic growth surface with xylene does not affect the quality of the preparation. The size of the Epon blocks also is such that they fit easily in the microtome.

The following two figures show some cultured cells processed for EM, where the attachment to the substrate is elucidated.

Figure 12.3 shows the junction between two human endothelial cells grown on a Falcon tissue culture dish. The cells are fixed and dehydrated *in situ* and then floated off the dish in propylene oxide, before embedding in Epon. The layer of extracellular material (arrowhead) which has coated the culture dish is no longer flat, indicating that the monolayer has been disturbed during its release from the dish.

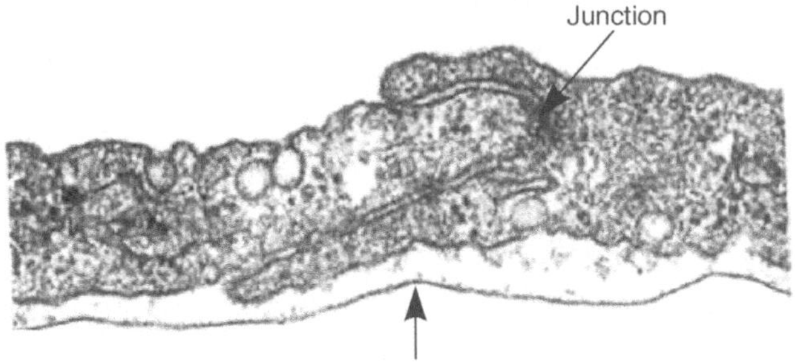

Figure 12.3 Junction between two human endothelial cells. (*Source:* Audrey, M. Glauert and Peter, R. Lewis. (1998). *Biological Specimen Preparation for Transmission Electron Microscopy.* Vol. 17. Princeton University Press, Princeton. p. 155, Fig. 5.6.)

Figure 12.4, shows a macrophage cultured on a layer of fibrin on a coverslip. The position of the coverslip, before its removal during embedding, is visible as a straight dense line (arrowhead), indicating that there has been no disturbance of the relationship between the cell and its substrate during processing.

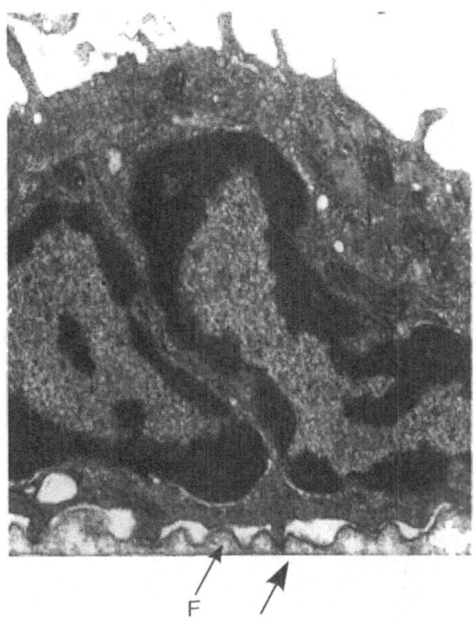

Figure 12.4 A macrophage cultured on a layer of fibrin (F) on a coverslip. (*Source:* Audrey, M. Glauert and Peter, R. Lewis. (1998). *Biological Specimen Preparation for Transmission Electron Microscopy.* Vol. 17. Princeton University Press, Princeton. p. 170, Fig. 5.8.)

Embedding of Cell Suspensions

In the case of cell suspensions, extreme dilution in a very large volume of fixative with agitation may improve preservation. But, it is better to centrifuge suspended cells into a compact mass and then to treat these pellets or small aggregates as solid tissue blocks. For some suspended specimens, repeated centrifugation at each preparative stage might prove harmful and may result in the loss of material. Then it is preferable to mix the warmed pellet (45°C) with a few drops of similarly warmed 2% agar. This can then be removed to a cooled microscope slide and cut into small cubes after the agar has set. These are then treated as tissue blocks.

A few illustrations of cultured cells received from students, which were processed, sectioned and screened in EM in our laboratory are given in Figures 12.5–12.14.

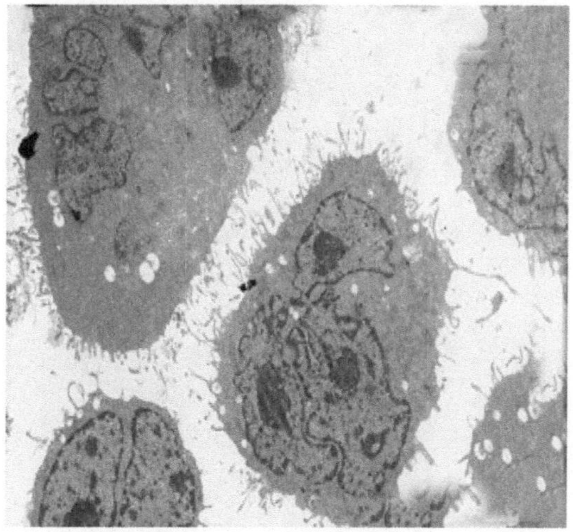

Figure 12.5 A few cells from acute leukaemic cell line differentiated into antigen-presenting cells (3000×)

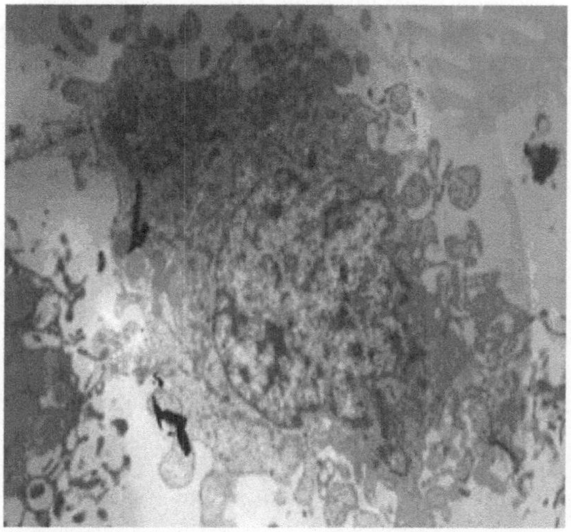

Figure 12.6 Cell from osteosarcoma cell line (5000×)

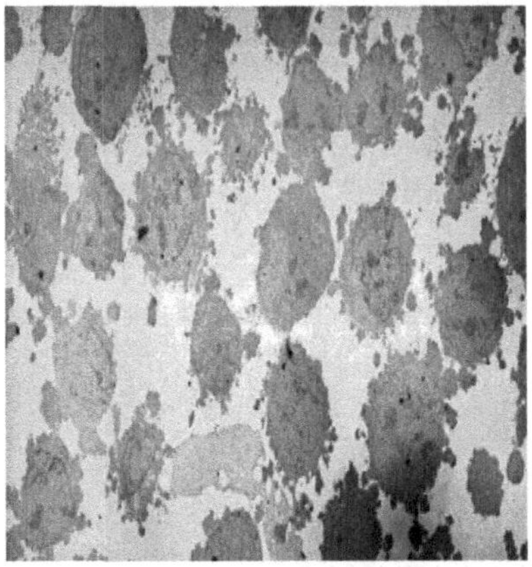

Figure 12.7 A few cells from Ewings sarcoma cell line (1000×)

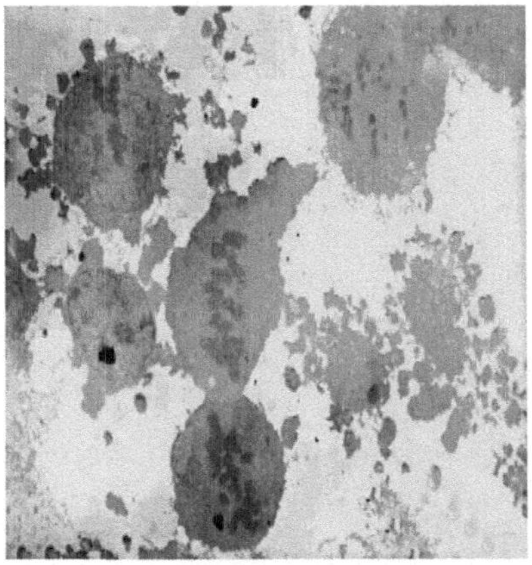

Figure 12.8 Mitotic cells from ES cell line (2000×)

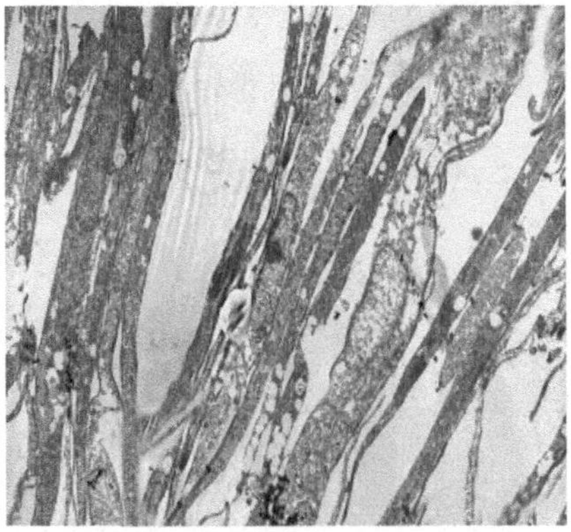

Figure 12.9 Undifferentiated mesenchymal cells (2000×)

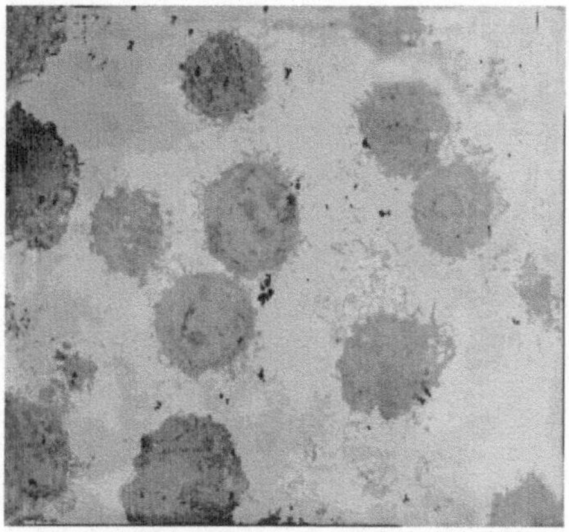

Figure 12.10 A few cultured epithelial cells (1500×)

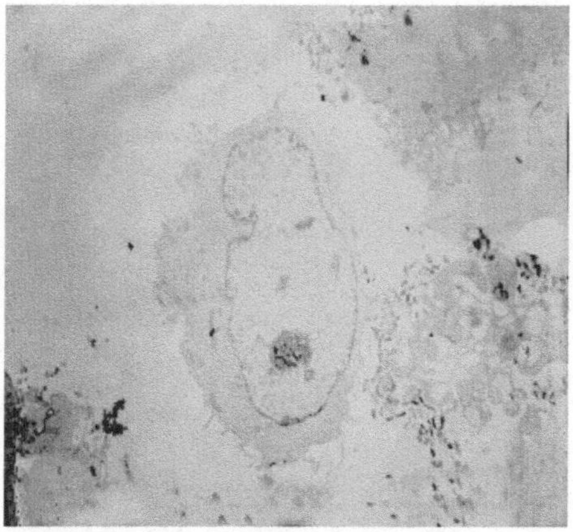

Figure 12.11 Cultured endothelial cell (3000×)

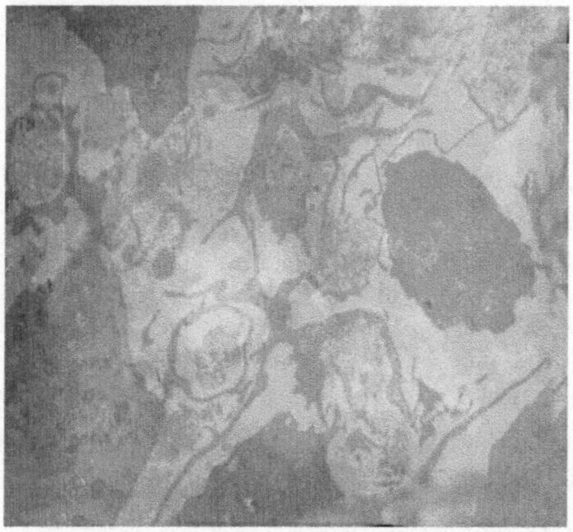

Figure 12.12 Dendritic cells cultured and suspended in RPMI medium (3000×)

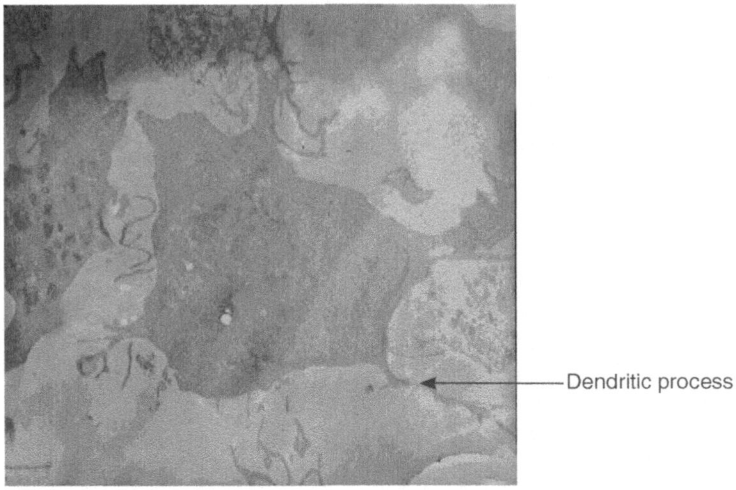

Figure 12.13 A single dendritic cell cultured and suspended in RPMI medium (4000×)

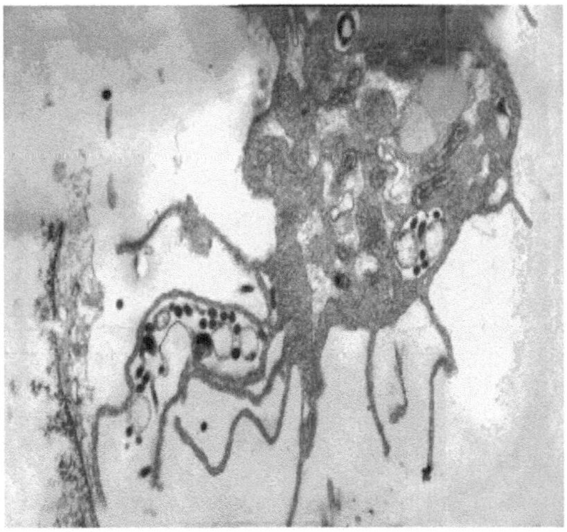

Figure 12.14 One mesothelial cell infected with *Mycobacterium tuberculosis* showing microvilli engulfing particles (10000×)

FINE NEEDLE ASPIRATES

For tumour diagnosis, wherever feasible, before attempting a biopsy, a fine needle is inserted into tumour tissue and cells are aspirated. This aspirated material can be used for EM work apart from the routine cytology. The aspirated cells are fixed in large volume of the fixative, centrifuged and the pellet treated as a tissue specimen for further processing.

Hagelquist described in 1973 a technique by which fine-needle aspiration material can be studied ultrastructurally. Kindblom (1983) elaborated a simple technique for using fine-needle aspiration biopsy specimens as small tissue pieces, for electron microscopic as well as conventional histology instead of preparing smears.

The procedure followed by him is given below.

The aspiration is carried out using the same type of instrument as used in routine cytology. After aspiration, the needle is detached from the syringe and a few millilitres of 2.5% glutaraldehyde in 0.1 M cacodylate buffer at pH 7.2 is aspirated into the syringe. The aspirated specimen with the glutaraldehyde is then immediately ejected with great care into a glass tube sealed at one end with a double ashless lens paper and unwaxed dental floss. The tube is then placed in a beaker containing glutaraldehyde. In this beaker the specimen is processed through all steps until finally embedded in Epon. This processing is performed according to a standard or rapid method. Before embedding in Epon the double ashless lens paper is opened. The osmium-stained tumour tissue adhered in fragments or fine threads is easy to handle for final embedding. Figure 12.15 illustrates the above.

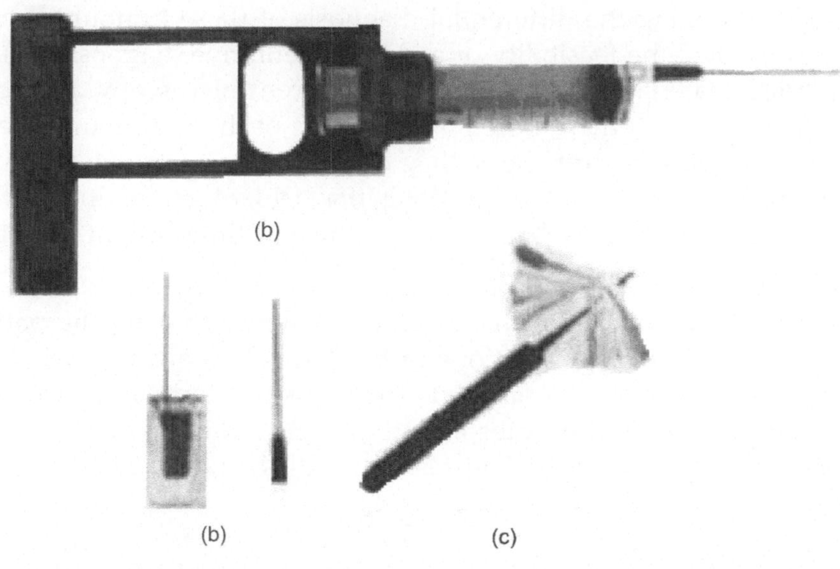

Figure 12.15 Processing of fine needle aspirates. (a) Instrument used for aspiration (b) Glass tube sealed with lens paper (c) Lens paper opened up

Figure 12.15 shows the instrument used for the aspirations. The aspirated material is ejected into a glass tube sealed at one end with lens paper; the tube is placed in a beaker in which the material is processed till the embedding in Epon. The lens paper is opened up to collect the tissue for final embedding.

Kindblom (1983) has evaluated the usefulness of fine-needle aspiration biopsy specimens for the pre-operative diagnosis of soft tissue and bone tumours. Since for these tumours pre-operative open biopsies are often not recommended, the use of fine-needle aspirations for both light and electron microscopy offers great advantages over routine cytologic smears.

- ✳ It gives a very well-preserved cellular morphology as seen both light and electron microscopically.
- ✳ Some conventional histologic staining can be applied giving a familiar appearance and additional information to the pathologist.
- ✳ The specimens can be saved in blocks and new sections made whenever required.
- ✳ EM studies help to confirm the LM diagnosis and sometimes may be necessary in order to arrive at a definite diagnosis.

Similarly, Mohammed Akhtar *et al.* (1985) have employed a combined light and electron microscopic approach for the fine-needle aspiration biopsy diagnosis of round cell malignant tumours of childhood. In the case of malignant round cell tumours, it is quite difficult to make a precise diagnosis at the light microscopic level. The use of electron microscopy for determining the differential diagnosis of these tumours is essential and the authors have verified the feasibility of making a definitive diagnosis using the easily accessible aspirated material for both light and electron microscopy and conclude that EM plays a definitive role in the differential diagnosis of these tumours as an aid to the light microscopic findings. Nevertheless, they caution that the use of aspirated material for EM requires more experience and expertise in this field. Moreover in less differentiated tumours, a prolonged search may be necessary which at times might not be fruitful.

We received the fine needle aspirated material from a thyroid tumour for electron microscopy. Ultrathin sections from this aspirated material showed the pathognomonic Birbeck granules at higher magnifications in the EM. Thus histiocytosis X of primary thyroid, a very rare tumour was confirmed. The biopsy material obtained subsequently also showed similar features thus substantiating the usefulness of FNAC material for diagnostic electron microscopy. (*See* Figures 20.27, 20.28 and 20.29.)

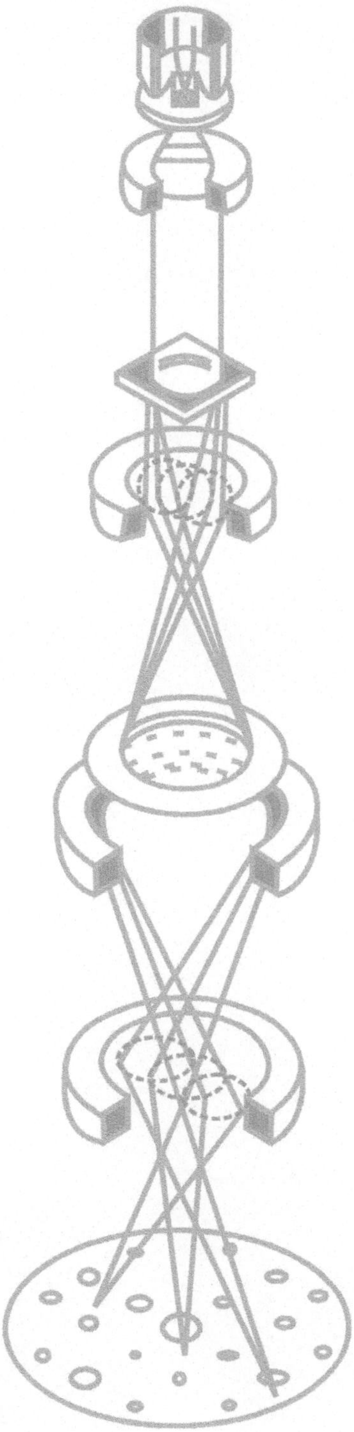

SECTION III

OTHER PROCESSING TECHNIQUES

- Rapid Processing and Replica Techniques
- Negative Staining
- Microwave Techniques in EM
- Cryotechniques in EM
- Immunoelectron Microscopy

13

RAPID PROCESSING, SHADOW-CASTING AND REPLICA TECHNIQUES

RAPID PROCESSING

Rapid processing is required mainly when there is an urgency for diagnostic confirmation, i.e., when EM is used as an aid in confirming a clinical and histological diagnosis. Also, it is practised when there is a need to reduce the processing time and fast results are essential. The entire processing, including sectioning and screening may be completed within 24 to 48 hours while the routine processing takes 4 days or more. For rapid processing, the specimens should have a maximum thickness of 0.1 mm in at least one dimension. Though this method gives optimal results and reliable preservation of fine structures, the rapid procedure may be insufficient for preservation of certain structures. Whenever possible, when adequate quantity of the specimen is available, it is advisable to keep a small bit of the specimen and process by the routine method also.

A Rapid Processing Schedule for Small Specimens

The specimen vials are placed on a rotator to provide constant agitation. This procedure is carried out at room temperature as given below:

- Fix by immersion for 15 to 20 min. in 2.5% glutaraldehyde in 0.1 M cacodylate or PIPES buffer, at pH 7.2 to 7.4, containing 2 to 3 mM calcium chloride.
- Wash for 5 min. each in 3 changes of the same buffer.
- Post-fix for 20 min. in unbuffered 2% osmium tetroxide, wash briefly in distilled water, treat for 20 min. in 1% aqueous uranyl acetate, and wash briefly in distilled water again.

- ✳ Dehydrate for 3 min. each in 70, 80, and 90% ethanol, and then for 2 × 5 min. in 100% ethanol and finally for 2 × 5 min. in propylene oxide.
- ✳ Infiltrate for 10 min. with a 1 : 1 mixture of propylene oxide and an epoxy resin embedding medium, followed by 15 min. with a 1 : 3 mixture, and then for 10 min. with the epoxy medium. Constant agitation is particularly important at this stage.
- ✳ Transfer to embedding moulds filled with fresh epoxy resin embedding medium and cure the resin by heating for 45 min. at 70°C and then for 45 min. at 95°C.

The blocks can be sectioned 15 min. after removal from the oven.

A Schedule for Rapid Processing of Biopsies

- ✳ Fix the biopsy for 5 to 10 min. in a mixture of 4% glutaraldehyde and 4% formaldehyde in 0.1 M cacodylate or PIPES buffer at pH 7.2 to 7.4, containing 2 to 3 mM calcium chloride.
- ✳ Gently lay the biopsy in buffer on dental wax and cut it into pieces about 0.5 mm thick under a stereomicroscope with a clean razor blade. Minimize the mechanical damage to the tissue.
- ✳ Cut out the area of wax carrying the tissue pieces and immerse it in the same fixative for 20 to 30 min.
- ✳ Transfer the specimens to the same buffer for 3 rinses of 5 min. each. Cut each piece of tissue into 2 or 3 slices, not more than 0.2 mm thick, if this can be achieved without much mechanical damage, and carefully detach from the wax.
- ✳ Post-fix, dehydrate and embed by stages as given above.

Rapid Processing by Microwave Irradiation

All stages of processing starting from the initial fixation to the final block preparation can be made rapid by irradiation with microwaves. The saving of time is particularly significant during primary fixation and also during the curing of the embedding medium.

To avoid exposure to the vapours of hot fixatives and solvents, the oven must be placed in a fume cupboard. Moreover, some organic solvents are potentially explosive at high temperatures. Therefore, it is advisable to use the oven only for the initial fixation and the final curing and not for the intermediate stages. For the intermediate steps, rapid processing method described above can be used.

An important requirement for using microwave ovens is the need to calibrate the oven before each fixation and embedding to identify the positions in the oven required to give the most even heating and to obtain reproducible results. Therefore, the time needed for calibration, though not long, must be taken into account for estimating the overall time required for the complete processing.

An outline of the schedule for rapid processing using microwave oven is given below:

- Calibrate the oven for the primary fixation, in order to identify the positions for even heating by microwave irradiation at full power.
- Soon, place the specimens (less than 10 mm^3 in size) immersed in the primary fixative at the positions identified, and irradiate. The fixation time should be about 9 sec. at 50–55°C.
- Remove specimens from the oven as soon as the oven switches off, to prevent further fixation from the hot fixative, and place them in the washing buffer.
- Further steps of washing, post-fixation, dehydration and infiltration are carried out using the rapid schedule outside the oven.
- Calibrate the oven again as before and place the embedding mould with specimen and the embedding medium in the oven for 15 min. at 50% power.
- Remove the mould from the oven as soon as it switches off and allow the blocks to cool to room temperature.

(*See* Chapter 15 for further details on microwave techniques.)

SHADOW-CASTING

In electron microscopy, shadow-casting means evaporating atoms of a dense material, usually a metal, from a point source at an oblique angle to the specimen. If the specimen is contoured, metal piles up on surfaces which face the source, but surfaces facing away from the source are shielded and receive no deposit, as shown in Figure 13.1. This is the shadow, which in a positive print appears white. However, in a negative print the shadows

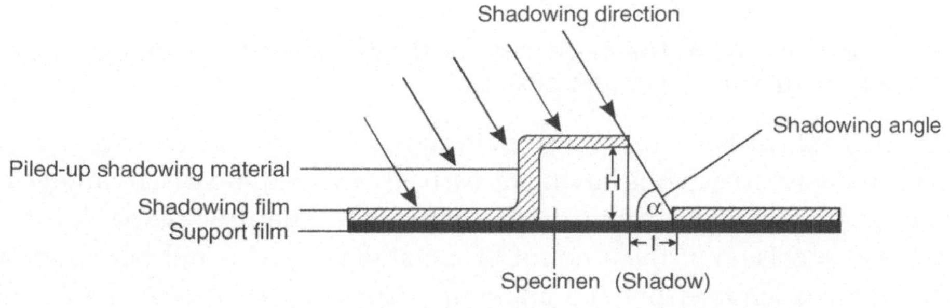

Figure 13.1 Enlarged representation of the specimen area

appear black while the metallized surfaces are light. Thus, the negative prints look like pictures of contoured objects illuminated from the side by light. If, the angle of shadowing θ

and the length of the shadow *l*, are known, the height of the object *h* which produced the shadow can be determined by using the formula, $h = l \cdot \tan \theta$. Thus, the third dimension of the objects can be ascertained.

Shadowing is important to define sharply the contours of small objects of relatively low density. Fine-order surface details also could be visualized with this method.

For shadowing, metal evaporation is done in vacuum. A vacuum evaporator is the instrument used for this purpose and which is shown schematically in Figure 13.2.

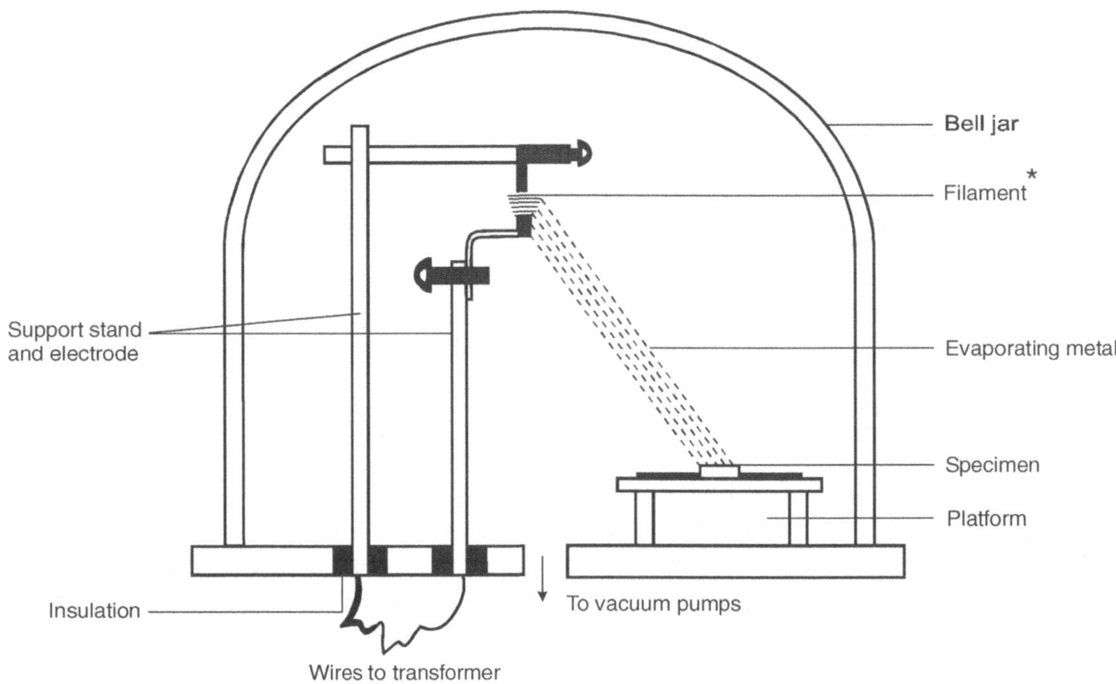

Figure 13.2 Representation of the mechanism of shadow-casting. (* Tungsten filament coated with melted metal (e.g., platinum))

Chromium, gold, nickel, palladium, platinum and uranium are some of the metals used for shadowing with each one having its own advantages and disadvantages. Platinum and carbon when evaporated simultaneously yield shadows capable of very high resolution. Great precision in the amount of metal deposited is not necessary and visual estimation is often adequate. But the angle of shadowing is of considerable qualitative importance. Very long shadows increase accuracy only for quantitative measurements. With long shadows, details of everything that happens to lie in shadowed areas are lost. For particulate matter of roughly spherical proportions, a shadowing angle of 20° to 30°

is sufficient; for moderately contoured surfaces which require greater obliquity, 15° will be better; for very shallow contours, angles as small as 10° prove useful.

The distance between the evaporating source and the specimen is not critical. However for biological specimens closeness to the heating filament may damage them. On the other hand, a long distance will make the source more effectively a point, and the shadows will be sharper.

Shadowing is of three types—unidirectional, bidirectional (two directions of 180° or 90°) or rotary, where the specimen is rotated at 60–120 revolutions per minute.

This technique is based on the principle that when a metal or carbon is heated in vacuum, it evaporates. The vapours when allowed to settle on a suitable substrate form a thin coating of metal/carbon film. Carbon coating on grids as support film is prepared using this technique. When the evaporated metal atoms fall on the specimen at an angle, it is referred to as "shadow-casting".

The same phenomenon occurs in the EM also. The high vacuum in the TEM column makes the carbon in biological specimens to evaporate when the specimen is irradiated by the electron beam. The evaporated carbon particles settling on various parts of the EM, particularly the apertures, cause contamination in the column. The reverse phenomenon is applied for cleaning the apertures in a vacuum evaporator. The contaminated apertures are heated in vacuum when the contaminants, mainly carbon, evaporates, thus making the apertures clean for further use in EM.

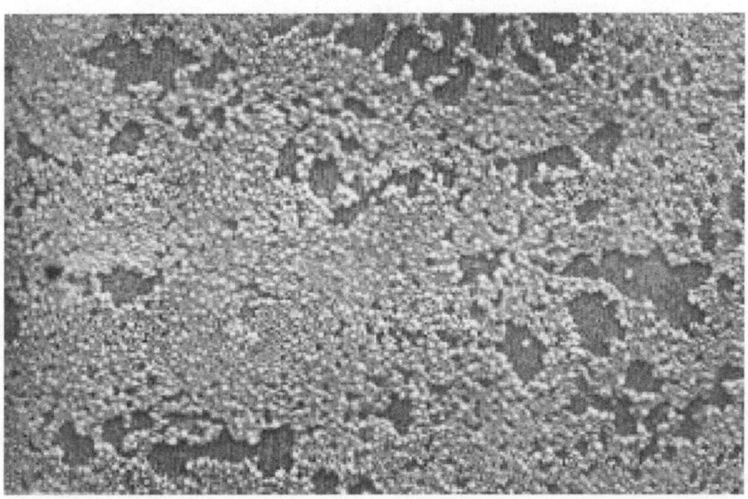

Figure 13.3 Layer of tobacco mosaic virus (TMV) shadowed with platinum (24000×). (*Source:* Cosslett, V.E. (1951). *Practical Electron Microscopy.* Butterworths Scientific Publications. p. 237, Fig. 119.)

A biological specimen, a viral specimen, prepared using this technique for electron microscopy is shown in Figure 13.3.

Recently, Virendra Kumar (2008) reported that shadow-casting technique using TEM gives more accurate counts of intact **leprae bacilli** of *Mycobacterium leprae* in patient specimens than standard optical microscopy. He visualized bacterial surface structures by thermal evaporation of platinum alloys under moderate vacuum and determined EM-MI (Electron Microscopic Morphological Index). This method according to him gives a high contrast image at relatively low resolution and is useful in correlating micro-morphology in a quantitative manner to early therapeutic effects of anti-leprosy drugs.

REPLICATION

Replication is the method used to study surface topography using TEM. Replicas of surfaces quite faithfully reproduce the smallest detail and are electron-transparent.

This technique is mainly employed for the following.

- Indirect examination of materials which cannot be studied directly by conventional scanning or transmission microscopy since they are degraded by the electron beam.
- Simple study of surface morphology at better resolution than offered by scanning electron microscopy.
- Examining and identifying small particles.

The important characteristics of a replicating material include the following:

- It should mould itself to follow completely the contour of the original specimen.
- It must be capable of being detached completely and cleanly from the specimen. In other words, it must be chemically stable to solvents for the original specimen.
- After detachment, the replica must be morphologically stable, that is, it must not flow spontaneously in which case fine structural details may be lost.

Types of Replicas

Replicas are of two types—single-stage and two-stage.

Single-stage replicas In the single-stage method a thin film is deposited on the specimen surface which can be detached for EM observation. It is prepared by flowing a solution of plastic on the surface to be replicated, by which a thin film of plastic is formed. The exposed upper surface of this film will be smooth and the other surface contoured corresponding to the contours of the specimen. This replica itself will be of variable thickness and will produce an electron image of varying contrast. But as the replica film gets thickened, the contrast is quickly lost and in most circumstances the replicated surface is observed after metal-shadowing.

The images produced by all single-stage replicas are negatives of the original surface which may not cause much difficulty in their interpretation. This method has proved quite useful in dealing with macromolecules, viruses, etc.

Pseudoreplicas If a shadowed single-stage replica is to be used, it is often best to metal-shadow the surface to be replicated first, and then build the replica film upon this. The metal will vigorously adhere to the replica, and usually can be removed with it. Such replicas are usually called **pre-shadowed replicas** or **pseudoreplicas**. The resolution of such replicas is limited only by the grain of the deposited metal. The method has found much application in virology and macromolecular research. It has an advantage over a single-stage replica, which is shadowed after replication, in that the shadow images are positives. Also, the angle of shadowing can be known exactly and employed quantitatively.

Two-stage replicas Single-stage replicas inevitably must be delicate objects, and are only practical when a thin film can be stripped easily from the surface to be replicated or when the specimen can be dissolved or digested leaving the replica behind. Two-stage replication is employed to circumvent these limitations. The primary replicas are most commonly made of plastics like Formvar. Evaporated carbon is almost always the material of choice for the final replica.

A schematic representation of both single-stage and two-stage replicas is given in Figure 13.4. Two-stage replication is advantageous when it is desirable to remove the replica by mechanical stripping and when the specimen is large enough so that these operations can be performed under a dissecting microscope. Hard tissues such as bone, teeth and even hair can be handled easily. Soft surfaces which can be dried, such as skin, also can be studied using two-stage replication technique. Surfaces of many botanical specimens have been studied with this method.

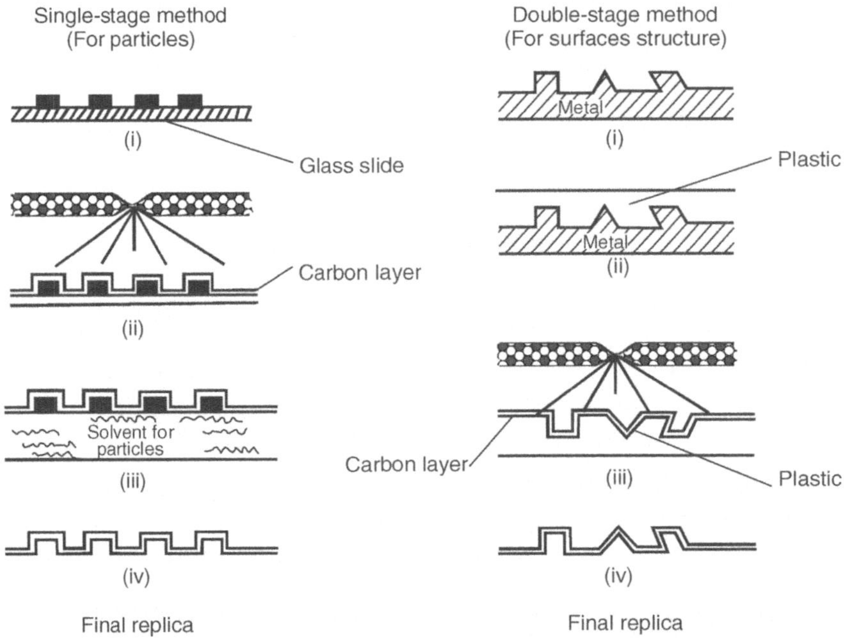

Figure 13.4 Single-stage and two-stage replicas. (*Source:* Boyd, J.D., Johnson, F.R. and Lever, J.D. (Eds.). (1961). *Electron Microscopy in Anatomy.* Edward Arnold Publishers Ltd., London. p. 34, Fig. 3.2.)

Preparation of Surfaces for Replication

Surfaces for replication require some preliminary preparation. If only mineral patterns on the surface are to be studied, as in bone or tooth, mild etching with weak acids may reveal details not otherwise visible. Freshly cleaved fractures uncontaminated by handling may also be employed as surfaces of choice for such specimens. Surfaces like skin or hair should be degreased using suitable organic solvents. Paint and wood surfaces have been studied after differentially extracting various components. If a primary replica is to be made by a plastic such as Formvar dissolved in an organic solvent like ethylene dichloride or chloroform. It is important that the solution wets the surface completely and does not trap bubbles. If a primary replica is to be made by evaporated carbon or other substance, the surface must be extremely clean.

Materials for Replication

Plastics and resins Solutions of Formvar in chloroform or ethylene dichloride have been most commonly used. Parlodion in amyl acetate also is suitable. Thick films are suitable as the first replica in a two-stage replication can be made with a thin sheet of polymerized methacrylate which is soaked in methacrylate monomer until its surface is softened and then laid over the surface to be replicated. If single-stage replication is attempted with a plastic replica film that must be stripped, it is usually better to transfer it directly to specimen grids.

Vacuum deposited materials Carbon, a light element is preferred but because of its brittleness, it can be used only when the underlying material can be dissolved or digested away. If a single-stage replica of particulate material is being sought, it is possible to deposit the sample directly upon specimen grids previously prepared with Formvar or Parlodion supporting films. Usually these would be metal-shadowed first, and then a carbon replica deposited on top. Evaporated carbon is the most common choice as the final replica in two-stage replication, when the primary replica is a thin plastic film which can be dissolved away leaving the intact carbon replica behind.

In the two-stage method an imprint of the surface is made on a relatively thick piece of plastic and a final replica is produced from this. Many materials are used for replica film but the most common are polymers, carbon and metal oxides.

Figure 13.5 Gold-shadowed replica of surface of earthworm cuticle (13500×) (*Source:* Cosslett, V.E. (1951). *Practical Electron Microscopy.* Butterworths Scientific Publications.p. 239, Fig. 121.)

For biological samples, pre-shadowed single-stage carbon replicas (positive) are commonly used. Various proteins have been illustrated using positive replication. Very few samples in biology require negative replication. Figure 13.5 shows the replicated surface of an earthworm cuticle which is metal-shadowed to increase the contrast.

Gaurav Sharma *et al.* (2005) have developed a new single-stage replication technique for petal surfaces of a flower using a cosmetic peel-off facial material. The surface morphology of a flower petal can be a significant marker for the characterization of its variety. The authors describe a material for making moulds with a single-step replica technique in which the step of applying pressure is avoided and thus the fragile flower petals are not subject to collapse. This technique is easily applicable for surface studies using SEM.

14

NEGATIVE STAINING

INTRODUCTION

Negative staining has been a useful specimen preparation technique for biological and medical electron microscopists. Huxley in 1956 first described negatively stained tobacco mosaic virus particles. The technique has undergone extensive modification and improvement within the past 50 years.

This is a very valuable and important technique for examination of fine surface detail of small particulate specimens. Basically the method consists of embedding the particles in a layer of dense material so that the specimen is viewed as a relatively light object against a dark background. Of course, the term "negative staining" is misleading, because staining in the proper sense of the word is not involved at all. The only "negative" feature is the dark-field effect of a light object against a dark background, giving a "negative contrast".

This is a most convenient method for rapid preparation for electron microscopic observation and can be applied to objects from macromolecules like proteins to organisms like viruses and bacteria. In this method, staining is achieved by heavy metal salts, which upon drying, form a "glassy coat" around the embedded objects. Metal salt solution is able to occupy the hydrated regions in and around the object. Positive staining involves direct interaction of stains with the cellular proteins, but negative staining does not involve protein–stain interaction.

The merits and drawbacks of negative staining are listed below.

Advantages

- technically simple method
- produces high-contrast images
- relatively insensitive to damage to the specimen by the electron beam

- easy and fast preparation of samples requiring mostly only 2–4 minutes
- requires only a small amount of sample (2–5 μl)
- the size and shape of the particles can be easily visualized

Drawbacks

- resolution limited to about 25 Å
- structure usually flattened owing to drying (air-drying for several hours to overnight air-drying is recommended), leading to introduction of drying artifacts
- uneven structure resulting in artifacts
- high ionic strength and non-physiological pH of the stains may alter the specimen
- a negatively stained image may not show all components of the specimen, including even those substances which might be layered on the surface.
- complex substances which might be hydrated and penetrated by the stain would remain unobserved

This method is applicable to:

- isolated viruses
- protein molecules
- macromolecular assemblies and fibrils
- subcellular membrane fractions
- cell fragments
- liposomes and artificial membranes
- synthetic DNA arrays
- polymer solutions

The shape and size of these specimens can be rapidly observed as the preparation technique is simple and less time-consuming.

Thus this technique can be readily employed to:

- identify human viruses in urine, faeces, blood serum, GI tract, CSF (cerebrospinal fluid), etc.
- detect viruses found in animal, plant, insect and bacterial specimens.
- study and compare viral nucleic acids.
- localize antigens within virus particles.
- study ribosomal subunits and protein macromolecules.

BASIC REQUIREMENTS OF THE SPECIMENS

There are certain requirements for the application of negative staining for biological materials. They are:

- Dimensions of specimen should not be less than the resolving power of the microscope
- Must be available at a minimum concentration
- Should not interact with the stain
- Must have affinity for the supporting materials
- Preparation must be free of detergents
- Specimen should be thin enough and have good transparency

In this technique, staining period is maintained very short, (30 secs. to 2–5 minutes). The specimen particles appear as electron-lucent structures in an electron-dense background. Hence the name "negative staining". This is because, the electron stain passes through the low electron density of specimens, but not through the metallic background. But specimen substructure can be revealed if the stain penetrates through holes or crevices of the specimen.

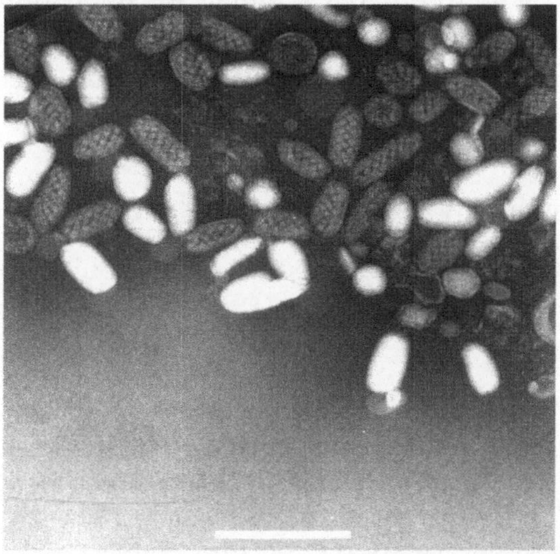

Figure 14.1 Negative staining of micronemes from *Cryptosporidium parvum*. *(Source: Microscopy and Analysis.* p. 6, Fig. 2.)

In Figure 14.1 the intact micronemes appear as electron-transparent rods, whereas micronemes with a damaged surface membrane allow stain entry. (Scale bar = 100 nm)

SPECIFICATIONS FOR NEGATIVE STAINS

1. The stains should not react with the specimen.
2. They should be of high density and good stability.
3. They should possess high stability under electron bombardment and high vacuum.
4. Atoms of the stain must be smaller than the details to be visualized.
5. The stains should be fairly soluble and should not crystallize.
6. The stain must be as dense as possible to achieve maximum contrast.

COMMONLY USED STAINS

The most commonly used stain for negative staining procedure is 1–2% phosphotungstic acid (PTA). To a solution of phosphotungstic acid, drops of concentrated sodium hydroxide is added till the desired pH is achieved. For viruses and isolated proteins a pH of 5–6 seems to give optimal results. But penetration of this stain is low because of formation of polyanion.

The cationic and slightly acidic 1 to 2% w/v aqueous uranyl acetate and solutions of several other anionic, heavy-metal-containing salts have also been routinely used for negative staining usually at neutral pH (e.g., tungstate, phosphotungstate, silicotungstate, molybdate, vanadate, etc.). By varying the stain concentration, different levels of stain density can be achieved. However, it has been found that after drying, the molybdate and vanadate negative stains impart lower mass density around the biological material than uranyl acetate.

Some of the most commonly used stains are listed below:

- Uranyl acetate (UA) having superior penetrability due to very small grains.
- Ammonium molybdate, mainly used for membranes and membrane-bound materials.
- UA–EDTA complex
- Uranyl oxalate
- Uranyl formate providing the highest contrast because of higher density
- Uranyl magnesium acetate having fine grain
- Sodium/potassium phosphotungstate
- Sodium/potassium silicotungstate
- Methylamine tungstate
- Methylamine vanadate (NANOVAN)

CRITERIA FOR NEGATIVE STAINING

- The specimen can be fragmented, dispersed or suspended.
- The specimen concentration should be optimal, i.e., should neither be too high which may lead to clumping, nor too low where very few scattered particles may be difficult to locate.
- The stain solution should be purified before use by filtration or centrifugation.
- Carbon-coated or formvar–carbon-coated grids are used for negative staining. These grids are now commercially available and can be used straightaway.
- The specimen suspension as well as the stain solution can be dropped on a stain mat/pad or a parafilm.
- Coated grids are floated on the drop of specimen suspension as well as on the drop of the stain with the coated side at bottom in contact with the solutions. This will prevent settling of debris or sediment on the grids/on the specimen. But the disadvantage of this method is that while picking up the grids from the drops, there are chances of the grid or the support film getting damaged. On the other hand, if the grid is inserted into the drop with the support film up, i.e., with the specimen in contact with the stain, the grid can be picked up easily.
- The staining period is kept very short (a few seconds to a few minutes).
- The specimen need not be washed after staining.
- The fragmented, dispersed or suspended particulate specimen, for example, protein solution, is dropped on the coated grid (grid allowed to float or inserted into the drop) and then allowed to stand for a brief period, say, a few minutes, before blotting the excess fluid, during which time, the particles get attached to the grid film. Once attached however, the particles will not come off unless they are dissolved or digested by subsequent treatment. Thus successive reactions can be performed upon the specimen. The specimen can be kept wet throughout all changes if this is desirable. The stain is added finally, and then the specimen is dried before viewing.
- While using isolated protein particles for negative staining, it is preferable to wet the hydrophobic carbon support film with a wetting agent like traces of albumin (0.01% enough) for easy penetration of the stain. But these proteins like albumin, used as wetting agents are not usually visualized by the commonly used negative stains. Smaller organic molecules like sucrose, etc., also remain invisible. Thus quite complex suspending media can be used without their components interfering with the final imaging of the specimens. This allows a flexibility which is not possible in shadowed specimens. (*See* Chapter 13.)

- In the case of labile specimens, the suspensions may be fixed initially with the commonly used fixative, buffered glutaraldehyde, to:
  - preserve morphology
  - remove potency if infective organisms are used
- After fixation the specimen may be washed with buffer.
- For viewing negatively stained specimens, usually low accelerating voltages and small objective apertures are used.

NEGATIVE STAINING METHODS

There are three methods of negative staining. They are:

1. Drop method
2. Spray method
3. Floating method

Drop Method

In this method we can follow one of the following procedures.

Simultaneous application of specimen and stain This method is useful when the specimen neither reacts with the stain nor suffers from extensive pH variations. The steps are as follows:

- Mix specimen and stain at optimal concentrations.
- Apply 10 μl of the mixture on coated grids.
- Leave for a few minutes to allow for the specimen to get adsorbed on the substrate.
- Drain most of the liquid by blotting with wedge of a filter paper. A thin film has to remain on the substrate.
- Air-dry and examine in the microscope.

Successive application of specimen and stain

- Place a drop of specimen on a coated grid and allow to be adsorbed for a few minutes.
- Drain off liquid with filter paper wedge until a thin film remains on the substrate but avoid drying completely.
- Rinse with water or a volatile buffer.
- Remove excess fluid with filter paper.
- As an optional step apply a drop of fixative and rinse as above.

- Place a drop of stain on the grid.
- Drain off liquid until a thin film remains.
- Air-dry and scan in the electron microscope.

(Fixative is used particularly, when the specimen is labile or contains infective organisms)

Spray Method

- Mix specimen and stain normally using equal volumes.
- Spray this mixture from a nebulizing device (atomizer spray) on to the coated grid in front of an absorbent screen.
- Place the grid horizontally and air-dry.

Floating Method

- Pipette 50–100 μl drops of sample, washing, fixing (optional) and staining solutions on to a parafilm or stain pad.
- Using sharp pointed scissors, cut a piece of about 2 × 3 mm from the carbon-coated mica. Leave the piece attached to the rest of the mica by a thin bridge.
- Holding one end with forceps, introduce the piece of mica into the specimen suspension at an angle of 30° to 45°. The carbon film will start floating but the other end of the carbon will be held in place on the mica with the forceps.
- Allow the specimen to adsorb on the carbon film.
- Remove the mica slowly. The carbon will follow and fall back to its original position. Drain off excess fluid from the forceps with filter paper.
- Using the same procedure, introduce the mica into the washing solution (water or volatile buffer).
- As an optional step, introduce the mica into fixative and wash again.
- Use forceps to break off the film when it is still attached and completely float off the carbon from the mica on to the staining solution.
- Using forceps, pick up a sticky grid and touch the sticky side to the surface of the carbon film.
- When the film is attached, pick up the grid and turn it around. Drain off the attached drop of stain with filter paper until a thin film evenly covers the carbon.

RECENT ADVANCES IN NEGATIVE STAINING

Negative Staining in the Presence of Trehalose

The anionic negative stains generally have a finer granularity and have less direct (charge-dependent) interaction with the biological material. Inclusion of the disaccharide, trehalose, as a biological protectant during negative staining has been shown to be beneficial. This di-glucose acts as a protectant and is uniquely beneficial for the preservation of biological materials during dehydration, cold-storage, freezing, high temperature, UV irradiation, etc. Trehalose in combination with ammonium molybdate is shown to give satisfactory results. The film of dried stain and trehalose is somewhat thicker than that of the stain alone, allowing the biological method to be subjected to a reduced flattening force. But higher concentration of the stain is required, because the presence of trehalose reduces the net mass density. Other negative staining salts such as phosphotungstate and silicotungstate can also be used in combination with trehalose. But samples stained with uranyl acetate and trehalose mixtures have been shown to be more sensitive to the electron beam. Glucose and sucrose more rapidly 'bubble' in the electron beam than trehalose and hence trehalose is preferred.

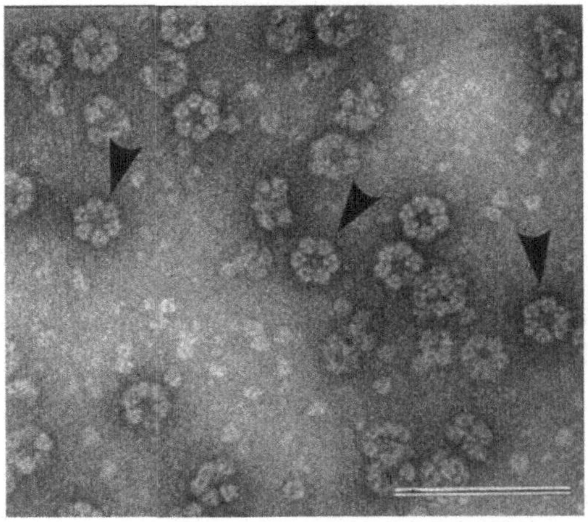

Figure 14.2 Negative staining of haemoglobin from the annelid *Nereis virens*, in the presence of trehalose. (Scale bar = 100 nm) *(Source: Microscopy and Analysis.* p. 31, Fig. 1.)

Figure 14.2 shows the haemoglobin negatively stained with 5% ammonium molybdate with 1% trehalose (pH 7.0) on a carbon support film, showing intact hexameric molecules (arrowheads), dissociating molecules and sub-components on the background.

Use of Holey Carbon Film

One of the most important disadvantages with negative staining is the "drying artifact". Since a supporting film is on one side of the specimen, the drying artifacts can be expected to be asymmetric. These drying artifacts can be avoided by using holey carbon film (a carbon film with multiple small holes in it). The holey carbon film also has the advantage of gain in contrast and resolution because of the elimination of a continuous support film. A grid coated with holey carbon film is commercially available and it is also used for objective lens astigmatism correction.

Samples are spread across glow-discharged holey carbon film on a grid and stained with negative stain and trehalose. The holes contain an even, thinly spread film of sample embedded in the negatives stain. When samples are spread across holes in the presence of trehalose alone, a thin film of dry trehalose is formed which is stable in the electron beam. This approach, which avoids the use of negative stain, has potential for biological specimens, polymer samples and nanoparticles only where inherent sample density is greater than that of the surrounding trehalose layer.

Cryonegative Staining

Computerized 3D image reconstruction was initially developed as a technique for the analysis of negatively stained biological macromolecules. But the use of negative stain has been superceded by cryoelectron microscopy—imaging of unstained hydrated specimens embedded in vitreous ice. Vitreous specimens are highly susceptible to radiation damage and require specialized electron microscopes to achieve high resolution.

The recent development of cryonegative staining by Adrian *et al.* alleviates some of these difficulties but at the same time retains the enhanced specimen preservation obtained through cryotechniques.

Figure 14.3 shows the comparison between conventional and cryonegative staining of a measles virus ribonucleoprotein. The superior image detail and specimen preservation in cryonegative staining can readily be noted.

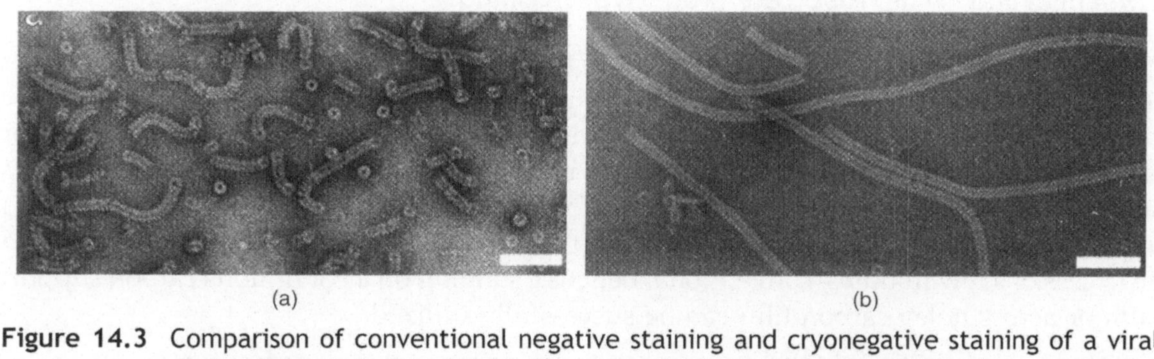

(a) (b)

Figure 14.3 Comparison of conventional negative staining and cryonegative staining of a viral ribonucleoprotein a) imaged in 2% w/v ammonium molybdate adsorbed to a continuous carbon support film b) imaged in ammonium molybdate cryonegative stain suspended across a hole in the carbon support. (Scale bar = 100 nm)*(Source: Microscopy and Analysis.* p.6, Fig. 3)

Cryonegative-staining requires a higher concentration of negative stain to attain adequate contrast as in the case of negative staining in the presence of trehalose. Typically a solution of 16–20% (w/v) ammonium molybdate at neutral pH is used. Approximately 50 μl of sample suspension at a concentration of 0.2 to 0.5 mg/ml is applied to a freshly glow-discharged or gold sputtered holey carbon support film. The sample is then washed for 5 to 60 seconds (depending on sample stability) in the stain solution. Finally the sample is blotted for 1–2 seconds to produce a thin aqueous film of sample in stain across the holey support film which is allowed to thin by air-drying for a further 1 to 2 seconds before the grid is plunged into a bath of liquid-nitrogen-cooled ethane slush. The vitrified sample is then transferred to the microscope and imaged under low electron dose conditions at liquid nitrogen temperature.

Many studies have demonstrated improvements in sample preservation using this method when compared to conventionally negative-stained material. The enhanced contrast and preservation of cryonegative stain brings considerable advantages to 3D reconstruction studies. Furthermore, cryonegative-stained samples have been shown to be less susceptible to radiation damage along with of course the added advantage of enhanced contrast. But this technique is limited by the grain size of the stain and is also restricted to analysis of the stain-excluded surface regions of the sample.

Negative Staining of Polymers and Colloids

Aqueous polymer solutions behave in an essentially similar manner to biological macromolecules and subcellular organelles. The potential and simplicity of air-dry negative staining for the study of polymers is gaining importance and it can be predicted that it will be increasingly used for polymer samples, alongside cryonegative staining and unstained cryoTEM.

Dynamic and Time-resolved Negative Staining

With negative staining, the drying of a thin layer of stain solution occurs over 1 to 2 minutes and this might be thought to impose a minimum time period for dynamic studies. But this is not so under conditions where the direct action of the negative stain or a fixative has the ability to trap biological material in a defined metabolic state. This approach has been successfully applicable for dynamic study of flexible molecules and myosin filaments. For any system where the time-dependent changes occur over a period in excess of a few minutes, conventional negative staining on a continuous carbon support film or across holey carbon film can be successfully utilized.

With cryonegative staining, sample pretreatment may be slow, but the possibility to treat a sample immediately before plunge freezing, such as by suddenly changing the pH, adding a metabolite or drug, or exposure to a temperature change, lighting conditions or gaseous environment could reduce the interaction time to the millisecond range.

There is great interest in many peptides that spontaneously form fibres in aqueous and physiological solutions, in particular the β-amyloid and 'tau' peptides involved in Alzhiemer's disease. Dynamic negative staining performed over a period of minutes, hours and days provides a system by which peptide oligomerization, protofibril and fibre formation and fibre aggregation can be assessed.

The time-dependent interaction of bacterial pore-forming toxins with biomembranes and artificial lipid systems can likewise be investigated using negative staining. It has been reported by Robin Harris *et al.* (2006) that over a period of a few minutes, the cytolysin from *Vibrio cholerae* formed oligomers attached to the bilayer edges of cholesterol microcrystals, but over a longer period of time (say, one hour) the planar surfaces of the microcrystals also become coated with oligomers as shown in Figure 14.4.

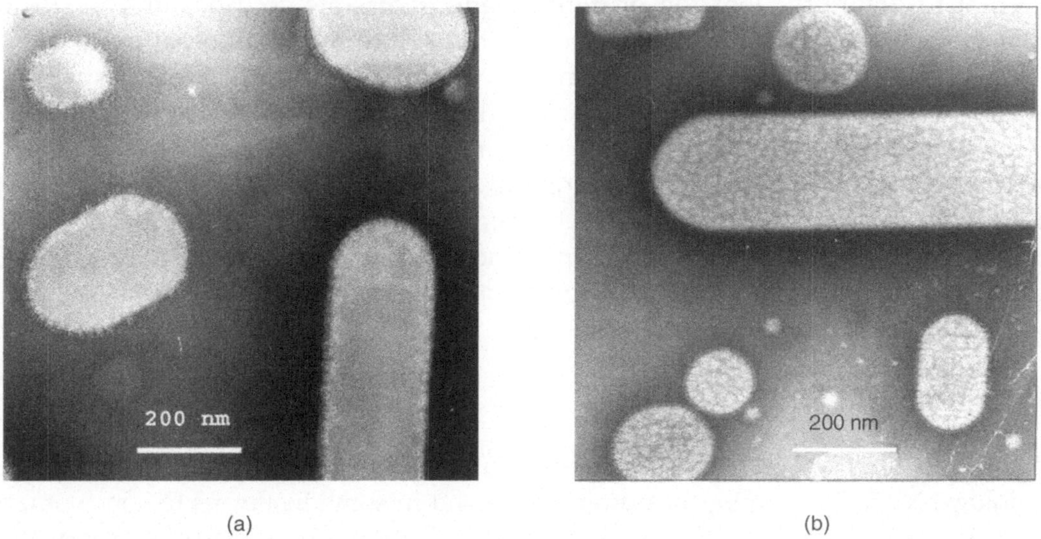

Figure 14.4 Cholesterol microcrystals with attached *Vibrio cholerae* cytolysin oligomers. a) After 15 minutes incubation b) After 1 hour incubation. *(Source: Microscopy and Analysis.* p. 32, Fig. 5.)

It can be seen in Figure 14.4 that after 15 minutes incubation, the pore-like oligomers are attached only at the edges of the planar cholesterol crystals. After 1 hour incubation, the whole of the surfaces are coated with oligomers.

2D Crystallization of Viruses and Proteins and Formation of Macromolecular Assemblies

Mica-spreading negative staining carbon film procedure which has been useful for preparation of 2D crystals of viral particles and protein molecules has been developed recently. Viruses and protein molecules have a tendency to produce 2D crystals. Induction of 2D crystal formation by viruses and protein molecules in the presence of ammonium molybdate and polyethylene glycol (PEG) is the underlying formative principle of the mica-spreading negative staining carbon film technique. Both negative staining and cryonegative staining can be used for crystal formation in the presence of PEG.

Negative Staining and Immunolabelling/Immunonegative Staining

In comparison with pre- and post-embedding immunogold labelling of thin-sectioned biological material, the combination of negative staining with immunolabelling has not received much attention. With the increasing availability of peptide sequence-specific polyclonal and monoclonal IgGs, Fab' fragments, and single chain variable (scFv) cloned antibody fragments, it is possible to perform immunolabelling at the molecular level, in an attempt to define the location of defined and accessible epitopes on the surface of macromolecules. Labelling with small gold probes (2 to 5 nm colloidal gold, nanogold and undecagold) can readily be performed in combination with negative staining and cryoelectron microscopy of unstained vitrified specimens. The density of the negative stain needs to be minimized, to avoid masking the small gold probes. Thus sodium vanadate or a low concentration of ammonium molybdate should be employed. Protein–protein interaction often can be studied by negative staining, significantly when one of the proteins possesses a fibrous nature and binds a soluble protein on to its surface. This technique has thus opened up the possibility of probing structurally the interaction sites of the two proteins by TEM.

Nanostructures

Alongside AFM, TEM has much to contribute in the fields of nanotechnology and nanobiology. Negative staining has often been used in some instances to complement the study of unstained samples. Systematic biological polymers and lipids (nanoparticles, nanovesicles, nanotubules and nanofibres) can all be studied by negative staining as long as aqueous or organic solvent suspensions or solutions are available at an appropriate concentration (i.e., overloading of an EM grid is worse than underloading).

The pharmaceutical industry, with its wide-ranging emphasis on drug delivery systems and biodegradable nanoparticles could have much to gain from the use of TEM during product development and quality assessment.

Figure 14.5 shows the PHA (polyhydroxy alkanoate) nanoparticles stained with 2% aqueous phosphotungstic acid. The isolated particle suspension is mixed with equal volume of the stain. A drop of this mixture is placed on the stain pad. The carbon-coated grid with the coating side on top was just dipped inside the drop and immediately lifted up. The excess stain was removed by blotting the edge of the grid with a filter paper wedge. The grid is then air- dried and examined in the TEM. Figure 14.6 shows the same as above but with the grid inserted into the drop for two minutes, removed and air-dried. The longer duration has resulted in better penetration of the stain.

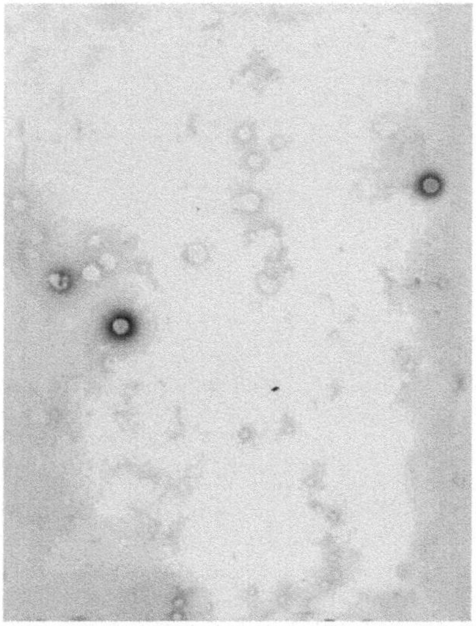

Figure 14.5 PHA nanoparticles negatively stained with PTA for a few seconds (grid just dipped in stain and removed) (15000×)

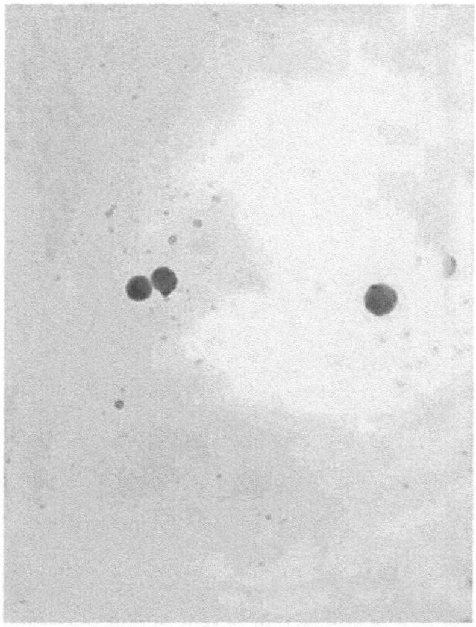

Figure 14.6 PHA particles negatively stained for 2 minutes (15000×)

15

MICROWAVE TECHNIQUES IN ELECTRON MICROSCOPY

WHAT ARE MICROWAVES?

Microwaves (MW) are a form of non-ionizing radiation that interact with dipolar molecules like water and polar side chains of proteins, and cause a 180° back and forth rotation of the molecules at the rate of over 2 billion cycles/second. As molecular agitation increases, the diffusion rate of all molecules in solution, including fixatives increases. In addition, organic reactions are enhanced by microwave irradiation. Heat is also generated during the process.

USE OF MICROWAVES IN EM PROCESSING

The use of MW radiation in biological tissue processing was first described as a heating source to reduce tissue fixation times from overnight to a matter of seconds. However fixation results were at times inconsistent and considered inadequate due to limitations of the MW technology being used.

The influential factors which limit acceptance and routine use of MW technology for biological tissue processing are:

- maximum sample temperature and heating time
- lack of energy uniformity within the MW cavity described as "hot and cold spots". Energy nonuniformity has been a significant cause of inconsistent results.
- lack of variable continuous power output of the magnetron, the source of microwaves.

All these led to a general lack of standardization in MW processing. But even without standardization, it was proved that MW methods could greatly reduce turnaround times for a wide range of fixation and tissue processing applications. In the earlier period there was no technology for the following.

- to mediate hot and cold spots
- to change continuous magnetron power output
- to maintain a constant sample temperature during MW use, i.e., during continuous irradiation

During 2004–2006, the following variables were identified as important to MW experimental design.

1. *Variable wattage* The technology to change the continuous power output of the magnetron and examine the role of energy in relation to processing results.
2. *Sample heating rate* The technology to control the heating rate of a sample during MW exposure without cycling off the magnetron.
3. *Sample temperature control* The technology to precisely control temperature without interrupting the MW radiation.
4. *Radiation uniformity* The technology to mitigate the uneven energy distribution in the MW cavity.

EVOLUTION OF MW TECHNOLOGY

Now, better control of the MW processing environment is achievable and this was first brought about by the following.

i. Development of variable/continuous power output (wattage) combined with a device to mitigate the influence of "hot and cold spots" during processing.
ii. Independent sample temperature control which allowed for continuous sample irradiation with precise temperature control.

With technology to control the above variables, a variety of protocols have been developed for tissue fixation, processing, staining and immunolabelling (immunoelectron microscopy).

Thus, with MW techniques, the turnaround times for EM and histology tissue processing are decreased to less than three hours, and single-antibody immunolabelling turnaround times are reduced to less than one hour.

New microwave tissue processor units are available now adding a new level of precise and extensive control needed for consistent quality tissue processing results. Recently, Leica Microsystems has produced the world's first automated microwave tissue processor for EM samples. These enable rapid specimen turnaround with consistent results for optimum imaging quality. Specific application kits have been configured to optimize EM tissue processing. The important features of the latest oven models are as follows:

- *Vented microwave cavity* Fumes from fixatives, solvents, stains and resins are effectively removed from the oven cavity. The exhaust fan remains 'on' for 30 seconds after the oven has completed its cycle, thereby reducing the possibility of any hazardous fumes remaining in the oven cavity when opening the door.
- *Automatic magnetron pre-warming* This can be a critical factor when time intervals are short say, less than 20 seconds. Normally MW ovens require manual magnetron prewarming. This is achieved by turning the oven on for 2 minutes at 100% power, with waterload in the cavity to pre-warm the electronics. If this is not done, the time countdown will begin but the magnetron will not produce full power. The lag time to full power can be up to 4 or 5 seconds. With automatic prewarming, the time does not begin until the magnetron is fully pre-warmed.
- *Programmability* This eliminates a number of steps during processing and greatly simplifies operation.
- *Temperature probe* The specially designed temperature probe which can be set in drops as small as 35 μl. The probe can be easily set to maintain any desired sample temperature, above ambient temperature, inside the microwave oven.
- *Load cooler* This device circulates the water load and will maintain its temperature at or below 43°C. This is important for run-to-run reproducibility. As water is heated, it absorbs less microwave energy. The load cooler also has a built-in air bubbler for sample mixing during a microwave procedure. This feature helps to mix fluids more evenly so that hot spots do not become a problem.
- *Computer attachment* The microwave ovens can be attached to a computer (PC) which allows the user to compare temperature data for quality control and set experimental parameters in a timely and efficient manner.

Whenever there is a microwave effect, there is a heating effect. It had been difficult to create an experimental design that could evaluate the radiation independent of a heating component. Giberson and Demaree (1995) reported that the fixation of tissues (both primary and secondary) using MW irradiation yielded preservation equal to or better than routine immersion fixation when a cold spot (as the sample location) and an ice-encased vial (most common microwave container) were used during MW fixation. Generally the temperature rise during fixation is the most common determinant for choosing the fixation time. Unlike water, ice is transparent to MW irradiation and easy to work with and requires no special equipment or modifications to the microwave oven.

IMPORTANT PARAMETERS FOR MICROWAVE (MW) FIXATION

- Exposure not more than 30 to 60 seconds.
- Temperature to be kept below 50–55°C.
- Specimen and container size should be maintained small.
- Proper location to place the specimen inside should be determined.
- A container of water to absorb excess microwave energy is to be included.
- After fixation, the container is to be shaken well for uniform temperature to be attained.

Although, the use of MW ovens for the fixation of tissue dates back to the early 1970s, the ability to get satisfactory consistent results has been difficult to achieve at the EM level. Major research efforts for the use of MWs for tissue preservation for EM was done by Login and Dvorak (1994).

Use of industrial or research-grade ovens are recommended over the conventional ones.

The advantages of the industrial or research-grade ovens are:

- Higher power with less specimen heating.
- High-volume ventilation ports to exhaust toxic fumes.
- Better control of temperature and power levels.
- Approximation of energy focal points where specimen should be located.

The oven is calibrated to accurately locate the energy focal points and also to determine the power settings and times.

APPLICATIONS OF MW TECHNIQUES

1. Reducing the duration of chemical processing of tissue for both TEM and LM.
   i. For various steps in tissue processing including
      - Fixation,
      - Dehydration,
      - Infiltration,
      - Polymerization and
      - Staining.

   ii. Particularly for sample preparation of plant tissues
2. Accelerating routine post-embedding immunoelectron microscopy.

The routine tissue processing protocol is given in Table 15.1.

Table 15.1 Tissue processing protocol

Treatment	Processing time			Temperature restriction set point
	Processing time in MW	Processing time outside MW	Total	
1. Fixation (Aldehyde)				
Microcentrifuge tubes	30 sec. to	4.5 min. to	5 min.	37°C
Cool fixative below 20°C	3 min.	2 min.		
2. Buffer rinse				
Microcentrifuge tubes	–	2 × 2.5 min.		
3. Fixation (Osmium)	30 sec. to	4.5 min. to	5 min.	37°C
	3 min.	2 min.		
4. Buffer rinse	–	30 sec.		
5. Acetone dehydration				
Change to processing baskets				
50%	2 × 40 sec.			45°C
70%	2 × 40 sec.			45°C
90%	2 × 40 sec.			45°C
100%	3 × 40 sec.			45°C
6. Infiltration				
1 : 1	15 min.			50°C
100% resin	3 × 10 min.			50°C
7. Polymerization				
95° C	15 min.			95°C
RT	60 min.			RT

Source: "Microwave Protocols" prepared for the workshop at the Tata Memorial Hospital, Mumbai, in 1996.

The temperature restriction checkpoints for fixation (both primary and secondary) are 37°C, for acetone dehydration, 45°C and for infiltration, 50°C. The temperature restriction checkpoint for polymerization is 95°C in the first phase and no restriction for the second phase.

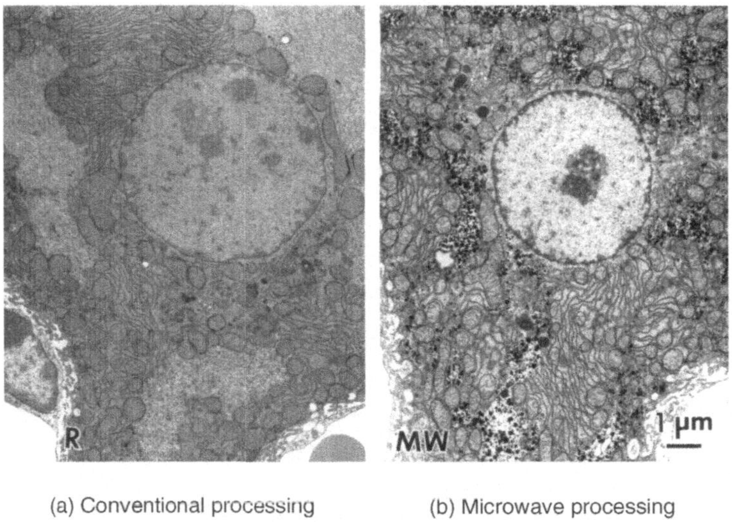

Figure 15.1 Comparison of microwave processing with conventional processing. (a) Conventional processing (b) Microwave processing (*Source:* "Microwave Protocols" prepared for the workshop at the Pathology Dept. of the Tata Memorial Hospital, Mumbai, by PELCO International, USA. p.23 (bottom figure).)

Figure 15.1a shows the mouse liver processed by using conventional method and compares it with microwave processing of the same specimen shown in Figure 15.1b.

Sample Preparation of Plant Tissues

The thick cellulose walls and high water content of plant tissues present a particular challenge for the design of methods for specimen preparation for electron microscopy. Typically, fixation, buffer washing, dehydration, resin infiltration and polymerization of resin-infiltrated leaf samples can take several days before ultrathin sections can be cut for TEM imaging. Zechmann *et al.* present a method for the rapid preparation of plant tissues for TEM that is assisted by MWs that reduced the preparation time from at least 3 days to about 5 hours. They compared the ultrastructure of different leaf materials which were prepared with a conventional well-established chemical fixation and embedding protocol and a commercially available MW tissue processor. Despite the massive reduction in sample preparation time, no negative effects on cutting properties of the blocks, stability of the sections in the electron beam, or contrast and ultrastructure of the cells were observed under TEM when samples were prepared with MW-assisted protocol.

Microwave-assisted Post-embedding Immunoelectron Microscopy

Microwaves have been used successfully for post-embedding techniques in immunoelectron microscopy. This is performed using the following steps.

1 to 12 grids can be simultaneously labelled in this procedure.

1. *Blocking step* The blocking solution contains Tris, sodium chloride and Tween 20. (Addition of BSA and /or normal goat/rabbit serum may cause background problems). At 37°C setting, this step is carried out for 2.5 minutes in the microwave oven.
2. *Rinsing step* After blocking, the samples are rinsed with a solution of Tris-buffer-containing NaCl. This step is carried out outside the microwave oven.
3. *Primary antibody treatment step* The dilution of the primary antibody can change when this step is done in the microwave oven compared to a normal incubation. This step is carried out for 2.5 minutes at 37°C in the oven.
4. *Rinsing step* This step is again done outside the oven using buffer solution, and the grids are blotted on filter paper.
5. *Gold conjugated secondary antibody treatment step* Incubation at 37°C for 2.5 minutes in the oven for this step.
6. *Rinsing step* This is again done outside the oven as before. Tris-buffer-containing NaCl is used for rinsing which is followed by distilled water.
7. The grids are finally blotted dry on filter paper and viewed in the electron microscope.

16

CRYOTECHNIQUES IN ELECTRON MICROSCOPY

To reduce the extraction of labile components by the fixative, dehydrating agent and embedding medium, specimens are prepared for electron microscopy at low temperatures. The simplest method is to carry out the standard schedule of processing at low temperatures in an ordinary deep-freeze cabinet. The other low-temperature methods, freeze-drying, freeze-substitution and cryoultramicrotomy involve an initial rapid freezing of the specimen and often this is the only means of fixation.

The tissue is frozen with extreme rapidity to minimize ice-crystal formation. Then the frozen tissue is dehydrated either by evaporative drying or by substitution. Freezing tissue fast enough to prevent or at least minimize ice-crystal formation demands that the size of the specimen should not exceed 0.2 mm in one of the dimensions.

Fernandez-Moran was one of the first to propose a low-temperature schedule for electron microscopy in 1961. Pieces of tissue were fixed at 0°C, dehydrated in a graded series of acetone or methanol with the temperature progressively decreasing from 0°C to –25°C, and then infiltrated with methacrylate at –25°C over a period of 3 to 5 days. The methacrylate was polymerized by ultraviolet (UV) irradiation at –10°C or at –55°C, the specimen temperature being below –10°C or –40°C respectively. Cope in 1968 developed a method of low-temperature dehydration and embedding in water-miscible methacrylates. The tissues were treated with a cryoprotective agent (dimethyl sulphoxide or dimethyl sulphoxide mixed with glycerol or ethylene glycol) and cooled to –20°C before fixation.

Various cryotechniques are used at various stages in the field of electron microscopy, starting from fixation, sectioning, scanning, including special techniques like immunotechniques, etc. Cryotechniques invariably involve rapid freezing, i.e., cryofixation.

To obtain information about the fine structure of organisms and their various tissues as close to their living state as possible, freezing and cryoprotection are important methods to be adopted.

Cryofixation means solidification of the aqueous specimen by rapid cooling with minimal displacement of its components. This can be achieved by rapid contact of the specimen with a cold copper surface or by rapid plunging of the specimen into a cold liquid such as propane or ethane cooled by liquid nitrogen. It requires rapid cooling of specimen to a temperature less than –140°C to prevent detrimental ice-crystal growth. At temperatures less than this, the mobility of water molecules is too slow to form ice crystals. Growing ice crystals displace solutes and macromolecular components by phase separation. Hence, it is essential to prevent, or rather, minimize their formation.

Cryofixation is shown to have more advantages over routine chemical fixation in certain aspects. It involves rapid freezing of fresh or pre-treated tissue. This initial rapid cooling is also referred to as "quenching". Necessary precautions to prevent ice-crystal formation are essential. For this purpose, cryoprotectants are used with or without pre-fixation. Specimens such as yeast, certain bacteria, seeds and fungal spores which have low water content may be frozen without cryoprotection. For specimens with high water content, addition of cryoprotectants before freezing is necessary.

The most commonly used cryoprotectants are DMSO, glycerol, ethylene glycol, sucrose and dextran. Dimethyl sulphoxide (DMSO) penetrates cells much more rapidly than glycerine and thus is a superior substance to prevent freezing damage to living cells. The rate of cooling also affects ice-crystal formation. Very high cooling rates, such as 10°C per second, prevent ice-crystal formation. Hence, most of the freezing methods employ rapid freezing techniques.

MERITS AND DEMERITS OF CHEMICAL AND CRYOFIXATION

The relative merits and demerits of both chemical and cryofixation can be summarized as follows:

Chemical Fixation

Advantages

- Well-established procedures
- Ease of specimen handling, embedding, sectioning, storing, etc.
- No transfer problems
- Image familiarity, so relatively easier to interpret

Disadvantages

- Modification or disruption of fine structures
- Loss of cell components
- Loss or displacement of elements
- Irreversible destruction of normal functions
- Usually unsuitable for immunocytochemistry, microanalysis, amorphous studies, etc.

Cryofixation

Advantages

- Instantaneous process
- Aqueous environment of biological sample is not disturbed
- Ultrastructure and chemical composition are preserved
- Enzyme activities preserved
- Well-suited for immunological studies and microanalysis
- Reduction of autolysis, post-mortem diffusion and cellular distortion.

Disadvantages

- Specimen size small
- Specimen handling, fixation, sectioning, storing, transfer problems, particularly embedding
- Thermal sensitivity of specimen
- Low contrast
- Ice-crystal formation. This is the main problem in these techniques
- *Complexity* Freezing methods are more time-consuming and expensive.
    - *Freeze-drying* In this, in addition to liquid nitrogen, apparatus capable of maintaining a vacuum at low temperature is essential.
    - *Freeze-substitution* In this, vacuum is not required, but substitution for several days is necessary.
    - *Freeze-thawing* This is the simplest method. It requires the same length of time for quenching and fixation.
- *Hazards* Handling of liquid nitrogen and risk of explosion.

TECHNIQUES INVOLVING CRYOFIXATION

The following techniques involve cryofixation

- Freeze-drying
- Freeze-substitution
- Freeze-thawing
- Freeze-fracture
- Cryosectioning
- Cryo-negative staining (*See* Chapter 14)
- Cryo-SEM
- Immunoelectron microscopy (*See* Chapter 17)

For freeze-fracture and cryo-scanning electron microscopy, mild chemical fixation usually with glutaraldehyde is used. However, for histochemistry and immunocytochemistry, it is necessary to choose the fixatives and conditions to obtain optimum fixation. A cryoprotectant, glycerol, which is most commonly used is also added. Then, the specimen of small size is mounted on a support and freezing is achieved by rapid immersion of the specimen and its mount into the cryogen.

CRYOGENS

The most commonly used cryogens are Freon and liquid nitrogen.

1. *Freon (Dichlorodifluoromethane)* It is used at its melting point (–158°C). It is liquefied by projecting a jet of the gas at the cold surface of a small condensing vessel held at LN_2 temperatures (–196°C).

   Freon 22 is preferred because:

   - it is safer to use
   - can be easily stored in a deep freeze
   - it can be reused after filtration at low temperature through filter paper

2. *Liquid nitrogen* The specimen is immersed directly into LN_2. For unfixed, fresh tissues, the following freezing methods are also used.

   i. *Metal contact rapid freezing* The tissue piece, mounted on a support, is guided to fall rapidly into contact with the metal (copper block) cooled with liquid nitrogen to –196°C or liquid helium to –296°C.

   ii. *Pressure freezing* The specimen is frozen under high hydrostatic pressure, thereby depressing the freezing point and lowering the rate of ice-crystal nucleation and growth.

iii. *Spray freezing* This method has been adopted for unicellular organisms. The specimen is sprayed in short bursts of about 1 sec. each, using an air brush (a very fine spray gun) into a small quantity of liquid propane held at liquid nitrogen temperature.

FREEZE-DRYING

In freeze-drying after initial rapid cooling, the tissue is rapidly transferred to a drying chamber which is maintained at a temperature below –30°C and at a high vacuum. Following drying, the tissue is embedded. It has been reported that the best method for preserving phospholipids and triglycerides is through cryofixation followed by freeze-drying.

Sjostand and Baker were the first to successfully adapt the freeze-drying technique to electron microscopy in 1958. This technique requires a costly apparatus, a part of which is usually custom-fabricated. The method depends on molecular distillation to achieve drying. The specimen is ordinarily kept at approximately –72°C (the temperature of a dry ice–alcohol bath) during dehydration. A cold trap chilled to about –196°C with liquid nitrogen (or liquid air) is placed as close to the specimen as possible. In the presence of a good vacuum, water then evaporates from the specimen area and condenses on the cold trap.

Tissue that is inadequately frozen and dried appears vacuolated. This damage may be due to "ice artifact", indicating probably that those cavities (vacuoles) earlier contained ice crystals. Similarly when the tissue is warmed to above –40°C while ice is still present, slow recrystallization occurs, which may produce damagingly large crystals even though the temperature is maintained much below 0°C. It has been found that tissue dried at –70°C had far fewer vacuoles than those dried at –40°C. Therefore, the temperature of the specimen should not exceed –70°C till drying is completed. The tissue can be fixed after drying and before infiltrating it with the embedment. Unstained frozen-dried sections must have sufficient thickness for contrast because staining involves treatment with aqueous solutions.

In the freeze-drying procedure the ice is removed from the frozen specimen under vacuum. Considerable damage is caused when the freeze-dried tissue is subsequently infiltrated with the liquid embedding medium, and consequently this method of preparation has generally been superseded by freeze-substitution or cryoultramicrotomy.

FREEZE-SUBSTITUTION

In this method, the solidified fluid of the quenched tissue is slowly dissolved or substituted at low temperatures in a fluid solvent, which may or may not contain a fixative. This method differs from freeze-drying. In this method, tissue specimens can be dehydrated and infiltrated with an embedding medium at a low temperature without presence of a gas phase, by using suitable intermediates. Thus, it offers possible advantages of exposing tissue to fixation, embedding and staining agents during the dehydration or infiltration process. The principle of the method is that ice is soluble in many solvents at temperatures far below its melting point. If the amount of ice is minute, and the volume of cold anhydrous solvent is large, escaping water molecules do not appreciably dilute the latter. In effect, there is never a free water phase present to damage the tissue during the substitution. The tissue pieces (minute in size) are glycerinated and rapidly frozen to prevent ice-crystal formation. Then the frozen tissue is moved to a precooled solvent bath. Some of the solvents include, acetone, absolute methyl and ethyl alcohols, propylene oxide, dioxane, etc. "The temperature can be as low as possible. The specimen can be fixed to prevent leaching of tissue components by adding osmium tetroxide to the chilled solvent. The tissue may be stained by adding heavy metal salts to the substituting fluid, although, it is reported that no staining is observed with osmium tetroxide unless the fluid is allowed to warm up. Alternatively, the sections are stained in the usual way.

The damage caused during the infiltration of dry tissues with the embedding medium can be avoided by using the method of freeze-substitution. Rapid freezing of the tissue is followed by solution ('substitution') of the ice at temperatures below 0°C. The frozen tissue is kept in the substituting fluid for a long period so that the ice dissolves slowly without distorting the structure of the tissue. After removal of the ice the specimen is infiltrated with an embedding medium, often at low temperature. The embedding medium is polymerized in the cold by UV irradiation or by heat in the usual way.

During dehydration by the freeze-substitution method, the specimen is rapidly frozen without chemical fixation (or after only light fixation) and then the ice in the specimen is replaced (substituted) by an organic solvent at temperatures of –80 to –90°C, which are low-enough to prevent recrystallization of the frozen water. This procedure results in a significant decrease in the extraction of specimen components as compared with standard dehydration methods. Following freeze-substitution, the specimen is infiltrated with an epoxy resin for ultrastructural studies, or with an acrylic resin at low temperatures for immunocytochemical investigations.

Solvents for Freeze-Substitution

The most widely used solvents for freeze-substitution are methanol and acetone, while other fluids, such as diethyl ether and tetrahydrofuran, have been found to have some advantages for X-ray microanalysis, since they are less likely to cause translocation of ions, but the preservation of ultrastructure is then very poor. The choice of fluid depends critically upon the type of specimen.

Acetone, when used as a solvent, has the following characteristics.

- It is a poor solvent for water at –85°C and so it is essential to pre-dry the acetone.
- Large volumes of fluid and large substitution times must be used.
- It acts more slowly and so is more gentle in its removal of water.
- It extracts considerably less lipid than methanol during freeze-substitution.

But substitution in methanol is much faster than with acetone and it substitutes ice in the presence of 10% water. Therefore the choice depends on the specimen.

Fixatives for Freeze-Substitution

The process of freeze-substitution in itself does not, of course, produce any chemical fixation of the specimen and introduces no contrast into the final electron micrograph. Fixatives, such as glutaraldehyde, osmium tetroxide and uranyl acetate, are therefore incorporated at appropriate stages into the freeze-substitution procedure. Significant cross linking of proteins by glutaraldehyde in acetone occurs only at temperatures above –45°C and so it is effective only in the later stages of freeze-substitution. But osmium reacts with the double bonds of unsaturated fatty acids at temperatures as low as –70°C. The addition of uranyl acetate to methanol reduces the extraction of phospholipids and glycolipids at –70 to –90°C.

Various combinations of the three basic fixatives have been suggested, depending on the substitution fluid used and the purposes for which the ultrathin sections are needed. Thus for immunocytochemical studies, fixation is kept to a minimum, although structural preservation is then no more than adequate for the identification of organelles. For ultrastructural studies, more than one fixative is often used. The most popular fixatives for freeze-substitution in acetone are 1 to 5% OsO_4 or a mixture of OsO_4 and uranyl acetate. Solutions of OsO_4 in organic solvents are unstable at room temperature and should be freshly prepared and then cooled immediately before use.

Freeze-Substitution Procedure

The procedure of freeze-substitution is carried out at very low temperatures and so all the safety precautions must be taken. Specimens to be dehydrated by freeze-substitution are cryofixed and stored under liquid nitrogen. A stock solution of the substitution fluid is then prepared and is kept in a closed container away since water is rapidly taken up by cold solvents. Open containers such as specimen vials are half-filled with frozen substitution medium and placed in Dewar flask containing liquid nitrogen. The volume of the fluid must be at least 1000 times greater than that of the specimen. The frozen specimens are transferred under liquid nitrogen to the surface of the frozen medium and screw caps, containing holes, are placed on the vials. The vials are then transferred to a substitution chamber and the temperature is raised to between –80°C and –90°C, so that the medium melts and the samples sink into it. The temperature of the substitution fluid is increased in stages until it reaches the temperature to be used for infiltration with the embedding medium. The vials should be agitated gently throughout the procedure and care must be taken that the specimen does not dry out during changes of fluid. Alternatively, specimens and the substitution fluid are placed in closed containers which are suspended in the coolant in a deep Dewar vessel. Substitution at very low temperatures takes many hours with methanol and some days with acetone. Consequently it is preferable to use one of the commercial units supplied by Leica or Balzers for freeze-substitution.

Limitations

i. The appearance of the tissue is very different from that observed after conventional fixation and embedding. In particular, membranes appear in negative contrast.
ii. Even with the best specimens, it is found that only a small proportion of the cells is free of damage due to ice crystals.
iii. The main disadvantage of the freeze-substitution method is the long period required in the substituting fluid. Periods of days or even weeks are required.

FREEZE-THAWING

This also requires initial rapid cooling/quenching of the tissue like freeze-drying and freeze-substitution. But it differs from these two techniques in that immediately following quenching, the tissue is rapidly warmed by being immersed in a chemical fixative at room temperature. The fixed tissue is then embedded in the conventional manner. Thus the tissue is subjected to two rapid changes of temperature. But when the thawing fluid is buffered osmium tetroxide, this technique proves to be more satisfactory than the other two (freeze-drying and freeze-substitution). Artifacts due to ice-crystal formation are less frequent and that too of smaller size than the freeze-drying and freeze-substitution procedures.

FREEZE-FRACTURE

This technique is useful to study the macromolecular architecture of biological membranes. When biomembranes are frozen to low temperatures such as –130°C or below they tend to be thermodynamically unstable and split along the middle of the hydrophobic bilayer. When such membranes are fractured with a fine knife, two halves of two monolayers are produced and these are referred to as the E face (extracellular half) and the P face (protoplasmic half) of the membrane. The original surfaces of the corresponding monolayers are known as PS (protoplasmic surface) and ES (extracellular surface). Figure 16.1 shows the freeze-facture process with a knife splitting the plasma membrane through its hydrophobic core.

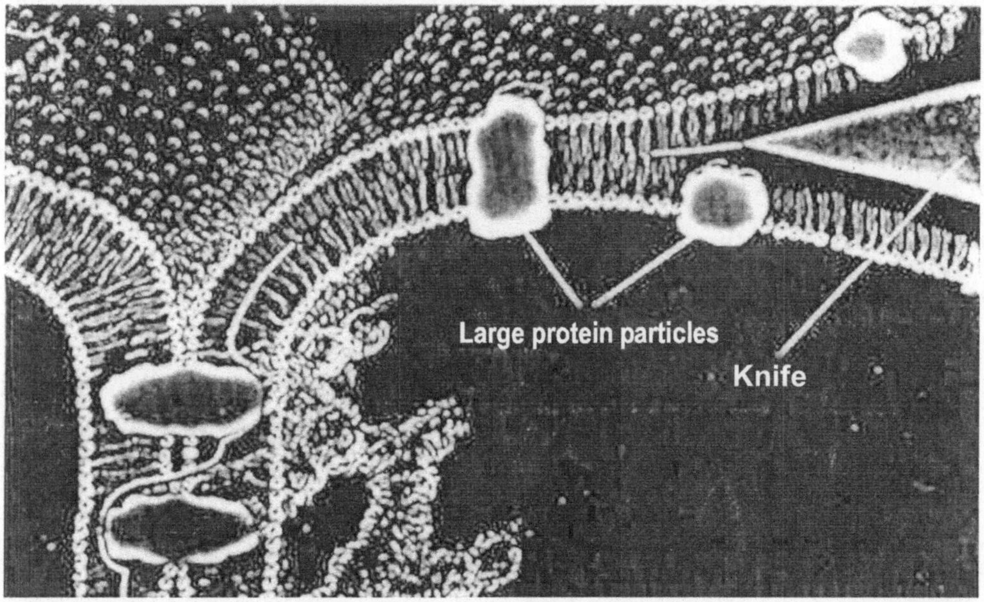

Figure 16.1 Freeze-fracture process. (*Source:* Jose Russo and Sheldon, C. Sommers (Eds.). (1988). *Tumour Diagnosis by Electron Microscopy.* Vol.2. Field and Wood Medical Publishers Inc., USA. p.8, Fig. 4.).

In freeze-fracture, the frozen and fractured specimen is immediately replicated with platinum and carbon to form an extremely thin metal replica, which is then removed and studied in the TEM. In freeze-etching or deep etching, after fracturing, the frozen solute is sublimated to expose macromolecules or membrane surfaces just below the fractured plane. For this, after fracturing, the specimen is elevated to higher temperatures (–100°C) and then replicated. Freeze-etching could be done in specimens which are not cryoprotected or cryoprotected in organic solvents like methanol.

Freeze-fracture method can be applied to study the earliest morphologic signs of cell injury and also to determine the composition of lipids in droplets and membranes. In the diagnostic side, it is useful in helping to establish cell types in poorly differentiated tumours and to distinguish a carcinoma (epithelial tumour) from a sarcoma (mesenchymal tumour) in histologically undifferentiated lesions. For diagnostic purposes, it can be performed on routine formaldehyde fixed sections also.

Johnson *et al.* (1988) reported that intramembranous particles observed in freeze-fracture replica correspond to integral membrane proteins or lipoprotein complex. These intramembranous particles are randomly distributed in the plane of the membrane, but many cell types have distinctive particle arrangements that indicate the cell of origin.

For example, the mitochondria-rich, carbonic anhydrase-positive, dark cells of the collecting duct of the kidney have been shown to contain rod-shaped particles in their plasma membrane. The plasma membrane of most cells has randomly dispersed intramembranous particles and pits.

In ultrathin sections, both tight junctions and gap junctions may simply appear as two closely apposed membranes. In freeze-fracture, however, they appear as strands of protein and lipid running parallel to the luminal surface (tight junctions) or as aggregates of 8 nm hexagonally packed particles in variegated arrays (gap junctions).

Thus freeze-fracture, they conclude, is a useful diagnostic tool.

This technique has certain limitations and artifacts as mentioned below:

- Deposition of materials like water, glycerol and freezing mixtures on freshly cleaned specimen prior to replication in the specimen chamber can contribute to contamination, causing artifacts. Artifacts can also be produced by antifreeze agents and by the freezing procedure itself. Insufficient cleaning of the replica also produces artifacts.
- Some organelles such as mitochondria tend to swell in antifreeze agents such as glycerol.
- Covalent bonds of proteins may break during fracturing depending upon the manner in which the proteins are anchored, and proteins may be subjected to plastic deformation at the time of fracture which leads to false results on the macromolecular distribution of membranes.
- The interpretation, as in microscopy, requires some expertise.

CRYOSECTIONING/CRYOULTRAMICROTOMY

Cryoultramicrotomy (CUM) was first reported by Fernandez-Moran in 1952. In 1964, Bernhard and Nancy formulated a procedure for it. As per their procedure, the specimens, after fixation and freezing in liquid nitrogen, were cut with a dry knife at –50°C. For sectioning, a conventional microtome was kept inside the cryostat chamber that had been maintained at –70°C. Temperatures below –70°C could not be achieved in these cryostats and ice crystal damage might occur when the temperature rises above –70°C. Instead, placing the ultramicrotome itself inside a cryostat, Christensen in 1972, designed an insulated chamber around the specimen and knife. The specimen area alone was cooled with liquid nitrogen and it was possible to maintain temperatures up to –170°C. CUM is most suited for immunoelectron microscopy. Moreover, for the elemental analysis of diffusible substances or ions by X-ray microanalysis, CUM is the best method available.

Cryoultramicrotomy allows the sectioning of vitrified biological samples. It produces ultrathin cryosections that are investigated in the cryoelectron microscope. High-pressure frozen biological samples are transferred to the cryoultramicrotome and cryosections are brought on an EM grid. These grids are mounted on a cryoholder for cryomicroscopy. All these are done at temperatures less than –140°C which is the recrystallization temperature of pure water.

The main difficulties in working at temperatures less than –140°C during the whole preparation are:

- cryosection transfer from knife edge to EM grid
- grid handling in the cryochamber
- grid transfer into the cryoholder of the EM
- ice-crystal contamination which can obscure structures of interest in the sections

The method applied to study cryosections of vitreous biological samples at a temperature less than –140°C is known as CEMOVIS (CryoElectron Microscopy Of Vitrified Sections) introduced by Al-Amoudi *et al.* (2004).

The difference between the cryosectioning technique and other specimen preparation methods in EM is the water status in the biological samples, which remains unchanged, reflecting the living structure of the cells at the time of freezing.

Advantages

- Relatively simple technique
- Is reproducible
- Yields high-quality sections with good preservation of enzymatic and immunological activities, because the cryosections remain fully hydrated during the immunolabelling process
- Cryosections from cryoprotected frozen specimens can be used for both LM and EM
- Sequential sections can be immunolabelled and used in correlative microscopy or as a substitute for confocal microscopy

Cryofixed tissue could be sectioned in specially designed cryo-attachments to microtomes. Cryoultramicrotomes offer ultrathin sections with excellent ultrastructural preservation. Though difficult, good sections can be obtained with this method.

The small tissue bits fixed in glutaraldehyde and washed are infused with sucrose (cryoprotectant). After trimming the block face to trapezoid shape, the bits are placed on a clean silver pin. Then the tissue pieces along with the carrier are frozen by direct immersion in liquid nitrogen. At –110°C, ultrathin sections are cut using dry diamond knife. Sections are picked up by using a sucrose drop at the end of a eyelash brush or with a platinum loop. The droplet is rapidly brought into contact with the section so that the sucrose does not freeze. The droplet with section is removed and brought to room temperature so that sections spread on the surface of the droplet. Then the sections are transferred on to the supporting film of the grid by touching it. The grid is then floated on the surface of the water droplet facing down to wash out the sucrose. Finally the sections are stained with uranyl acetate and screened in the EM.

In some methods, the tissue is fixed and then embedded in a water-soluble medium such as gelatin, before freezing and sectioning while in others the specimens are unfixed and unembedded. The latter, technically much more difficult, method is necessary in studies on diffusible substances and is of great value in cytochemistry.

Cryoelectron microscopy of frozen hydrated specimens is a technique for determining the "native" 3D ultrastructure of individual organelles and cells. Two techniques are available, namely, "stereo-pair imaging and electron tomography, the latter providing full 3D information about the specimen. A resolution of 4–5 nm can be currently obtained with cryotomography. Frozen hydrated specimens are very radiation-sensitive and have low contrast because they lack heavy metal stains. So careful choice of imaging conditions, use of a sensitive CCD camera system, and microscope automation are important requirements for conducting cryoelectron tomography.

Takizawa and Robinson (1993) combined immunocytochemistry and enzyme cytochemistry to localize two different proteins on the same ultrathin cryosections. In this method, the immunocytochemical localization is visualized with colloidal gold probes and the enzyme cytochemical detection is achieved with cerium as the capture agent. The immunocytochemistry is conducted first so that any potential adverse effects of the enzyme cytochemical procedure will not alter the antibody-binding properties of the cryosections.

Cryoprocedures for immunoelectron microscopy are discussed in Chapter 17.

CRYOSEM

This refers to a SEM provided with a temperature-controlled cryostage in it, with which the specimens can be examined at very low temperatures (less than –140°C). This cold stage is located in the EM specimen chamber and is indirectly cooled (thermally insulated) from an external source.

Modern cryosem equipment consist of a high-vacuum cryopreparation chamber directly attached to the SEM, while in some systems, the cryopreparation chamber stands separately from the SEM.

The cryopreparation chamber is an evacuated chamber which contains a cold stage cooled by liquid nitrogen, enabling the specimen to be kept at a controlled temperature. The specimen may be retained intact or fractured and kept either fully frozen hydrated, partially freeze-dried (etched) or fully freeze-dried. The samples may be coated for subsequent observations in the SEM. The chamber is actually connected to the SEM by a high-vacuum valve. The sample can be transferred to the SEM by opening the gate valve. When the preparation chamber is separated from SEM, specimen transfer has to be carried out externally.

CryoSEM has been applied to examine the frozen hydrated plant tissue. This technique enables to visualize the state of the cytoplasm and also water distribution in the plant tissues. The specimen is first frozen in liquid nitrogen. The frozen sample is examined in cryoSEM after freeze-fracturing and metal coating or cryosectioning. Since cryosectioning for botanical samples is difficult, freeze-fracturing is preferred. Cryofixation is fast enough to retain the original distributions of inorganic elements of tissues sufficiently for microanalytical studies. Thus elemental distribution in plant tissues can be studied using EDX in the cryoSEM. This approach has been expected to have broad applications for various types of localization of relevance to plant physiology, environmental pollution and plant–microbe interaction.

17

IMMUNOELECTRON MICROSCOPY

As the name implies immunoelectron microscopy is the integration of immunology and electron microscopy. In other words, it is an immunological technique at the ultrastructural level. With this technique, the antigen can be localized at the subcellular level. For example, a cytoplasmic antigen can be traced to the organelle to which it is attached; if it is a mitochondrial antigen, it is possible to identify, whether it is attached to the outer membrane or the inner membrane or present in the matrix. Thus in TEM, immunolocalized molecules can be accurately correlated to the large amount of visible subcellular components in the sections. But this technique is used mainly for research purposes.

Immunocytochemistry is a widely used technique for the identification and localization of antigens or tissue components *in situ.* Localization of a cell constituent can be achieved by using specific antibodies against it. The antibodies are tagged with electron-opaque particulate probes to enable easy identification of the reaction products, and thus confirm the presence as well as subcellular localization of the cellular antigens.

In immunoelectron microscopy, apart from the fine structural morphology, the antigenicity of the antigen has also to be preserved. In TEM, localization of cellular components requires a balanced fixation of the cells under study. While antigenic determinants must not be denatured during fixation, the ultrastructural morphology of the cell should also be well-preserved. Moreover markers (probes or tags), which render antibodies visible at the EM level are needed to identify antigens within or on the surface level. The kind of tissue processing required for studies in electron microscopy imposes certain limitations primarily because of the generally labile nature of antigens and steric hindrances imposed by the antibodies during the staining procedure.

Localization of the antigen can be done prior to sectioning or after sectioning. These are referred to as, pre-embedding immunoelectron microscopy, and post-embedding immunoelectron microscopy, respectively.

In the former, the antigen–antibody reaction precedes the specimen-processing steps of embedding, sectioning, etc. Immunostaining is carried out on a relatively thick unembedded tissue sections. Only after immunostaining, the tissue slices are embedded in resins, sectioned and screened. The advantage of this method is that the antigenicity of the component is preserved. But the penetration of antibody poses a big problem which can be overcome to some extent by mild fixation which will preserve the ultrastructure. The amount of antibody required may also be large making this method less cost-effective. This procedure is generally good for surface antigens. Pre-embedding procedure is useful in situations where antigen is present in low concentrations or where the antigen is particularly sensitive to embedding. It provides excellent localization of external antigens but can cause poor penetration of probes into the tissue. It also has limited applications because it permits the labelling of only one antigen in any one tissue block.

In the post-embedding method, on the other hand, the immunological reaction is carried out on ultrathin sections after processing. Here the problems of antibody penetration are avoided and the ultrastructure is also well-maintained, but antigenicity of the constituent is sometimes compromised upon. Hence, one has to be careful about the choice of fixatives, resins used and its curing conditions. This procedure is good for intracellular antigens. The post-embedding technique may be employed only for those antigens which are stable and can withstand the drastic processing steps. Unlike the pre-embedding method, the post-embedding technique can be applied for the detection of all antigens throughout a sample or for the staining of many different antigens within a tissue. Ultrathin sections are collected on grids, incubated with a series of immune reagents, washed, contrasted, dried and then viewed. The major drawback in this technique is that resin embedding renders many potentially reactive targets inaccessible to probes and antibodies. For example, cytoskeletal proteins will be largely hidden in the resin. Cryotechniques are more suitable for this. Many antigens are very sensitive to osmium particularly and so this is normally avoided in post-embedding procedures. However, osmium can be added after pre-embedding immunostaining. Therefore, the pre-embedding procedure is normally the method of choice in situations where osmium is necessary in order to obtain adequate membrane fixation and contrasting. For pre-embedding procedures, copper grids can be used. But for post-embedding, unsupported nickel or gold grids have to be used because copper may get oxidized during the immunological staining procedure.

Apart from these two major types, there are some special methods as mentioned below:

- Cryoimmunoelectron microscopy
- Scanning immunoelectron microscopy
- Negative immunoelectron microscopy

In the cryotechniques, immunocytochemical reactions are performed on cryoultrasections or freeze-fractured surfaces. These techniques allow direct labelling of the constituents without any compromise on antigenicity or penetration of antibody and thus, artifacts introduced are also minimized.

If the antigen to which antibody is directed is within the cell, post-embedding will give better results. If the target molecule is a surface antigen, negative or pre-embedding protocol may be followed. Thus if the structure and general localization and stability of the antigen is known then the appropriate protocol can be selected which will optimize its detection.

PRINCIPLES OF IMMUNOLOGICAL TECHNIQUES

The antigen–antibody reaction is traced with the help of an electron-dense probe attached to the antibody. There are two methods of performing this technique. Direct and indirect methods are illustrated in Figure 17.1.

In the direct method, the tissue antigen in the given species is treated against the tagged primary antibody (attached with an electron-opaque probe) from another species. The tag facilitates ready identification of the antigen. The disadvantage of this method is the necessity for raising an antibody for every target antigen.

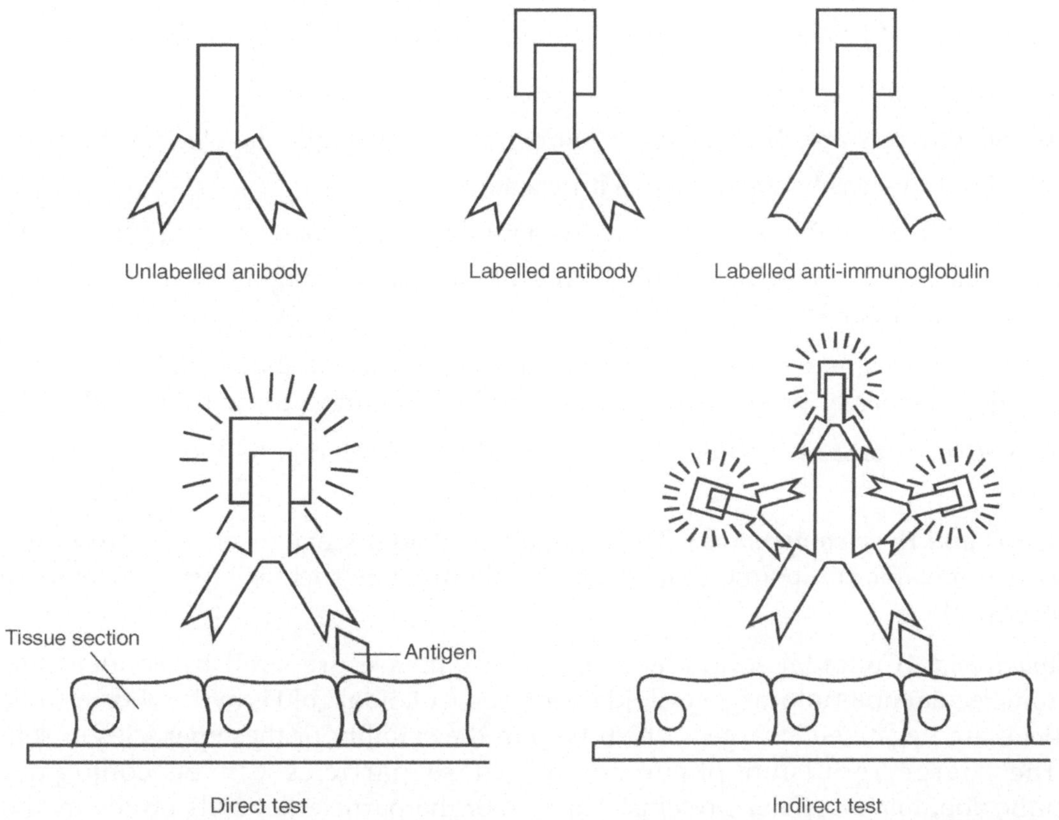

Figure 17.1 Direct and indirect methods of immunolabelling. (*Source:* Connie, R. Mahon, Donald, C. Lehman and George Manuselis. (2007). *Textbook of Diagnostic Microbiology,* 3rd edn. Elsevier, New Delhi, India. p.251, Fig. 10.20 (a) & (b).)

In the indirect method, on the other hand, the tissue antigen under study, from species 1 is exposed to the primary antibody produced in species 2 that has been made to bind the antigen. Then a tagged secondary antibody, raised in the third species is exposed to the antigen–antibody complex. Actually, the secondary antibody produced reacts against all immunoglobulins of the second species and also contains a tag. The result is a two-layered antibody sequence with an antibody tag, for recognition of the antigen. This indirect method is more sensitive than the direct method.

REQUIREMENTS FOR IMMUNOELECTRON MICROSCOPY

This technique has specific reagents and requirements like probes (tags), carrier proteins, etc.

Probes (Electron-opaque Tags) for Immunocytochemistry

For the fine structural localization of antigens, only certain probes are useful. The probes can be used either through carrier proteins or other specific chemical compounds.

Characteristics of probes The probes should bear the following characteristics:

- Should be electron-dense with a specific shape.
- Should be stable both at room temperature and under the electron beam.
- Should be easily conjugated with carriers.
- Should preferably be non-toxic to minimize cell damage during incubation.
- Same probe, if possible should be useful for various imaging modes (on size basis, e.g., gold particles).

All the probes may not have all the characteristics listed above, but, however, there are a large number of probes available for routine immunocytochemical studies.

Classification of probes Probes can be broadly classified into five distinct groups. They are:

1. ***Metals and their compounds*** These include colloidal metallic suspensions of gold or silver and metallic compounds especially metalloproteins such as ferritin, haemocyanin, ferrocene, etc.

Colloidal gold Colloidal solutions of metallic salts are prepared by condensation of micromolecular units into larger colloidal particles. Colloidal gold is hydrophobic. Colloidal particles are negatively charged which help in the stability of these particles in solution by the charge repulsion phenomenon. These particles can be conjugated to immunoglobulins or carrier proteins. The size of the particle depends on the method of preparation and varies from 2–3 nm to 150 nm in diameter.

Ferritin Ferritin, a protein of molecular weight 750,000, contains 2000 to 3000 atoms of iron per molecule which give enough electron density to be visualized under the electron microscope. Besides ferritin, other iron-containing compounds such as ferrocene and hemocyanin are also used. Compounds of uranium, mercury and osmium are also employed to make antibodies electron-dense.

Ferritin conjugation or immunoferritin techniques—the ferritin-containing core is electron-dense and also large enough, facilitating visualization at relatively low magnifications in EM. Being a particulate marker, it allows for quantitative analysis, provided the ratio of antibody to ferritin is known.

To obtain an effective, specific labelling of antigenic sites in cells and tissues, high-titre antisera or preferably purified antibodies are required for conjugation to ferritin. In the direct technique, highly purified antibodies should be used. In the indirect technique, the specimen is first incubated with unlabelled specific antibody or unfractionated antiserum, the excess washed off and then the sections are treated with ferritin-conjugated antibody to the γ-globulin of the animal species from which the unlabelled antibody has been prepared. Hybrid antibodies can also be used to circumvent the undesirable aspects of the chemical coupling of ferritin to antibody. The hybrid antibodies contain two different specificities united into one molecule. In theory, a hybrid antibody in which one specificity is directed to ferritin and the other to the antigen to be marked, is an excellent reagent for the purpose. In practice, hybrid antibodies with the dual-specificity antiferritin/anti-immunoglobulin of a given animal species have been used in an indirect, sandwich technique to locate cell-surface antigens. Hybrid antibodies can also be used in conjunction with any other marker which is antigenic, e.g., with enzymes such as HRP (horseradish peroxidase) and certain viruses.

2. *Enzymes* Enzyme-labelled antibodies are used in both light and electron microscopic techniques for localization of intracellular antigens. The suitability of a given enzyme as probe depends on a number of factors.

- enzymes should not act on sample under investigation
- the reaction product should be insoluble and electron-dense
- should precipitate directly at the site of reaction

Acid phosphatase, horseradish peroxidase, glucose oxidase, tyrosinase, alkaline phosphatase, etc. are covalently linked to mono-specific antibodies or to their fragments by bifunctional compounds. The antigenic binding site of the antibody should remain free on conjugation. The probe is made to react with the cells and the site of enzyme is visualized by formation of electron-dense products of the enzyme reaction.

Peroxidase Peroxidases are most widely used probes in immunocytochemical studies with H_2O_2 as the substrate. The chromogen for this enzyme, DAB (diaminobenzidine), gives an electron-dense reaction product. Sensitivity of the immunoperoxidase depends on conditions of fixation and tissue processing.

Some of the disadvantages of peroxidase are as follows:

- ✳ There is need to inhibit endogenous peroxidase activity which may be achieved by the use of phenylhydrazine.
- ✳ Purity and toxicity of the chromogen used (DAB) has to be considered.

Alkaline phosphatase Unlike peroxidases, in the case of alkaline phosphatases, the background staining is negligible due to the absence of endogenous enzymes. Lead salts are used as the chromogen.

Many viruses, enzymes, hormones and immunoglobulins have been localized in tissue sections by the enzyme method.

3. *Radioisotopes* Iodine (^{125}I) is the most commonly used radioisotope. It is introduced into the tyrosine moieties of antibodies using chloramine T or lactoperoxidase. Tritium (^{3}H) labelling of antibodies is done by the help of tritiated anhydride. Carbon (^{14}C)-labelled antibodies have also been used by some workers. Intracellular antigens form complexes with isotopically labelled antibodies. These can be traced by autoradiography both at the light and electron microscopic levels.

4. *Viruses* Viruses with well-defined shape could be conjugated to antibodies to be used as markers. Viruses have also been used as labels in immunocytology. It gives excellent results for the location of surface antigens especially by scanning electron microscope. When cells are exposed to the virus–antibody conjugate, the antibody along with the virus binds to the surface antigen. The viruses can easily be identified, because of their definite shapes, in the scanning electron microscope.

5. *Synthetic beads* Synthetic materials with definite shape and size can also be used as probes. Many synthetic beads, microspheres and such materials which can be conjugated to carrier proteins and immunoglobulins are used as probes. Polystyrene latex spheres, co-polymer microspheres and silica spheres are the common examples.

Carrier Proteins

Probes are attached either by chemical fixation or physical binding to certain proteins which recognize either the antigen or antigen–antibody complex. These proteins are called "carrier proteins". The probes are attached to carrier proteins at one particular site. The other site recognizes the specific antigen and thus probes can trace antigenic sites. Isolation, purification and characterization of carrier proteins are very important steps in the whole process of localization of antigens.

When probes are attached to specific antibodies, the immunocytochemical reactions are performed in a single-step antigen–antibody reaction. This "direct method" is not very popular because it is difficult to get probe-conjugated specific antibodies for every antigen individually. Probes are generally attached to anti-immunoglobulins or certain proteins which specifically recognize antibodies for indirect labelling of antigens. Immunocytochemical reaction is performed in many steps. The antigen–antibody complex of the first step is recognized by the probe-conjugated second antibody or probe-conjugated carrier protein.

Antibodies Antibodies raised against an antigen would recognize and bind to the antigen and hence antibodies are excellent carrier proteins. Antibodies raised against antibody (primary antibody) or a portion of antibody is called "secondary antibody" which generally serves as carrier for the probe and is used in indirect methods for localization of antigens.

Polyclonal antiserum comprises a range of antibodies directed against a collection of antigenic determinants. The reaction is less specific but more sensitive (because of amplified signal). Monoclonal antibodies (MAbs), on the other hand, comprise a single class of antibody directed against a single epitope. They are more specific but less sensitive. Aldehyde fixatives such as glutaraldehyde produce strong cross-linking of proteins which in turn may result in direct modification of, or conformational changes in the epitope(s). These effects may be pronounced in monoclonal antibodies where modifications of one antigen site, may affect the accessibility and binding of the antibody. Another important factor for consideration is the source of antibody. When indirect methods are used, it has to be ensured that the antibody from one species will interact with that from another. In the case of protein A- or protein G-complexed gold, only the complexes compatible with the specific antibodies should be used. Secondary antibodies are commercially available.

Types of antibodies

- Horseradish peroxidase–DAB conjugated antibody
- Antiferritin–ferritin conjugated antibodies
- Enzyme-labelled antibodies
- Immunogold antibodies(IGS)

Gold-labelled antibodies are the most commonly used ones and can be used for SEM also.

Protein A and Protein G Several proteins capable of binding immunoglobulins have been identified and isolated from different strains of bacteria.

- Protein A from staphylococci
- Protein G from streptococci
- Protein L from peptostreptococci

These proteins are located at the surface of bacteria. Among these, protein A is the most commonly used protein.

Protein A Protein A produced by most strains of *Staphylococcus aureus* is linked to their cell walls. Protein A consists of a single polypeptide chain with a molecular weight of 42,000 kDa. It has high affinity towards immunoglobulin type G by binding to the Fc fragment. It is functionally a bivalent molecule having two IgG binding sites. A low-affinity interaction of this protein and the Fab fragments of IgG and IgE has also been reported. It interacts with IgG of several mammalian species. It exhibits good stability towards changes in pH, temperature and also against several denaturing agents. Since protein A–IgG reaction does not involve antigen-binding site, it does not affect antigen–antibody reaction. Protein A can be conjugated to colloidal gold or enzymes.

Protein G Like protein A, protein G is an immunoglobulin-binding molecule with high affinity for IgG of several mammalian species through its binding to the Fc fragment. It also binds to the Fab fragment and is a bivalent molecule having two binding sites. It has higher affinity for IgG molecules from goats, mice, rats and cows and binds to a wider range of subclasses of IgGs than protein A. In addition it has good affinity for monoclonal antibodies.

Colloidal gold marker The colloidal gold cytochemical marker system was first introduced by Faulk and Taylor in 1971, when they adsorbed gold particles to a rabbit anti-salmonella antibody. Indirect labelling techniques were introduced by Romano *et al.* in 1974 who subsequently in 1977 reported the use of gold-labelled protein A for the detection of primary immunoglobulins. Colloidal gold has proven to be one of the best electron-dense markers in cytochemistry. It has more advantages than other markers such as ferritin and peroxidase.

- Because of its spherical particulate nature and high electron density, very accurate identification and delineation of the labelled structures is possible.
- Being one of the smallest markers (up to 1 nm), it allows for the best resolution in cytochemistry.
- Multiple labelling of various binding sites in the same section is possible because it can be easily prepared in different sizes from 1 to 100 nm.
- Quantification is also easy.
- Can be used for both TEM and SEM.
- Non-toxic and easy to use.
- Unlike other markers (fluorescent labels), gold particles do not fade with time and thus give permanent label.
- There is no interference with cellular enzymes.

Colloidal gold is a negatively charged hydrophobic sol, formed by electron-dense metallic particles. Its binding to macromolecules (IgG, Protein A, or G) occurs by noncovalent electrostatic van der Waals forces between the charged surface of the particles and the protein. The important point to note is that the adsorption and formation of the protein–gold complex do not affect the biological activity of the tagged macromolecule. Thus they are ideal for use as biological markers.

The size of the gold probe (range 1–100 nm) influences the resolution and intensity of labelling. Resolution (distance from gold marker to the site of antigen–antibody interface) is influenced by the selection of the immunolabelling procedure (direct, indirect or bridge) and/or on the size of the gold probe. The best resolution is obtained with the direct labelling procedure which incorporates a small gold probe-complexed antibody. Probe size also affects the intensity of labelling, as smaller probes penetrate the tissue better.

Colloidal gold immunostaining is applicable for pre-embedding, post-embedding and cryotechniques as well as silver enhancement.

Danscher in 1981 first reported the immunogold silver enhancement technique, IGSS, (Immuno gold silver staining). Silver ions are reduced to metallic silver by reducing agents catalysed by gold. This technique is particularly useful when only a few cells in a mixed cell suspension possess the antigen in question, such as bone marrow aspirates or lymph node cell suspensions. The different procedures using gold probes are represented in Figure 17.2.

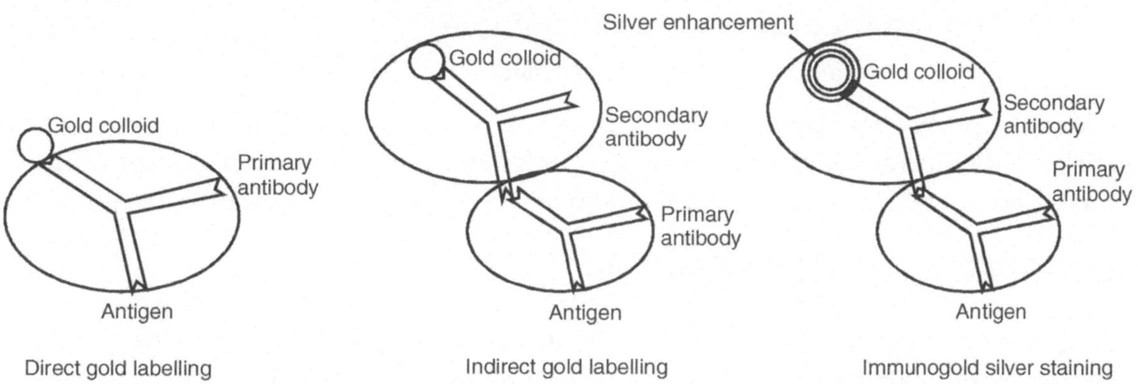

Figure 17.2 Staining methods using colloidal gold probes

Colloidal gold immunolabelling allows multiple labelling, co-localization and quantification. By relating colloidal gold particles to cell ultrastructure, it is possible to know

- the distribution of gold particles between compartments like organelles, membranes or filaments/tubules, within a given group of cells.
- whether the particles within a group are randomly distributed between compartments.

* whether the patterns of compartment labelling are the same in different groups of cells.

Thus the gold particle counts, combined with simple stereological and statistical tools, provides efficient ways of comparing the antigen under study in different compartments within a group of cells or in the same compartments in different groups of cells. Terry Mayhew (2005) reports a method of quantifying immunogold labels in ultrathin sections.

Lectins Lectins are proteins/glycoproteins of non-immune origin and derived mostly from plants but some of them are also obtained from animals. Lectins bind specifically to carbohydrate groups.

CONJUGATION

Probes and carrier proteins can be conjugated by using glutaraldehyde which can bridge them together. Successful conjugation depends on the molecular proportions of the carrier proteins and the probe.

Criteria for the preparation of conjugates:

* Neither the carrier proteins nor the probes become inactive.
* Very large aggregates are not formed.
* Homopolymers should not be formed.

Metallic probes are conjugated to the carrier proteins under appropriate conditions of pH, temperature, concentration of reactants, etc. since they do not form normally covalent bonds with proteins but get adsorbed or bound with weak electrostatic forces.

NON-SPECIFIC INTERACTIONS

During the immunological reaction, i.e., the reaction between the antigen and the added antibody, some non-specific interactions also may take place which might interfere with antigen localization, apart from increasing the background noise. The most common causes of these non-specific interactions and the ways to minimize them are given below:

1. If for fixation of the tissue, higher concentration of the aldehyde fixative is used, say 2–5% glutaraldehyde for 40–60 min., then free aldehyde group will give non-specific labelling. Therefore, the free aldehyde must be quenched by NH_4Cl (ammonium chloride).
2. To prevent non-specific binding of antibodies to "reactive group" within the specimen, the specimen is incubated either with buffer of high ionic strength or

with competing inert proteins, referred to as **blockers.** These blockers are dissolved in either phosphate buffered saline (PBS) or tris buffered saline (TBS).

The most routinely used blockers are:

- ✳ Ovalbumin (1%)
- ✳ Bovine serum albumin (BSA) (1%)
- ✳ Gelatin, including fish gelatin (1–5%)
- ✳ Defatted milk (skimmed milk) (2–4%)
- ✳ Acetylated BSA (0.01%)

3. Use of Triton X-100 or Tween-20 (0.05–1%) will reduce the non-specific interaction by reducing the charge attraction of antibody.

PROCESSING FOR IMMUNOELECTRON MICROSCOPY

In pre-embedding techniques, immunostaining is carried out on relatively thick unembedded tissue sections. Only after immunostaining, the tissue slices are embedded in resins, sectioned and screened. In contrast, post-embedding involves the immunostaining of ultrathin plastic embedded, grid-mounted sections. The penetration of antibodies into thick, fixed tissue slices is slight and this is the major disadvantage with pre-embedding staining. But the pre-embedding procedure is useful in situations where antigen is present in low concentrations or where the antigen is particularly sensitive to embedding.

An Outline of the Steps

Pre-embedding (where immunolabelling precedes tissue sectioning)

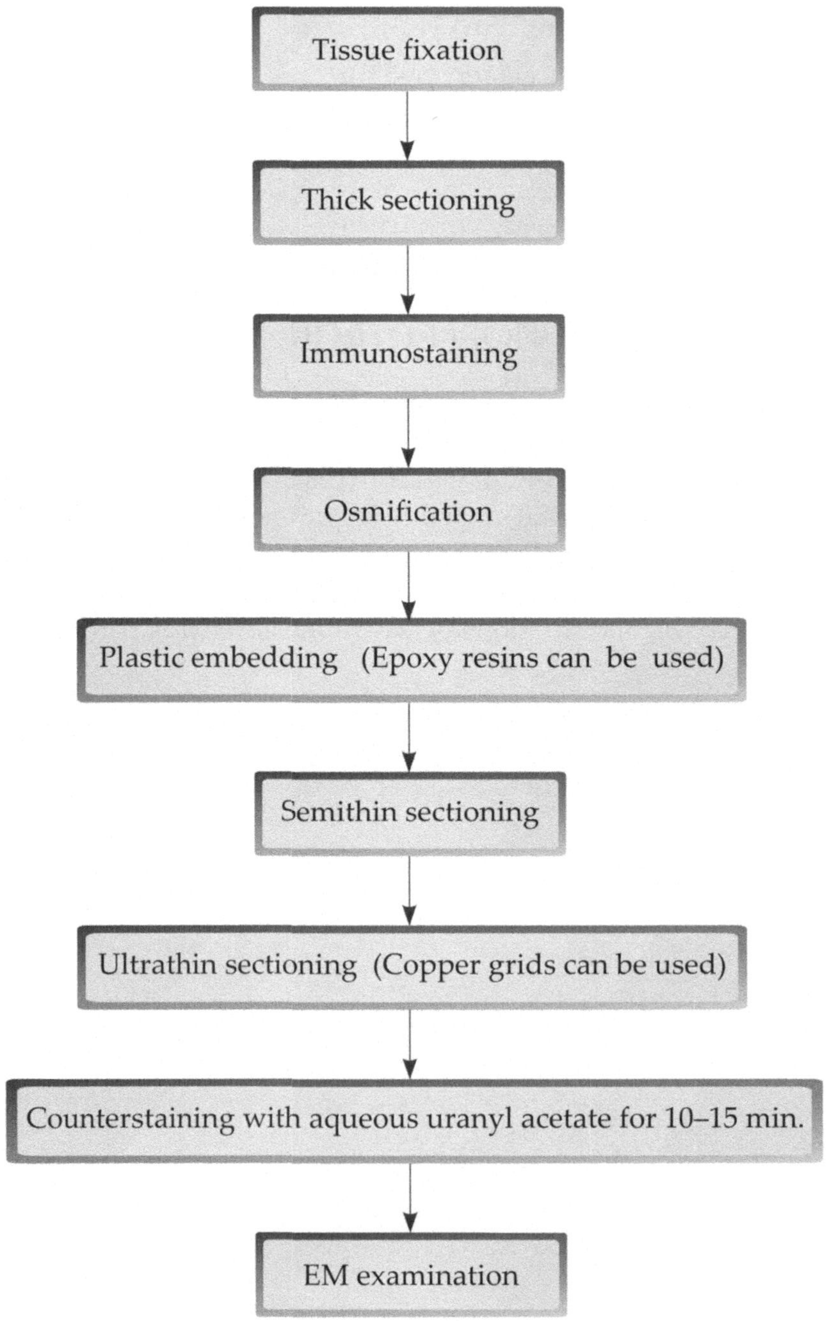

Post-embedding (done on ultrathin, grid-mounted sections)

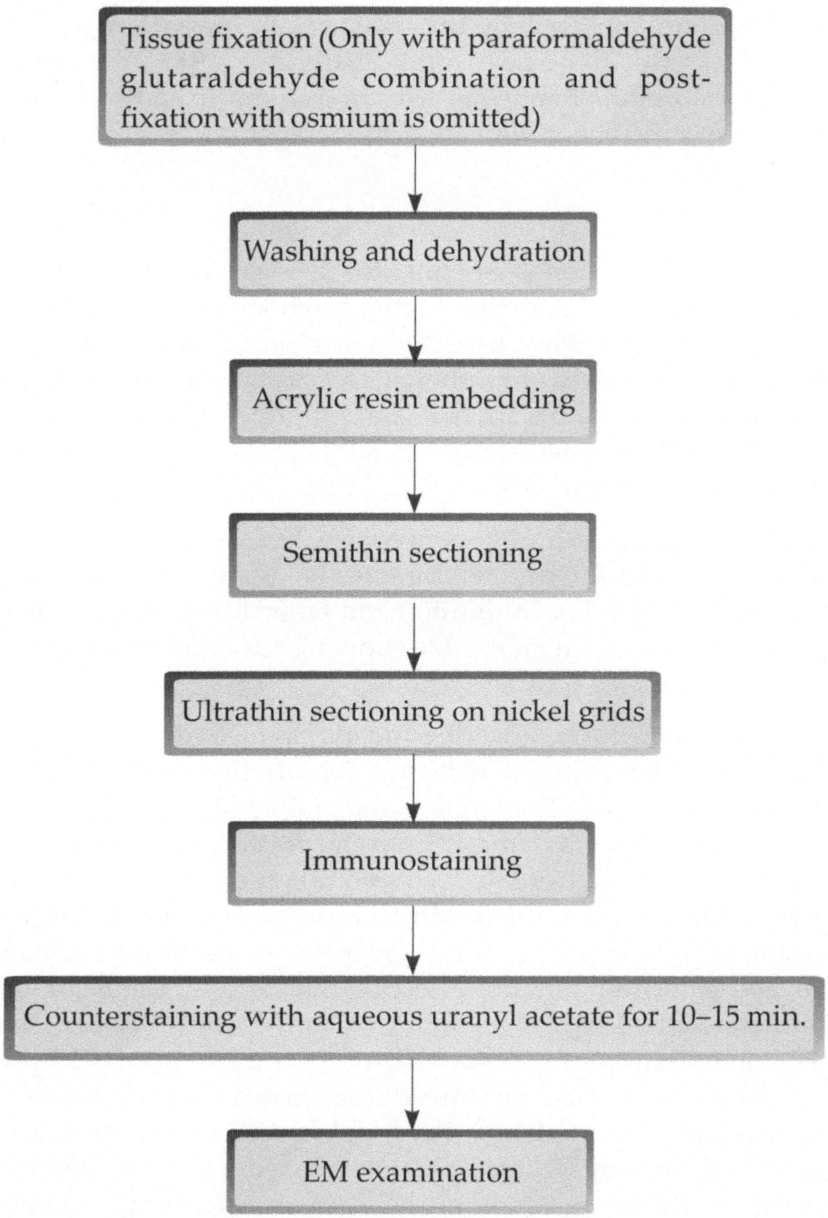

Post-embedding method for osmicated tissues Osmium normally almost completely inhibits immunolabelling for many proteins and hence osmium post-fixation is generally omitted in post-embedding immunolabelling techniques. But strong oxidizing agents have been found capable of unmasking protein antigenic sites on glutaraldehyde-fixed, post-osmicated tissues. Sodium metaperiodate has been shown to restore protein antigenicity. Thus, routinely processed tissues which are post-fixed with osmium can also be used for immunolabelling after ultrathin sectioning and treatment with metaperiodate. After immunolabelling, the sections can be stained with uranyl acetate followed by lead citrate. Bendayan and Zollinger (1983) described a method for ultrastructural localization of antigenic sites in osmium-fixed tissues using protein A-gold technique after pretreatment with sodium metaperiodate.

Fixation

As we have already seen, fixation is essential for preservation of the ultrastructure and also to render the antigen insoluble. But fixatives, while maintaining the ultrastructure, may have a deleterious effect on antigen reactivity. In glutaraldehyde-fixed tissue, generally, the antigen sensitivity is inversely proportional to the concentration of the glutaraldehyde and the time for which it is used. Fixation in concentrations higher than 1% for periods of time exceeding 15 min. have been shown to be deleterious to the antigenic response. The loss in antigenicity is irretrievable.

For pre-embedding techniques, a mixture of 2–4% paraformaldehyde and 0.5% glutaraldehyde is used for 30 to 60 minutes at room temperature. Dehydration is carried out with ethanol only. For infiltration and embedding epoxy resins can be used. Ultrathin sections can be collected on routine copper grids and staining can also be milder for a shorter duration than for routine sections.

For post-embedding methods, lowering the glutaraldehyde concentration (0.1–0.2%) and processing at room temperature with full dehydration of the tissue also does not improve the antigenic response. The tissue is thus minimally stabilized and more prone to extraction and conformational damage from the effects of complete dehydration and heat polymerization. Therefore using low concentration of fixative at room temperature followed by partial dehydration at room temperature or Progressively Lower Temperature dehydration technique (PLT) along with the use of acrylic resins for embedding produces good results in post-embedding immunoelectron microscopy. The fixation is carried out using 0.1–1% glutaraldehhyde with 4% paraformaldehyde for 1–8 hours. Osmium fixation is not used. For dehydration only ethanol is used and not acetone. Epoxy resins are hydrophobic and hence not used for embedding. Instead acrylic resins are used. The ultrathin sections are collected on nickel grids and never on copper grids for immunolabelling. McLean and Nakane (1974) developed a fixative containing periodate, lysine and paraformaldehyde for immunoelectron microscopy. Theoretically the carbohydrates are oxidized by periodate and cross-linked by lysine. The fixative can preserve antigenicity by formaldehyde as well as ultrastructure by glutaraldehyde. Using this fixative and the peroxidase-labelled antibody technique, basement membrane antigen was localized within the cisternae of the endoplasmic reticulum of parietal yolk sac cells and in extracellular basement membranes which was not possible with conventional fixation.

Eldred *et al.* (1983) have used the following procedure for fixation of cells of the central nervous system and retina for localization of various antigens. Cells are fixed for one hour in 4% paraformaldehyde and 0.2% glutaraldehyde (pH 7.4) followed by overnight fixation 4% paraformaldehyde (pH 10.4). Treatment with sodium borohydride after glutaraldehyde fixation restores much of the immunoreactivity that would otherwise be undetected. The penetration of immunological reagents can be greatly increased by using either a buffered ethanolic series or by freezing the tissue after adequate cryoprotection. This method may prove useful for localization of many different antigens in a variety of tissues.

Embedding

Epoxy resins are not preferred for immunoelectron microscopy because they form heavy cross links and also show very low water adsorption due to the presence of many hydrophobic groups. This resin, however, can be made more hydrophilic by treatment of the tissue with oxidizing agents like potassium permanganate, hydrogen peroxide or periodic acid. By oxidizing the hydrophobic alkane side chains to hydrophilic alcohols, aldehydes and acids, this treatment increases the overall hydrophilicity of the resin. But the antigenic moiety may also get oxidized and thus lose its property. The heavy cross links formed by epoxy resins are predominantly ester cross-linking which can be broken by etching the sections with sodium methoxide or ethoxide added to the resin. But these drastic treatments may affect the ultrastructural details of the specimen. Therefore, it is best to avoid epoxy resins and use only acrylic resins for immunotechniques.

Acrylic resins are more hydrophilic and less densely cross-linked than the epoxides and therefore there is no need for peroxide or etch treatment. The curing condition can be milder than epoxide, as curing can be carried out at low temperature using UV radiation or at room temperature using accelerators.

The most commonly used acrylic resins are:

- LR white
- LR gold
- Lowicryls

LR white resin

- This resin monomer has the viscosity similar to water enabling rapid penetration into the tissue.
- The monomer is partially water-miscible and can polymerize in the presence of up to 12% by volume of water, enabling unosmicated tissue to be embedded which is

incompletely dehydrated. Thus the extraction of tissue components by treatment with organic solvents can be reduced.

- LR white has good sectioning properties.
- Sections cut are hydrophilic and freely permeable to all aqueous solutions for staining.
- It is available in three grades—hard, medium and soft. The hard-grade resin is the one most preferred because this can be considerably under-polymerized enabling easier penetration of antibodies into sections and also retaining good sectioning properties.
- Three methods can be used for polymerization, namely, heat, chemical catalysis and UV light. The choice depends on the extent of tissue fixation tolerated by the antigen under study. Large blocks of heavily fixed tissues, and also osmicated tissues should be heat-polymerized. Osmicated tissues should never be exposed to UV light. Smaller blocks of delicately preserved tissue, i.e., tissue treated with low concentration of fixative, retain their immunoreactivity and structural integrity if polymerized more rapidly with chemical catalysts or UV light in the cold.

LR gold LR gold is similar to LR white in many aspects. But LR gold can be used for embedding "unfixed" tissue at low temperature in order to retain histochemical reactivity in enzymes which are denatured by aldehyde fixation. Further this resin shows good preservation of lipid-rich structures. But due to many associated problems, this resin is not used much. Like LR white, LR gold can be polymerized using UV light at temperatures near –20°C, but both resins freeze at about –27°C. So they cannot be used at the sub –30°C temperatures.

Lowicryls This group of acrylic resins has low viscosity at temperatures of –35°C and below and hence can be polymerized using UV light at these very low temperatures which can preserve the ultrastructure as well cytochemical and immunocytochemical reactivity. The Lowicryl resins commercially available are Lowicryl K_4M, HM20, $K_{11}M$ and HM23.

Lowicryl K_4M and $K_{11}M$ are hydrophilic like LR Gold and LR White and therefore need partial dehydration only. Lowicryl K_4M is the most hydrophilic, followed by $K_{11}M$, LR gold, LR white and HM20.

The two most commonly used Lowicryls for immunoelectron microscopy are K_4M and HM20. Lowicryls $K_{11}M$ and HM23 have been more recently introduced for cryosubstitution techniques where $K_{11}M$ at –60°C and HM23 at –80°C can be polymerized with UV light.

Polymerization temperatures of polar (hydrophilic) acrylic resins:

✳ Lowicryl K_4M	Polymerizes with UV at –35°C
✳ Lowicryl $K_{11}M$	Polymerizes with UV at –60°C
✳ LR White	Polymerizes at room temperature with added accelerator
	At 55°C without accelerator
	By UV light at –10 to –20°C
✳ LR Gold	Polymerizes at –25°C with UV

Polymerization with hydrophobic (nonpolar resins):

✳ Lowicryl HM20	Polymerizes with UV at –70°C
✳ Lowicryl HM23	Polymerizes with UV at –80°C

There are four ways to polymerize acrylic resins over a temperature range of 60°C to –80°C, using,

- ✳ heat
- ✳ chemical catalysts/accelerators
- ✳ ultraviolet (UV) light
- ✳ blue light

Of these, heat is most damaging to the tissue particularly when the tissue is treated with low concentration of the fixative. Chemical, UV and blue light polymerization can be carried out at low temperatures (0°C or lower) to improve retention of immunoreactivity and to minimize extraction of cellular components. For example, the heat produced during exothermic UV polymerization reaction can, at room temperature, result in a temperature rise up to 12°C. But when the reaction is carried out at low temperature, say around –35°C, the rise in temperature is not more than +2°C.

At low temperatures, polymerization is slow. Therefore, resin components have to be mixed very carefully so as not to introduce oxygen from the atmosphere. Mixing is best done by bubbling dry nitrogen through the resin solutions.

The cross linking of acrylic resins can be hastened by the addition of certain catalysts, also called as initiators. For heat polymerization dibenzoyl peroxide is added as the catalyst. The same catalyst can be used for chemical polymerization at room or low temperatures also but requires an additional accelerator/activator, usually an amine compound as the activity of the catalyst is temperature-dependent. For UV, an ethyl or methyl ether is used as an initiator.

Labelling Procedures

Direct, indirect and bridge techniques can be used in immunoelectron microscopy. These have already been outlined.

CONTROLS

It goes without saying that adequate control preparations are absolutely essential for the success of the technique of immunoelectron microscopy. The use of controls depends on the experiment and the availability of the material.

Some of the control procedures are listed below.

1. ***Non-specific binding*** Parallel staining experiments should be performed using unconjugated probes. This type of control experiments are useful where ferritin, colloidal gold, viruses and synthetic materials are used as probes.
2. ***Blocking the binding sites of carrier proteins*** Treatment of the conjugate with excess of antigen on a solid-phase immunoadsorbent can block the site for specific binding of the antigen before the conjugate comes in contact with the tissue. This is a very effective control when antibodies are employed as carrier proteins.
3. ***Blocking the antigenic sites on cells*** The specific blocking of the antigenic sites can be achieved by the treatment of tissues/cells with homologous antibodies or specific inhibitors of the antigen.
4. ***Removal or inactivation of antigen*** Sometimes it is possible to remove antigens from the target cells by chemical treatment. For example, flagellar antigens of *Salmonella paratyphi* B can be dissolved by ethanol treatment. Protein antigens can be denatured by processes such as heat treatment, so that they no longer bind to carrier proteins.
5. ***Penetration of conjugates*** Probes conjugated with carrier proteins may not be able to penetrate inside the cells because of their large size. It becomes important, therefore, to ascertain whether a negative reaction in the preparations is due to non-accessibility of the conjugate or due to the absence of antigen. Poor penetration may lead to false-negative results. To ensure that the conjugate penetrates the cells, the permeability control normally exercised by plasma membrane should be neutralized. This can be achieved by fixation of the cells prior to incubation with the conjugate. Sometimes, intracellular antigens may also be denatured by such a procedure.
6. ***Non-target tissue control*** Treatment of non-target tissue, in a manner which is exactly parallel to the treatment of target tissues for tracing a particular antigen, provides valuable information as a negative control.

7. ***Endogenous activity*** Sometimes the probe material is already present in target tissues and therefore a false positive reaction may occur. Checking of endogenous activity is essential when enzyme or ferritin probes are used. Histochemical procedures for probe enzymes should be carried out on tissues which are not incubated with the conjugate.

SPECIAL METHODS

Cryosectioning for Immuno-gold Labelling

Cryo-ultramicrotome (CUM) is a procedure for cutting ultrathin frozen sections, especially for use in immunohistochemical techniques. It has several advantages over conventional ultramicrotomy.

1. In conventional ultramicrotomy, biological tissues are embedded in resins that are polymerized at higher temperatures. The resins often mask the cellular antigens. Further, the dehydration step may also lead to loss and/or redistribution of cellular substances.
2. In CUM on the contrary, no dehydration and embedding steps are involved and hence, the antigenicity is well-preserved and loss or distribution of substances from cells is also minimized.

In 1973, Tokuyasu devised a method for preparing ultrathin frozen sections which is routinely used for producing cryosections for immunohistochemistry and it is briefly outlined below.

The Tokuyasu method involves infiltrating chemically fixed specimens with a sucrose solution followed by freezing and then sectioning on a dry knife while the specimens are still frozen. The sections are removed from the knife surface, thawed and placed onto a coated specimen grid. They are then immunolabelled at RT before examination in TEM.

- The tissue sample is fixed in a chilled solution of 2% paraformaldehyde and 0.2% glutaraldehyde for 15–30 min. at RT.
- It is washed in cooled phosphate buffer containing 0.1 M glycine to inactivate any residual aldehyde groups.
- Sample is immersed in 2.3 M sucrose in 15–20% polyvinyl pyrrolidone in PBS overnight.
- Sample is mounted onto metal holders or pins and trimmed in such a way that they have a wide base close to the pin and a pointed tip. A tiny droplet of sucrose is used to hold the sample to the holder.
- The mounted sample is quickly frozen by plunging the holder with the specimen into liquid nitrogen very rapidly.

- The block is then transferred to the cryochamber of the ultramicrotome.
- Sections are cut using cooled glass or diamond knives.
- As dry sections are cut at a temperature of –90 to –100°C, they will collect along the knife edge.
- Sections are retrieved by using a 1–2 mm wide metal loop that has already been dipped in 2.3 M sucrose such that a droplet of sucrose remains in the loop.
- The droplet is allowed to thaw and the sections are transferred to carbon-coated grids, washed with PBS and then further processed for immunolabelling.

Figure 17.3 shows the colloidal gold distribution within the secretory granules in a tumour sample from a case of medullary thyroid carcinoma where the specimen is immunolabelled for the protein hormone, calcitonin, using the protein A-gold technique by post-embedding immunoelectron microscopy. The specimen after immunolabelling by post-embedding method is counterstained with the routine uranyl and lead salts.

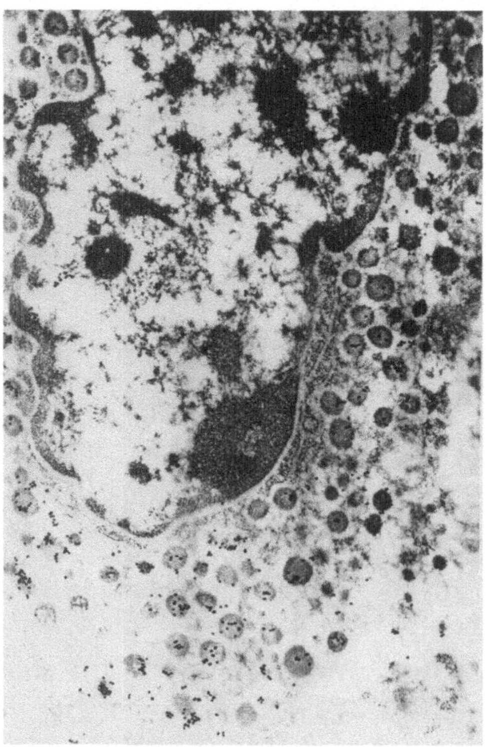

Figure 17.3 Medullary thyroid carcinoma immunolabelled for calcitonin with the protein A-gold technique and counterstained with uranyl and lead salts. Gold particles are present within the secretory granules (30,000×). (*Source*: Henry, A. Azar. (Ed.).(1988). *Pathology of Human Neoplasms. An Atlas of Diagnostic Electron Microscopy and Immunohistochemistry*. Raven Press, New York. p. 158, Fig. 19.)

Negative Contrast Immunoelectron Microscopy

This technique is mainly useful for virological and bacterial applications. But, however, applicable to any specimen which is particulate and can be adsorbed to an EM grid, particularly, surface antigens of particulate samples. Coated nickel grids are glow-discharged to make them hydrophilic, and used. The particulate material after adsorption on the coated nickel grid, is treated with the primary and secondary antibodies and then negatively stained with 2% phosphotungstic acid (PTA) or 1% ammonium molybdate or uranyl acetate.

Scanning Immunoelectron Microscopy

The immunological methods for antigen detection can be carried out using the scanning electron microscope. Gold particles of different sizes can be employed as probes in this method also. The smaller (5–15 nm) probes can only be detected by backscattered imaging and HRSEM (High Resolution SEM). But, larger probes (20–40 nm) can be detected using standard SEM or low-vacuum SEM, both in the secondary electron and backscattered electron modes.

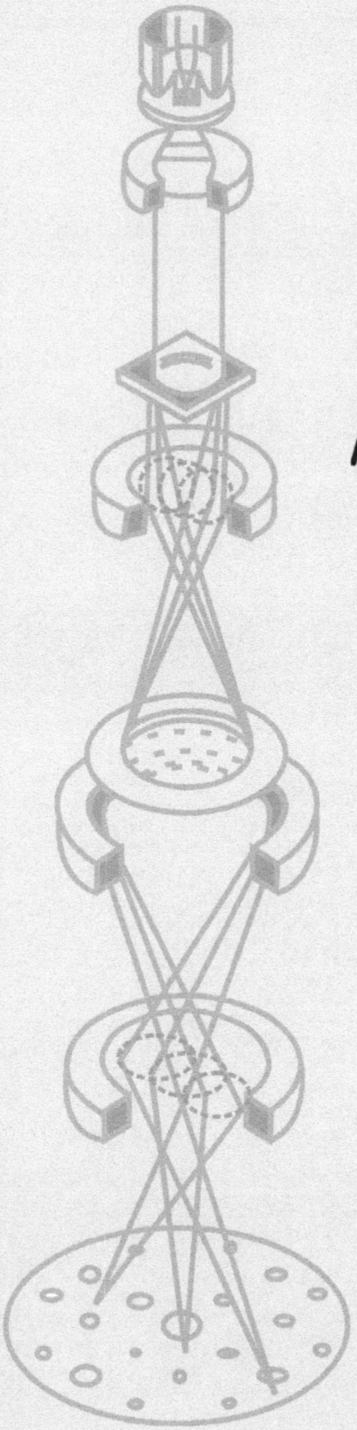

PART III

APPLICATIONS OF TEM

- Artifacts in EM
- EM in Biological Research
- EM in Tumour Diagnosis
- EM in Diagnosis of Diseases other than Tumours
- EM in Microbiology
- EM in Oral Pathology
- Applications in Nanoscience

In biological arena the applications of electron microscopy are innumerable. As we have already seen, the cell organelles were identified only after the advent of electron microscope. Earlier, the cell fractionation was carried out using the ultra centrifuge by which the microsomal, mitochondrial, nuclear fractions with distinct functions could be separated. But, organellar structure, their numbers, their disposition and distributions within the cell and also with respect to other organelles and inclusions could not be deciphered with this method. Electron microscopy by virtue of its high magnification and high resolution came as a boon in this aspect. Organelles thought to be artifacts were later on confirmed as genuine structures by using different methods of processing.

The structure can be correlated with the functions as we know both are intimately interlinked. For example, presence of secretory granules in a cell confirms the secretory function of the cell. Thus, structure-function relationship can also be studied using EM. (*See* Chapter 22 also)

The study of normal structure can naturally be extended to cellular variations both physiological and pathological. In cellular pathology EM finds applications in diagnosis of:

- Certain genetic and metabolic diseases, especially storage diseases.
- Certain renal glomerular diseases
- Certain skeletal muscle diseases
- Certain peripheral nerve diseases
- Certain infectious diseases, particularly viruses.
- Most important—Tumour diagnosis.

Though surgical biopsy specimens form the majority of samples which require diagnostic electron microscopy, there are a number of other types of specimens such as sputum, feacal matter, urinary sediments, cytology samples, cerebrospinal fluids, bone marrow biopsies, peripheral blood, synovial fluid, stones, bronchial lavage and others where electron microscopy can yield significant diagnostic information of value not only to pathologists, but also to haematologists, virologists and microbiologists.

There are certain instrumental parameters which are of significant value in diagnostic pathology apart from the routine transmission electron microscopy. They are:

- Specimen tilt—an underutilized technique
- Selected area or microbeam diffraction
- Backscattered electron imaging
- Scanning transmission imaging (STEM)
- Analytical electron microscopy

Like any other technique, EM has its own limitations and disadvantages. Some of the crucial ones are listed below:

LIMITATIONS

- Artifacts can be introduced at each and every stage of processing (*See* Chapter 19).
- Plane of sectioning varies for each section (*See* below and chapter 18).
- Interpretation of 2D images of 3D sections requires experience.
- The specimen is cut into very small and ultrathin sections. Therefore semithin sections have to be screened to look for representative material and then the material trimmed accordingly.
- Sometimes prolonged scrutiny has to be made to identify even a single-diagnostic feature from several sections and also from several blocks.
- High cost of the instrument apart from running and maintenance cost.
- Time taken for processing and screening in EM.
- Requirement of expertise and skills.

IMPORTANCE OF PLANE OF SECTIONING

Thin sections prepared for TEM are usually 400 to 600 Å thick, so this is the maximum thickness of any profile of the organelles that can be seen in a section. For organelles or features larger than the section thickness, either a 400–600 Å profile or a thinner slice of the structure, depending on how much of the structure is encountered by the plane-of-section and retained in the section will be seen. For organelles smaller than the thickness of the section, it is possible that the entire structure, or any part of the structure, may be contained within the section. When an organelle has an ordered structure, the plane-of-section can intersect the organelle at different angles, and the appearance of the structure in the section will depend on its orientational plane. In other words, the identification of the feature from any part of its structure that may appear in a section is influenced by its orientation in the profile, by the plane-of-section of the organelle, and by the amount of the structure that is represented in the profile.

This creates effective artifacts which may mislead interpretation. An experienced user with a prior knowledge about structures cut at different planes would be able to judiciously account for the features that appear differently depending on how they are included in the plane of the section as seen on the screen/micrograph and conclude the diagnosis. An example this type of artifact is given in Figure III.1 a,b and c.

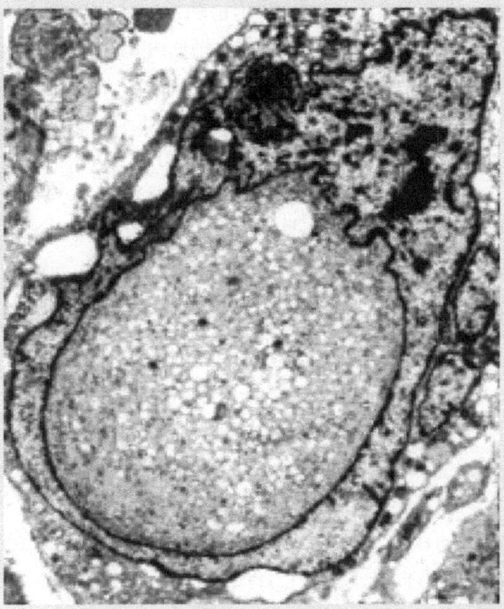

(a)

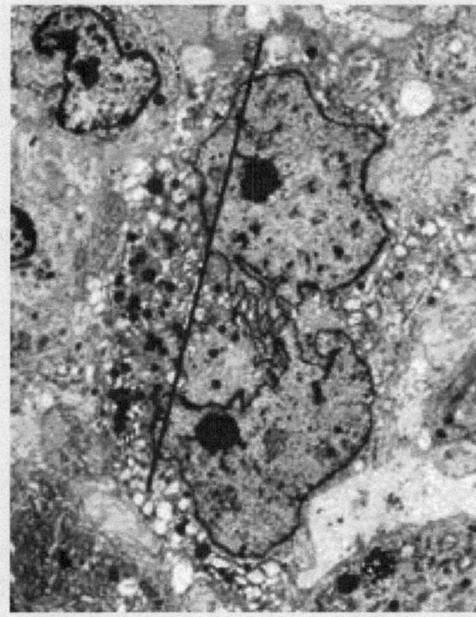

(b)

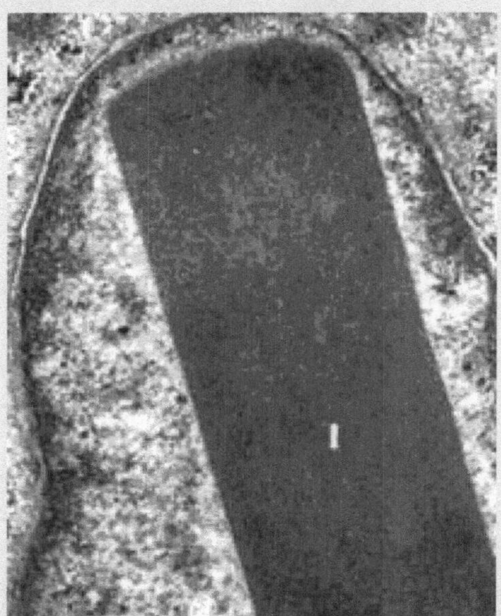

(c)

Figure III.1 Planes of sectioning artifact

Figure III.1a shows a feature which appears as an intranuclear inclusion, enclosing mitochondria, endoplasmic reticulum, filaments and other cytoplasmic organelles in it. The inclusion is surrounded by a nuclear rim, which confirms that it is not within the nuclear space and thus is a plane of section artifact. Because, a true nuclear inclusion will have no membrane separating it from the nuclear chromatin. In Figure III.1b the nucleus is very irregular, and has numerous invaginations of cytoplasm into the irregularities. The solid line demonstrates one plane-of-section that would yield the artifact shown in Figure III.1a. A true intra nuclear inclusion is seen in Figure III.1c and there is no rim of nucleus or nuclear membrane surrounding it.

Some of the applications in the fields of biological research, tumour diagnosis, microbiology, oral pathology and nanoscicence are illustrated in the following chapters. Most of the micrographs, unless otherwise stated, are from the work carried out at Cancer Institute (WIA), Chennai with JEOL JEM 100SX Transmission Electron Microscope at an accelerating voltage of 80 kV using samples given by various students for their research/ dissertation work. The semithin and ultrathin sections were cut using Leica Ultracut R Ultramicrotome with glass or "Diatome" diamond knives. The processing of these samples for ultrastructural study and screening the sections in TEM including photography were carried out in our laboratory.

18

ARTIFACTS IN EM

At the outset, every electron micrograph is in a sense an artifact. At each stage, from fixation to photography, changes in ultrastructure are inevitable. These cellular changes occur when

- materials are extracted
- dimensions are altered
- molecular rearrangements occur

The best way to get good micrographs is to keep these changes to a minimum. The identification of artifacts in a particular electron micrograph is not always made directly. It is mostly learnt from previous work on closely related problems for which knowledge of types of artifacts found by other workers is essential. The best test is a comparison of the same or similar specimens prepared in the same or different ways.

WHAT IS MEANT BY ARTIFACTS?

Artifacts are actually features which appear as facts but may not be facts at all. The *American Heritage Dictionary*'s definition of the word "artifact" in biology is, "A structure or substance not normally present but produced by some external agency or action." This should also include structures normally present but which are removed by some external agency or action and not seen by microscopy. The features seen in the electron micrographs, which are photographs taken in the electron microscope, or seen on the fluorescent screen itself, are expected to represent those in the sections loaded in the EM which again arise from the specimen from which these ultrathin sections are cut. In reality it may not be so. Because, cells undergo sudden and drastic changes in morphology when the blood supply is cut off, or when they are removed from the organism or body, or whatever. These changes can be overcome by perfusion fixation where, as we have already seen, the fixative is perfused into the animal under anaesthetic conditions when the animal is still alive. But this method of fixation is impossible for human tissues and can be carried out only in animals which can be sacrificed. The only possible fixation method for human samples is immersion fixation in which the tissue of interest is removed and immediately immersed into the fixative (mostly, 2.5% buffered glutaraldehyde).

Early post-mortem changes occur between the time of resection of the tissue and its immersion into the fixative, however short the duration is, that is, even when the fixative is kept as close as possible to the specimen to be fixed such that the time gap between resection and immersion is minimal. The best example to visualize this concept is as follows. Fast growing aggressive tumours have been reported to show minimal mitotic figures when screened in the TEM contrary to the fact that mitosis is more common in these tumours. This reverse observation can only be explained on the assumption that the fast growing tumour cells complete their cell cycles before the tissue is put into the fixative even though the fixative is kept ready to minimize the time gap. Thus, the final micrograph may not be a true representation, which, in other words means that we cannot see "How a cell is in its living state."

On the other hand, only after the advent of EM have many of the cellular structures, as listed below, been identified.

- cellular membranes
- interior of plasma membrane (by freeze-fracture method)
- Golgi complex
- presence of DNA in mitochondria
- lysosomes which could be distinguished from mitochondria
- viruses

How to identify facts and artifacts Discovery of structures normally precede knowledge of their function. So, many features were thought to be initially artifacts rather than true structures. For example, Golgi and endoplasmic reticulum were considered as artifacts initially. When viewed in the electron microscope, even when the processing techniques vary, these features were persistent indicating that they were not artifacts. Similarly, by comparing the specimen under different processing techniques, some structures which were thought to be genuine turned out later on to be only artifacts. For example, mesosomes in gram-positive bacteria which appeared along the membrane and at sites of septum formation differed in their size and number with different techniques. By using cryotechniques they have been proved to be only artifacts arising from glutaraldehyde fixation and not true structures. Thus, when different processing techniques are performed on the specimen, the features consistently present in all types of processing techniques tend to be facts, at least near facts.

ORIGIN OF ARTIFACTS

Biological specimens have to be killed and processed to endure the stresses of:

- further treatments with chemicals, sectioning, etc.
- high vacuum in the microscope column
- bombardment of the electrons

So artifacts can be introduced at each and every stage of processing starting from collection of samples itself.

STAGES OF INTRODUCTION OF ARTIFACTS

The various stages of processing at which artifacts can arise are enumerated below:

- Collection of samples
- Fixation and washing
- Dehydration
- Infiltration and embedding
- Sectioning
- Staining
- Screening in the electron microscope
- Photography

Therefore it is mandatory to take precautionary measures to minimize artifacts at each and every stage of processing, whatever be the technique employed.

TYPES OF ARTIFACTS

Broadly, the artifacts can be classified into two types. They are,

- Inherent artifacts
- Introduced artifacts

Inherent Artifacts

The inherent artifacts are those which occur due to the type of technique employed for processing the sample. This depends on the nature of the processing methods used and therefore unavoidable as long as that particular technique is chosen. The best way to deal with these inevitable artifacts is to recognize them in the image with prior knowledge and take them into consideration while interpreting the images and micrographs.

Introduced Artifacts

The artifacts which arise due to the actual way of carrying out the chosen technique of processing are known as the introduced artifacts. So they are introduced probably by improper way of executing the methodology. The only way to prevent them is by paying proper attention to the technique and strictly adhering to the methods and preparative conditions for specific materials while processing the specimens.

REASONS FOR OCCURRENCE OF ARTIFACTS

- Autolysis before a tissue is fully fixed
- Incorrect osmolarity of the fixative
- Absence of divalent cations from the fixative
- Extraction of lipids during dehydration
- Lack of uniformity in the curing or polymerization of the embedding medium
- Instability of the embedding medium under the electron beam

The above reasons indicate that the procedures learnt from literature and experience should be modified suitably, if need be, depending on the nature of the specimen and the field of interest. Apart from that, the stringent conditions must be followed strictly. Only then, can the inherent artifacts be minimized and the introduced artifacts be avoided.

INDICATORS OF ARTIFACTS

There are certain indicators of artifacts which have been recognized and which can be utilized for their identification in the micrographs. The important ones are enumerated below:

1. Disparity in appearance of adjacent cells of same type (Figure 18.1)
2. Loss of continuity of any membrane (Figure 18.2)
3. Any distortion or disorganization of organelles (Figure 18.3)
4. Presence of empty spaces in cytoplasm of cell (Figure 18.4)
5. Swollen or empty spaces between the two nuclear membranes (Figure 18.5)
6. Irregularities in the apposition of adjacent cells
7. Sharp bends or curves in filamentous structures which are usually straight, such as microtubules

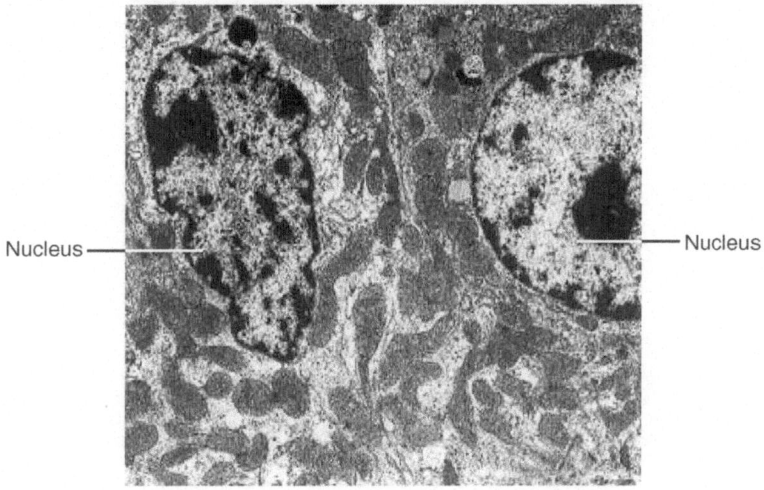

Figure 18.1 Disparity in appearance between adjacent cells. (*Source*: Audrey, M. Glauert and Peter, R. Lewis. (1988). *Biological Specimen Preparation for Transmission Electron Microscopy.* Vol. 17. Princeton University Press, Princeton/Portland Press Ltd., London. p. 7, Fig. 1.3)

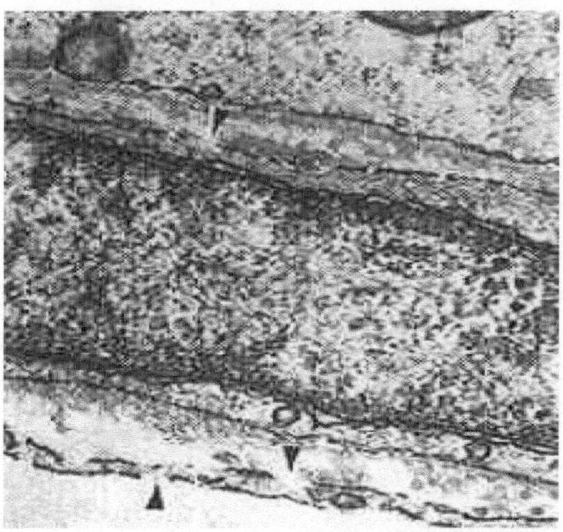

Figure 18.2 Discontinuity of membranes (*See* arrows). (*Source*: Audrey, M. Glauert and Peter, R. Lewis. (1988). *Biological Specimen Preparation for Transmission Electron Microscopy.* Vol. 17. Princeton University Press, Princeton/Portland Press Ltd., London. p. 8, Fig. 1.4a.)

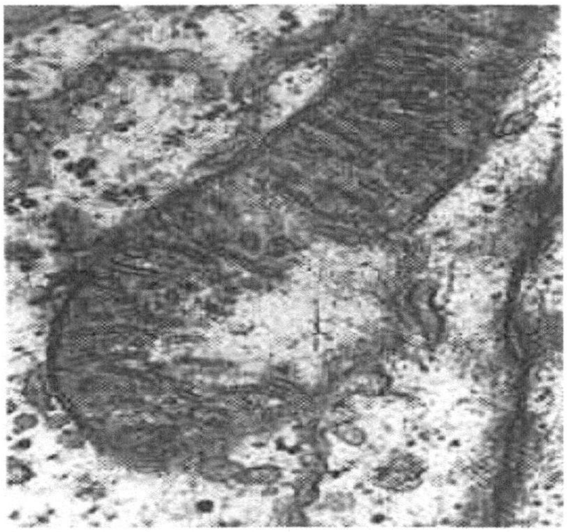

Figure 18.3 Distortion of organelles (Mitochondria). (*Source*: Audrey, M. Glauert and Peter, R. Lewis. (1988). *Biological Specimen Preparation for Transmission Electron Microscopy.* Vol. 17. Princeton University Press, Princeton/Portland Press Ltd., London. p. 8, Fig. 1.4b.)

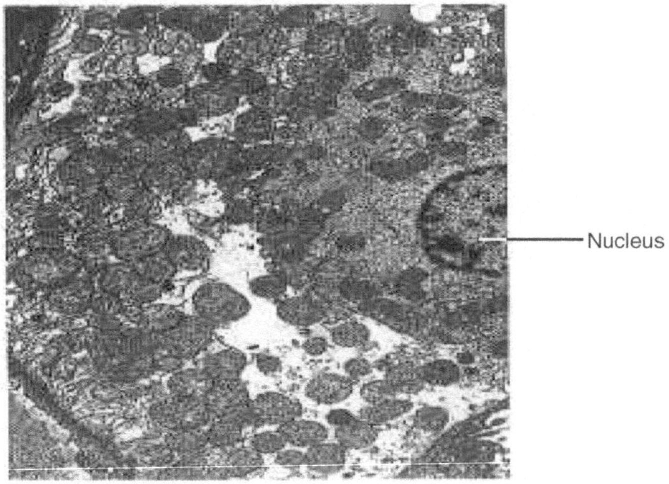

Figure 18.4 Empty spaces in cytoplasm. (*Source*: Audrey, M. Glauert and Peter, R. Lewis. (1988). *Biological Specimen Preparation for Transmission Electron Microscopy.* Vol. 17. Princeton University Press, Princeton/Portland Press Ltd., London. p. 8, Fig. 1.4c.)

When the specimens have been fixed by different types of chemicals, such as oxides (osmium tetroxide) or cross linkers (aldehydes), or by non-chemical means such as freezing, freeze-drying, etc., some structures appear consistently which could be considered as facts rather than artifacts. And an experienced observer is quite familiar with subcellular organelles and their structural relationships.

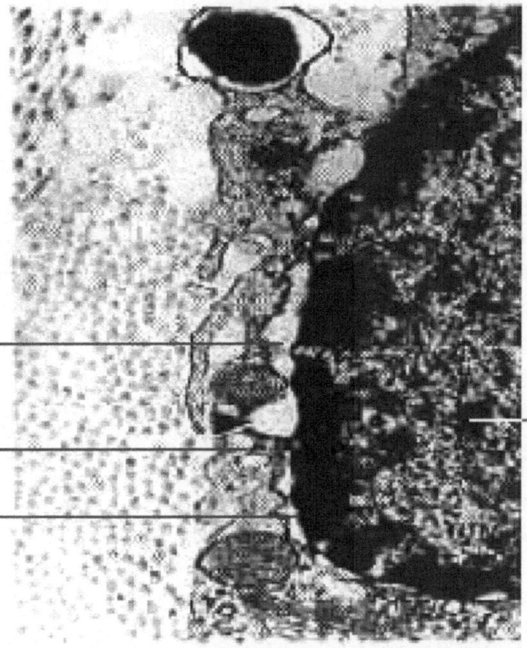

Figure 18.5 Swollen, empty spaces between nuclear membranes. (*Source*: Audrey, M. Glauert and Peter, R. Lewis. (1988). *Biological Specimen Preparation for Transmission Electron Microscopy.* Vol. 17. Princeton University Press, Princeton/Portland Press Ltd., London. p. 8, Fig. 1.4d.)

Since specific artifacts can be introduced by certain procedures, such artifacts can be considered as consistent with the structure or even useful in the recognition of a structure. Heavy metal stains are not part of natural cell conjunction, but without their use it would not be possible to delineate structures with clarity and precision. Similarly, dehydration is known to remove certain diffusible components but that is essential to get a clear image of the cell's interior. Therefore, one must be able to identify artifacts inherent with various steps and then try to arrive at facts.

FIXATION ARTIFACTS

Fixation preserves and stabilizes structures and chemical constituents of the cell, and thus, morphology. Therefore this step should be fast and perfect to minimize artifacts. The methods as we have already seen, are chemical and physical methods. The chemical method is the routinely used mode of preservation, though it tends to be a little harsher on the material as it introduces certain artifacts inherent to its chemical nature. Thus the fixation step gives rise to **inherent artifacts**. Fixation does not "fix" or stop all processes instantly. The so-called "myelin figure" which results from lipases that are not inactivated by glutaraldehyde is a good example of this type of artifact. Heat inactivation of enzymes by microwave irradiation or cryotechniques may be better options in this regard but these techniques may produce their own artifacts.

During the fixation many labile structures disappear and sometimes artifactual structures may appear. By comparative techniques, as pointed out in earlier paragraphs certain natural structures ultimately turned out to be artifacts and vice versa. So, while establishing a technique each sample could be preserved with more than one fixation protocol, using different fixatives or a combination of them. At the same time, it is not wise to totally depend only on the nature of chemical reactions involved during fixation to choose the fixation protocol. Potassium permanganate was getting preference over osmium tetroxide because of its ability to fix lipids, its oxidizing nature and its ability to enhance membrane contrast. Later it was realized that potassium permanganate does not preserve better, rather extracts most of the other cytoplasmic constituents enhancing the membrane contrast. Potassium permanganate is hence not used as a fixative now.

Fixation artifacts may arise by:

- Method of fixation (immersion, perfusion, perfusion being a better preservative than immersion) (Figure 18.6)
- Nature of fixative (Figure 18.7)
- Buffer vehicle and other additives used (Figure 18.8)
- Concentration of the fixative
- pH of fixative solution
- Temperature of fixation
- Duration of fixation

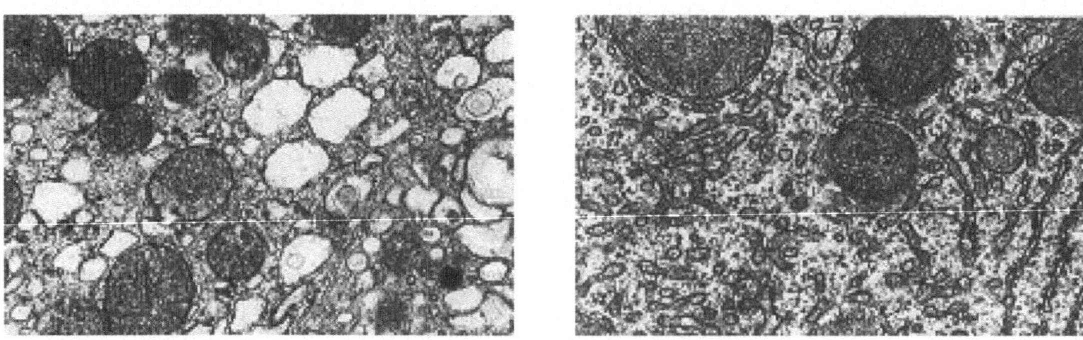

(a) (b)

Figure 18.6 Comparison of 2 fixation methods a) Immersion fixation; b) Perfusion fixation. The smooth endoplasmic reticulum (SER) is well-preserved after perfusion fixation. (*Source*: Audrey, M. Glauert and Peter, R. Lewis. (1988). *Biological Specimen Preparation for Transmission Electron Microscopy.* Vol. 17. Princeton University Press, Princeton/Portland Press Ltd., London. p. 80, Fig. 3.2.)

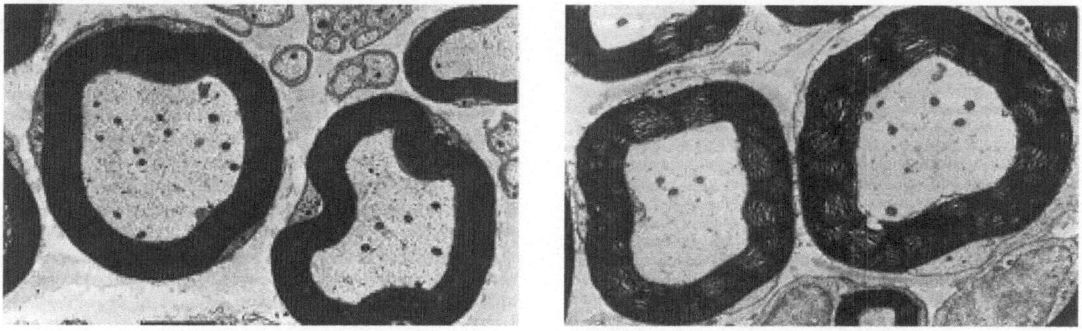

(a) (b)

Figure 18.7 Comparison between 2 different fixatives a) 2.5% glutaraldehyde; b) 4% HCHO. Specimen: Myelinated nerve. With glutaraldehyde the myelin layers are better preserved than with formaldehyde. (*Source*: Audrey, M. Glauert and Peter, R. Lewis. (1988). *Biological Specimen Preparation for Transmission Electron Microscopy.* Vol. 17. Princeton University Press, Princeton/Portland Press Ltd., London. p. 28, Fig. 2.1a and 2.1b.)

The appropriate fixation conditions and method can be chosen by observing the gross changes brought about by a fixative in a phase-contrast microscope. The correct fixation procedure can be selected by

- Trial and error methods
- Previous experience
- From available literature and suitable modification to suit the sample under study and purpose of the study

Liver cells are the most common test objects when exploring new preparatory methods in biological electron microscopy; the proximal tubule of the kidney is also suitable. This is because, these are very sensitive to variations in preparation procedures including type of fixative, osmolarity, mode of application of fixative, hydrostatic pressure during perfusion, handling during immersion fixation, dehydration, etc. Moreover, they are also suitable for correlation between structure and function.

In general, human biopsy materials exhibit more preparation artifacts than samples from experimental animals.

Physical methods of preserving a sample involve cryofixation and recently microwave methods are also used. There are generally no major "inherent artifacts" associated with these methods and the "introduced artifacts" could be avoided by taking appropriate precautions. The major artifact introduced in cryomethods is ice-crystal formation which damages the cytoplasmic and nuclear structures (Figure 18.9).

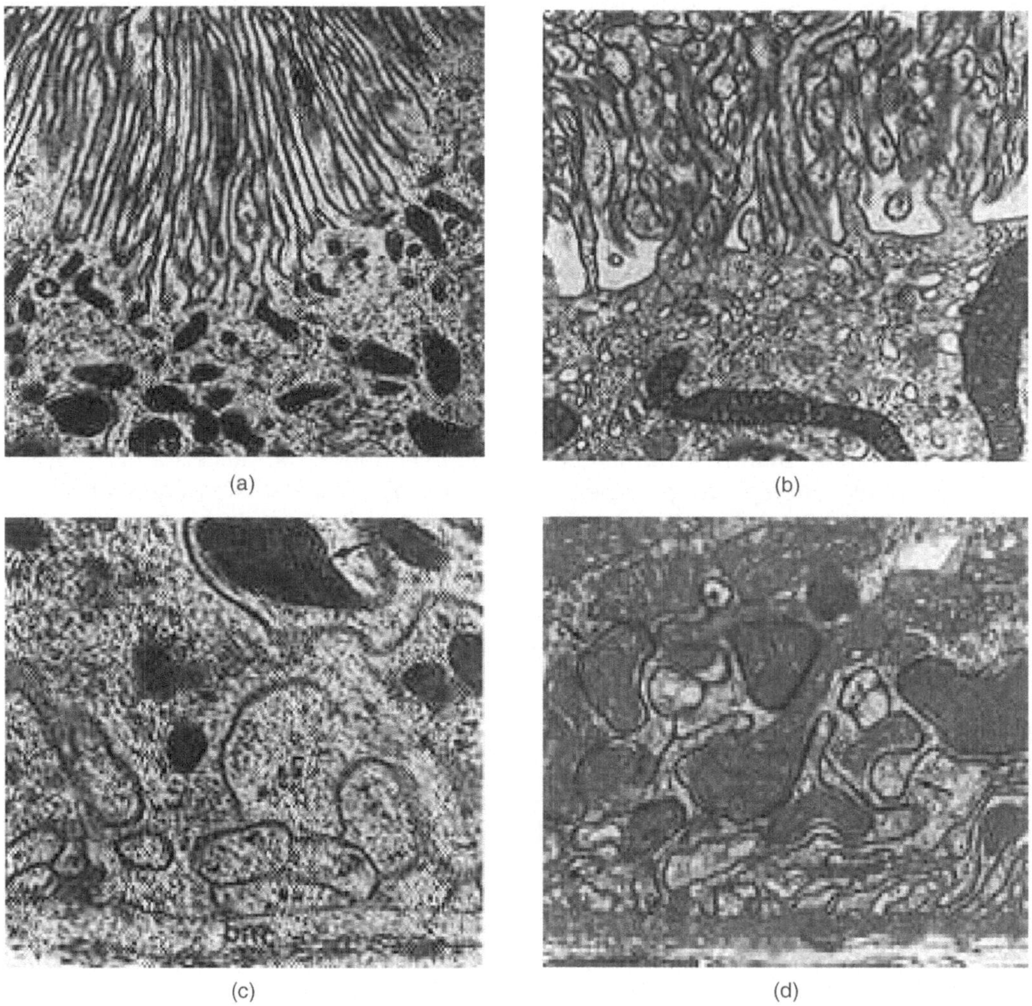

Figure 18.8 Use of additives. Potassium ferricyanide added to Osmium fixative (a) and (c) without additive; (b) and (d) with additive. Improvement in membrane fixation, contrast and better preservation of mitochondrial cristae are seen with addition of ferricyanide. (*Source*: Audrey, M. Glauert and Peter, R. Lewis. (1988). *Biological Specimen Preparation for Transmission Electron Microscopy.* Vol. 17. Princeton University Press, Princeton/Portland Press Ltd., London. p. 133, Fig. 4.2.)

This could be minimized by vitrification of water during freezing. By increasing the rate of cooling (plunge cooling) and pretreating the specimen with cryoprotectants, the ice-crystal formation can be prevented and thus the introduced artifacts can be minimized. Certain inherent artifacts like differential freezing of certain chemical constituents have been, in fact, utilized to develop certain techniques. For example, in freeze-fracture technique a fracture can be introduced in a predictable manner along the lipid bilayer and thus the inner face of the membrane can be observed.

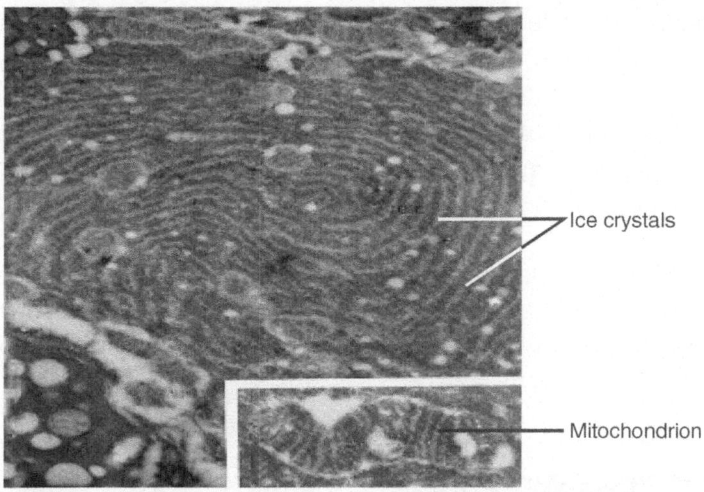

Figure 18.9 Cryofixation artifacts (ice-crystal formation). (*Source*: Daniel, C. Pease. (1964). *Histological Techniques for Electron Microscopy*, 2nd edn. Academic Press, New York & London. p. 7, Fig. 6.)

There are certain identified features of good fixation which could be used as guidelines for confirming minimal artifacts during fixation procedures. They are exhibited by well-preserved organelles as mentioned in Table 18.1.

Table 18.1 Appearance after good fixation

Cellular element	**Appearance after good fixation**
Mitochondria	Few swollen or empty looking mitochondria; both membranes smooth; no gross distortion in shape
Endoplasmic reticulum	Membranes intact; especially parallel when stacked; cisternae arrangement uniform
Golgi	Intact and vesicles of various dimensions appearing in circumscribed area
Plasma membrane	Intact and smooth around the entire cell
Cytoplasmic matrix	Fine precipitates seen; no empty spaces generally present
Nuclear envelope	Both membranes intact (except at pores); both membranes parallel to each other
Nuclear contents	Finely granular with denser masses both in the interior and periphery of the nucleus

DEHYDRATION ARTIFACTS

Dehydration is an essential step. Though water in the tissues is only replaced and not removed from the cells, this step is called dehydration. When water-miscible resins are used, this dehydration step is omitted. Generally acetone or alcohol is used for dehydration. Removal of lipids and other unfixed material is unavoidable during this step and this gives rise to inherent artifacts. If this step is improperly carried out, it could result in **introduced artifacts** which could even affect impregnation and polymerization of resins and thus subsequent sectioning of blocks. But these could be avoided by proper precautions during dehydration. Longer duration of the dehydration step may lead to shrinkage of cells and also undue extraction of lipids. Therefore partial dehydration is also suggested and tried. But this may cause problems in impregnation of resins and therefore not followed routinely. Figure 18.10 illustrates the better preservation of lipids by partial dehydration method.

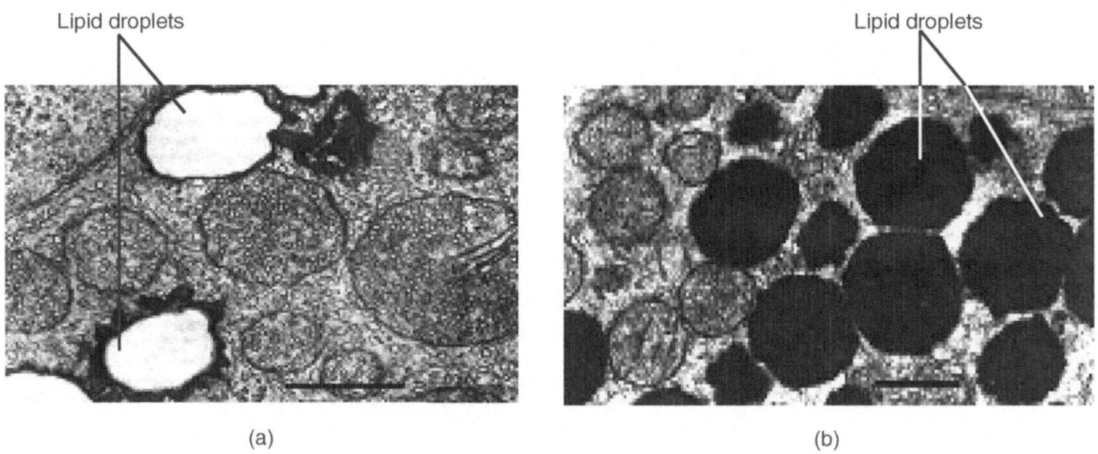

Figure 18.10 Comparison of preservation between partial dehydration and conventional dehydration a) Normal dehydration where lipid droplets are extracted; b) Partial dehydration where the tissue is transferred from 70% ethanol directly to 1 : 1 mixture of ethanol and Epon. The lipid droplets appear dense. (*Source*: Audrey, M. Glauert and Peter, R. Lewis. (1988). *Biological Specimen Preparation for Transmission Electron Microscopy.* Vol. 17. Princeton University Press, Princeton/Portland Press Ltd., London. p. 141, Fig. 4.3.)

INFILTRATION AND EMBEDDING ARTIFACTS

Like the above-mentioned steps, these steps also could result in both inherent and introduced artifacts. The dehydration and infiltration cause some amount of differential shrinkage or swelling depending on the resin formulation. Individual components of the resin, (resin, hardeners, plasticizers, catalyst for embedding) may penetrate at different rates and cause differential infiltration and polymerization of blocks, as well as shrinkage or swelling of certain organelles. These defects could be serious while doing morphometric studies. Figures 18.11, 18.12 and 18.13 indicate the differences in preservation of organelles with different resin formulations. Discussion on various resin formulations is given in Chapter 8.

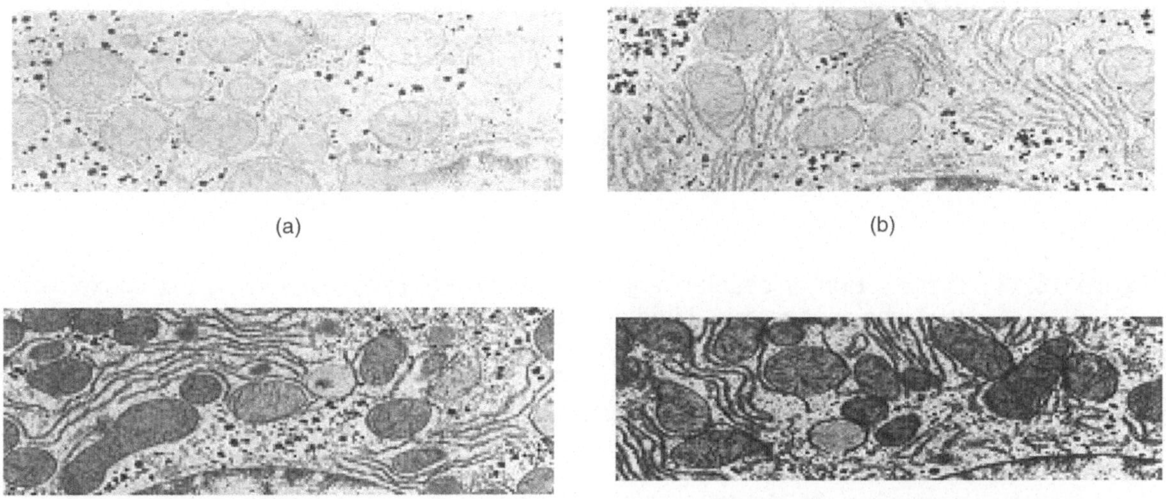

(a) (b) (c) (d)

Figure 18.11 Contrast comparison of different resin formulations (a) Araldite; (b) DER 736; (c) SPURR; d) Quetol 651. (*Source*: Audrey, M. Glauert and Peter, R. Lewis. (1988). *Biological Specimen Preparation for Transmission Electron Microscopy.* Vol. 17. Princeton University Press, Princeton/Portland Press Ltd., London. p. 217, Fig. 6.7.)

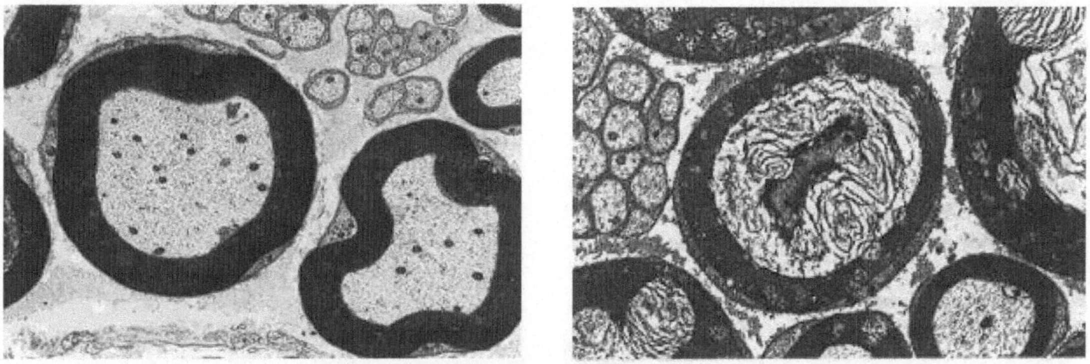

(a) (b)

Figure 18.12 Preservation of myelin by epoxy resin and LR white resin (heat polymerization) (a) Epoxy resin which shows only slight separation of the myelin layers; (b) LR white (heat polymer). The myelin sheaths are disrupted and vesiculated. (*Source*: Audrey, M. Glauert and Peter, R. Lewis. (1988). *Biological Specimen Preparation for Transmission Electron Microscopy.* Vol. 17. Princeton University Press, Princeton/Portland Press Ltd., London. p. 241, Fig. 7.4.)

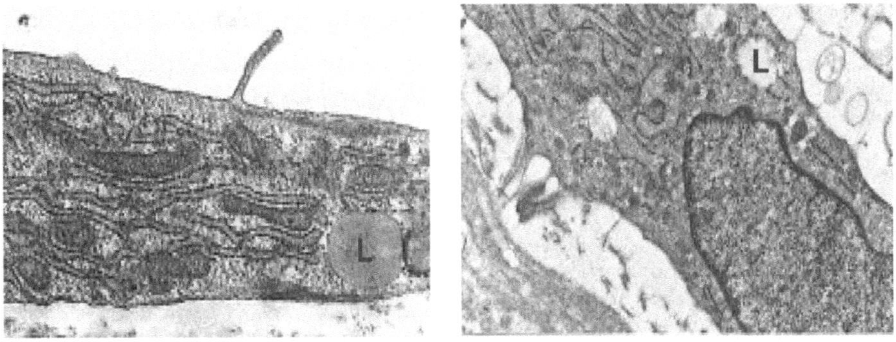

(a) (b)

Figure 18.13 Preservation of cellular organelles using different resins (a) using Epoxy resin; (b) using acrylic resin. Lipid droplets (L) appear extracted and the membranes of the rough endoplasmic reticulum are less distinct after acrylic resin (Lowicryl HM20) embedding. (*Source*: Audrey, M. Glauert and Peter, R. Lewis. (1988). *Biological Specimen Preparation for Transmission Electron Microscopy.* Vol. 17. Princeton University Press, Princeton/Portland Press Ltd., London. p. 252, Fig. 8.1.)

Polymerization Damage

This damage is seen with methacrylate embedding. It is thought to result from nonuniform polymerization which disrupts the fine structure of some tissues under certain circumstances. This damage may be difficult to be distinguished from poor fixation or from decomposition from electron bombardment. By comparing a single tissue embedded conventionally using methacrylate with samples embedded using prepolymerized methacrylate or using epoxy or other resins, polymerization damage can be demonstrated in methacrylate-embedded tissues.

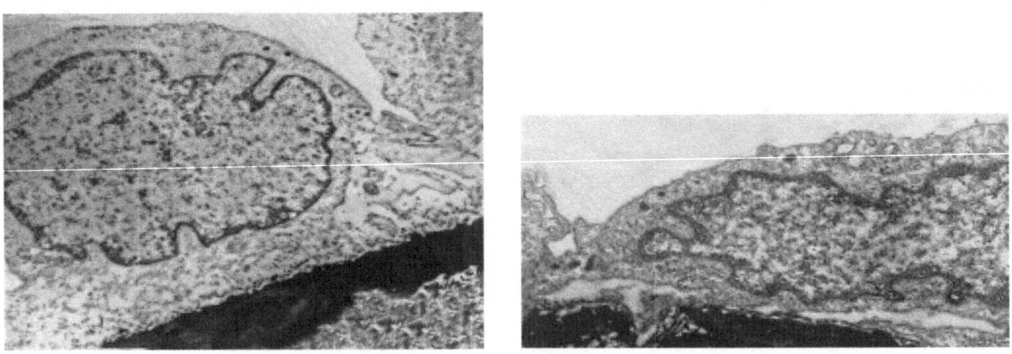

(a) (b)

Figure 18.14 Polymerization damage (a) Affected cell; (b) Normal cell. (*Source*: Daniel, C. Pease. (1964). *Histological Techniques for Electron Microscopy*, 2nd edn. Academic Press, New York & London. p. 103, Fig. 8.)

Polymerization damage can be correlated with "explosion" because there is always some swelling (Figure 18.14). In this figure, the apparent size of the affected cell is almost twice the size of the well-preserved one. The membranes of all sorts are interrupted. Cytoplasmic and nuclear contents are dispersed and have a flocculent appearance. This simulates poor fixation. It is obvious that the whole cell is poorly preserved. Polymerization damage can be a graded phenomenon so that varying degrees of explosion can be recognized. Exposed cells on the surface of tissue blocks are the ones particularly prone to be damaged. This damage is also tissue-specific.

SECTIONING ARTIFACTS

Blunt knives, contaminated knives, nicks in the knives all may cause introduced artifacts in the sections. Orientation of features in the sections and the plane of sectioning are also factors to be taken into consideration during interpretation of micrographs.

Figure 18.15 shows the mottled and unsatisfactory appearance of image of the sections which are seen generally in sections cut from Epon blocks using glass knives. Such appearances are not visible when diamond knives are used for sectioning. Presumably it results from a disturbance or displacement of the surface of the section as it is cut. It is as though the surface is scraped rather than properly cleaved. (*See* Chapter 8.)

Figure 18.15 Mottled appearance seen on sections from Epon blocks cut with glass knife. The specimen is skeletal muscle. (*Source*: Daniel, C. Pease. (1964). *Histological Techniques for Electron Microscopy*, 2nd edn. Academic Press, New York & London. p. 114, Fig. 11.)

When a membrane is sectioned transversely, it appears as a line. When it is sectioned in the horizontal plane, it appears as a smudge. Oblique sections of membranes look like pale ribbons. The inexperienced operator may overlook the smudge, and interpret the ribbon as an out-of-focus image, or, perhaps as an image spoiled by drift. Figure 18.16 shows numerous cytomembranes at high magnification. In many places the lines are sharp and obviously in focus. The rest of the micrograph does not appear so. Regions of apparent local fuzziness and lack of contrast also appear. This cannot be accounted for by difference in focusing. Because in EM with high depth of focus, uniform focusing can be achieved easily over the entire thickness of the specimen which is usually ultrathin. When the membranes are viewed vertically in the section, as a result of being sectioned transversely, their substance provides a maximum total density, and the boundaries appear sharp. When the membranes are viewed obliquely, and particularly tangentially in the horizontal plane, there is less scattering material in the path of the electrons, and hence a reduced density. Thus their images appear relatively pale and without sharp delineation. Therefore orientation becomes a matter of considerable importance in electron microscopy and it is often necessary to search diligently for a preparation which shows detail adequately.

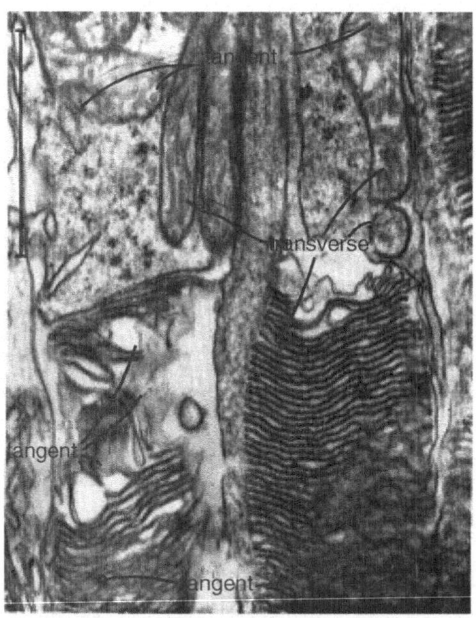

Figure 18.16 Features of vertical and transverse sectioning. (*Source*: Daniel, C. Pease. (1964). *Histological Techniques for Electron Microscopy*, 2nd edn. Academic Press, New York and London. p. 322, Fig. 39.)

The above can be better illustrated in Figure 18.17 where a group of serial sections of a three-dimensional nucleus depicts the artifacts arising due to differences in plane of sectioning. The 3D nucleus is shown in (f). The semiserial sections shown in (a) to (e) depict the features in different planes of sectioning. Thus it is obvious that the apparently multinucleated cell as seen in these serial sections are in reality features of a single nucleus cut at varying planes.

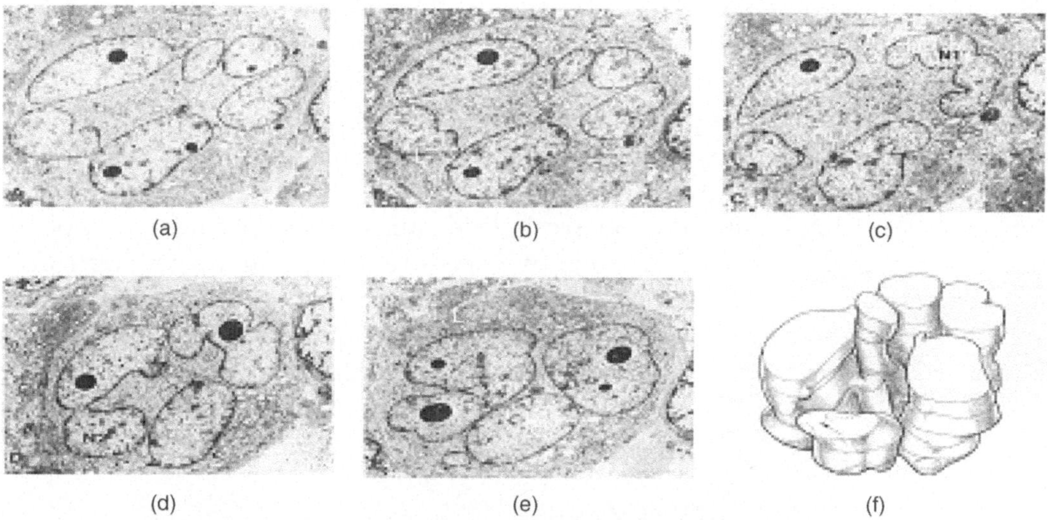

Figure 18.17 Plane of sectioning. (Source: Douglas, W. Henderson and John, M. Papadimitriou and Mark Coleman. (1986). Ultrastructural Appearances of Tumours, 2nd edn. Churchill Livingstone. p. 22, Fig. 1.2

The apparent "hole" in the nucleus in Figure 18.18 is due to artifacts arising from differences in plane of sectioning.

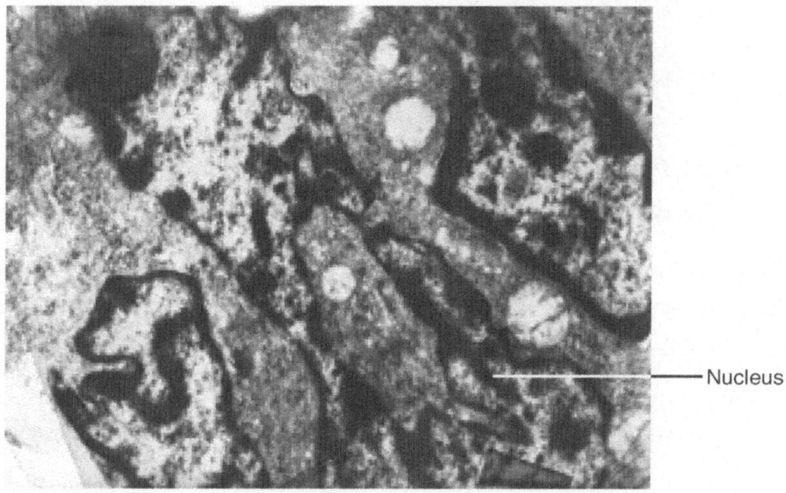

Figure 18.18 Plane of sectioning artifact appearing as a hole in nucleus

STAINING ARTIFACTS

Artifacts could be introduced by interaction of resin with stains. Most common effect introduced is "pepper effect" due to lead salts. Routine staining procedure, which is uranyl acetate followed by lead citrate in tissues embedded in different embedding media revealed that pepper effect is more pronounced with Spurr and Quetol resins than with Epon or Araldite resins. Improper staining procedures like, not using sodium hydroxide crystals surrounding the grids during lead staining may result in lead carbonate crystal deposition as shown in Figure 18.19. Figure 18.20 depicts the effect of electron bombardment for a longer duration on excess lead stains in the specimen. It appears as if the tissue is "burned" in the microscope by exposure to too intense an electron beam. Tissues heavily stained with lead are particularly vulnerable to this artifact, apparently because the leaded areas can easily become heated to liquefaction, whereupon surface tension causes coalescence and produces fine granular deposits as seen in Figure 18.19.

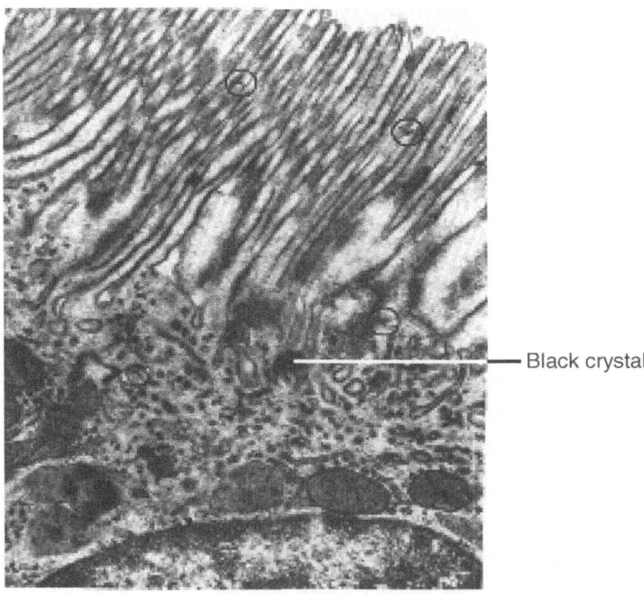

Figure 18.19 Deposition of lead carbonate crystals. (*Source*: Daniel, C. Pease. (1964). *Histological Techniques for Electron Microscopy*, 2nd edn. Academic Press, New York and London. p. 112, Fig. 10.)

Another artifact, stain exclusion, has been described by Meek and Chapman (1985). Stain exclusion is presumably caused by bulky polymeric complexes at cross-linking sites.

It is advisable to use freshly prepared stains especially for lead staining. Old preparations and improper storage all can result in introducing staining artifacts.

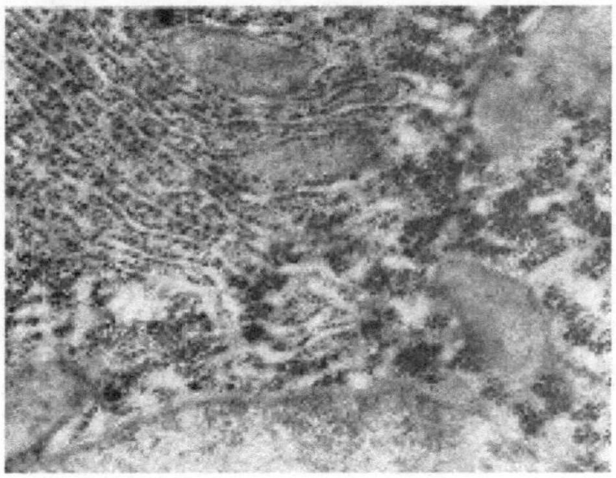

Figure 18.20 Effect of electron bombardment on excess lead. (*Source*: Daniel, C. Pease. (1964). *Histological Techniques for Electron Microscopy*, 2nd edn. Academic Press, New York and London. p. 220, Fig. 25.)

Negative Staining

Specific components in the samples are shown to have different reactions with various stains and thus appear different with negative staining procedures. For example, lipids rich in unsaturated fatty acids undergo phase transition when stained with silicotungstate stains and appear filamentous, while the preparations with phosphotungstic acid (PTA) or uranyl acetate appear as sheets or vesicles. Similarly, artifactual stacking of lipids occurs in tubulin preparations when stained with PTA but uranyl acetate maintains their vesicular form. Therefore, the possibility of such reactions must be borne in mind while employing negative staining methods and while interpreting micrographs of negatively stained preparations. Negative staining with more than one stain should be tried to rule out possibility of such staining artifacts.

Histochemical Staining

When histochemical or immunoctytochemical reactions are performed at the ultrastructural level, care should be taken that the products are formed rapidly with enough mass so as to provide accurate localization with good detectability. The reactions should be very specific for the component of interest. For example, to demonstrate and localize enzymes ultrastructurally, specimens are incubated with suitable substrate maintaining the optimum temperature and pH using an appropriate buffer along with a reagent to tag the reaction product for its detection in the EM. This tag is usually a heavy metal like lead for alkaline phosphatases. If enzyme concentration is high and concentration of lead ions is small, hydrolysed phosphate ions could diffuse from site of activity and there will be precipitates at distant site and an overall diffuse reaction. This artifact could be overcome by increasing the concentration of metal ions but at the same time the toxic nature of these metals have to be taken into consideration while handling such metals at high concentrations. The favourite site for such artifactual precipitates has been identified to be the nucleus. Other factors like duration, temperature, etc. have also to be carefully chosen.

IMAGING ARTIFACTS

These are generally instrumental/operational artifacts. The most important of these are those which arise due to lens aberrations.

Lens Aberrations

Spherical aberrations, chromatic aberrations and astigmatism are the most common aberrations as encountered in optical microscopes.

i. Spherical aberration
   - In objective lens—causes spread of each image point.
   - In projector lens—leads to variation in magnification across the image as a whole, with centrally high magnification and peripherally low magnification.

ii. Chromatic aberration—causes local changes in focus and magnification.

iii. Astigmatism
   - In objective lens—each spot elongated first in one direction and then the other.
   - In projector lens—image shows different magnifications in two different directions at right angles to one another.

iv. Distortion in lenses; charging effects due to accumulation of charges and drift of the image either because of movement of specimen or improper mounting of specimen result in focusing artifacts.

v. Contamination anywhere in the microscope causes loss of resolution and focusing artifacts.

vi. Vibrations (waves, undulations, chatter, chatter marks) in which the section displays periodic wavy changes in thickness (Figure 18.21), whereby the alternating thicker and thinner zones run parallel to the knife edge and perpendicular to the knife marks (*See* Chapter 9). Vibrations are produced by periodic oscillations
   - of the building,
   - of the entire ultramicrotome,
   - of the components of the ultramicrotome, the knife and the specimen,
   - in the microarea of the knife edge,
   - on the surface of the specimen while cutting.

The electron microscopes, therefore, should be installed in vibration-free areas preferably in the ground floor or basements away from elevators, compressors, heavy motors, etc.

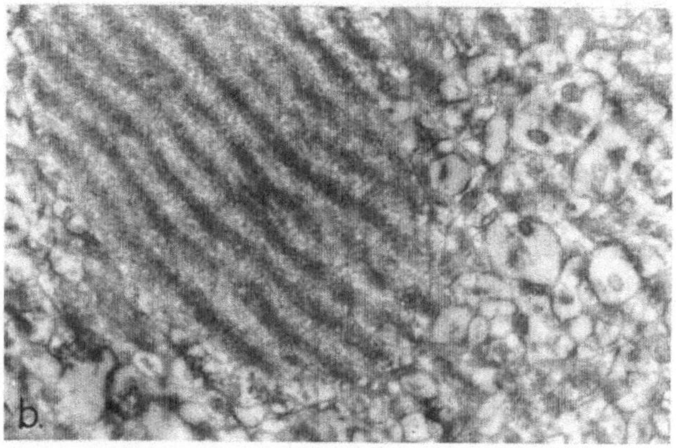

Figure 18.21 Chatter effect. (*Source*: Daniel, C. Pease. (1964). *Histological Techniques for Electron Microscopy*, 2nd edn. Academic Press, New York & London. p. 180, Fig. 21b.)

vii. Magnification displayed on the control panel of the electron microscope may not be accurate always. For quantitative measurements the magnification has to be calibrated and suitable correction applied. Possibility of error in magnification may arise due to the following.

- ✳ drift in accelerating voltage over time
- ✳ change of objective lens current after moving a specimen to another area
- ✳ hysteresis of the lenses
- ✳ build-up of charges in imaging lens due to dirt or contamination

Magnification calibration is carried out using standard specimens (*See* Chapter 4).

PHOTOGRAPHIC ARTIFACTS

These artifacts arise mainly from defects in exposure of photographic emulsion and mechanical handling of films. In modern instruments, with CCTV cameras, most of these artifacts become insignificant. Most artifacts arise because of apparent contrast of a specimen or photographic recording of a 3D image on a 2D film or screen and could lead to major misinterpretation of shapes, sizes and spatial relationships of structural features.

Image focused by electrons on the screen is projected directly on the photographic plate in TEM. Any error in focusing or contrast could be the operator's fault when autofocusing facility is not available. Mechanical vibrations during exposure while recording the image or during photographic imaging could produce artifacts in image.

In older instruments, the rotary pump for creating vacuum was installed within the instrument and hence it was necessary to switch it off while photographing the features and switch it on again after the exposure was over. But in later microscopes the rotary pumps are located outside the microscope column.

TEM films are inserted in the photographic chamber of the microscope and are exposed to electrons. So film emulsion must be able to withstand vacuum and produce less noise. Noise is produced by electrons resulting from random fluctuations in electron beam. These electrons contribute to granularity or electron density. For printing also an enlarging paper of suitable contrast should be chosen such that only limited areas, lines, etc., are black in a properly exposed print. Light source of enlarger during printing also is a matter of concern in the production of final image. A point source is better than diffuse illumination.

Reasons for Unsatisfactory Pictures

Summarizing the above, the unsatisfactory micrographs could be due to the following reasons:

- Unduly thick supporting films—fairly thick sections on a thick supporting film may look deceptively good on the fluorescent screen. It will have most of the contrast properties of a thin section on a thin film, but will, of course, lack the resolution of the latter.
- Specimen drift during picture taking.
- Changes in focus during preparations for photography—when an unduly bright beam is used for accurate focusing, and afterwards the intensity is greatly reduced, then, a change in the focal plane of the objective lens may occur so that the image that is photographed would be out of focus. If the pictures are taken rapidly, this can be avoided.
- Objective lens astigmatism—limits the resolution of the pictures.
- Microscope instability—most commonly caused by contamination of the objective aperture, but dirt on the specimen or the specimen holder may cause problems. Circuit instability also can produce unsatisfactory pictures.
- Characteristics of the specimen.
- Sectioning artifact—the fault may be with the embedding material, knife angle, block size or cutting speed.

Therefore, it is mandatory for the electron microscopist to choose the proper methodology and stringently carry out the procedure in order to achieve good pictures. Needless to mention, the instrumental conditions including contaminations and vibration are to be taken care of during screening the specimens.

19

EM IN BIOLOGICAL RESEARCH

The structural details offered by the electron microscope at very high magnifications, (in the order of a few lakhs), and high resolution (in the order of a few angstroms) make the technique, EM, a very valuable and powerful tool in morphological research of both biological and non-biological systems. In biological systems, the structure and functions are closely interrelated, elevating the importance of the structural study in these systems. For example, the presence of Golgi complex and secretory granules ultrastructurally are indicators of secretory function of the cell.

The facility of actually looking at the features directly, in spite of the various artifacts that may arise, attracts the researchers from many fields. To cite an example, viral infection can be studied by identifying the viral DNA in the tissue extract using molecular techniques. But the viral particles can be visualized/ localized directly within the tissue intracellularly, whether intranuclear or intracytoplasmic, using the EM. Thus, both the techniques are complementary. The viral DNA detection by molecular techniques may also be due to contamination and similarly, when the viral DNA is integrated into the host DNA, the visualization using EM is not possible. The morphological features of the infected cells in the tissue as well as the clinical symptoms confirm the infection.

Presently, for publication of research work, ultrastructural study, wherever applicable, has become mandatory, even for nanoparticles.

The following are few examples emphasizing the importance of ultrastructural studies in biological research. Ultrastructural comparative study of the test and control systems clearly outline the significance of the study. The specimens prepared elsewhere were processed; ultrathin sections were cut using the Leica Ultracut R Ultramicrotome and were screened in the JEOL JEM 100 SX Transmission Electron Microscope at an accelerating voltage of 80 kV in our laboratory.

EFFECT OF PETROLEUM HYDROCARBONS ON SOME MARINE ANIMALS

The study undertaken was to explore the probable damage to marine animals exposed to petroleum hydrocarbons, in case of accidental oceanic oil spills. For this purpose, a marine species, *Perna viridis*, was exposed to hydrocarbons in different concentrations and for different time intervals. Primarily, the hepatopancreas was chosen as the organ of study. Figures 19.1–19.5 show the ultrastructural features after exposure. The rupture of membranes including nuclear membrane, autophagic vacuoles, proliferation of ER (which may be considered as a step in the detoxification mechanism), distortions in basal lamina and loss of cell inclusions and organelles are some of the important features identified in the EM.

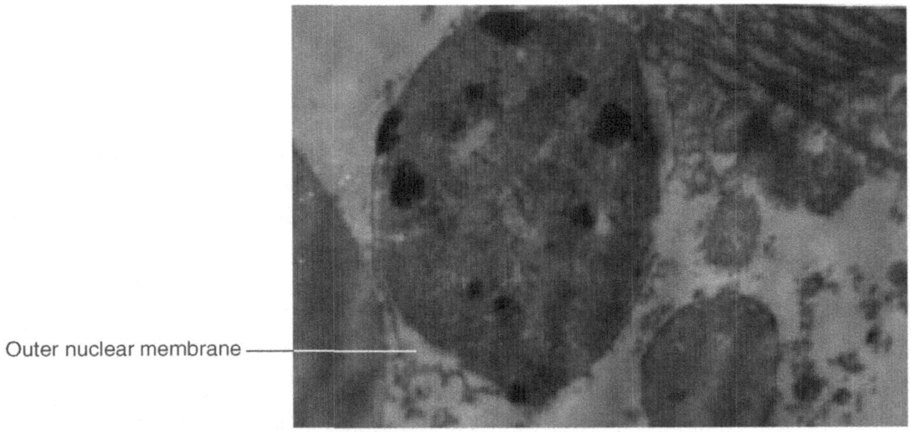

Figure 19.1 Damage of outer nuclear membrane (15,000×)

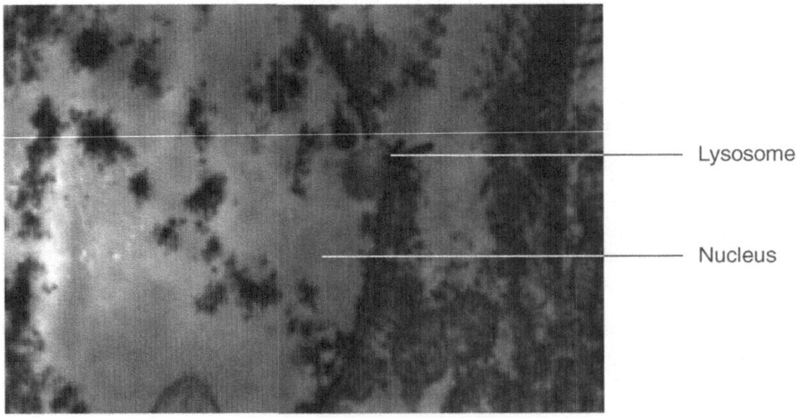

Figure 19.2 Entry of a lysosome through the ruptured nuclear membrane (20,000×)

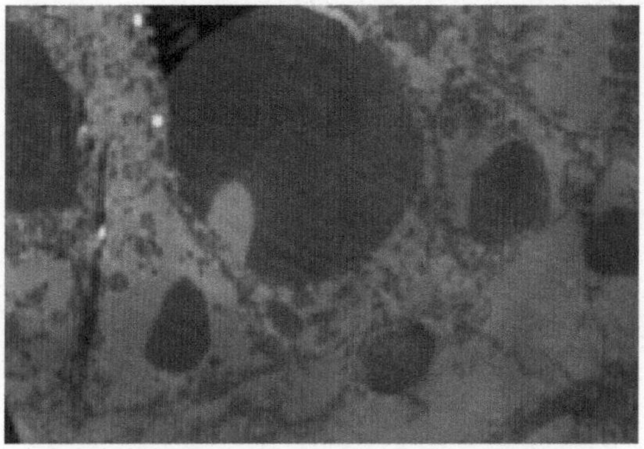

Figure 19.3 Autophagic invagination by the lysosome (30,000×)

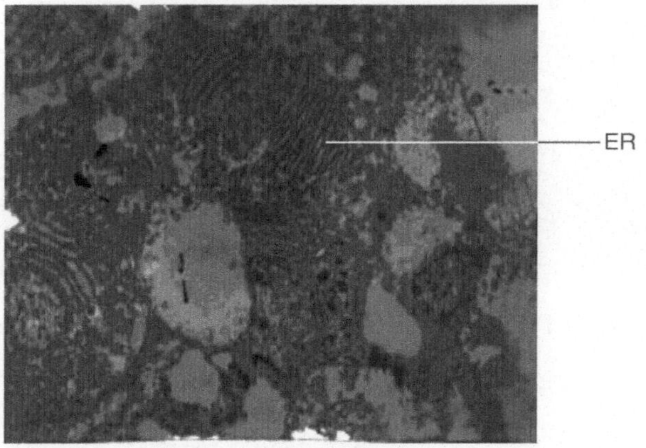

Figure 19.4 Proliferation of ER (Endoplasmic reticulum) (40,000×)

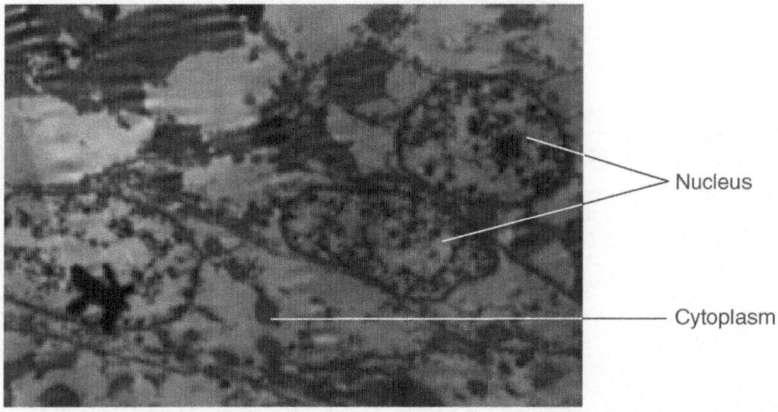

Figure 19.5 Distortions in basal lamina and loss of cytoplasmic organelles (8000×)

INDUCED ARTHRITIS IN WISTAR RATS

Ultrastructural study on induced arthritis in Wistar rats (*Rattus norvegicus*) shows hypertrophy of chondrocytes (Figure 19.6) and also infiltration of joint cartilage by inflammatory cells including lymphocytes and macrophages. The extracellular matrix (ECM) surrounding the inflammatory cells appear to be completely digested (Figure 19.7).

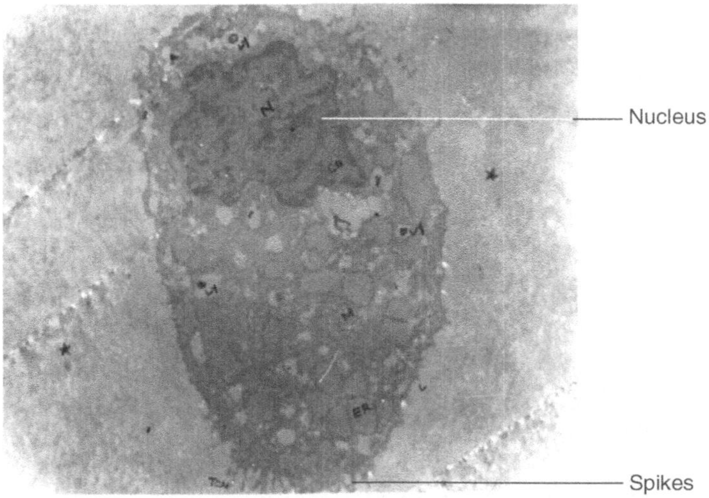

Figure 19.6 The proliferating chondrocyte (5000×)

The proliferating chondrocyte is seen with its characteristically scalloped cell membrane showing flaps and spikes protruding into the prominent extracellular matrix (Figure 19.6) The cytoplasm is mainly occupied by RER, particularly, extended cisternae filled with proteinaceous material.

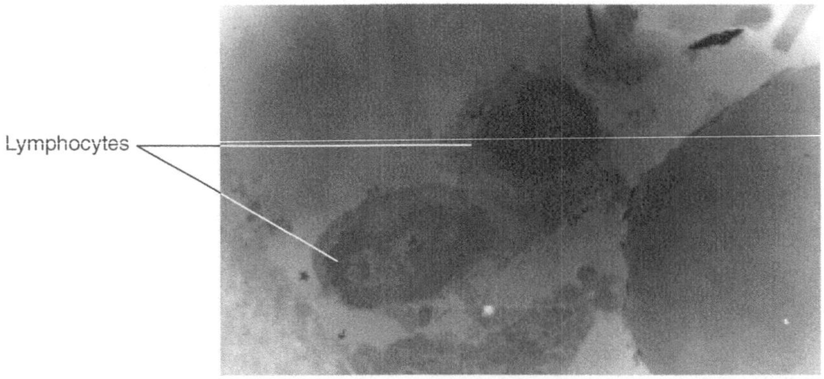

Figure 19.7 Two lymphocytes identified at the joint cartilage. Surrounding the lymphocytes the ECM appears to be completely digested (8000×)

EFFECT OF THE INDIAN DRUG "TRIPHALA" ON INFECTED WOUND IN A LABORATORY ANIMAL

The management of infected wound by an Indian herbal drug "Triphala" in the form of ointment and triphala-incorporated collagen dressing has been studied. The collagen fibres at higher magnifications in the EM appear more packed and presumably better aligned in the Triphala group than the groups without Triphala, which probably suggests the beginning of macromolecular organization of granulation tissue (Figures 19.8 and 19.9).

Figure 19.8 Rat skin collagen fibres—normal (20,000×)

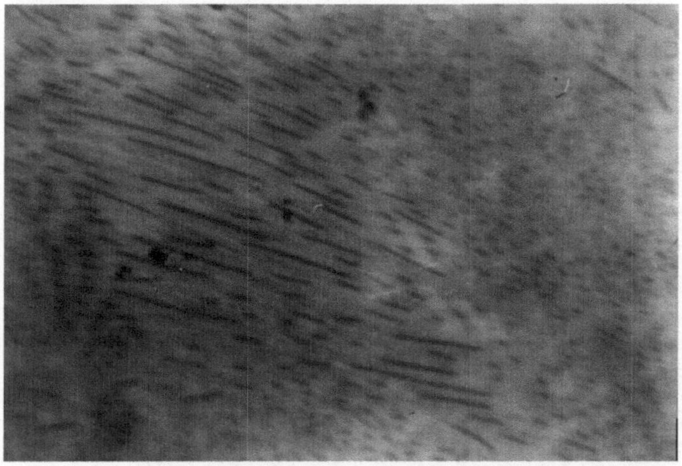

Figure 19.9 Collagen bundles after application of Triphala ointment (20,000×)

CHEMOPREVENTIVE ACTIVITY OF CAPSAICIN

Capsaicin, an effective natural antioxidant derived from capsicum fruit extracts, is known to exhibit chemopreventive activity against benzo(α)pyrene-induced experimental lung carcinogenesis. Ultrastructurally, the lung tissue of benzopyrene-treated animals (male Swiss albino mice) shows closely packed pleomorphic cells with irregular nuclei. But the lung tissue of benzo(α)pyrene along with capsaicin-treated animals shows features of both early and late apoptosis characterized by nuclear clumping and chromatin condensation (Figures 19.10–19.13).

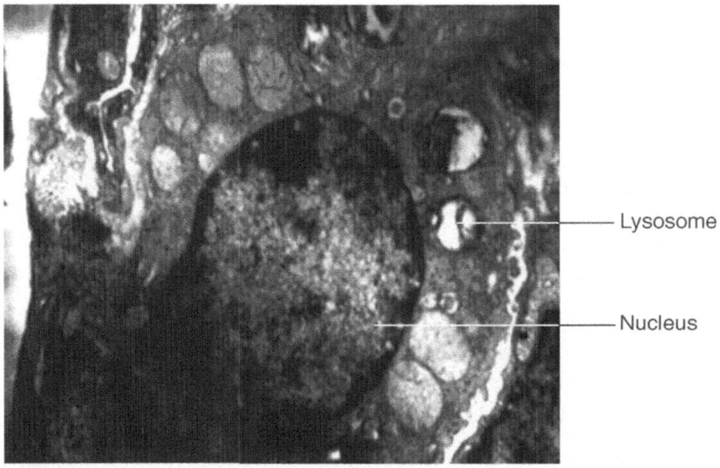

Figure 19.10 One cell from a lung tissue of a control animal showing normal cell structure with intact nucleus and cytoplasm (10,000×)

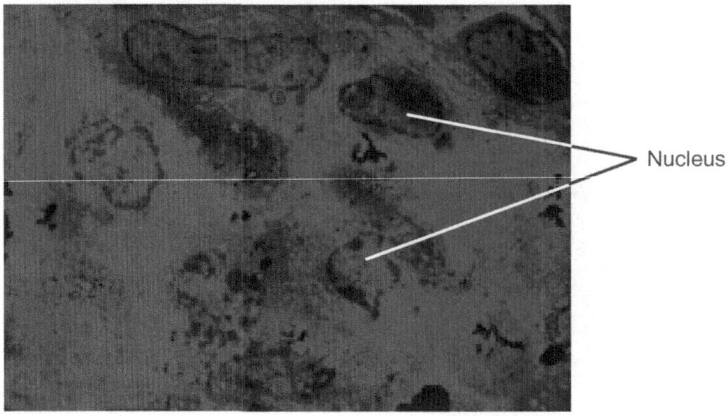

Figure 19.11 Lung tissue from benzopyrene-induced lung cancer showing closely packed pleomorphic cells with irregular shaped nuclei (4000×)

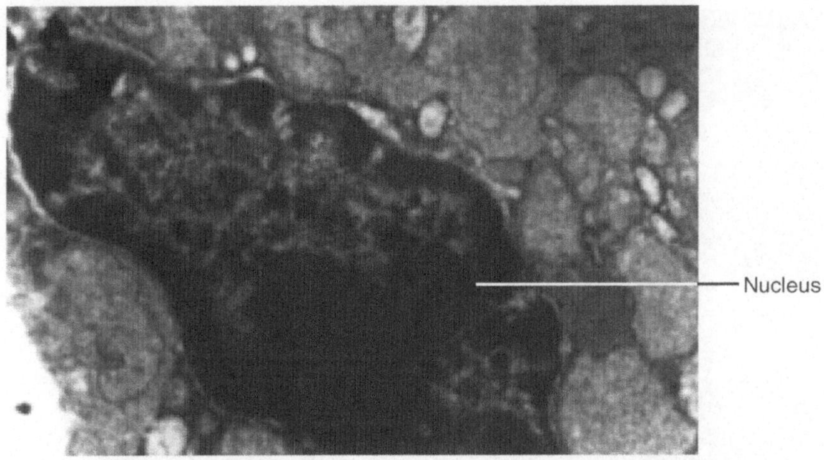

Figure 19.12 Cell from sample treated with benzo(α) pyrene + capsaicin—early apoptosis (20,000×)

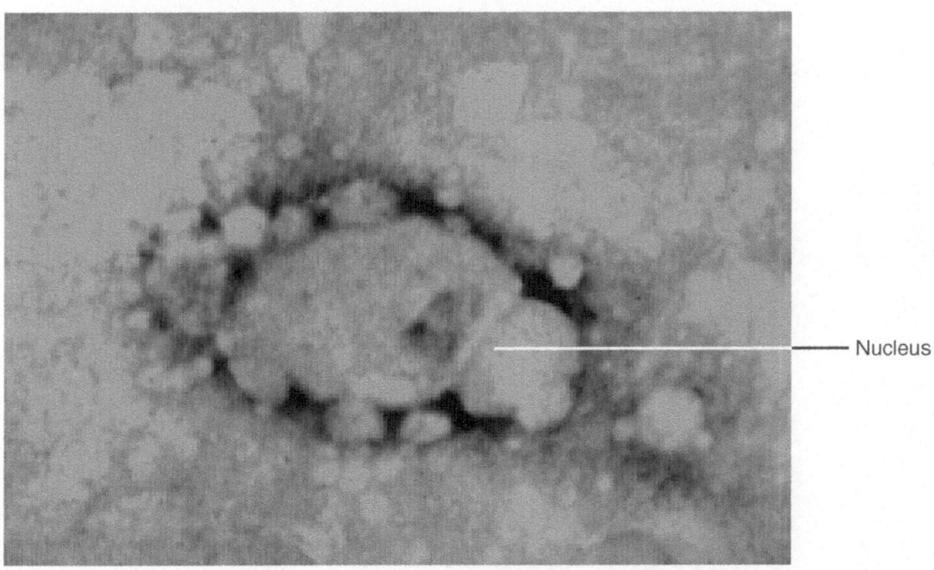

Figure 19.13 Benzopyrene along with capsaicin-treated lung tissue showing an apoptotic cell (20,000×)

The study also investigates whether lysosome is a target in benzo(α)pyrene-induced oxidative-stress-mediated lung cancer in Swiss albino mice and the plausible role of the phytochemical substance capsaicin in mitigating lysosomal damage. Ultrastructural studies shows the defensive action of capsaicin against the lysosomal damage caused in benzo(α)pyrene-induced lung cancer (Figures 19.14–19.16).

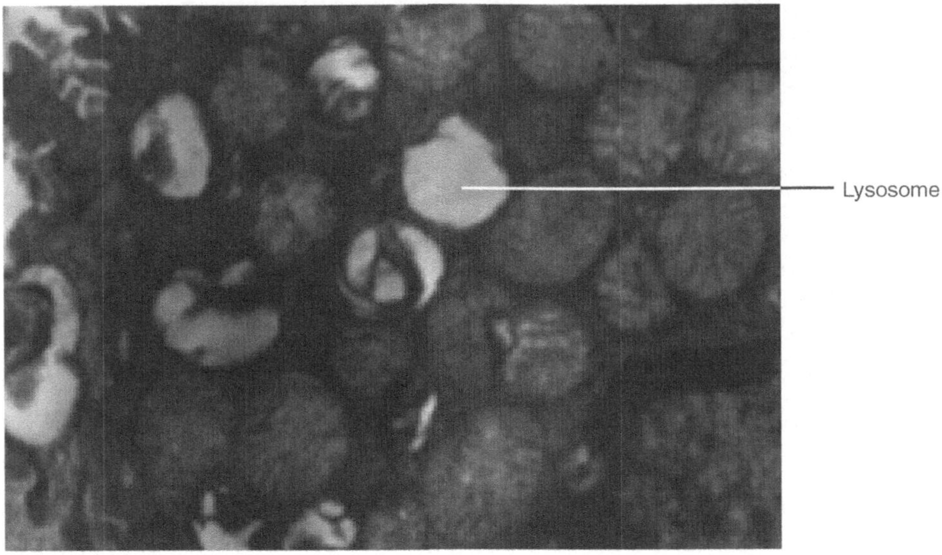

Figure 19.14 Cell from the control group showing the normal architecture of the cytoplasmic lysosomes (15,000×)

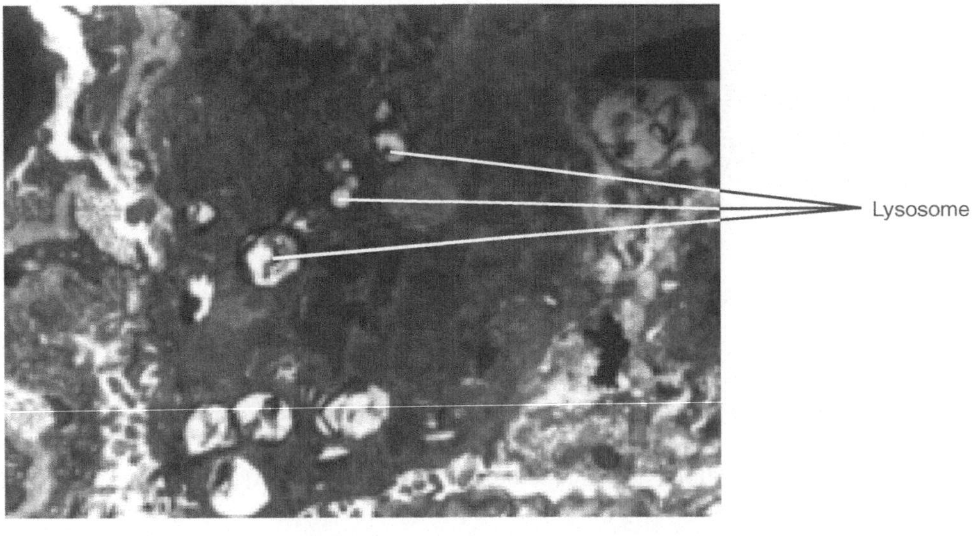

Figure 19.15 The cytoplasmic lysosomes in the sections from benzo(α)pyrene induced lung cancer show irregular shape, with small and large amorphous densities, and aggregation of intramembranous engulfed particles revealing loss of lysosomal membrane integrity (10,000×)

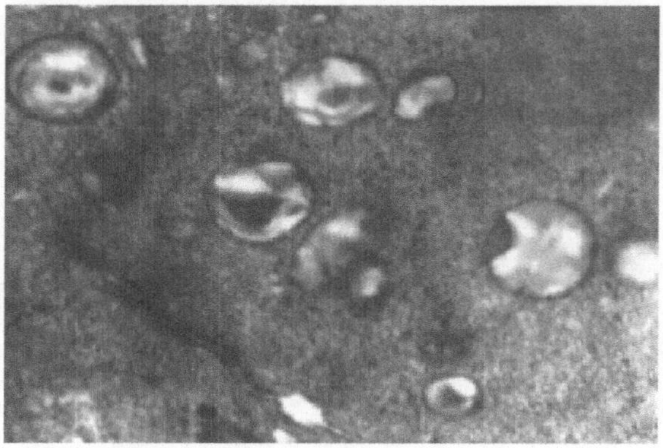

Figure 19.16 Benzo(α)pyrene along with capsaicin administration showing restoration of lysosomes to near normal architecture (20,000×)

MYCOBACTERIUM-INFECTED CULTURED MESOTHELIAL CELLS

Mesothelial cell line (Met5A cells) infected with *M. tb* (*Mycobacterium tuberculosis*) for different periods of time is studied. Though uptake of *M. tb* by non-immune cells like epithelial cells, fibroblasts, adipocytes, etc. has been reported, this is claimed to be the first report on the internalization of *M. tb* bacilli by mesothelial cells. The uninfected cells reveal the presence of numerous microvilli as surface projections which is the characteristic feature of mesothelial cells. In addition, the cells show a high nuclear–cytoplasmic ratio, scanty mitochondria and lysosomes in the cytoplasm (Figure 19.17). But the infected cells show prominent, elongated mitochondria, increase in fat vacuoles and more prominent smooth and rough endoplasmic reticulum, which are the characteristic features of activated mesothelial cells. The membrane changes and the engulfing of the particles within membrane-bound vacuoles could be easily made out (Figures 19.18–21).

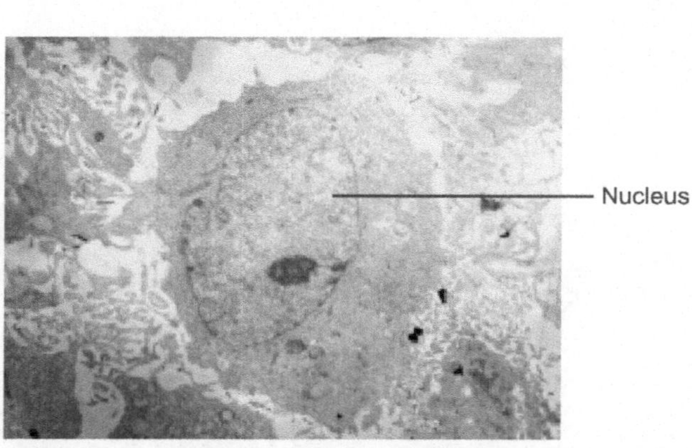

Figure 19.17 One mesothelial cell from cultured mesothelial cell line (3000×) Control, uninfected cell shows microvilli and cytoplasmic organelles, nucleus and nucleolus.

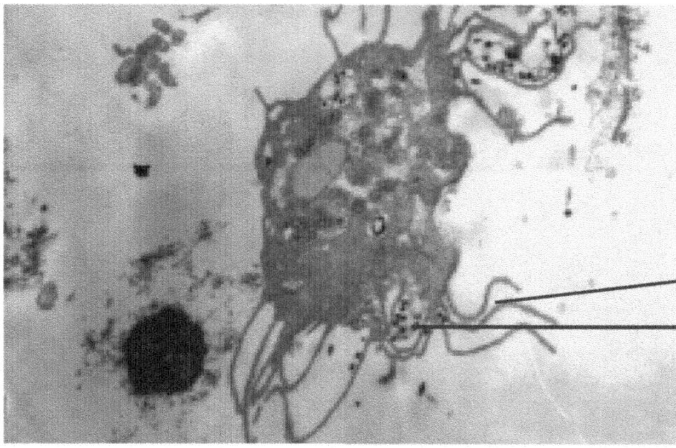

Figure 19.18 One mesothelial cell infected with *M. tb* showing microvilli and engulfed particles (6000×)

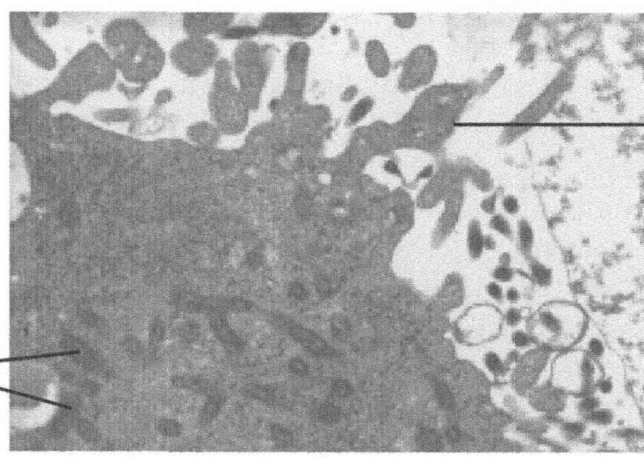

Figure 19.19 Blunt and thickened mv in a mesothelial cell infected with *M.tb* (15,000×)

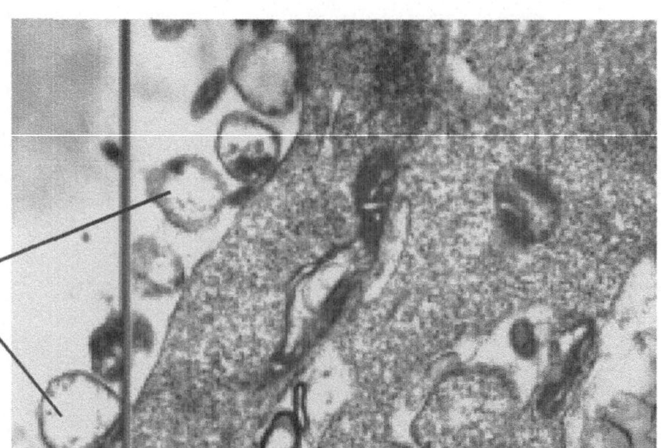

Figure 19.20 Membrane bubbles in a mesothelial cell infected with *M.tb* (25,000×)

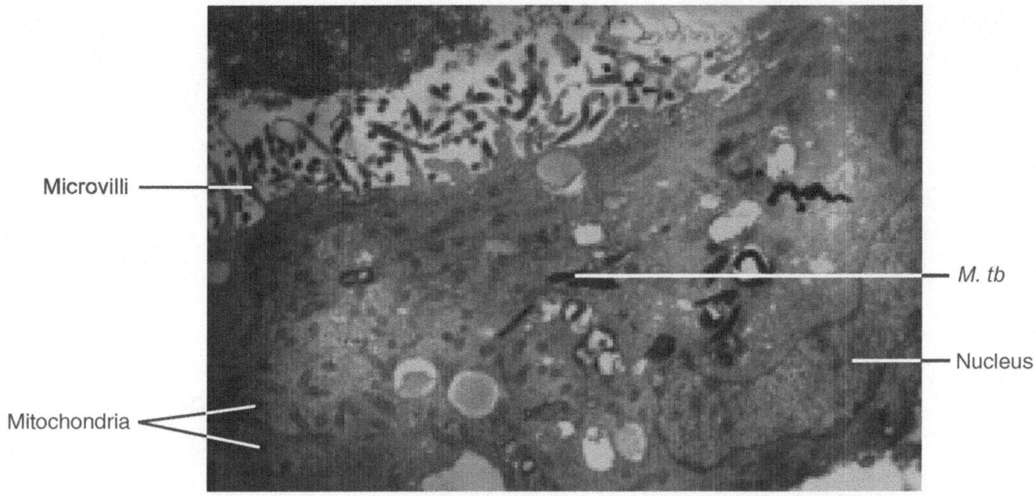

Figure 19.21 Rod-shaped *M. tb.* in cytoplasm of mesothelial cell infected with *M.tb* (8000×)

ULTRASTRUCTURAL FEATURES OF 1-DAY- AND 4-DAY-OLD FISH LARVAE

The 1-day-old and 4-day-old fish larvae are compared for their morphological variations including development of cellular organelles (Figures 19.22–19.26).

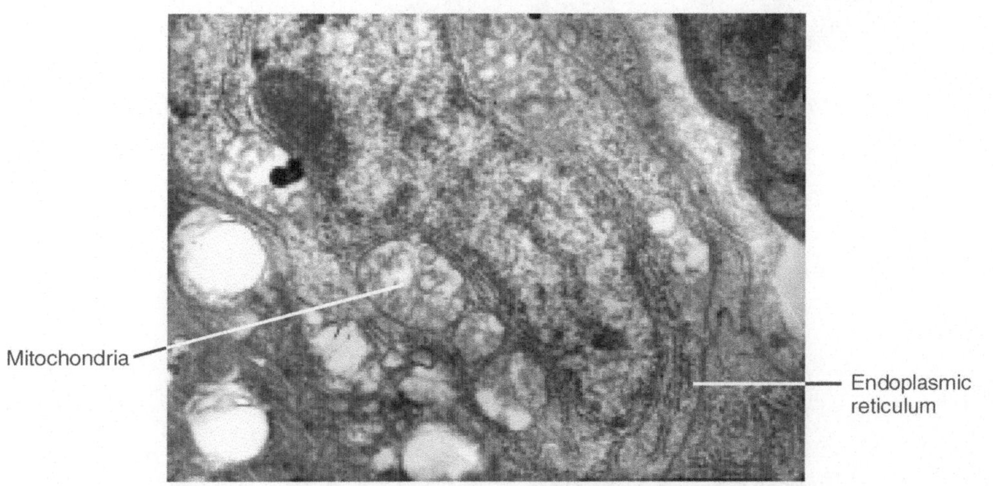

Figure 19.22 Cell shows cellular organelles including developing mitochondria in a 1-day-old larva (15,000×)

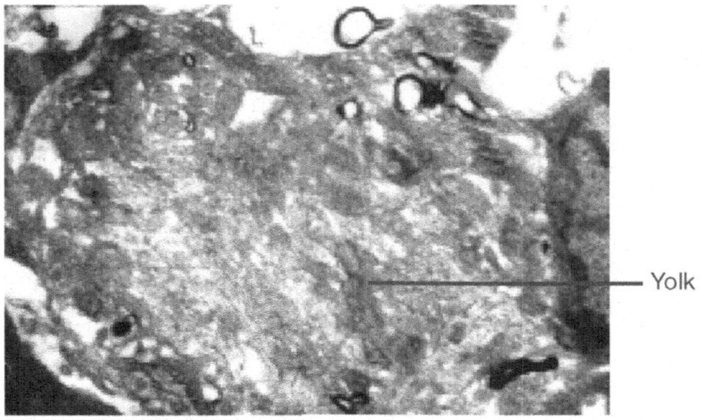

Figure 19.23 Yolk material in a 1-day-old larva (15,000×)

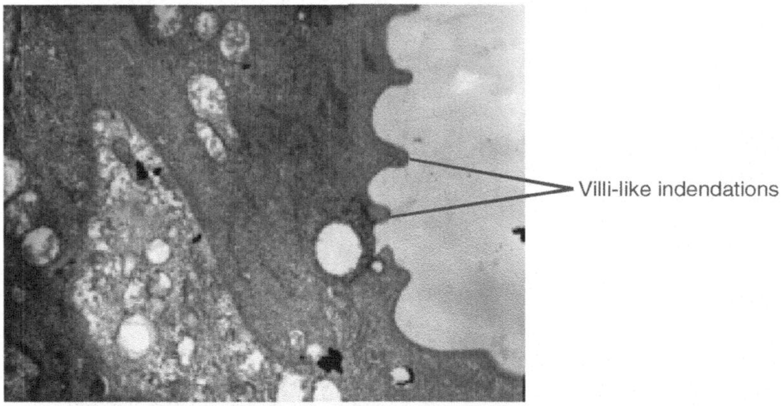

Figure 19.24 Villi-like indentations in a 1-day-old larva (10,000×)

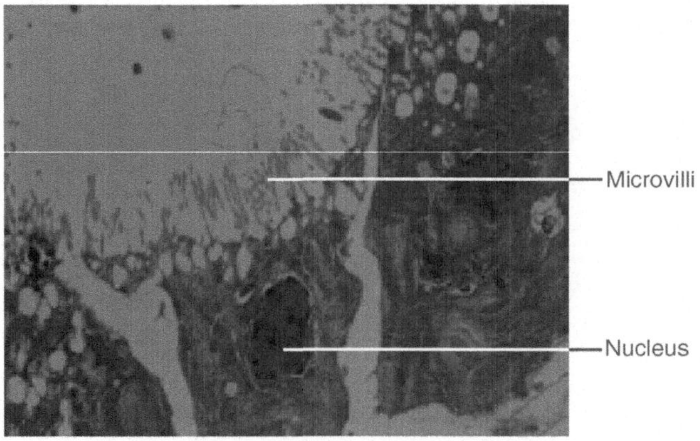

Figure 19.25 Cells from 4-day-old fish larva showing microvilli (6000×)

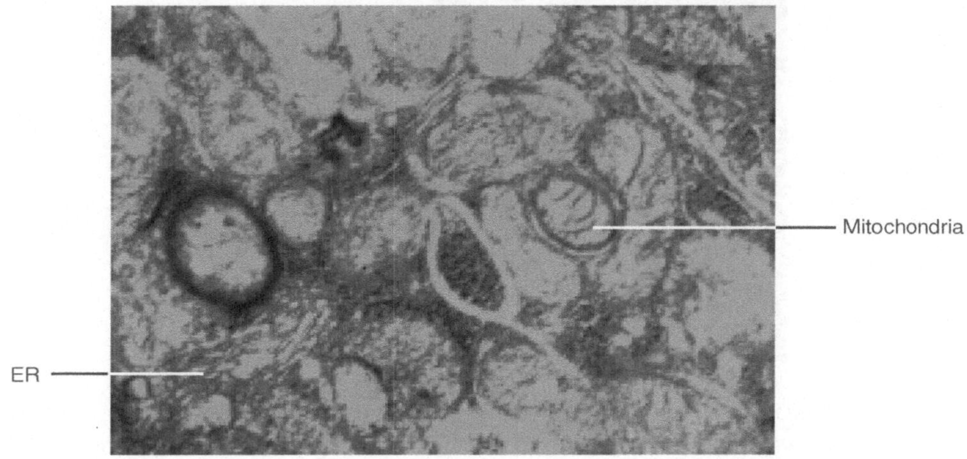

Figure 19.26 Cell organelles in a 4-day-old larva (80,000×)

DIFFERENTIATION OF MESENCHYMAL STEM CELLS INTO CARDIOMYOCYTES AND HEPATOCYTES

A. Undifferentiated mesenchymal cells (mesenchymal stem cells) are isolated from bone marrow and made to differentiate into cardiomyocytes by exposure to heart extract (Figures 19.27–19.31).

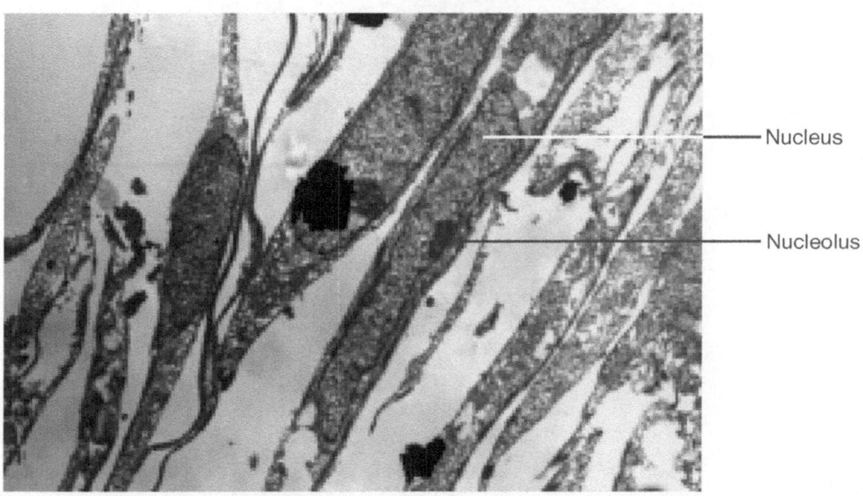

Figure 19.27 Undifferentiated mesenchymal cells from bone marrow showing primitive features such as spindle cells with high nuclear cytoplasmic ratio, spindle nucleus with a small nucleolus (4000×)

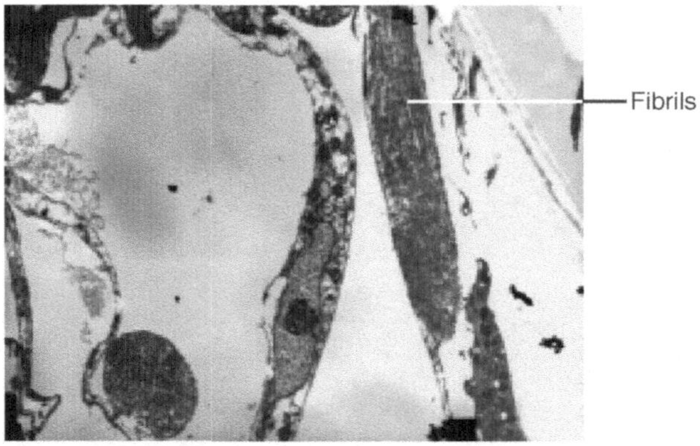

Figure 19.28 Differentiating mesenchymal stem cells showing more cytoplasmic organelles. One cell shows cytoplasmic fibres probably indicating the differentiation into a myocyte (3000×)

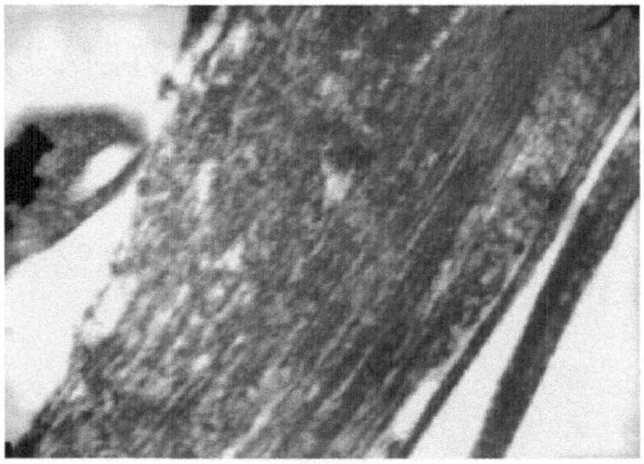

Figure 19.29 Differentiating mesenchymal cells showing myofibres (30,000×)

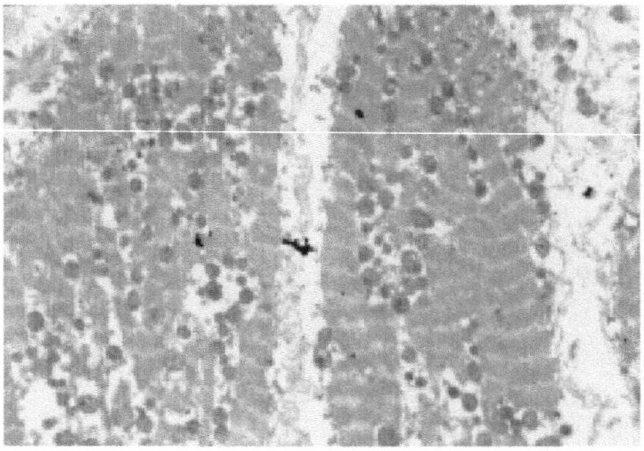

Figure 19.30 Myofibrils (myotubes) and mitochondria in a more differentiated area (4000×)

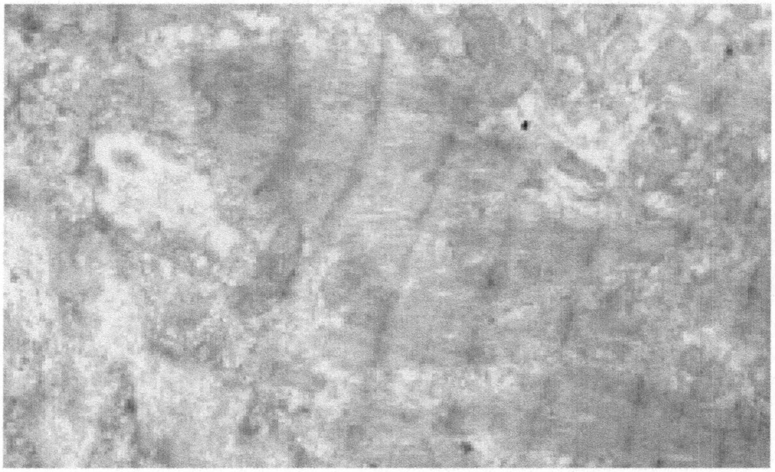

Figure 19.31 Normal cardiomyocyte (Myofibres) (10,000×)

B. Mesenchymal stem cells from bone marrow differentiate into hepatocytes by treatment with jaundiced serum. The ultrastructure of the differentiating cells show features like spherical cell with round nucleus, and abundant cytoplasm; cytoplasmic organelles including Golgi, and also membrane projections resembling microvilli. These features are indicators of the differentiation into epithelial cells (Figures 19.32–19.34).

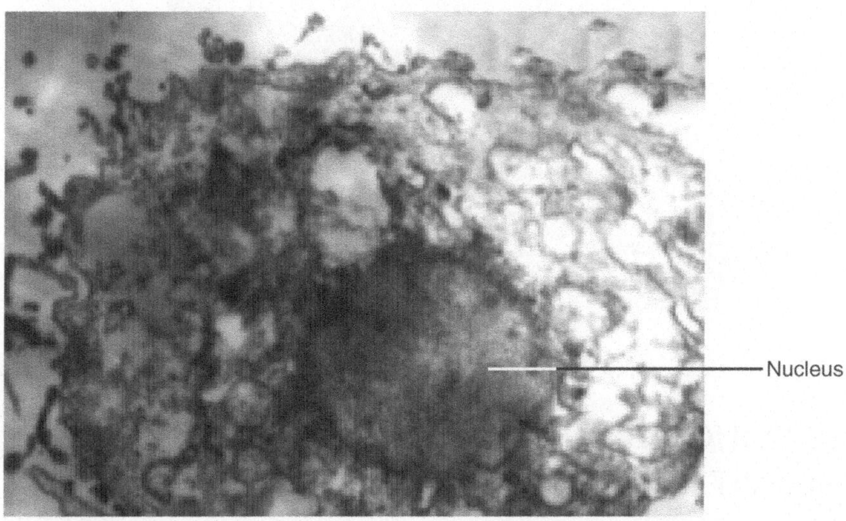

Figure 19.32 One cell showing a round nucleus, and plenty of cytoplasmic developing organelles (8000×)

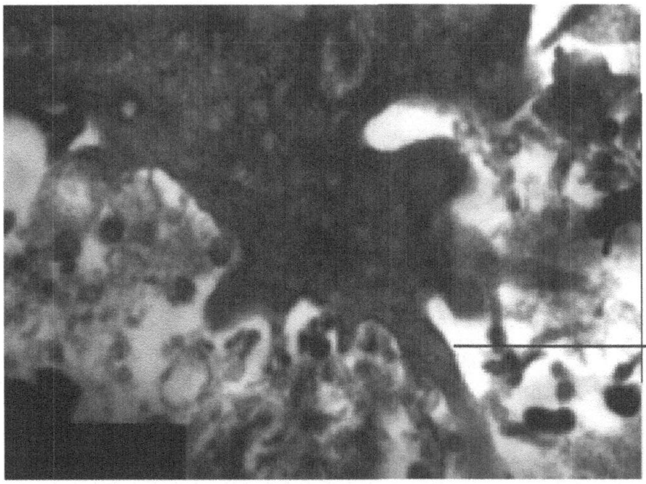

Figure 19.33 One cell showing membrane projections resembling microvilli (20,000×)

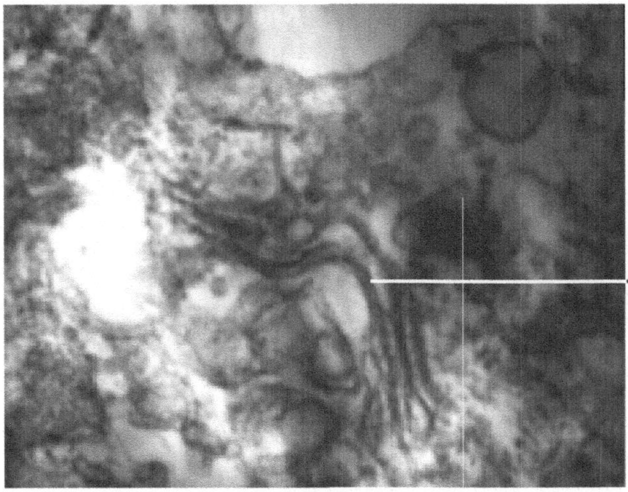

Figure 19.34 Cell showing Golgi complex and other cytoplasmic organelles (50,000×)

MALE REPRODUCTIVE SYSTEM OF MARINE HERMIT CRAB

The ultrastructural studies on the ultrathin sections from the male reproductive system of a **marine hermit crab** shows several interesting features (Figures 19.35–19.41).

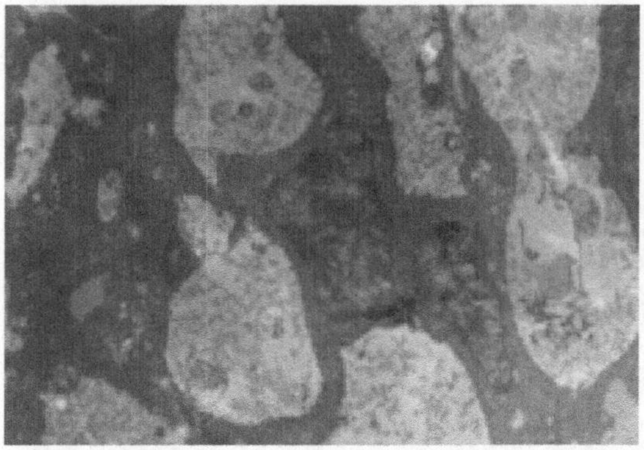

Figure 19.35 Spermatocytes and nourishing cells (3000×)

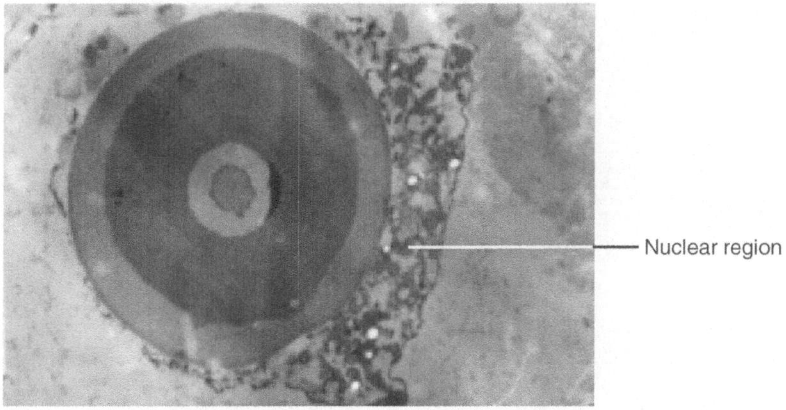

Figure 19.36 Cross-sectional view of a spermatocyte showing nuclear region (8000×)

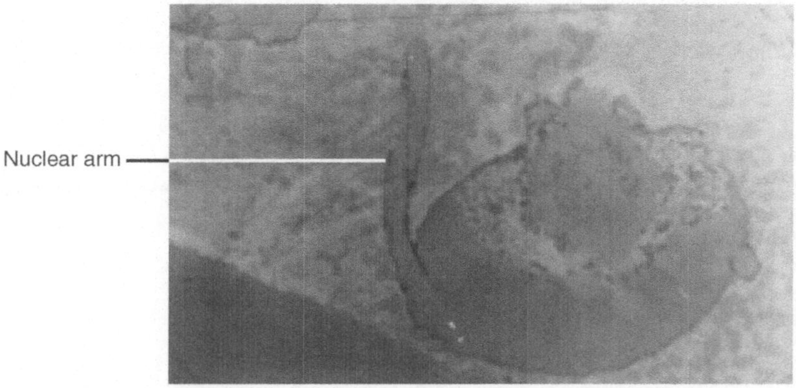

Figure 19.37 Nuclear arm of a spermatocyte showing microtubules (10,000×)

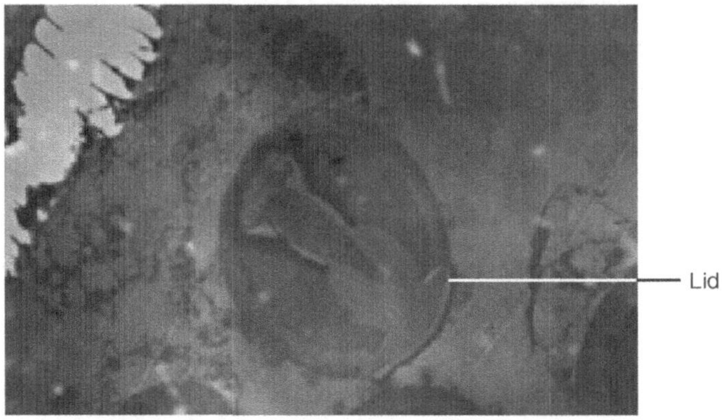

Figure 19.38 Longitudinal section of a spermatocyte showing the operculum (lid) (5000×)

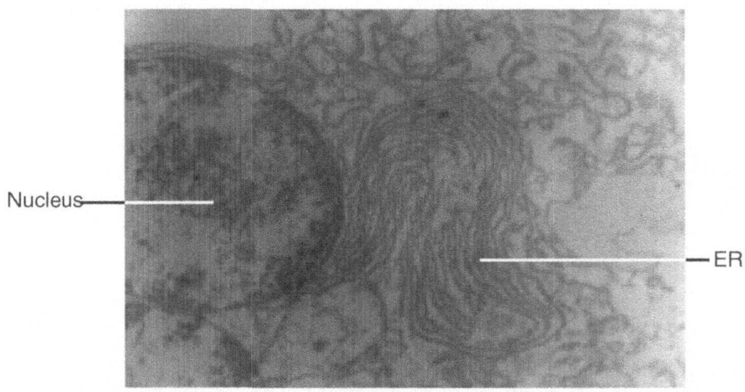

Figure 19.39 Nuclei and cytoplasmic organelles in the spermatocyte (6000×)

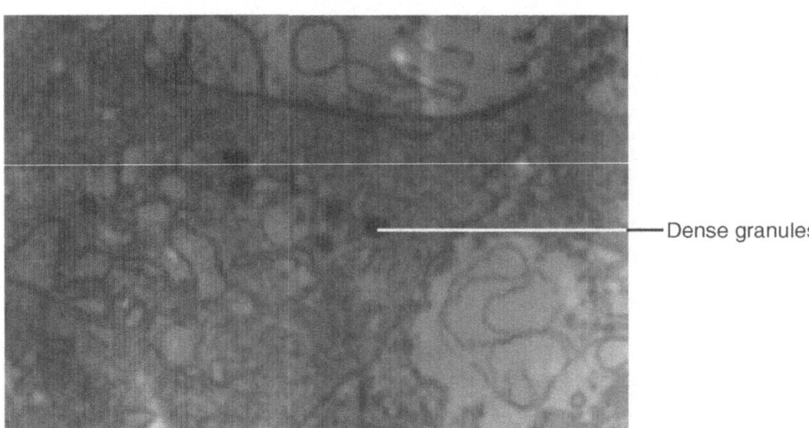

Figure 19.40 Membrane invaginations and a few granules with dense core (15,000×)

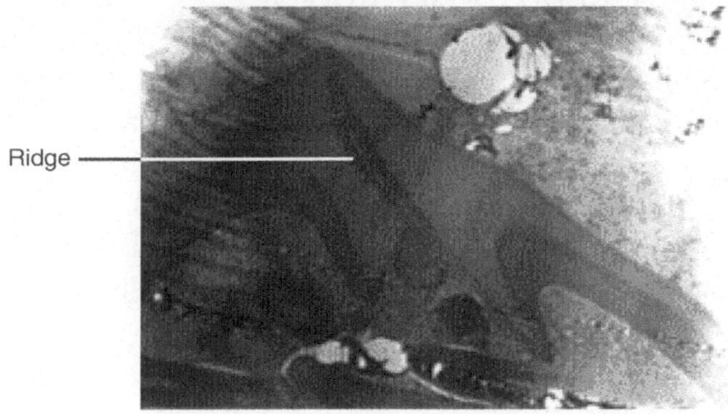

Figure 19.41 A ridge through which the sperms may be released (10,000×)

RECONSTITUTED COLLAGEN

Reconstituted collagen after negative staining with uranyl acetate shows the characteristic bands as seen in rat tail tendon collagen as evident in Figures 19.42 and 19.43.

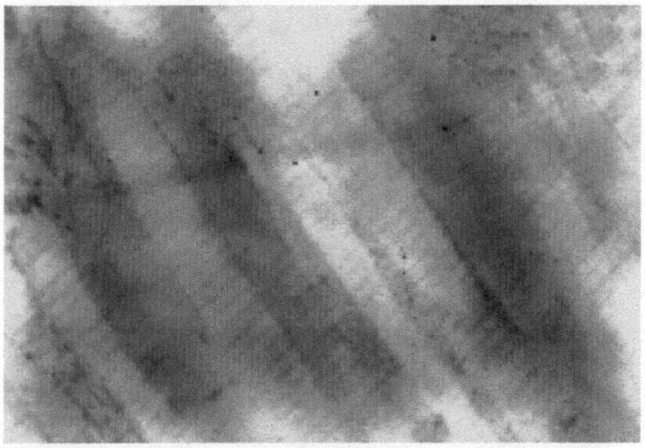

Figure 19.42 Banded collagen fibres from rat tail tendon (40,000×)

Figure 19.43 A few fibres of reconstituted collagen with bands (50,000×)

STUDY ON ENDOTHELIAL AND EPITHELIAL CELL LINE

The efficacy of the antibacterial activity of a phytochemical drug on human endothelial cell line (ECV 304) and human epithelial cell line (EHEp2) infected with *E. coli* (E0157:H7) is studied ultrastructurally. The structural damage caused by the bacterial infection is alleviated to some extent, when the cells are pretreated with the protective drug. Figures 19.44 to 19.48 show features in cultured endothelial cells while Figures 19.49 to 19.56 show ultrastructural features in cultured epithelial cells.

1. Endothelial cell line

   Control (Figure 19.44)

   Exposed to *E. coli* (Figures 19.45 and 19.46)

   Pretreated with protective drug prior to exposure to *E. coli* (Figure 19.47 and 19.48)

2. Epithelial cell line

   Control (Figures 19.49 and 19.50)

   Infected with *E. coli* (Figures 19.51 to 19.53)

   Pretreated with drug and then infected with *E. coli* (Figure 19.54–19.56)

Endothelial cell line

a. Control

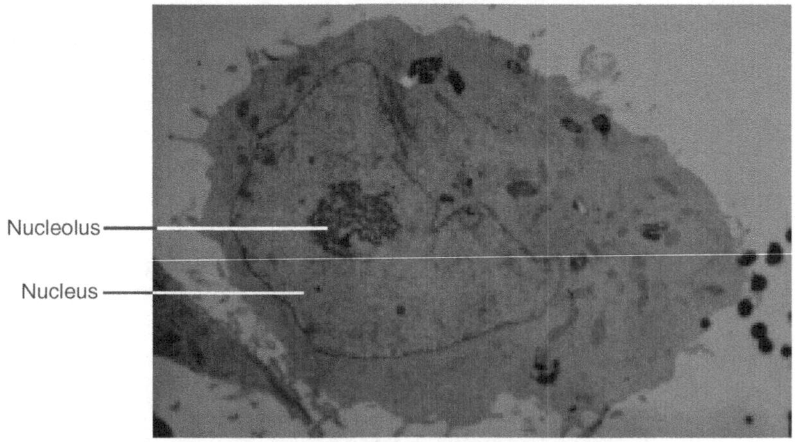

Figure 19.44 One intact cell in the control sample (5000×)

b. Endothelial cells exposed to E. coli

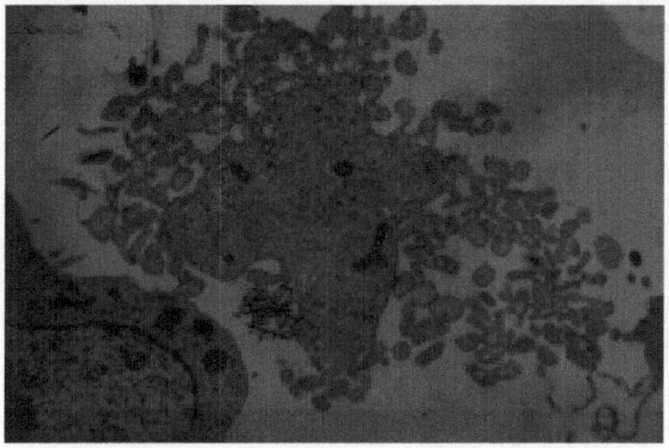

Figure 19.45 *E. coli*-induced damaged cell (6000×)

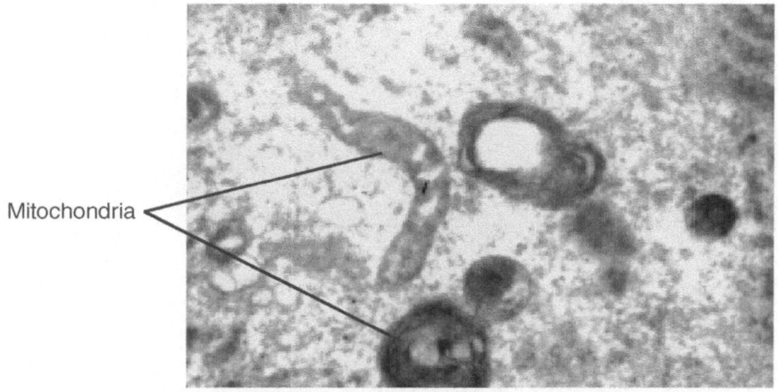

Figure 19.46 One cell infected with *E. coli* showing mitochondrial alterations (30,000×)

c. Endothelial cells exposed to E. coli but protected with protective drug

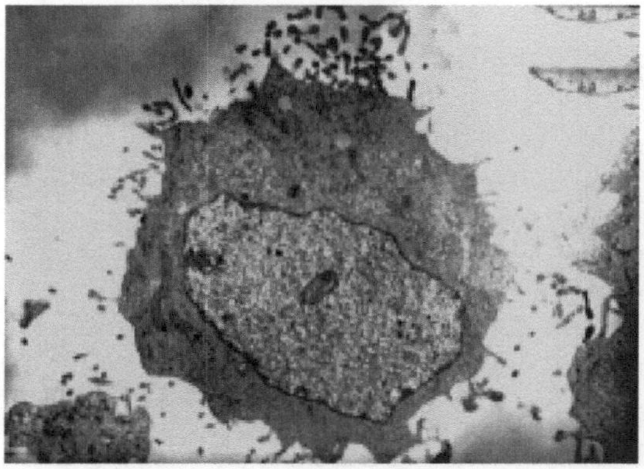

Figure 19.47 One cell with microvilli

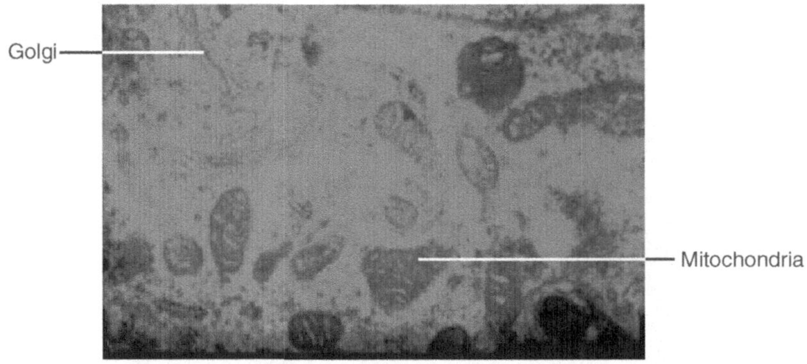

Figure 19.48 Mitochondria with more cristae and Golgi (25,000×)

2. Epithelial Cell Line

a. *Control sample*

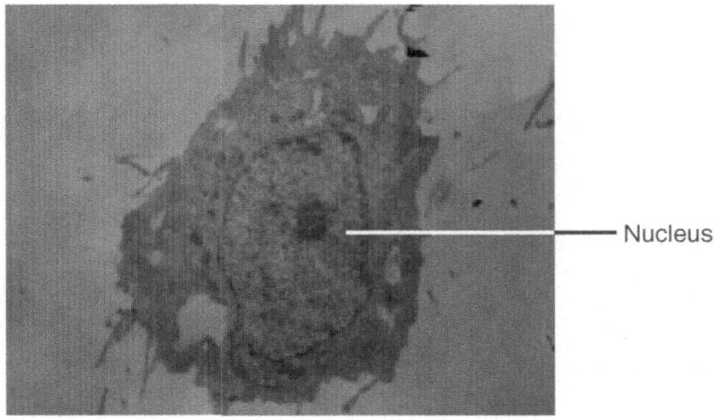

Figure 19.49 One cell in the control sample (5,000×)

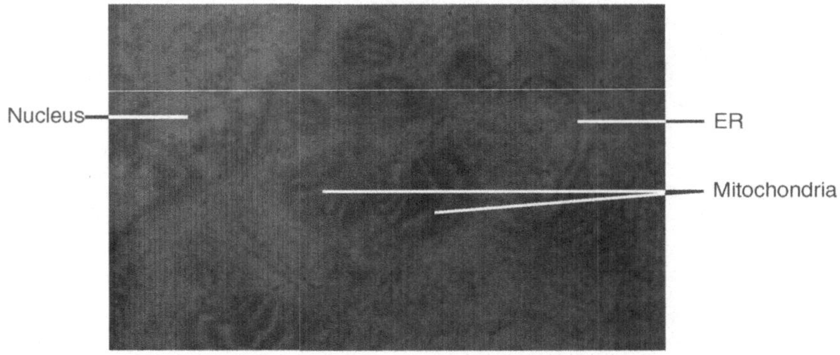

Figure 19.50 A few mitochondria and a small portion of nucleus (40,000×)

b. *Epithelial cells infected with E. coli*

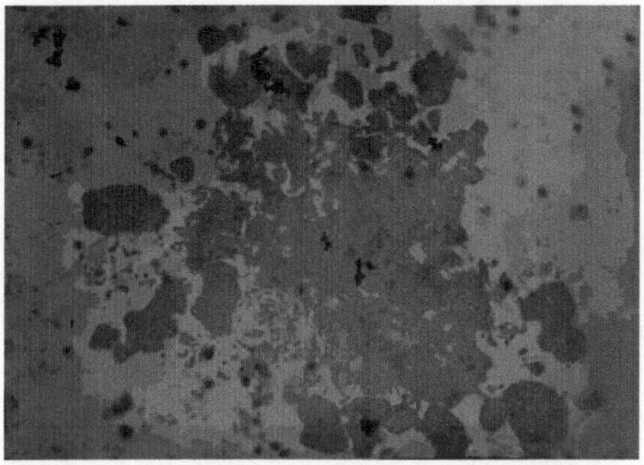

Figure 19.51 A completely damaged cell (4000×)

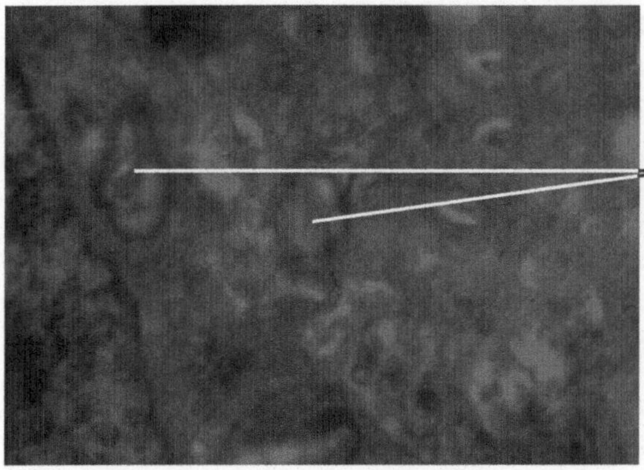

Figure 19.52 Damaged cell showing mitochondria with loss of cristae (30,000×)

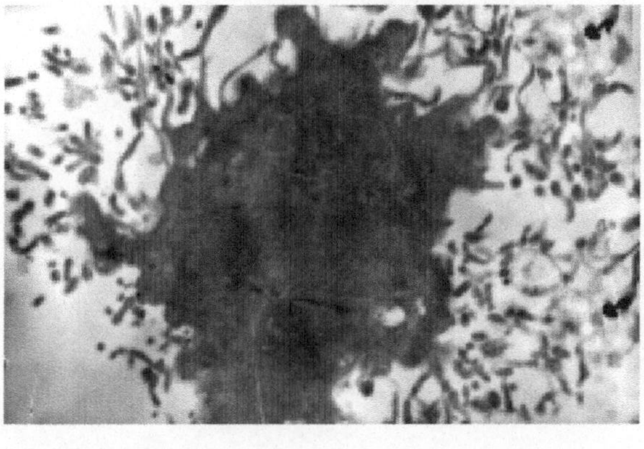

Figure 19.53 Shredding of mv (10,000×)

c. Epithelial cells infected with E. coli but pretreated with protective drug

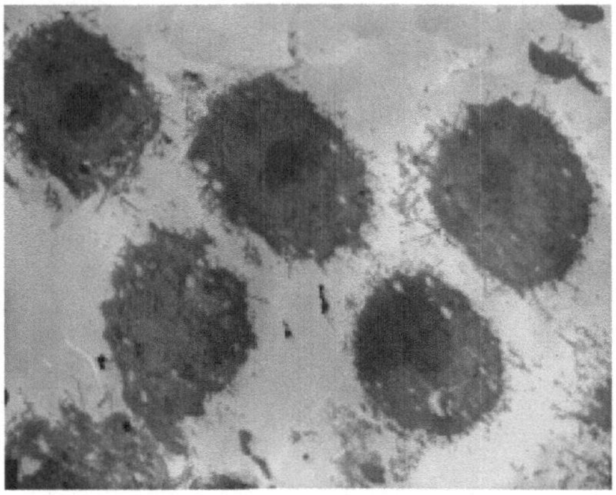

Figure 19.54 A few cells with restoration of size, shape and mv (2000×)

Figure 19.55 One full cell with restoration of mv (5000×)

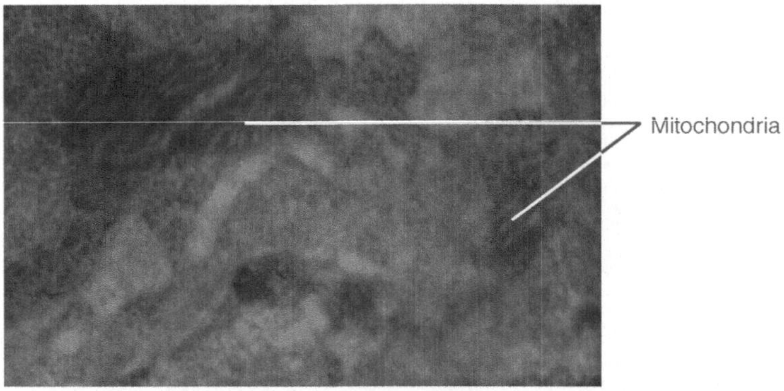

Figure 19.56 Mitochondria with more cristae (30,000×)

SPERM AND OVARY OF MARINE ORGANISMS

Mature sperms of a shrimp wild type, those cultured in ponds and in laboratory tanks are shown in Figures 19.57 to 19.59.

Figure 19.57 Mature sperm. Wild type (10,000×)

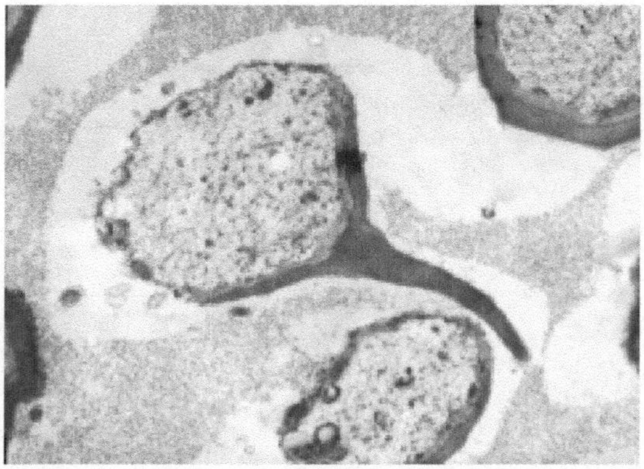

Figure 19.58 Mature sperm from pond (10,000×)

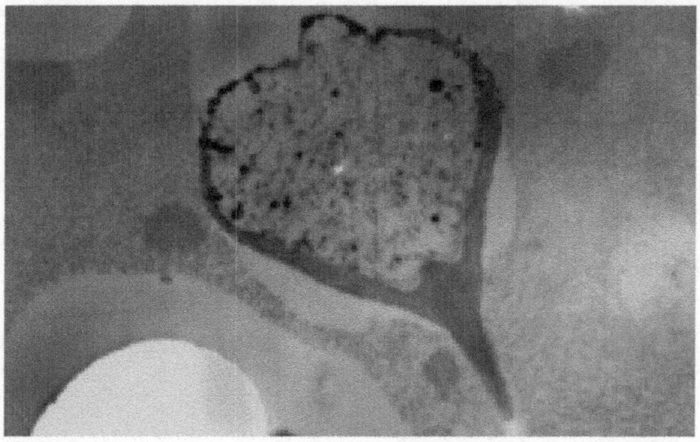

Figure 19.59 Mature sperm from laboratory tanks (10,000×)

The features seen in the ovary of a shrimp are shown in Figures 19.60 and 19.61.

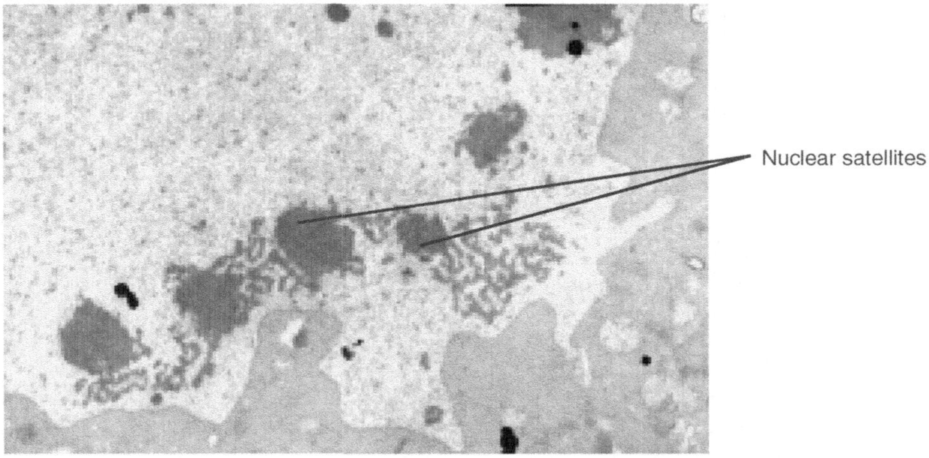

Figure 19.60 Nuclear satellites in an ovary of a shrimp (5000×)

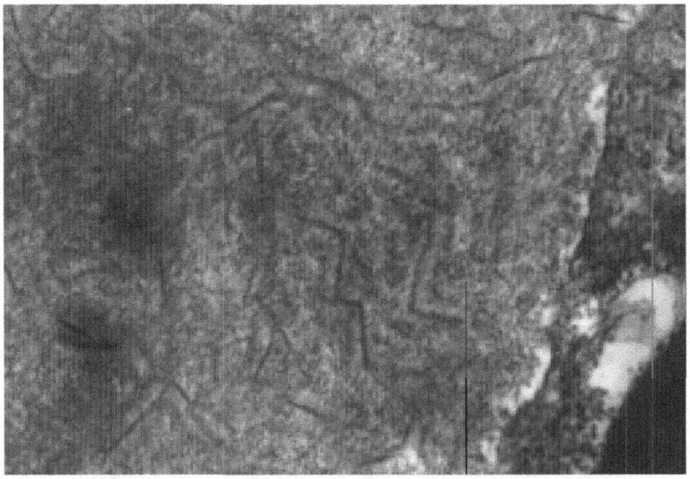

Figure 19.61 Corticle rods in a mature ovary (30,000×)

SPERMS OF A MARINE ORGANISM

Sperms from another marine organism are shown below at higher magnifications (Figures 19.62–19.64).

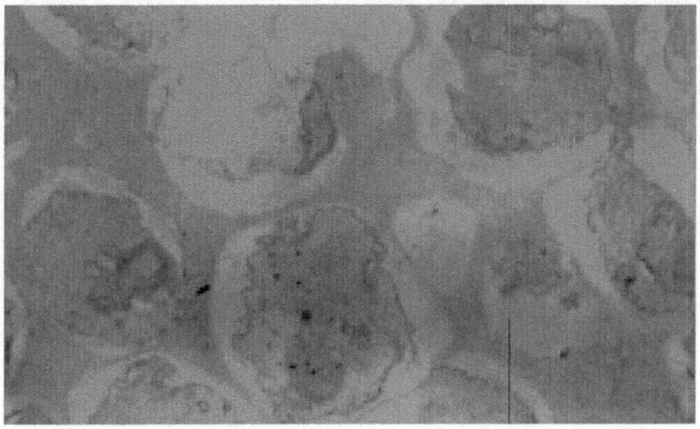

Figure 19.62 A few immature sperms (4000×)

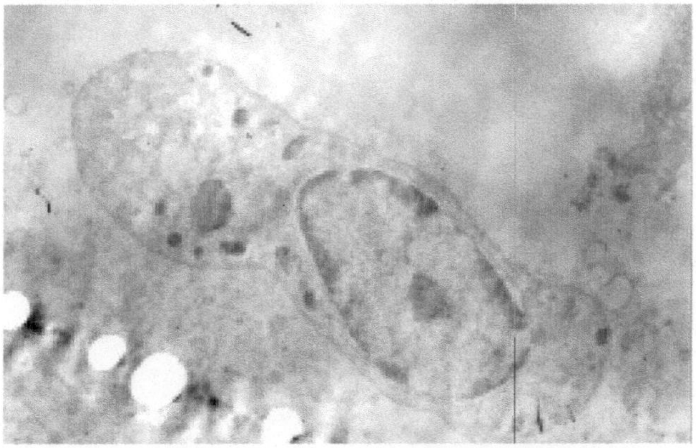

Figure 19.63 Spermatid (6000×)

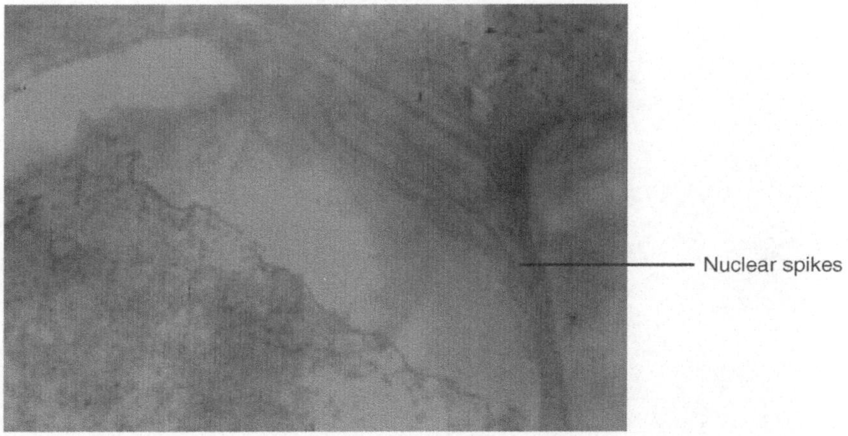

Figure 19.64 Nuclear spikes in a sperm (10,000×)

ULTRASTRUCTURAL STUDIES ON HEART TISSUE OF WISTAR RATS

The electron micrographs shown in Figures 19.65–19.71 are from a study on heart tissue of the laboratory animal, wistar rat. Heart tissue from the following three groups of animals is collected and processed for ultrastructural study.

1. Control sample (Figures 19.65 and 19.66)
2. Animal treated with isoproterenol (ISO) which is an inducer of myocardial infarction (Figures 19.67–19.69)
3. Animal pretreated with a protective herbal drug followed by treatment with ISO (Figures 19.70 and 19.71) to elucidate the protective effect of the drug.

1. *Heart tissue from a control animal*

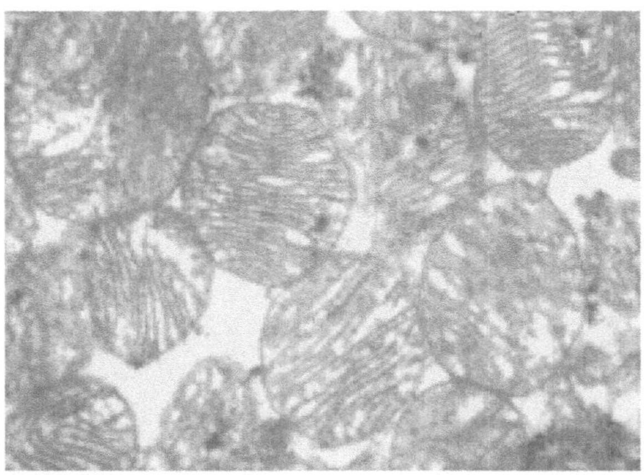

Figure 19.65 A few mitochondria (20,000×)

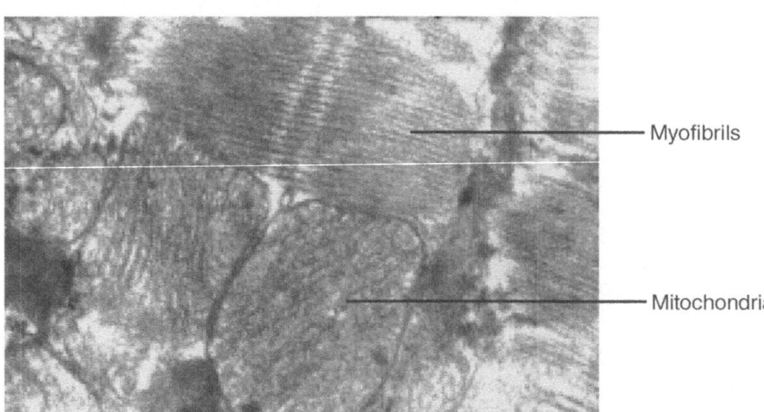

Figure 19.66 Myofibrils and mitochondria (40,000×)

2. *Heart tissue from an animal treated with ISO*

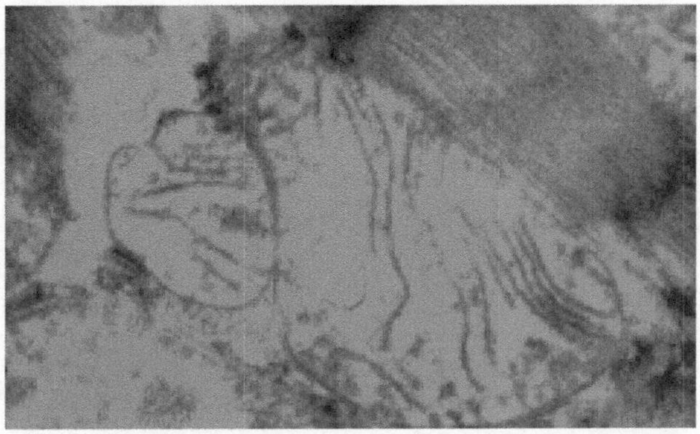

Figure 19.67 Pleomorphic mitochondria with loss of cristae (40,000×)

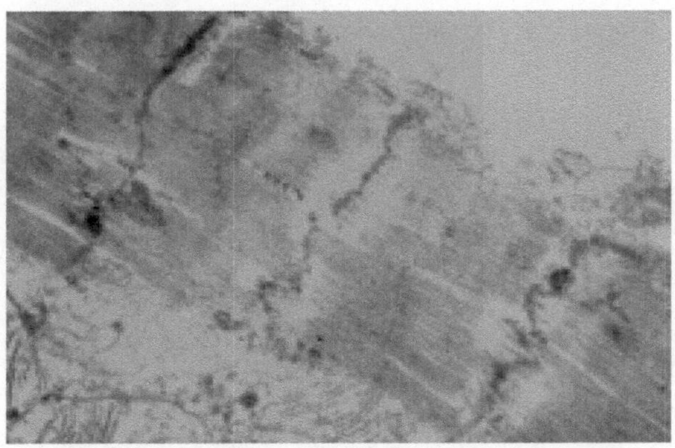

Figure 19.68 Fibrillar damage (15,000×)

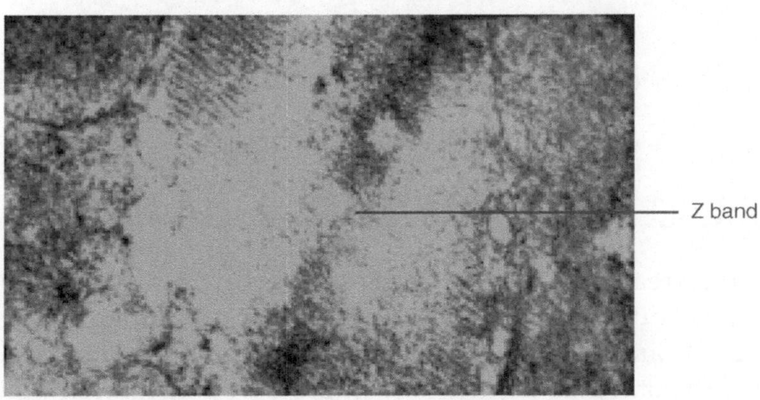

Figure 19.69 Z band damage (40,000×)

3. Heart tissue from an animal pretreated with drug followed by ISO Mitochondria and myofibrils show near normal architecture.

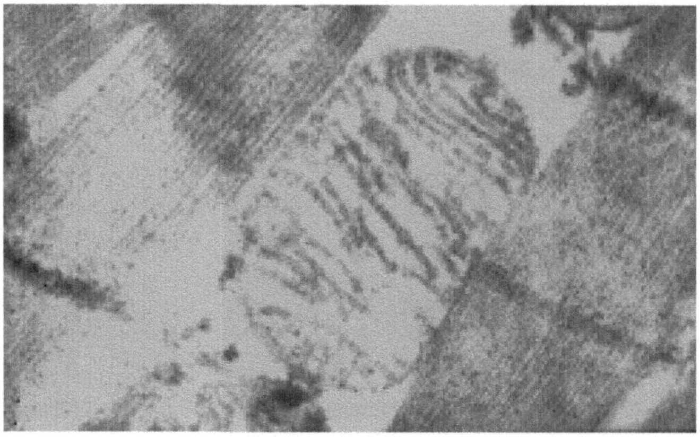

Figure 19.70 Mitochondria + fibrils (40,000×)

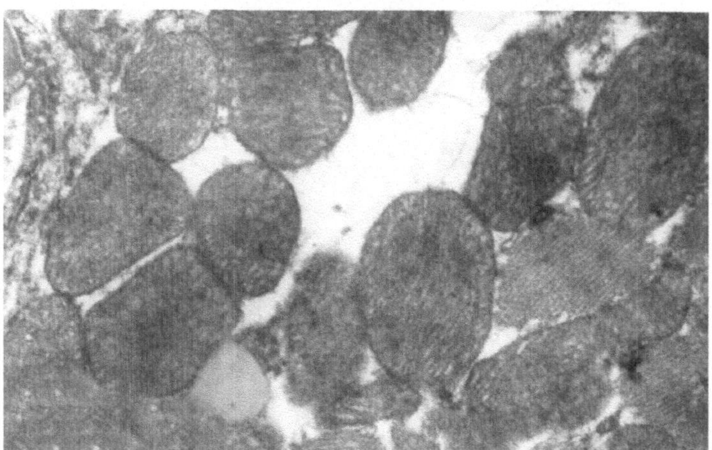

Figure 19.71 A few mitochondria with cristae (50,000×) (drug in maximal dose given)

FIBROBLAST CELL LINE TREATED WITH CHEMICALS

Fibroblast cell line treated with two different chemicals, a metal, chromium, and benzimidazole shows different ultrastructural features of cellular disintegration as seen in Figures 19.72a, b and c. The chromium-treated cells exhibit nuclear marginal condensation of chromatin and organelles with preserved membrane while those treated with the organic chemical show cytoplasmic vacuolization and bulging of the organelles.

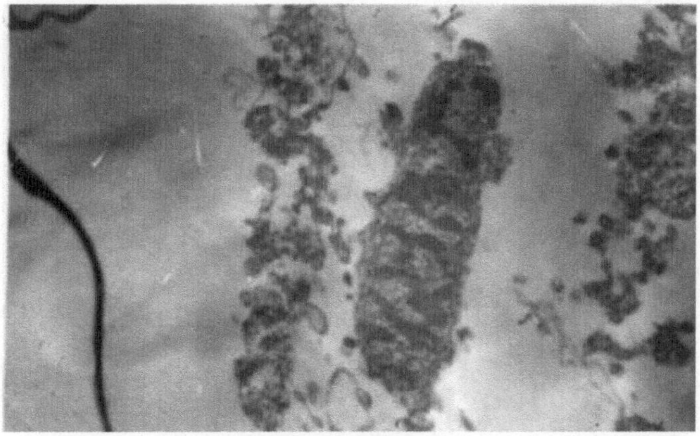

(a) (6000×)

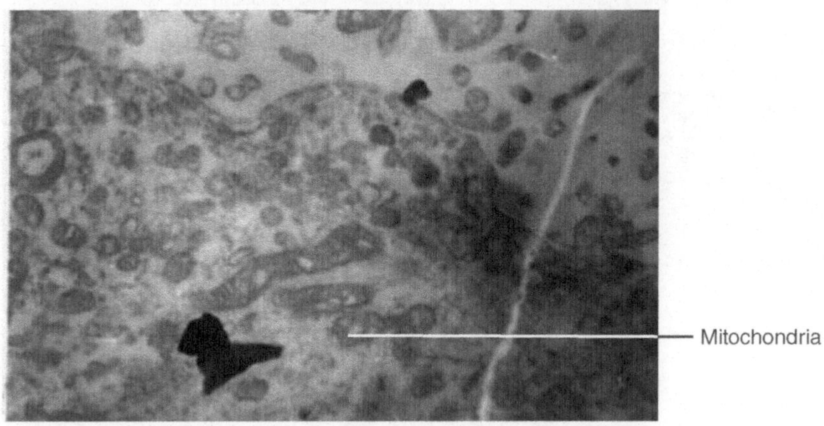

(b) (15,000×)

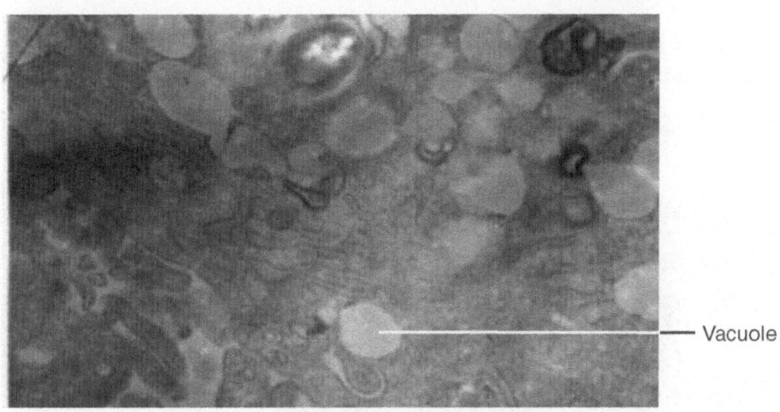

(c) (15,000×)

Figure 19.72 Fibroblast cell line. a and b—treated with chromium and c—treated with benzimidazole

EWINGS SARCOMA CELL LINE

Ewings sarcoma cell line treated with thalidomide to elucidate the effect of the drug on these cells is shown in Figures 19.73–19.75.

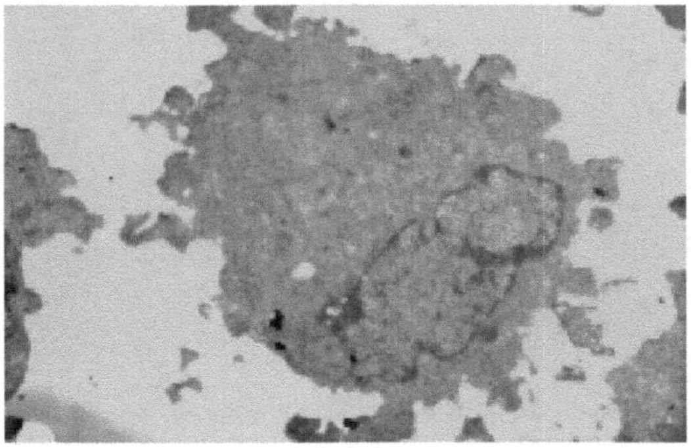

Figure 19.73 ES cell line trypsinized. One cell showing membrane projections. (6000×)

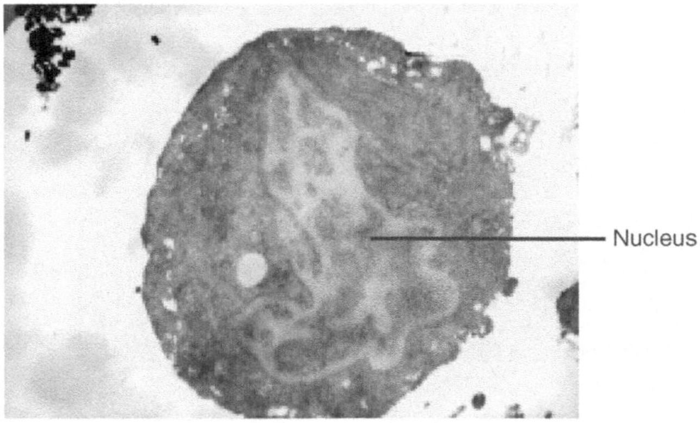

Figure 19.74 ES cell line treated with thalidomide. A cell showing vacuolated cytoplasm, ground glass appearance of nucleus and loss of cytoplasmic projections (8000×)

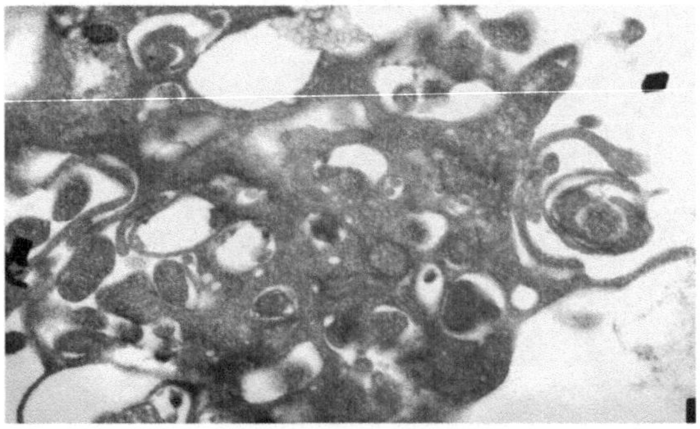

Figure 19.75 ES cell line + thalidomide. The cytoplasm of a damaged cell showing vacuoles with electron-dense contents (20,000×)

LIVER FROM WISTAR RATS

Liver tissues from the laboratory animal, Wistar rats, were treated with different chemicals, that have been listed below, and screened in the EM. Some of the interesting features observed are shown in Figures 19.76–19.94.

1. Animal treated with alcohol alone (Figure 19.76)
2. Animal treated with a carcinogen, dimethyl nitrosamine (Figures 19.77–19.79)
3. Animal treated with ethanol and carcinogen (Figures 19.80–19.83)
4. Animal treated with ethanol + carcinogen but pretreated with a herbal drug (Figures 19.84–19.86)
5. Animal treated with a toxicant, hydrazine, which is known to cause fatty liver (Figures 19.87–19.89)
6. Animal treated with ethanol and toxicant (Figures 19.90–19.92)
7. Animal treated with ethanol + toxicant but pretreated with a protective drug, an antioxidant, silymarin (Figures 19.93 and 19.94)

1. *Treatment with alcohol alone*

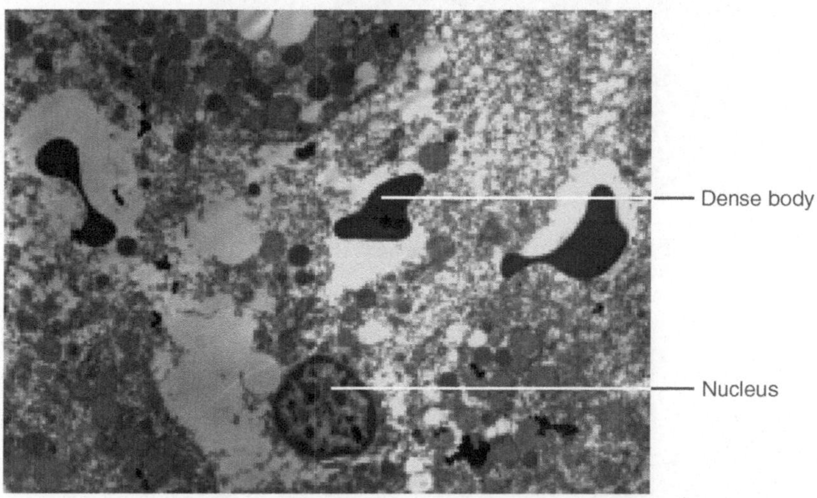

Figure 19.76 Paranuclear vacuoles and dense bodies (3000×)

2. *Treatment with carcinogen (dimethylnitrosamine) alone*

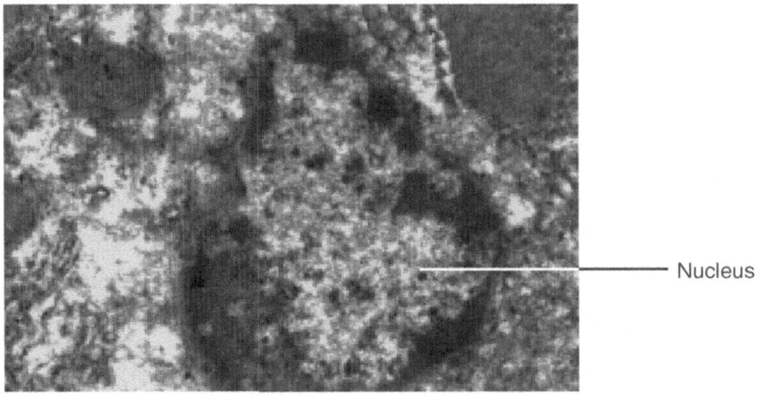

Figure 19.77 Nucleus with peripherally condensed chromatin (20,000×)

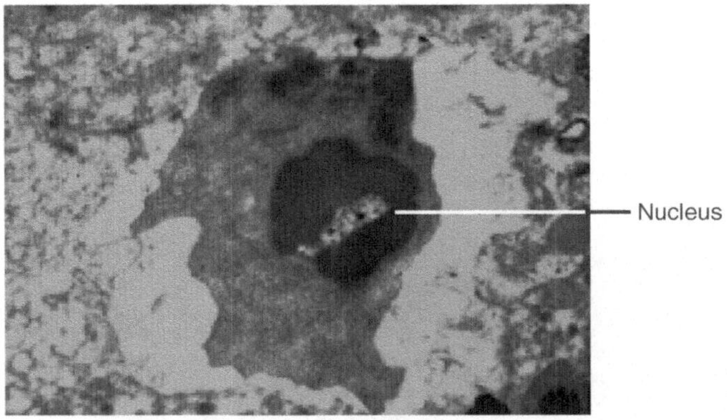

Figure 19.78 Condensed nucleus (10,000×)

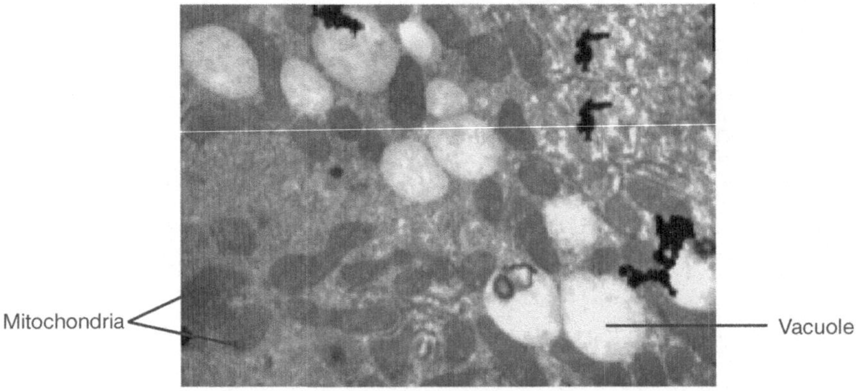

Figure 19.79 Mitochondria with condensed cristae and vacuoles (10,000×)

3. *Treatment with ethanol and carcinogen*

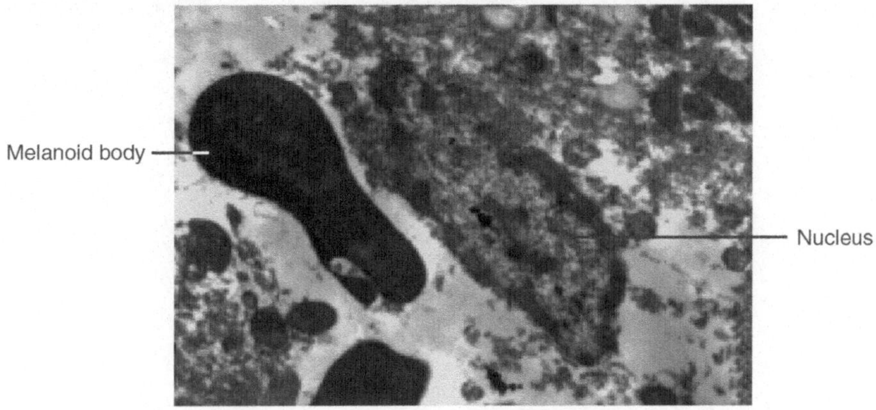

Figure 19.80 Nucleus with peripherally condensed chromatin, disrupted nuclear membrane and melanoid body (10,000×)

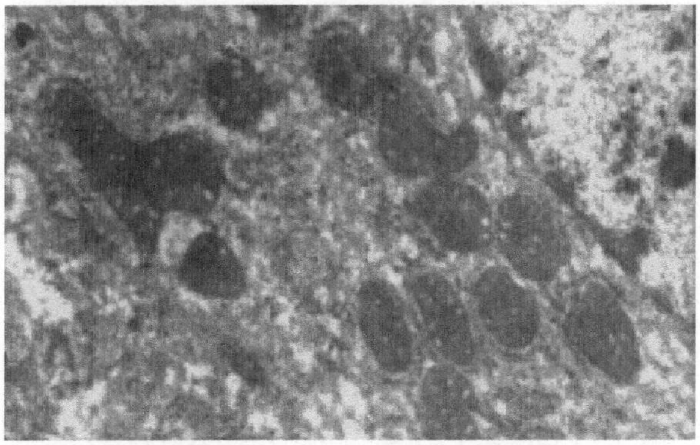

Figure 19.81 Pleomorphic mitochondria (15,000×)

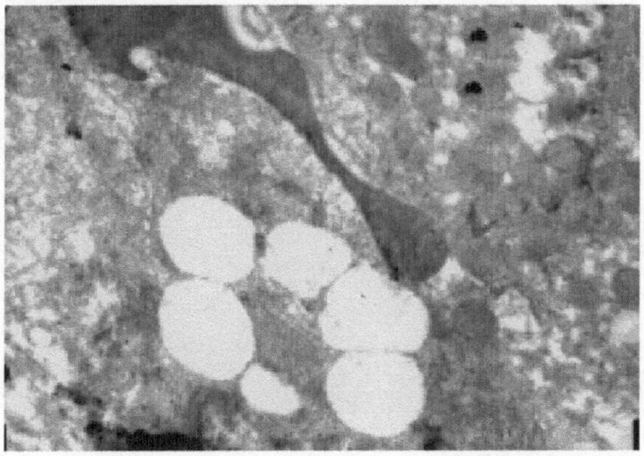

Figure 19.82 Microvesicular bodies and melanoid body (10,000×)

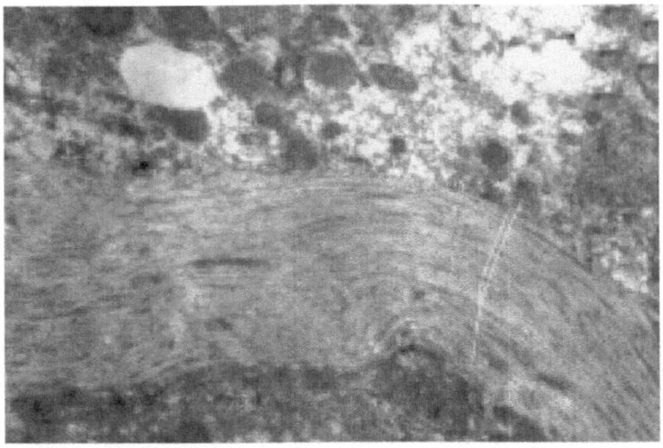

Figure 19.83 Intercellular fibres (10,000×)

4. *Treatment with ethanol + carcinogen + drug*

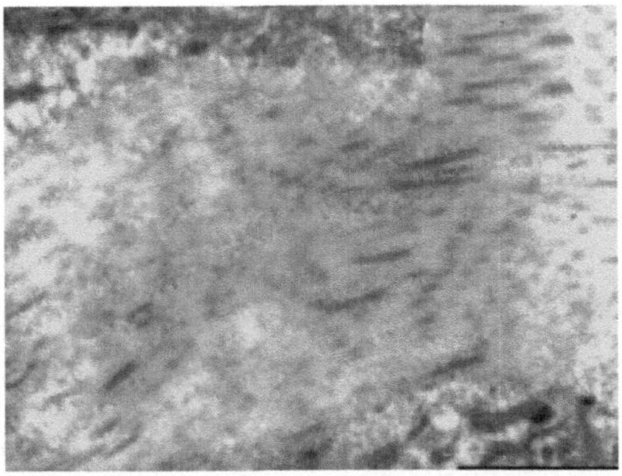

Figure 19.84 Fibrosis (20,000×)

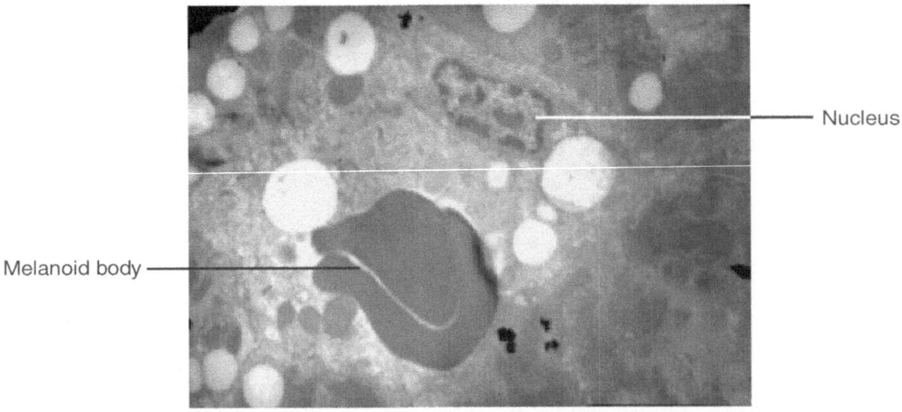

Figure 19.85 Melanoid body, small nucleus with condensed chromatin and cytoplasmic vacuoles (6000×)

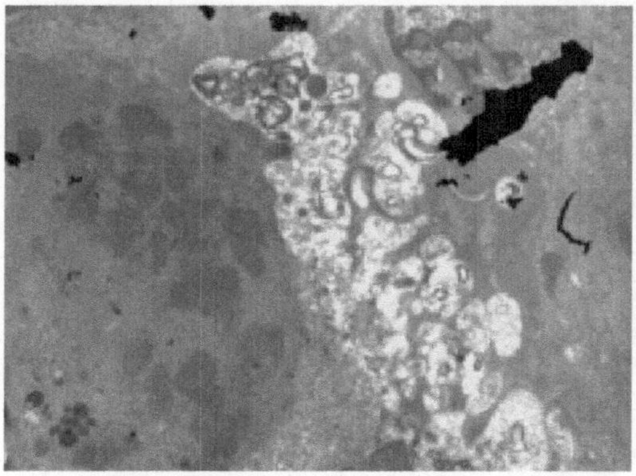

Figure 19.86 Intercellular area with contents (5000×)

5. *Treatment with toxicant (hydrazine which causes fatty liver)*

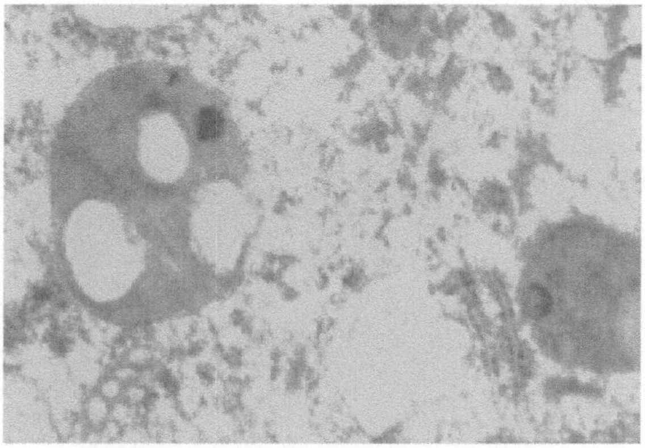

Figure 19.87 Lipid bodies (40,000×)

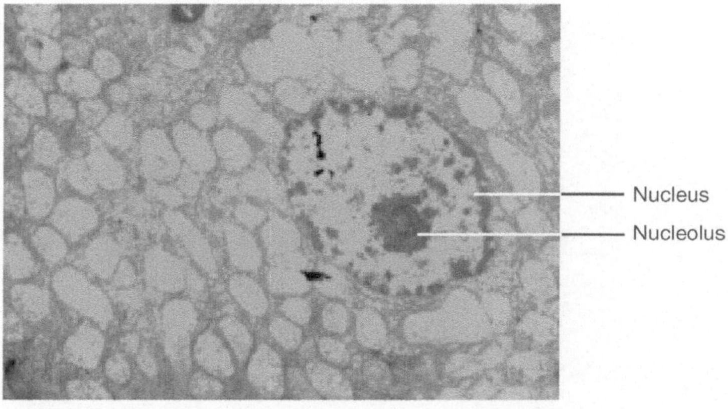

Figure 19.88 One cell with nucleus showing a prominent nucleolus (6,000×)

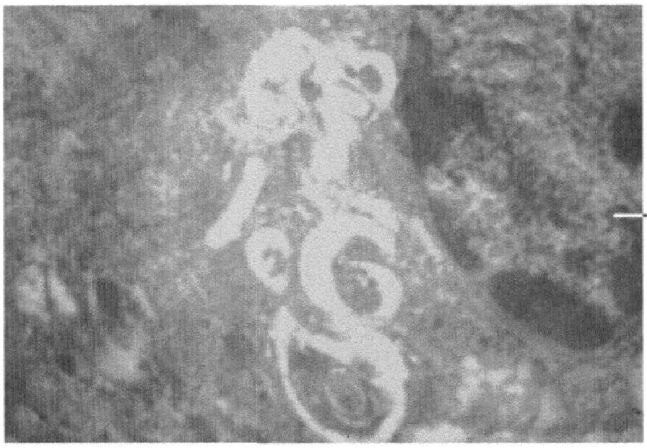

Figure 19.89 Lamellar bodies close to nucleus (30,000×)

6. *Treatment with ethanol + toxicant*

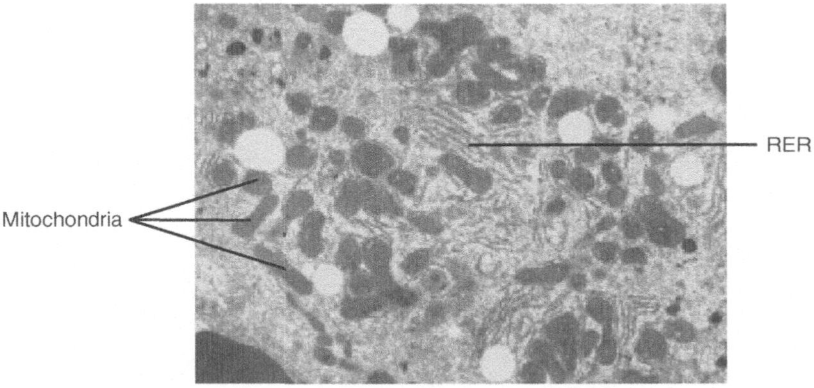

Figure 19.90 Pleomorphic mitochondria, RER and few vacuoles (6000×)

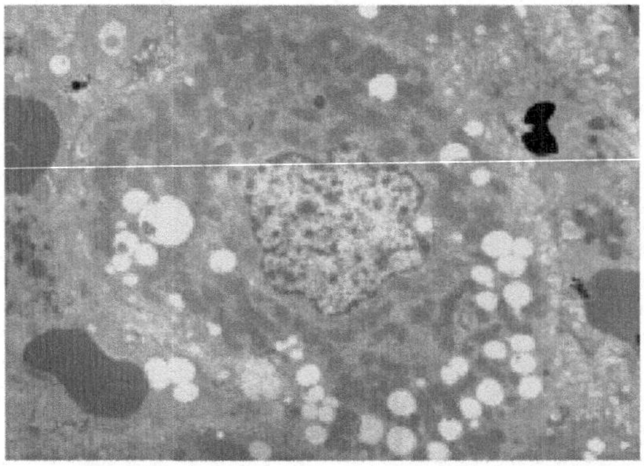

Figure 19.91 One cell showing cytoplasmic vacuoles, probably, lipid (3000×)

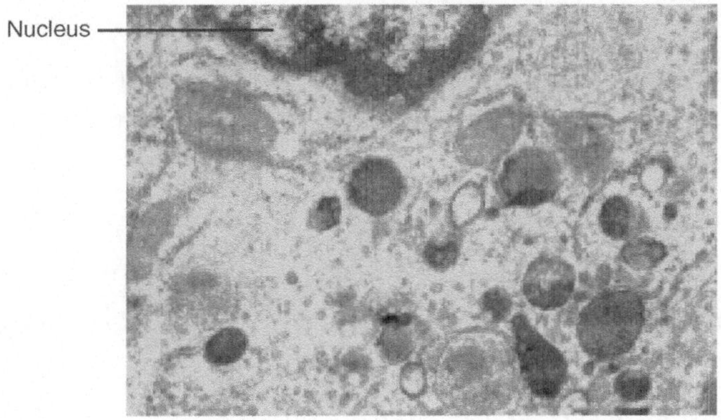

Figure 19.92 Part of the nucleus and dense particles (20,000×)

7. *Treatment with ethanol + toxicant + protectant (silymarin)*

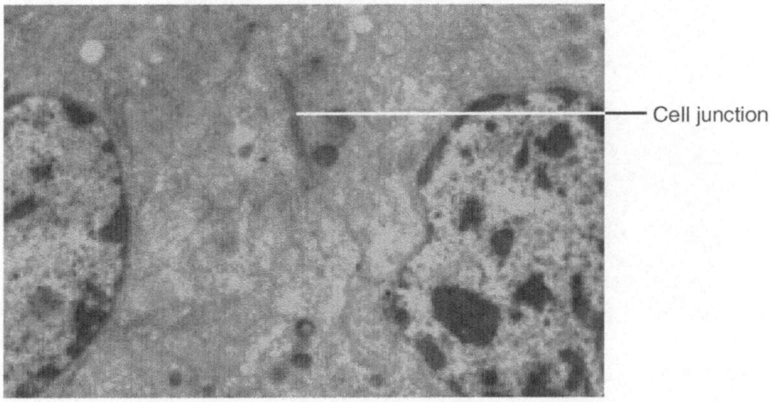

Figure 19.93 Two cells with junction in between (8000×)

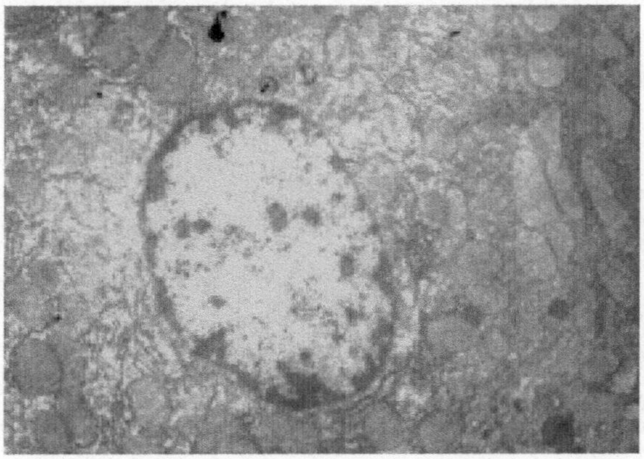

Figure 19.94 One cell with nucleus and cytoplasmic organelles (8000×)

SPINAL CORD OF A WISTAR RAT

Electron microscopic studies are done on ultrathin sections of spinal cord of a Wistar rat after induced injury and also inflicted injury followed by immediate treatment, which show the features as seen in Figures 19.95–19.103.

1. Spinal cord from an animal in the control group (Figures 19.95–19.97)
2. Spinal cord after induced injury to the animal (Figures 19.98–19.100)
3. Spinal cord from an animal treated immediately after induced injury (Figures 19.101–19.103)

1. *Spinal cord from an animal in the control group*

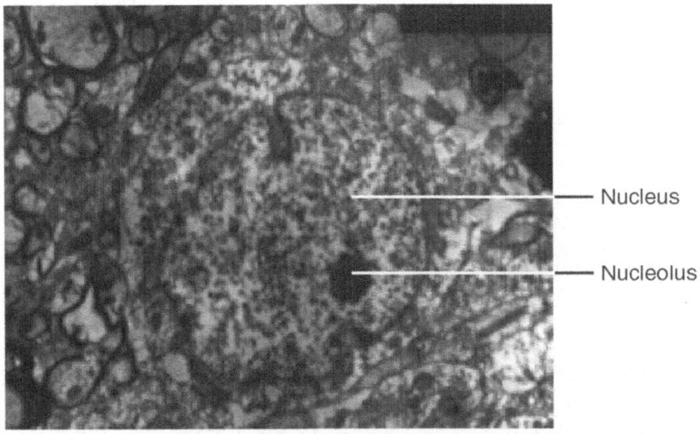

Figure 19.95 One cell with nucleus and prominent nucleolus (6000×)

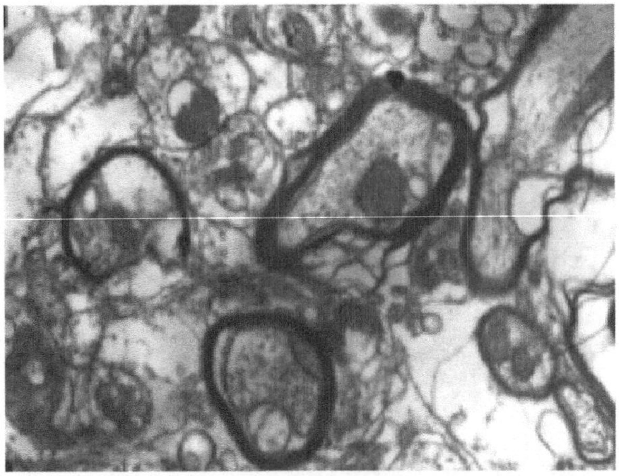

Figure 19.96 Myelinated nerve fibres showing neurofilaments and tubules (15,000×)

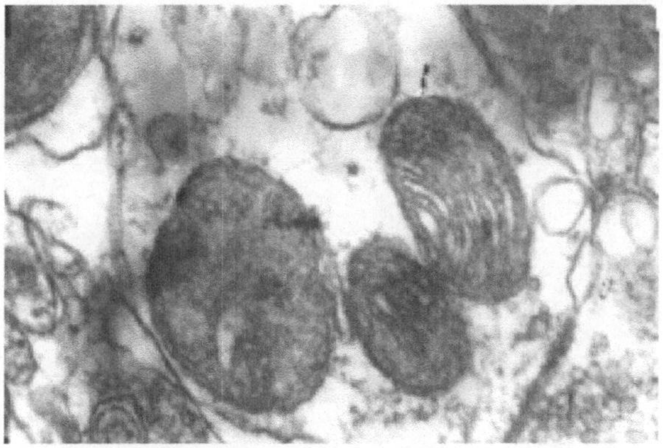

Figure 19.97 A few mitochondria (50,000×)

2. *Spinal cord after induced injury to the animal*

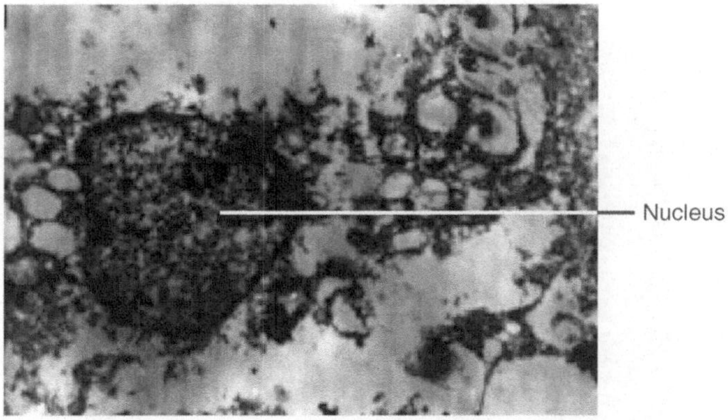

Figure 19.98 One cell showing nucleus with condensed chromatin and cytoplasm with vacuolated spaces and scanty organelles (6000×)

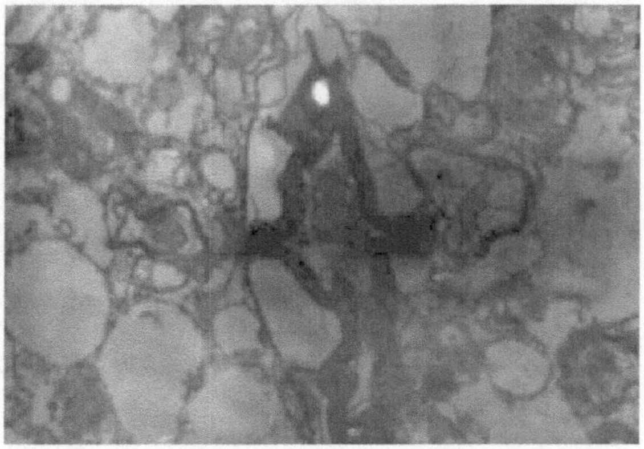

Figure 19.99 Crushed nerve fibres (8000×)

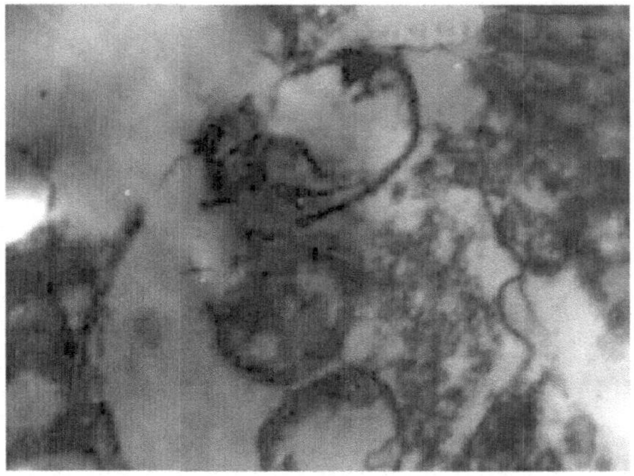

Figure 19.100 A few mitochondria showing loss of cristae and vacuolization (40,000×)

3. *Spinal cord from an animal treated immediately after induced injury*

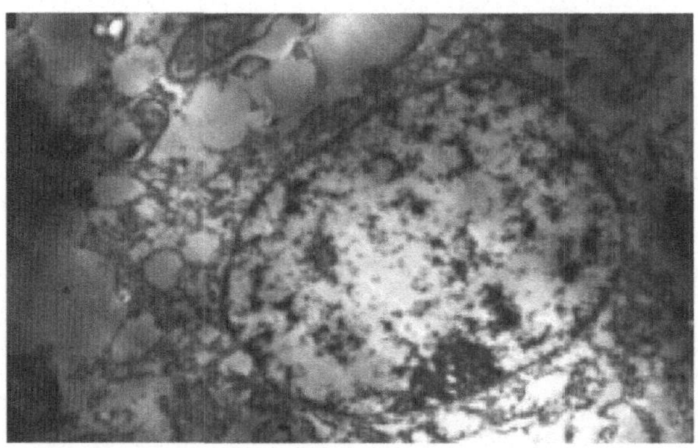

Figure 19.101 One cell showing more organelles than the one after injury (6000×)

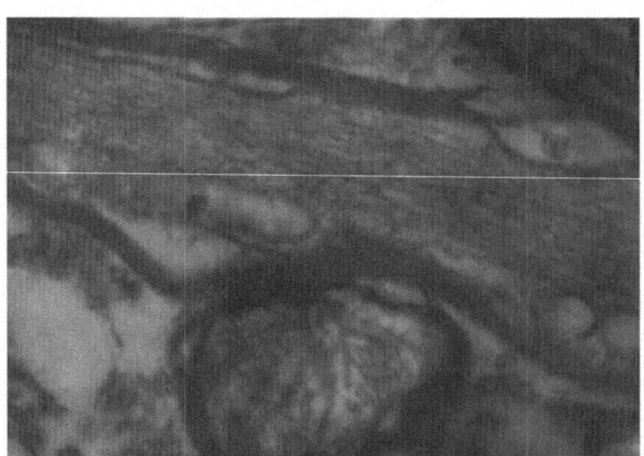

Figure 19.102 Myelinated nerve fibres showing neurotubules (30,000×)

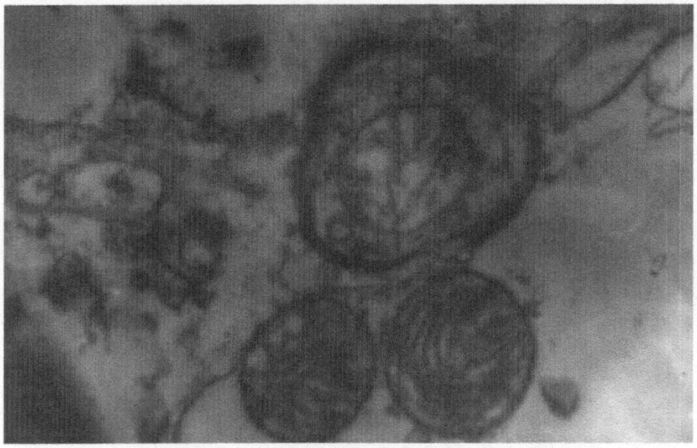

Figure 19.103 A few mitochondria showing more cristae and more regular shape (25,000×)

HPV IN CERVICAL CARCINOMA

The HPV (Human papilloma virus) particles within the cell of an advanced case of cancer of uterine cervix in one of our patients is shown in Figure 19.104. The presence of tonofibrils (cytokeratin bundles) indicate the epithelial nature of the infected cell. When the patient is given a dose of dendritic cell vaccine, more of necrotic and apoptotic cells are seen in the tumour tissue in the post-vaccinated sample, probably denoting the efficacy of the vaccine in controlling the tumour cell proliferation, as seen in Figures 19.104 to 19.106.

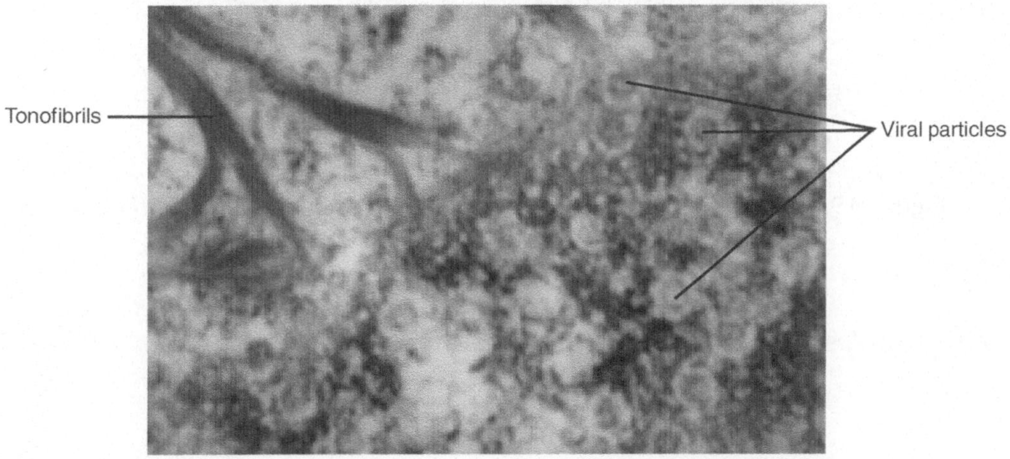

Figure 19.104 HPV in an epithelial tumour cell of cancer uterine cervix (40,000×)

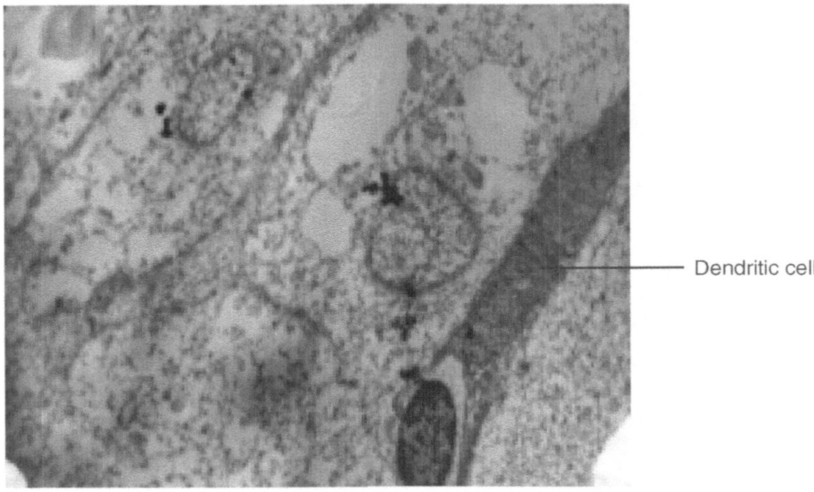

Figure 19.105 The dendritic cell activating the lymphoid cell (40,000×)

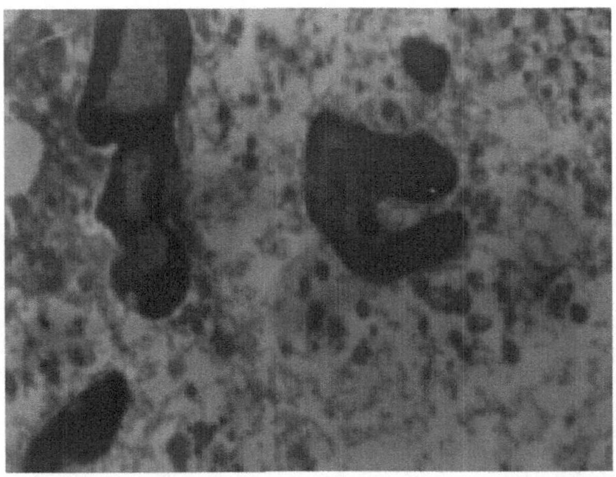

Figure 19.106 Few apoptotic cells in the post-vaccinated sample (8000×)

INFECTION IN A LYMPH NODE BIOPSY SAMPLE

The biopsy samples of infection in a lymph node are shown in Figures 19.107 and 19.108.

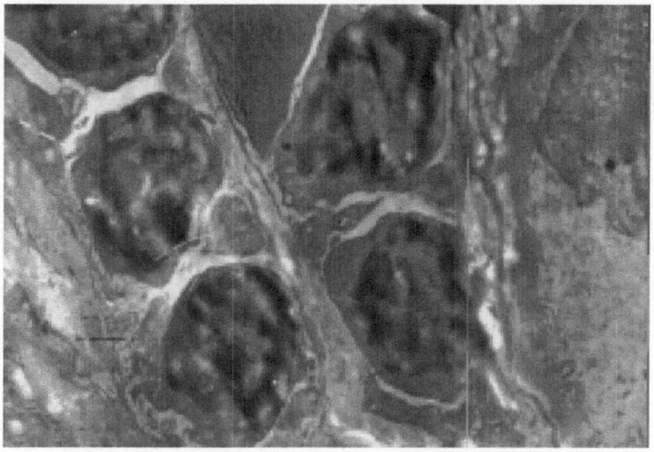

Figure 19.107 A few infected cells in a supraclavicular lymph node biopsy (5000×)

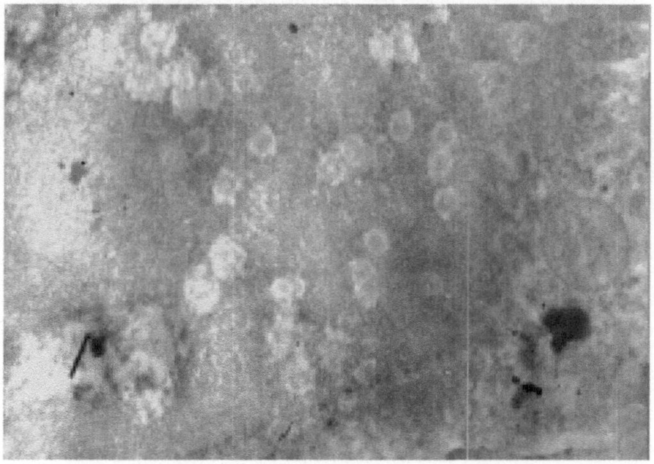

Figure 19.108 A few viral-like particles in the nucleus of one of the above cells (20,000×)

LOOSE SHELL SYNDROME (LSS) IN BLACK TIGER SHRIMP *PENAEUS MONODON*

The affected shrimps exhibit predominantly muscular wasting leading to death. The significant etiologic factor has been suggested as virus. Viral-like particles have been identified in the affected animals in tissues like hepatopancreas, lymphoid organs, etc. The viral DNA has been isolated from the animals. The isolated infectious agent from the affected animals on negative staining also shows similar viral-like particles (Figures 19.109–19.111).

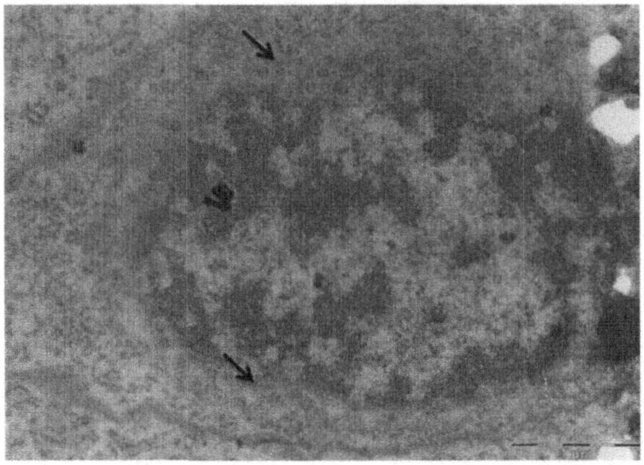

Figure 19.109 Viral particles especially at the nuclear membrane (20,000×)

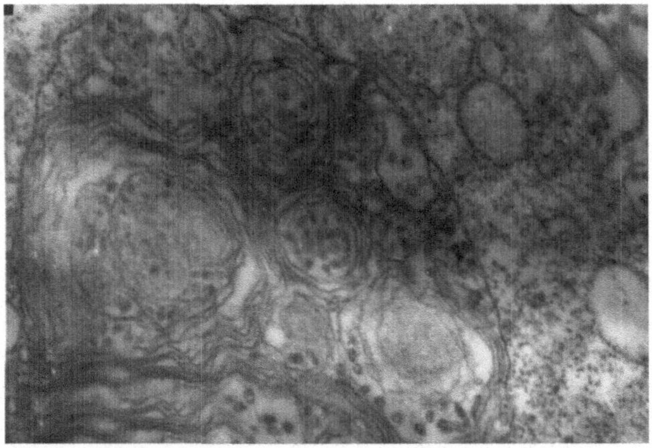

Figure 19. 110 Multilayered vesicles in the cytoplasm showing trapped particles (25,000×)

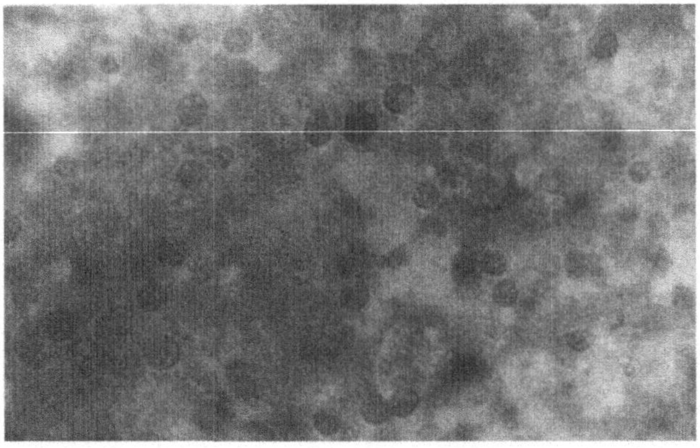

Figure 19.111 Purified viral particles negatively stained with PTA (80,000×)

DEVELOPMENTAL STAGES OF CELL WALL FORMATION DURING XYLOGENESIS IN THE BRANCHES OF *LEUCAENA LEUCOCEPHALA*

The specimen is vascular cambium and its xylem derivatives from the branches of *Leucaena leucocephala.* Figure 19.112 shows the low-magnified image of the entire cambial zone which includes meristematic cell and xylem derivatives. In Figure 19.113, the initial stages primary of cell wall formation which includes cell plate formation and deposition of cell wall materials, like cellulose and hemicellusose is shown. The wall materials are brought from the cytoplasm to the cell plate by Golgi vesicles. The fusion of vesicles into the cell plate is more clearly visible in Figure 19.114, which is a higher magnification of Figure 19.113. The cell wall undergoes further special modification to become xylem cell or woody cell. This involves deposition of three layers of cell wall into the existing primary wall, collectively known as secondary wall. In Figure 19.115, the cell wall shows two layers of deposition of secondary wall in the fibre cell which is one type of xylem cell.

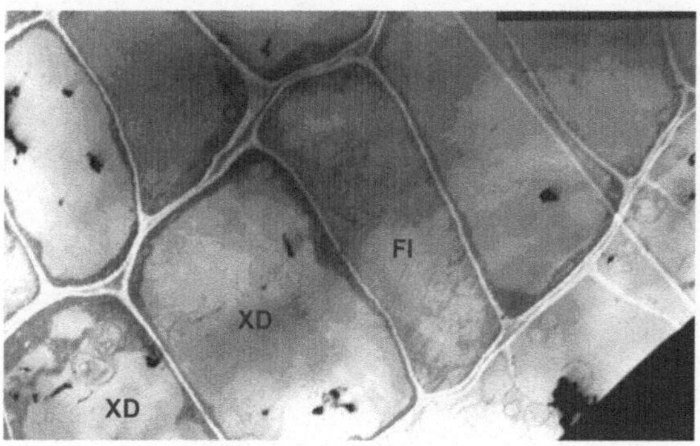

Figure 19.112 Cambial zone showing fusiform initials (FI) and xylem derivatives (XD) (3000×)

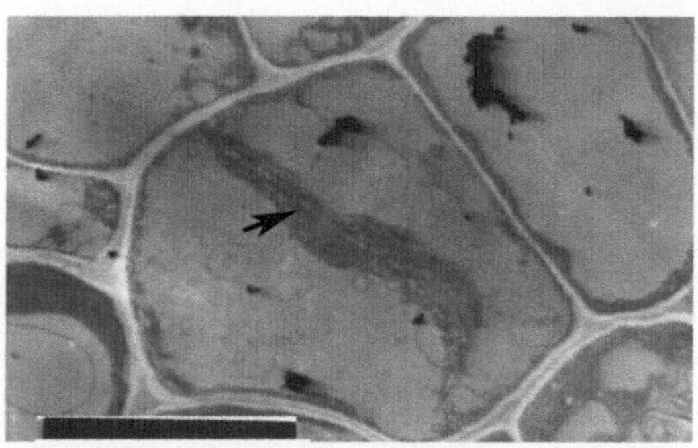

Figure 19.113 Formation of primary wall (arrow). Note the deposition of cell wall materials into cell plate through Golgi vesicles (4000×)

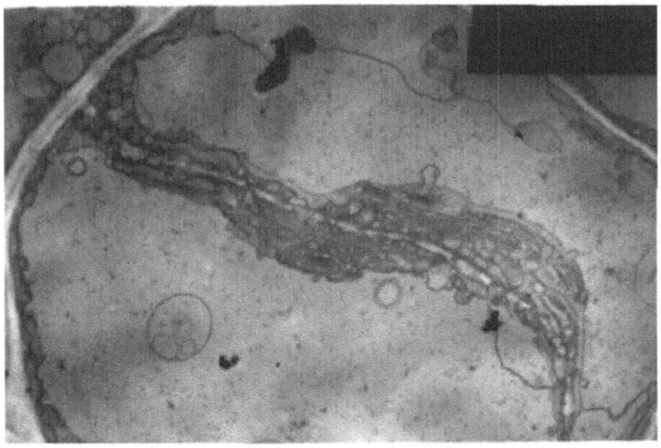

Figure 19.114 Deposition of cell wall materials at higher magnification (8000×)

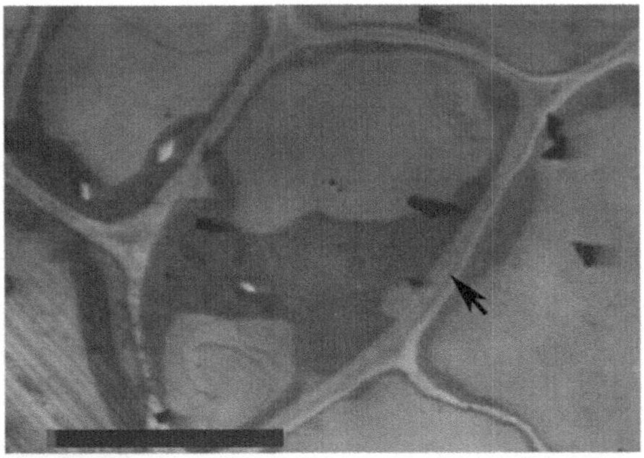

Figure 19.115 Initial stages of secondary wall deposition in the xylem fiber (arrow) (4000×)

20

ELECTRON MICROSCOPY IN TUMOUR DIAGNOSIS

In surgical pathology, indications for the use of TEM are limited to areas that include renal biopsies and certain other non-neoplastic specimens. But the study of tumours has proved to be the most extensive and rewarding role of diagnostic electron microscopy. The fundamental role of EM in tumour diagnosis is the assessment of differentiation within cells and tissues as expressed by cytoplasmic characteristics and cell relationships—to detect and evaluate aberrant submicroscopic morphology. Familiarity with normal ultrastructure is essential before abnormal fine structure can be interpreted. Although a few subcellular structures such as Weibel–Palade bodies and Langerhans' cell granules appear to be specific for certain cell lines, there is probably no single structure pathognomonic of a particular tumour when taken in isolation.

This field of tumour diagnosis is very vast by itself and complete coverage of this topic is beyond the scope of this book. Only the most common and salient features are outlined in this chapter.

Tissue diagnosis is not required for all diseases. For example, the most common condition, Diabetes Mellitus requires only a few ml of blood and probably a few ml urine for diagnosis as well as for prognosis and maintenance. Only in chronic, advanced stages when diabetic nephropathy or retinopathy sets in, tissue diagnosis may be required. But there are certain diseases where use of tissue/cells is essential for diagnosis. One such is tumours/cancer.

To understand the importance of tissue diagnosis in cancer we should know what cancer is. As we all know, cancer is defined as uncontrolled proliferation of cells. These cells stem from our own body. Due to some mutation or genetic changes, the control for cell division of a particular type of cell is lost. This, resulting in the proliferation of that particular cell uncontrollably, leads to cancer. Therefore, cancer is monoclonal though, very rarely multiple primaries are reported.

When we say, for example stomach cancer, all stomach cancers are not the same. Stomach is an organ containing different types of cells/tissues. It has epithelial tissue, both glandular and squamous, fibrous tissue, blood vessels and blood cells, lymphoid tissue (gut-associated); moreover, stomach itself is a muscular organ, so muscle tissue also. Any one of these may acquire malignant potential, depending on the mutation or other factors involved, and start proliferating. So it is essential to know the cell of origin or histogenesis of the tumour cell or rather, the cell of differentiation the tumour exhibits. This is a prerequisite to adopt any particular treatment modality for that cancer. The biological profile of each cell type is different so also the treatment response. Hence it is essential to identify the tissue type or its histogenesis or differentiation to initiate any effective treatment by the oncologist.

The cell of origin or of differentiation of the tumour can be assessed by using the structural and morphological features of the tissue, by the immunological features or by important molecular or cellular markers. The tumour cells are aspirated using a fine needle wherever feasible, a smear prepared and viewed under the light microscope. If the specimen collected is a representative sample, then it is possible to assess whether the tumour cells belong to malignant or benign category. Moreover, if the tumour cells are differentiated enough, probably the tumour cell type or rather the cell of origin can also be identified.

Most often, a biopsied material from the tumour is required for diagnosis as well as for confirmation of cytological opinion. In quite a few instances, additional inputs from electron microcopy and immunohistochemistry (IHC) are necessary. Electron microscopy by virtue of its high magnification and high resolution definitely has an important role to play in tumour diagnosis. EM identifies features beyond the resolution of LM. EM is indispensable in a few cases and in some tumours, it may be the only means of obtaining essential information. In this respect, TEM is more useful than SEM because cellular details can be visualized in TEM. Thus, along with LM and IHC it has a major role in tumour diagnosis. Both modalities of investigation namely, EM and IHC are complementary and their combined use can serve towards better diagnosis and patient care. However, the prognostic value of EM is insignificant.

EM is not required and not possible in all cases of tumours, unlike LM, due to

- its high cost, including maintenance and consumables cost, etc.
- time taken for processing and screening
- need for specialized ancillary equipment
- need for skilled technical support
- expertise required in interpretation of the images

In 1947, Porter and Thomson gave the first ultrastructural description of a neoplastic cell and in 1969, Juan Rosai and Hector Rodriguez studied four tumours in which TEM helped in their differential diagnosis.

NATURE OR TYPES OF SPECIMENS

Different forms of specimens can be used for diagnostic EM for tumour diagnosis. They are:

- Tissue specimens—biopsies and surgical specimens
- FNACs (fine needle aspirates)
- Cell suspensions
- Paraffin block sections

When necessary, for immediate diagnostic confirmation, **rapid processing** techniques can be employed where the screening of the specimen sections in the EM and diagnostic confirmation may be completed within 48 hours or less.

There are certain **limitations** in using EM for tumour diagnosis. Apart from the general limitations mentioned in the previous pages, the ones specific with respect to tumours are given below:

- In poorly differentiated tumours, quite often, prolonged search has to be made to identify even a single diagnostic feature.
- Hyperplastic or neoplastic distinction is not usually possible.
- Invasive or non-invasive nature of the tumour also cannot be determined because only very small bits of the specimen are processed.

In essence, the EM does the following

- It clarifies and amplifies light microscopic features.
- It helps in reaching an exact diagnosis in problem cases.
- It helps in confirmation of certain tumours where the diagnostic possibilities are reduced to a few, light microscopically.

EM has a significant role to play in the differential diagnosis of,

- Poorly differentiated (p.d.) tumours—to identify the major cell types such as epithelial, mesenchymal, lymphoid (carcinoma, sarcoma, lymphoma) and even identify subtypes occasionally.
- Large cell neoplasms—This includes most carcinomas, melanomas, mesotheliomas, lymphomas, some sarcomas, represented by large, round to polygonal cells.
- Malignant round cell tumours—rhabdomyosarcoma, Ewings sarcoma, neuroblastoma, lymphoma, neuroendocrine carcinoma, etc.

- ✳ Malignant spindle cell tumours—malignant fibrous histiocytoma, leiomyosarcoma, rhabdomyosarcoma, schwannoma, etc.
- ✳ Anterior mediastinal tumours—thymoma, lymphoma, germinoma, thymic carcinoma, etc.
- ✳ Various types of testicular tumours, adrenal tumours, ovarian tumours, etc.

IMPORTANT ROLE IN OTHER TUMOURS

- ✳ There are certain neoplasms which exhibit some characteristic ultrastructural features. Some of them include mesothelioma, chondrosarcoma, hairy cell leukaemia, carcinoid, alveolar soft part sarcoma, granular cell tumours, etc.
- ✳ Characteristic nuclear contours of mycosis fungoides and sezary syndrome are clearly well observed in EM.
- ✳ The intracytoplasmic crypts in endodermal sinus tumour and "Reinke" crystals in Leidig cell tumour are also clearly evident ultrastructurally.
- ✳ Histiocytosis X, otherwise known as Langerhans cell histiocytosis (LCH) is confirmed by Birbeck granules identified ultrastructurally.
- ✳ Similarly, melanoma especially amelanotic melanoma shows stage II and III premelanosomes, the pathognomonic features even in metastatic lesions.
- ✳ EM may indicate possible source of metastasis of unknown origin. For example,
  - ✳ Presence of glycocalceal bodies is an indicator of GI adenocarcinoma
  - ✳ Cytoplasmic neolumina are typical of some mammary carcinomas
  - ✳ Presence of melanosome, particularly stage II and III premelanosomes indicate a primary malignant melanoma
- ✳ EM established certain neoplasms as distinct entities, e.g., Merkel cell carcinoma

Viral etiology also can be confirmed using the EM. EM can resolve and directly image the structural details of particles of viral dimensions. It confirms the presence of viruses with identification of their size and shape.

EM in conjunction with LM and IHC plays a distinct role in tumour diagnosis. IHC may not be conclusive all the times. In such instances, EM comes in handy to the pathologists. Since interpretation of ultrastructural details needs some expertise, pathologists normally resort to IHC rather than EM. Even the resin-embedded semithin sections prepared in EM laboratory are very useful because the section thickness is less than the routine histopathological sections. But the drawback here is that regular H&E stains cannot be applied to resin sections directly.

STAGES OF ULTRASTRUCTURAL ASSESSMENT OF TUMOURS

While screening the ultrathin sections in the EM, the following features have to be taken into consideration to arrive at any conclusion:

- Topographic cellular relationships
- External lamina
- Cell contours
- Intercellular junctions
- Cytoplasmic granules
- Cytoplasmic filaments
- Cytoplasmic vacuoles and vesicles
- Nuclear and nucleolar morphology
- Stroma

Some of the common types of tumours and their characteristic diagnostic ultrastructural features which can aid in their differential diagnoses are outlined in the following pages in this chapter.

ULTRASTRUCTURAL CHARACTERISTICS OF MAJOR TUMOUR TYPES

Cell types can be broadly classified into three major categories, namely, epithelial, mesenchymal and lymphoid, the tumours of which are referred to as carcinoma, sarcoma and lymphoma respectively. There are many sub- and subsub-classifications under each category. When these tumours are well-differentiated, the morphological features for tumour typing can be recognized at the light microscopic level itself. When the differentiation is poor as in aggressive tumours, ancillary techniques are essential to arrive at a conclusive diagnosis. Immunohistochemistry and electron microscopy are the most useful techniques in this regard. The comparative ultrastructural features in the differential diagnosis of the major tumour types, namely carcinoma, sarcoma and lymphoma in tumours designated as poorly differentiated light microscopically are tabulated in Table 20.1.

Table 20.1 Comparative ultrastructure of major tumour types particularly in poorly differentiated tumours

Features	Carcinoma	Sarcoma	Lymphoma
Cell shape	Rounded to polygonal	Elongated, bipolar	Rounded
Disposition of cells	In groups or sheets	Singly with matrix	Forms sheets
External lamina	Present, variable	Variable	Absent
Distribution of EL	Surrounds cell groups	Around individual cells	Absent
Junctions	Present, desmosomes, tight junctions, etc.	Variable; desmosomes absent	Absent
Cell processes	Variable	May be present	Poorly developed
Filaments	Tonofilaments, cytokeratin	Intermediate, microfilaments	Absent
Remarks	Cell disposition + shape + desmosomes + tonofilaments; tight junctions, lumen, granules, microvilli	Cell disposition + shape + cytoplasmic filaments + absence of desmosomes	Cell disposition + shape + sparse organelles + absence of junctions and filaments

CARCINOMAS/EPITHELIAL TUMOURS

Desmosomal junctions are almost always indicative of epithelial differentiation. Even in undifferentiated tumours, where they are often incompletely developed, recognition of junctions, however tenuous they are, is diagnostic of epithelial differentiation. Desmosomal junctions are characteristically seen in squamous cell carcinomas. Tonofibrils, i.e., bundles of keratin filaments, some even attached to the desmosomes, are typical features of carcinomas, particularly squamous cell carcinomas. Keratohyalin granules distinguished by their electron density and oblong shape may be seen in well-differentiated tumours but usually undetectable in poorly differentiated tumours. Desmosomes decrease in invasive tumours but increase along with keratin filaments in squamous cell carcinomas induced after irradiation. Basal lamina usually is present around aggregates of squamous cells and may show attachment to cells by hemidesmosomes. Thus the minimal criteria are desmosome–tonofilament complexes and cytoplasmic tonofilaments.

Squamous cell carcinomas generally show close morphological similarities to each other, irrespective of their site of origin. EM is extremely valuable in establishing the diagnosis in poorly differentiated tumours like nasopharyngeal carcinoma, large cell undifferentiated carcinoma of lung, undifferentiated carcinoma of cervix, etc.

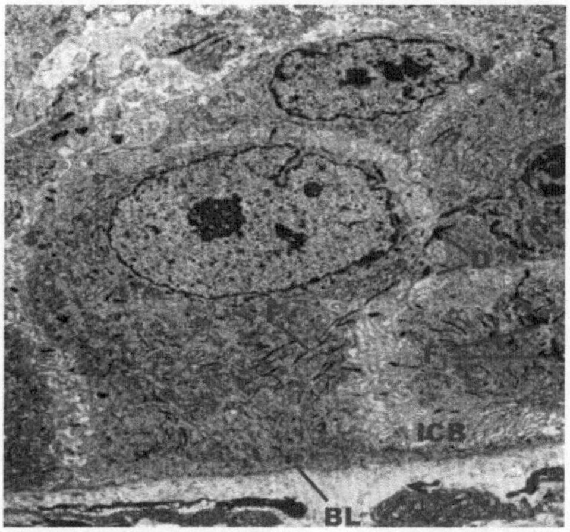

Figure 20.1 Well-differentiated carcinoma. F—Filaments (tonofilaments), D—Desmosomes, ICB—Intercellular bridges (12,150×). The basal lamina BL is seen surrounding groups of cells. (*Source:* Richard Dickersin, G. (2000). *Diagnostic Electron Microscopy—A Text/Atlas,* 2nd edn. Springer-Verlag Inc., New York. p. 66, Fig. 3.38)

Figure 20.1 illustrates the characteristic diagnostic features of a well-differentiated carcinoma, the tonofibrils, desmosomes and the basal lamina surrounding several cells in a group. Figure 20.2 represents the desmosomal junctions and Figure 20.3a and b demonstrates the desmosomal junctions in well-differentiated and poorly differentiated squamous cell carcinomas.

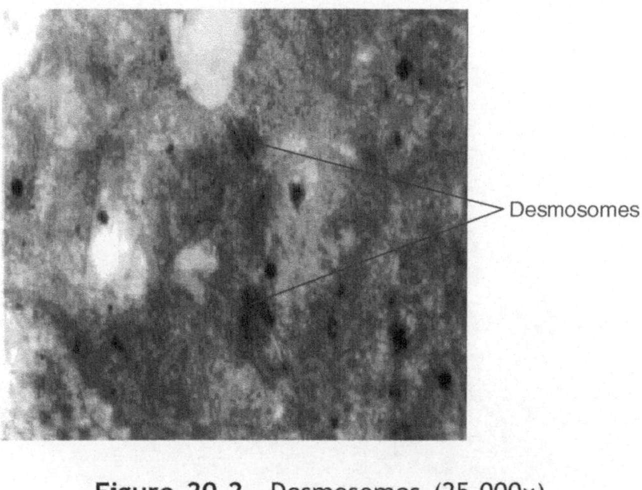

Figure 20.2 Desmosomes (25,000×)

There are two closely placed parallel lines centrally in Figure 20.2, one is the cell membrane of one cell while the other is that of the neighbouring cell. At certain regions the closely apposed membranes get thickened and the keratin filaments get attached to them at right angles. This constitutes a desmosome.

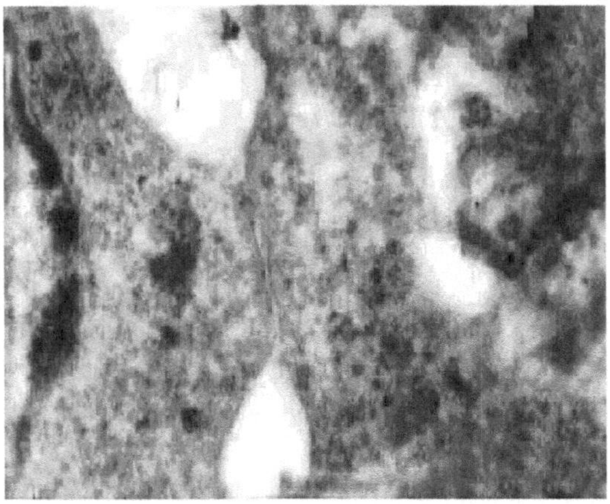

Figure 20.3 Junctions in poorly differentiated carcinomas (30,000×)

In well-differentiated SCCs (squamous cell carcinoma) the desmosomes are numerous. On the other hand, in Figure 20.3, the apposed membranes of the two neighbouring cells show focal thickening which cannot be considered as a junction. This type of thickening may be seen even among lymphoid cells where absence of desmosomes, rather cell junctions is a diagnostic criterion. In tumours which have been classified as poorly differentiated tumours, histopathologically this thickening can be considered to represent poorly differentiated junction and a diagnosis of carcinoma is more appropriate than a sarcoma or a lymphoma.

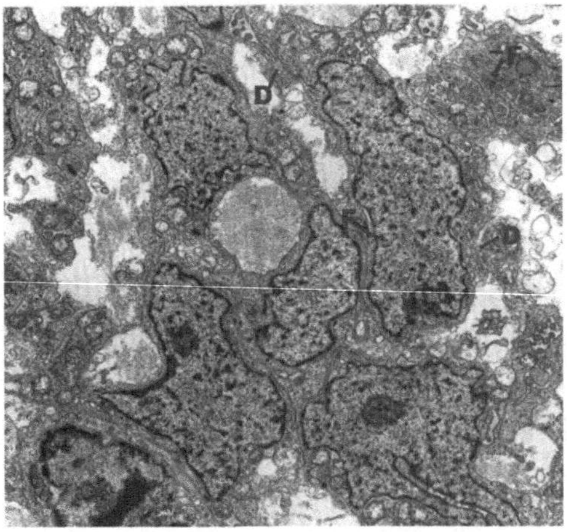

Figure 20.4 Adamantinoma (D—Desmosomes, F—Filaments) (10,400×). (*Source:* Richard Dickersin, G. (2000). *Diagnostic Electron Microscopy—A Text/Atlas*, 2nd edn. Springer-Verlag Inc., New York. p.74, Fig. 3.46.)

In the case of adamantinoma of long bones as seen in Figure 20.4, the cells appear elongated or spindle-shaped resembling mesenchymal cells. But the presence of filaments and junctions confirm the diagnosis of tumour of epithelial origin.

In tumours of glandular epithelium known as **adenocarcinoma**, the lumen, microvilli, luminal junctions, tight junctions, secretory granules, prominent Golgi, etc. are the important distinguishing features. Cilia are supposed to be the diagnostic markers for ependymomas and choroid plexus neoplasms. Figure 20.5 shows a well-differentiated adenocarcinoma whereas Figure 20.6 shows a moderately differentiated and Figure 20.7, a poorly differentiated adenocarcinomas.

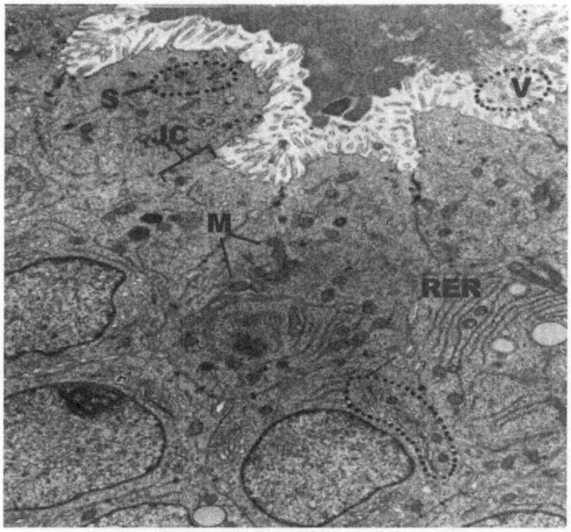

Figure 20.5 Well-differentiated adenocarcinoma (G—Golgi, S—Secretory granules, V—Villi, M—Mitochondria, RER—Rough endoplasmic reticulum, JC—Junction) (11,700×). (*Source:* Richard Dickersin, G. (2000). *Diagnostic Electron Microscopy—A Text/Atlas,* 2nd edn. Springer-Verlag Inc., New York. p.31, Fig. 3.4.)

In Figure 20.5, the lumen, the luminal microvilli and the junctional complexes are evident apart from the other cytoplasmic organelles.

But in some moderately differentiated tumours, the lumen is not well-defined and the microvilli are not prominent. But the presence of secretory products in the lumen and secretory granules in the cytoplasm of the cells suggest the glandular nature of the tumour cells (Figure 20.6).

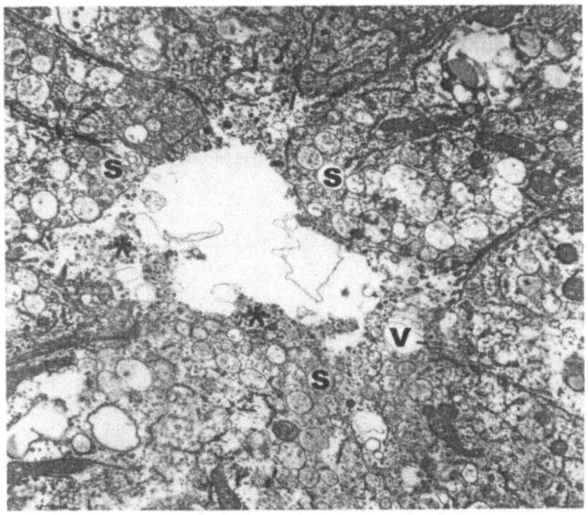

Figure 20.6 Moderately differentiated adenocarinoma (S—Secretory granule, V—Microvilli, *—particles of secretion) (13,000×). (*Source:* Richard Dickersin, G. (2000). *Diagnostic Electron Microscopy—A Text/Atlas,* 2nd edn. Springer Verlag Inc., New York. p.59, Fig. 3.34.)

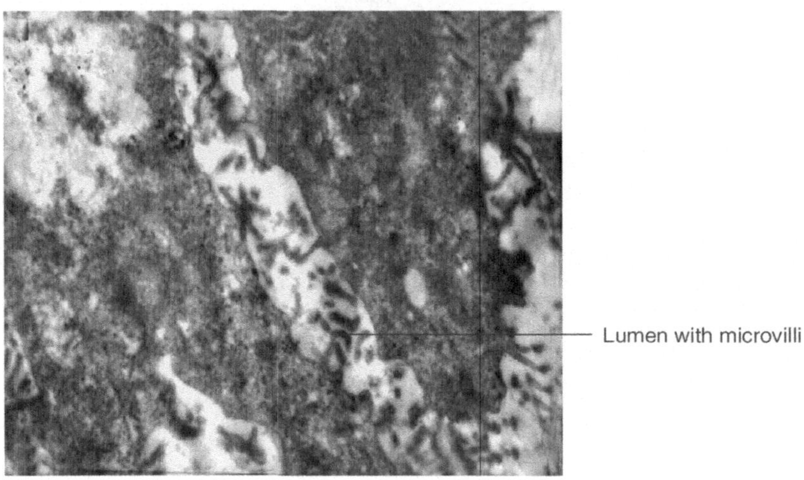

Figure 20.7 Poorly differentiated adenocarcinoma (10,000×)

In poorly differentiated cases (Figure 20.7), proper lumen formation may not be evident. Presence of microvilli though only a few, on the cell membrane is suggestive of the glandular nature of the tumour.

Another aberrant feature found in adenocarcinomas is intracytoplasmic crypts or neolumina which is lined by blunt microvilli. Secretory products may sometimes accumulate in these crypts. Figure 20.8 shows such a feature detected in an adenocarcinoma.

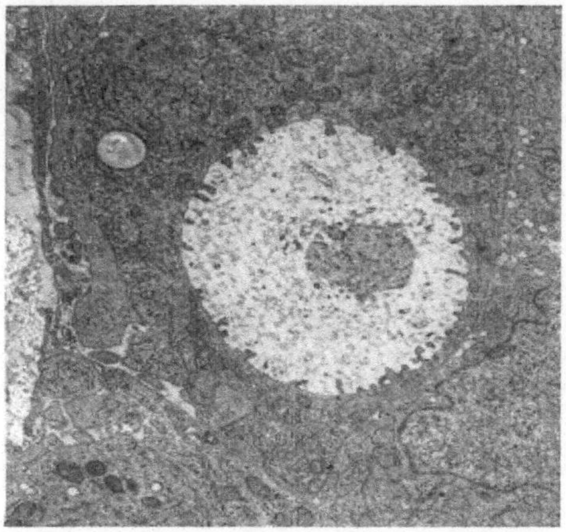

Figure 20.8 Intracytoplasmic neolumina–adenocarcinoma (27,300×). (*Source:* Richard Dickersin, G. (2000). *Diagnostic Electron Microscopy–A Text/Atlas,* 2nd edn. Springer Verlag Inc., New York. p.34, Fig. 3.7.)

Figure 20.8 shows a collection of microvilli within the cytoplasm referred to as intracytoplasmic neolumina, which is an aberrant feature because microvilli are seen only on the luminal surface of the epithelium. These structures indicate glandular epithelial tumour and while in a metastatic set-up suggests probable breast primary. If in intracytoplasmic lumen, tight junctions are present, then their presence is indicative of a pseudolumen caused by invagination of the cell by extracellular matrix (ECM).

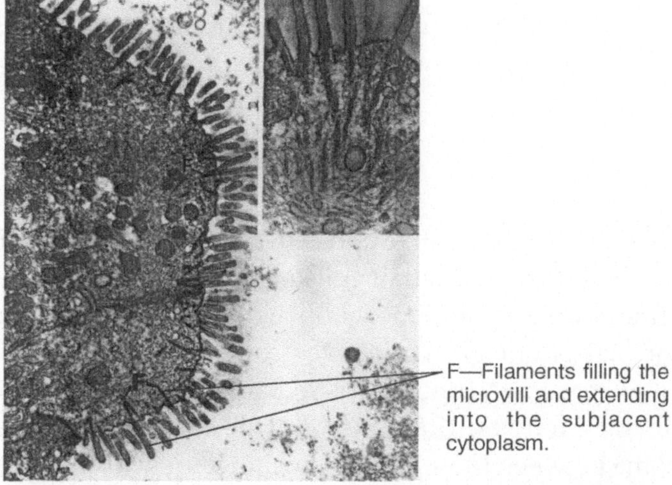

Figure 20.9 Terminal web in GI adenocarcinoma (22,300×). **Inset:** Higher magnification of the apical surface of a cell illustrates more clearly the filaments in the microvilli and cytoplasm (47,100×). (*Source:* Richard Dickersin, G. (2000). *Diagnostic Electron Microscopy–A Text/Atlas,* 2nd edn. Springer-Verlag Inc., New York. p.35, Fig. 3.8.)

In gastrointestinal adenocarcinomas, the microfilaments in the microvilli extend into the cytoplasm anchoring the microvilli to the cytoskeleton. These "terminal webs" are the distinguishing features of adenocarcinomas of the gastrointestinal tract (Figure 20.9). Similarly, the microvilli projecting into the lumen are usually covered by glycocalyx. The combination of membrane-associated glycocalyceal bodies and microvillar filamentous cores penetrating the underlying cytoplasm identifies the tumour that differentiates as intestinal-type epithelium.

MALIGNANT ROUND CELL TUMOURS (MRCT)

EM is an essential tool in the differential diagnosis of malignant round cell tumours, also called small round cell tumours. Under this category, most commonly, neuroblastoma (a nerve tumour), rhabdomyosarcoma (a striated muscle tumour), Ewings sarcoma (a bone tumour), small cell carcinoma (an epithelial tumour), lymphoma (a lymphoid tumour) are included. Many of these tumours occur predominantly in infancy and childhood. Light microscopically, small, uniform, round cells in diffuse sheets are visible. The differential diagnosis may not be possible at times even with special stains. IHC though very helpful may not be always conclusive. The diagnostic ultrastructural features are very useful. Moreover, ultrastructurally identified atypical and "hybrid" appearances have also been reported in some of these tumours.

But poorly differentiated tumours, particularly extremely poorly differentiated neuroblastomas and rhabdomyosarcomas may still pose a diagnostic dilemma as their features are sporadic and poorly developed. Prolonged search may be necessary to identify neuritic processes with neurosecretory granules or myosinlike or myosinlike myofilaments. In some instances, occasionally a precise diagnosis cannot be made ultrastructurally.

Ewings Sarcoma

In Ewings sarcoma (Figure 20.10), small round cells closely apposed with tenuous junctions most often occurring as infrequent small thickenings of apposed membranes are seen. The nucleus shows fine chromatin and the cytoplasm exhibits large aggregates of glycogen rosettes (rosettes) or "lakes" of glycogen. Most often the glycogen is depleted during routine processing and glossy, glycogen-depleted areas are noticeable. Cell processes and microtubules are inconspicuous; filaments and neurosecretory granules (NSGs) are typically absent. Stroma between tumour cells is characteristically sparse and lacks collagen fibres but may contain glycogen. Figure 20.10 shows a few closely apposed round tumour cells in Ewings. (*See* Figure 20.37 also).

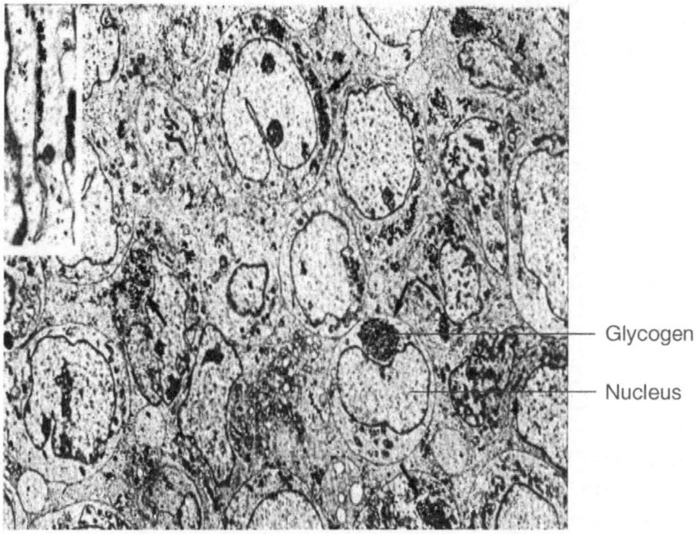

Figure 20.10 Ewings sarcoma—a few round to oval cells. Glycogen "lakes" (arrows) seen in the cytoplasm (3700×). (*Source:* Douglas, W. Henderson, John, M. Papadimitriou & Mark Coleman. (1986). *Ultrastructural Appearances of Tumours*, 2nd edn. Churchill Livingstone. p. 269, Fig. 25.7.)

Rhabdomyosarcoma

In rhabdomyosarcomas (Figure 20.11) (*See* also under striated muscle tumours), apart from small round cells, larger cells with abundant cytoplasm can be made out. These cells show roughly parallel arrays of thick and thin filaments with electron-dense zones reminiscent of abortive Z lines characterizing these tumours with striated muscle differentiation. The aggregated parallel thick and thin filaments in the small round cell tumours indicate rhabdomyoblastic differentiation, even in the absence of definite sarcomeric arrangements. The nucleus is most often irregular with multiple profiles and the nuclear chromatin is condensed and often marginated unlike Ewings and neuroblastoma. External lamina may be prominent in highly differentiated tumours, but is usually fragmentary and is sometimes absent in poorly differentiated ones. Cell processes, microtubules and intercellular junctions are inconspicuous. Cytoplasmic glycogen is often prominent but usually monoparticulate. Intracellular collagen fibres may be found which are not found in other round cell tumours.

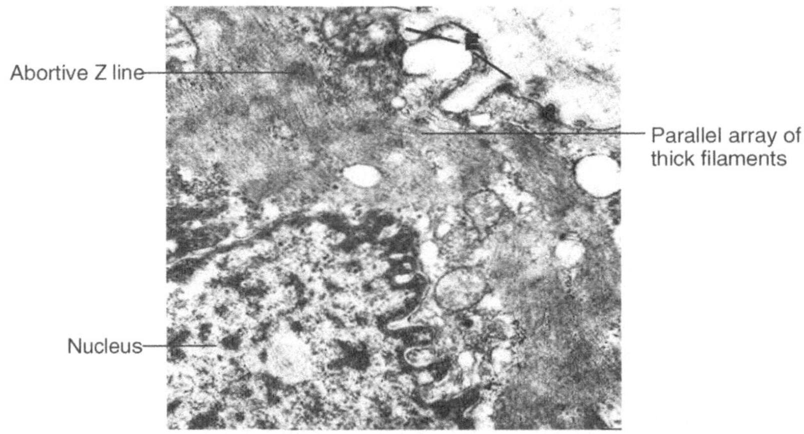

Figure 20.11 Rhabdomyosarcoma (26,400×). E—External fragmented lamina. (*Source:* Douglas, W. Henderson, John, M. Papadimitriou and Mark Coleman. (1986). *Ultrastructural Appearances of Tumours*, 2nd edn. Churchill Livingstone. p. 245, Fig. 22.7.)

Lymphoma

Lymphomas are even considered as diagnostic wastebaskets as they do not exhibit any characteristic features. In fact, absence of any significant unique feature is indicative of lymphoid nature. The cells round to oval mostly show scanty cytoplasm with sparse organelles (Figure 20.12). Junctions, basal lamina, filaments are typically absent.

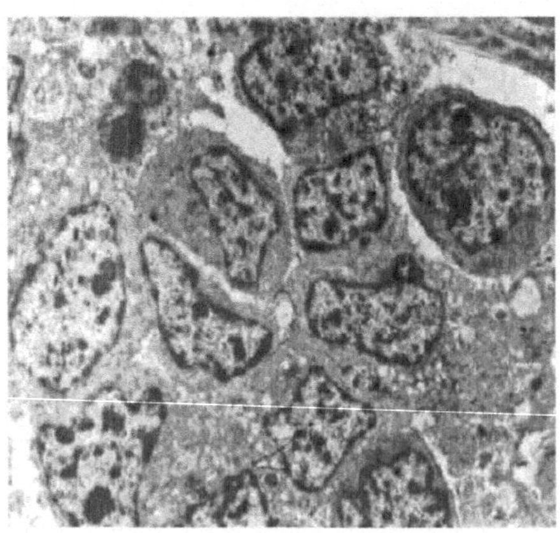

Figure 20.12 Non-Hodgkins lymphoma—a few cells (2500×)

Neuroblastoma (NBS)

Neuroblastomas are predominantly childhood tumours, though adult types also occur. The ultrastructure of neuroblastomas is characterized by cells with elaborate cytoplasmic processes in which microtubules, neurofilaments and neurosecretory granules are concentrated. Intercellular junctions may include synapses apart from the desmosomes. Varying numbers of primitive embryonic cells devoid of distinctive features are also present and they may predominate. However, EM permits a diagnosis in the majority of cases, and has contributed to the recognition of neuroblastoma in adults.

Smooth nucleus with fine chromatin, long slender processes forming bundles showing neurotubules and neurosecretory granules (NSGs) are the contributory factors towards the diagnosis of neuroblastoma. The neurosecretory granules have a distinct morphology. The granule has a limiting membrane, central dense core with a halo in between. Figure 20.13 shows a few cytoplasmic processes with microtubules and NSGs while Figure 20.14 shows the neurosecretory granules in the processes and the typical morphology of these granules is depicted in Figure 20.15.

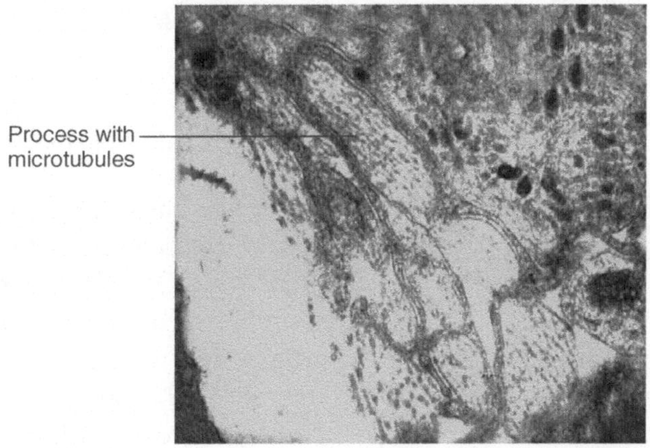

Figure 20.13 NBS—The processes with microtubules and granules (30,000×)

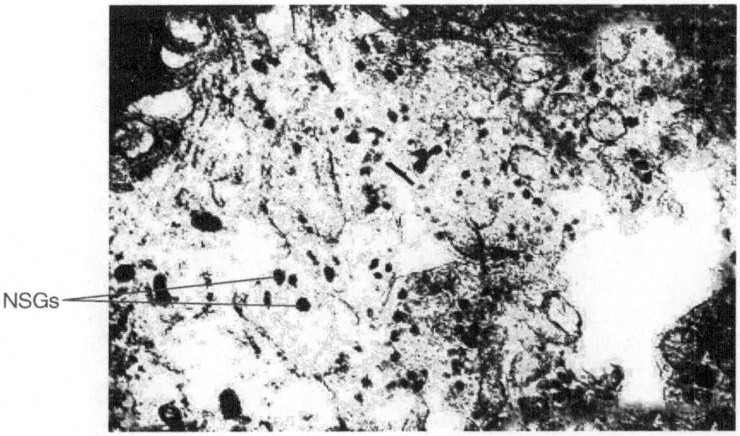

Figure 20.14 NBS—Neurosecretory granules in processes (20,000×)

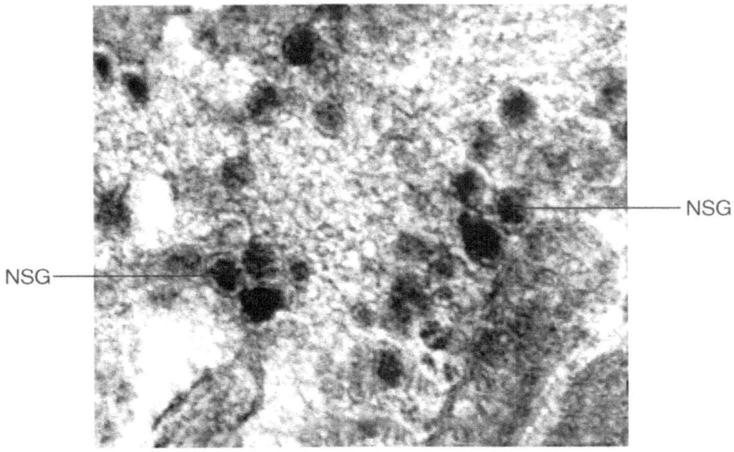

Figure 20.15 NBS—Neurosecretory granules at higher magnifications showing typical morphology (80,000×)

Neuroendocrine Carcinoma of Skin (Merkel Cell Carcinoma)

This tumour has been classified into a separate entity based on EM characteristics. In the past, these tumours have been reported as trabecular carcinoma, small cell neuroepithelial tumour of the skin and peripheral adult neuroblastoma. A definitive diagnosis is facilitated by electron microscopy. The cells are classically polyhedral and loosely cohesive with sporadic intermediate-type junctions. Moderate numbers of rounded neurosecretory granules constitute the cytoplasm. These granules are usually at the periphery, often forming linear, sub-plasmalemmal arrays, but they also occur in cytoplasmic processes. Paranuclear aggregates of intermediate filaments also typify these tumours. Figure 20.16 shows a few round, oval to polyhedral cells and Figure 20.17 shows the NSGs in the cytoplasmic processes.

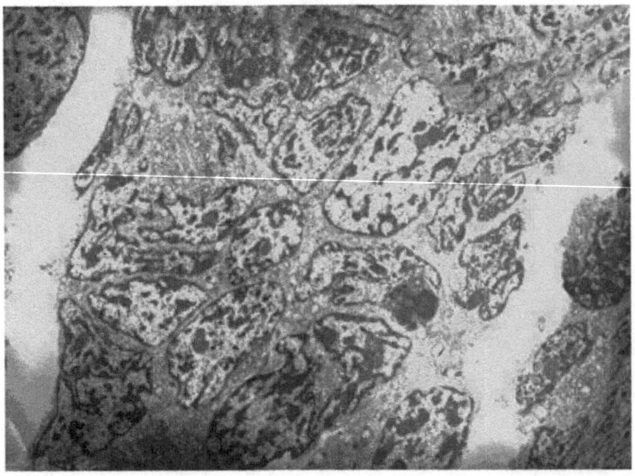

Figure 20.16 Merkel cell carcinoma—a few round to oval tumour cells (2500×)

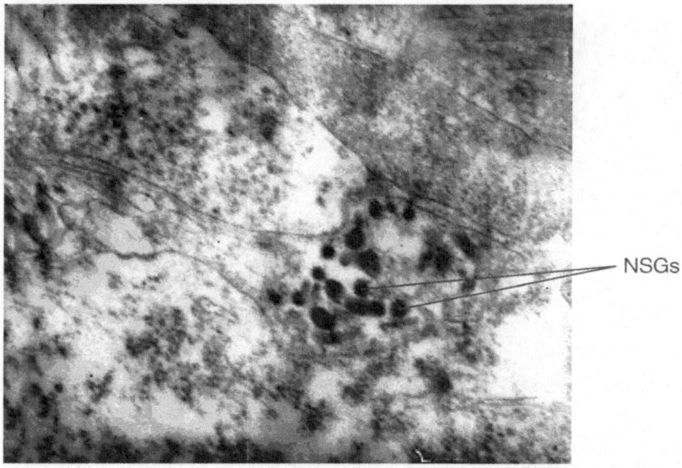

Figure 20.17 Neurosecretory granules in cytoplasmic Merkel cell carcinoma (30,000×)

DIFFERENTIAL DIAGNOSIS OF ANTERIOR MEDIASTINAL TUMOURS

Similarly, in the differential diagnosis of anterior mediastinal tumours, the ultrastructural features reveal some unique distinct characteristics. Thymoma, lymphoma, germinoma/seminoma and thymic carcinoid are the most common tumours occurring in the anterior mediastenum. The comparative ultrastructural features are summarized in Table 20.2.

Table 20.2 Comparative ultrastructure of anterior mediastinal tumours

Features	Thymoma	Lymphoma	Germinoma	Carcinoid
Ext. lamina	Prom.multilayered	Absent	Absent	Present-focal
Cell processes	Prom.interleaving	Incons.	Incons.	Incons.
Desmosomes	Numerous, well developed	Absent	Infrequent	Occasional
Tonofilaments	Prom. in bundles	Absent	Absent	Incons.
NSGs	Absent	Absent	Absent	Numerous
Glycogen	May be present	Incons.	Typically prom.	Not prom.
Nucleolus	Not prom.	Not prom.	Prominent nucleolonema typically in ribbons	Not prom.
Nuclear projections	Incons.	Frequent	Incons.	Incons.

Prom.—Prominent; Incons—Inconspicuous

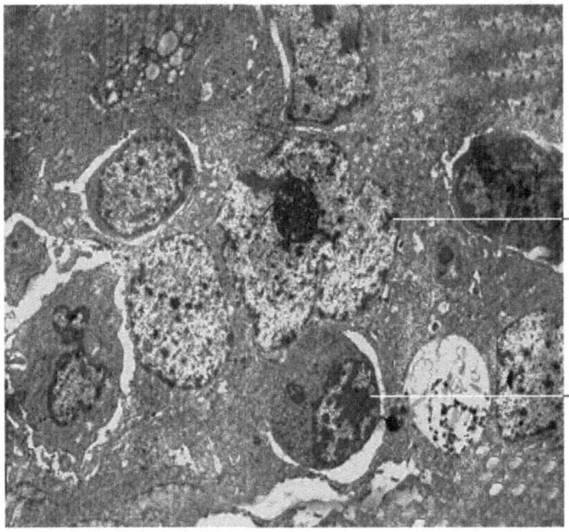

Figure 20.18 Epithelial and lymphoid cells in thymoma (2500×)

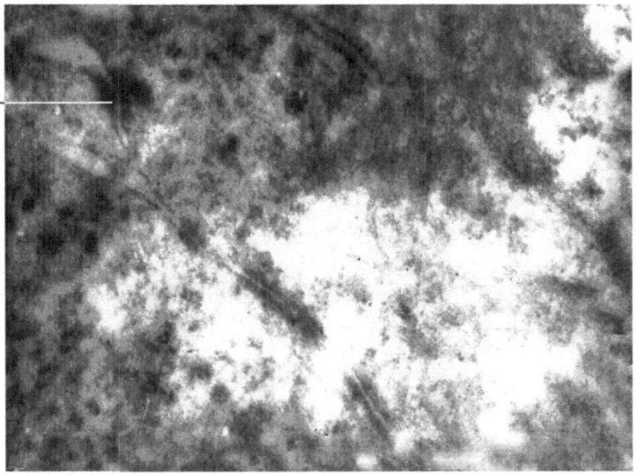

Figure 20.19 Desmosomes between epithelial cells in thymoma (40,000×)

Thymomas, the thymic tumours, are characterized ultrastructurally by the presence of epithelial cells with cytoplasmic tonofilaments, numerous desmosomes and basal lamina at the epithelial–stromal interface along with lymphocytes. Even in lymphocyte-rich thymomas, the epithelial component is readily apparent on ultrastructural examination.

In thymomas, the lymphocytes are seen surrounding the epithelial cells and in close proximity with these cells (Figure 20.18). The tonofilaments in the larger cells and junctions between them (Figure 20.19) are suggestive of epithelial differentiation. In thymic carcinoids, composite, (both) squamous and neuroendocrine differentiation is reported. Both glycogen and lipid are absent on ultrastructural examination.

Germinomas/seminomas are characterized by coiled nucleolonema and glycogen as described in Figure 20.25.

ABDOMINAL (PAEDIATRIC) TUMOURS

In abdominal tumours, especially in childhood tumours, the most common neuroblastomas and nephroblastomas (Wilms tumour) may be differentiated by using the fine structural features as mentioned below.

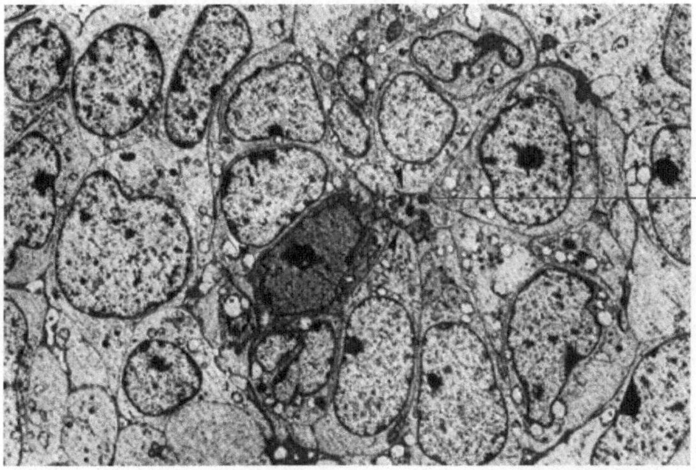

Figure 20.20 Wilms tumour (Nephroblastoma)—tubule formation (3700×). (*Source:* Douglas, W. Henderson, John, M. Papadimitriou & Mark Coleman. (1986). *Ultrastructural Appearances of Tumours*, 2nd edn. Churchill Livingstone, New York. p. 338, Fig. 30.10.)

The tubular formation with central lumen is suggestive of Wilms tumour, whereas the cytoplasmic processes, neurofilaments and neurosecretory granules particularly in processes are indicative of neuroblastomas as we have already seen in Figures 20.13–20.15.

The cell populations of Wilms tumour can be subdivided into four main compartments, namely, blastema cells, tubular formation, glomeruloid bodies and stroma. In Figure 20.20 the cells comprising the rudimentary tubule are radially arranged around a central point where there are intercellular junctions (arrowheads) but a tubular lumen is not evident. A well-developed layer of fuzzy external laminal material surrounds the tubule.

DIFFERENTIAL DIAGNOSIS OF ADRENAL TUMOURS (CORTEX AND MEDULLA)

The adrenal cortical and medullary tumours show some distinctive clinical features but there are times when definite diagnosis is doubtful because of overlapping features. The cortex secretes corticosteroids and hence the steroid-secreting cell characteristics can be visualized in these tumours, namely numerous smooth endoplasmic reticulum vesicles, stacks of rough endoplasmic reticulum and mitochondria with tubular cristae. Pleomorphic mitochondria with plentiful matrices and mitochondrial gigantism may also be encountered in these tumours. Mitochondria sometimes contain dense materials. Similarly mitochondria with tubular cristae or lamellar cristae can also be seen. On the contrary, the medullary tumours, often called **Pheochromocytoma,** secreting adrenaline and noradrenaline show multiple neurosecretory granules. The granules of pheochromocytoma are abnormally large. The NSGs have characteristic dense core, limiting membrane and a halo in between. In the case of adrenaline and noradrenaline the halo is large and therefore these granules are referred to as **halo granules**. The granules with central less dense cores and uniform thin submembranous halos depict adrenaline and dense comma-shaped eccentric core granules represent noradrenaline as seen in Figure 20.21. Figure 20.21a is the cortical tumour showing the smooth endoplasmic reticulum vesicles and mitochondria with tubular cristae, while Figure 20.21b shows neurosecretory granules with central as well as eccentric cores. Figure 20.22 shows the stacked lamellae of RER in cortical tumour.

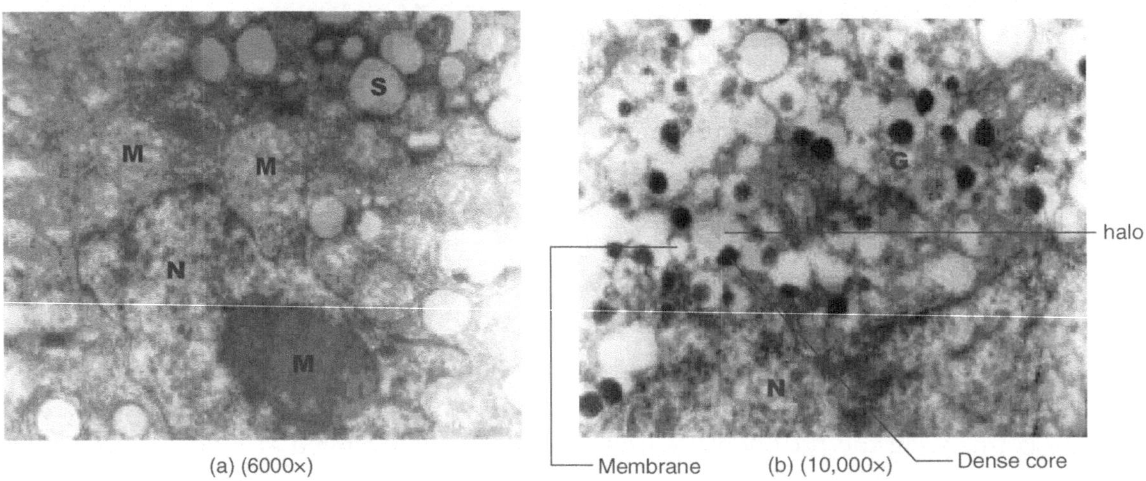

Figure 20.21 Adrenal tumours (cortex and medulla) (N—Nucleus, M—Mitochondria, S—SER vesicles, G—Granules

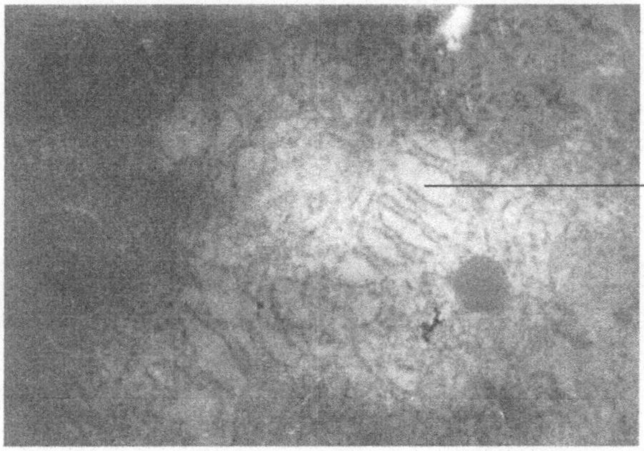

Figure 20.22 RER in stacks in cortical tumour (15,000×)

Similarly in pancreatic islet tumours also, the NSGs show classical granule morphology in tumours secreting insulin, gastrin and glucagon which appears to be diagnostic but the diagnostic value is limited by the occurrence of nonspecific granules.

Insulin-secreting pancreatic islet tumours secrete granules which contain typical crystalline cores and some also contain large dense spherules separated from the limiting membrane by a distinct halo. The glucagon-secreting tumours show round to oval granules in which only a small peripheral halo separates the electron-dense material from the limiting membrane. In gastrin-secreting tumours, large lucent granules predominate.

MESOTHELIOMAS

Mesothelial neoplasms are characterized by attenuated mesothelial cells with epithelial features, either enclosing luminal spaces or situated on the surface of cellular aggregates. Innumerable long, slender microvilli generally protrude from the cell surface, which can be demonstrated both by TEM and SEM.

The mesotheliomas and mucin-secreting adenocarcinomas require special techniques for their differential diagnosis. Though junctional complex, desmosomes, acini, external lamina are seen in both adenocarcinoma and mesothelioma, terminal web filaments and mucinlike granules are typically absent in mesotheliomas. Microvilli are numerous, long (with length : breadth more than 10 : 1), sinuous and complex in mesotheliomas while in adenocarcinomas they are relatively fewer and shorter. The other two characteristics of mesotheliomas are cytokeratins often plentiful and microvilli directed towards collagen in the stroma. Glycogen is often copious in mesothelioma. Difference in the organization of intercellular junctions in both as revealed by freeze-fracture technique is also reported.

The three most important distinguishing features namely, 1) cytokeratin filaments forming probably bundles of tonofibrils, 2) complex, long microvilli and 3) collagen–microvilli interactions are clearly identifiable in the Figure 20.23a, b and c. The ultrathin sections were cut from paraffin-embedded thick sections and not fresh tissue biopsy. The paraffin-embedded sections were reprocessed for electron microscopy after removal of the paraffin using xylene.

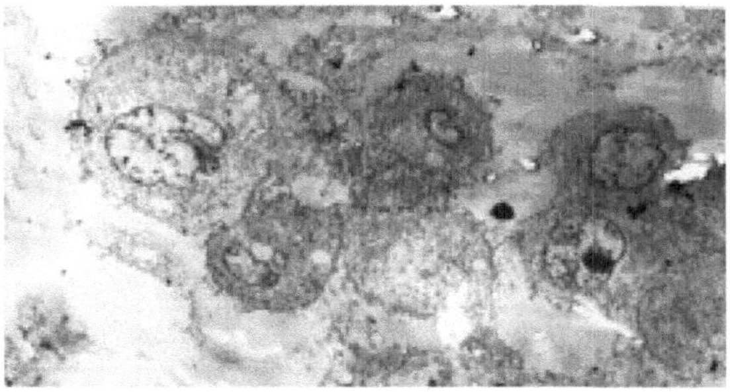

(a) A few cells in mesothelioma (2000×)

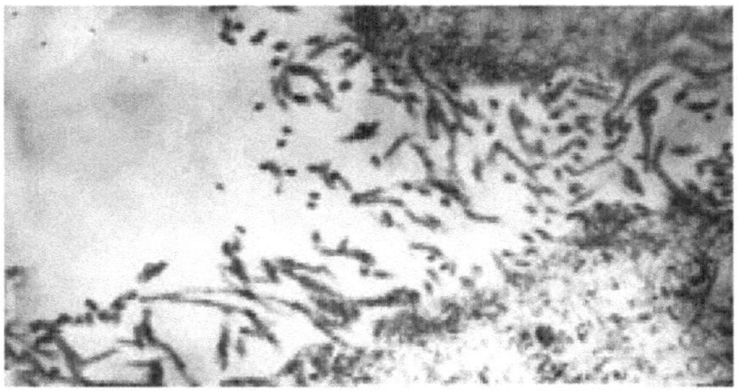

(b) Long complex microvilli (15,000×)

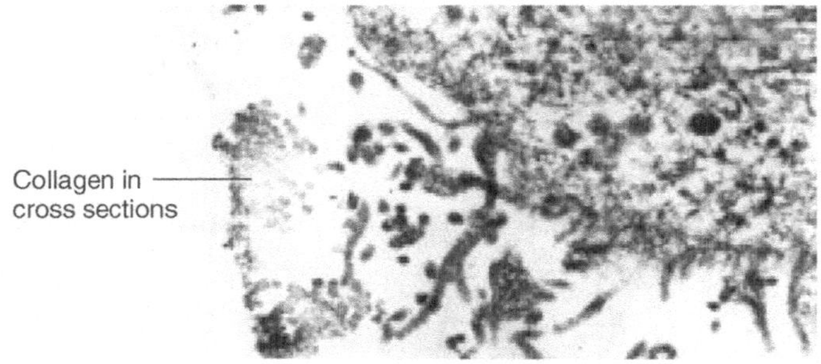

(c) Collagen–microvilli interaction (20,000×)

Figure 20.23 Mesothelioma–diagnostic features. a) A few cells in mesothelioma (2000×) b) Long, complex microvilli (15,000×) c) Collagen–microvilli interaction (20,000×)

Figure 20.23a shows a few tumour cells which exhibit microvilli on their surfaces. Figure 20.23b shows long, complex microvilli, while Figure 20.23c shows the collagen–microvilli interaction. (Collagen is seen in cross sections.)

MALIGNANT MELANOMA

Melanoma is another condition where ultrastructural characterization is highly useful. EM definitely serves to narrow down the differential diagnosis. The melanocytic nature of the neoplastic cells may be confirmed by the detection of melanosomes—the organelles characteristic of melanocytic differentiation in situations where the production of melanin is very minute that they cannot be detected by routine LM techniques. The melanosomes consist of membrane-bound vesicles with a distinctive internal structure. Depending on the stage of development, these may be parallel lamellar, helical or zig-zag structures. Melanin synthesis takes place in 4 stages, the first a non-diagnostic vacuole, the second and third stages are diagnostic referred to as stage II and stage III premelanosomes. Stage II premelanosomes are larger oval/elliptical vesicles with pathognomic internal lamellar structure while those in stage III are partially pigmented structures. Stage IV is again fully pigmented granule which is again non-diagnostic. The presence of stage II and stage III premelanosomes is indicative of tumour of melanin-synthesizing cell, i.e., melanoma. Similarly, if these premelanosomes are identified in a metastatic node, then the primary is melanoma. Cytoplasmic intracisternal arrays of microtubules and seen but are nonspecific. Apart from this, EM aids in identifying the variants of melanomas, namely,

- malignant melanoma of soft parts
- desmoplastic malignant melanoma
- neutropic melanoma—amelanotic melanocytes with Schwaanian differentiation

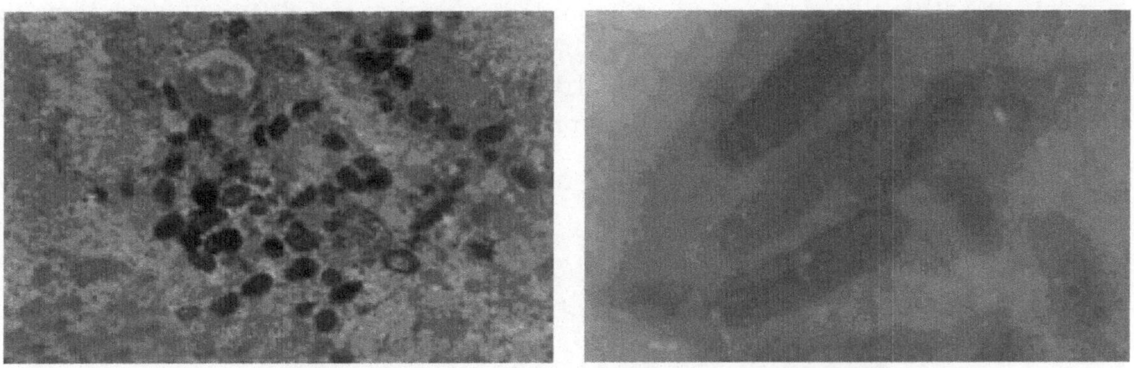

(a) (20,000×) (b) (80,000×)

Figure 20.24 Malignant melanoma

Straight intracisternal tubules with a fuzzy coating are occasionally present in the RER forming geometric aggregates. These coated parallel tubules (CPT) have been most often found in malignant melanoma and have diagnostic value particularly in the absence of demonstrable melanosomes.

Compound melanosomes have no diagnostic significance, because they may be found in many non-melanocytic cells especially macrophages.

Figure 20.24a shows melanosomes at various stages of melanization; Figure 20.24b shows stage II premelanosomes.

LYMPHOMAS

Lymphoma, the malignant tumour of the lymphoid tissue, is an area where EM has not much to contribute. The T cell, B cell and other subclassifications as well as the differentiation between non-Hodgkins and Hodgkins lymphomas are best carried out using immunohistochemical techniques including Facscan, so EM is required only when IHC is inconclusive or negative. The resin-embedded semithin sections are also highly useful. Moreover, some of the stromal components may also be of diagnostic significance, namely

1. intercellular amyloid fibrils in some plasmacytomas
2. prominent vascular reaction in immunoblastic lymphadenopathy
3. infiltration of vessel walls by malignant cells may be seen in the leukaemic phase of some lymphomas

But there are certain situations where EM is considered to be very significantly useful. They are listed below:

* In the differential diagnosis of anaplastic carcinoma/ diffuse lymphoma, the diagnosis of lymphoma can be established on ultrastructural examination by the absence of the indicators of epithelial differentiation, namely intercellular junctions, BL, acinus formation and cytoplasmic tonofilaments.

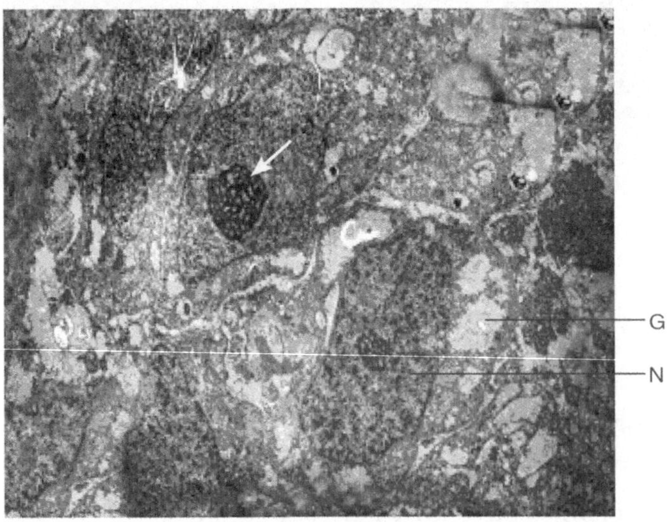

Figure 20.25 A few cells in seminoma (Arrow—Nucleolus as coiled nucleolonema, G—Glycogen, N—Nucleus) (3000×)

* EM plays a role in the diagnosis of testicular tumours, particularly, in the differential diagnosis of poorly differentiated primary germ cell tumours like seminoma and some lymphoid tumours. The neoplastic cells in seminoma are characterized by prominent nucleolar spiremes, often referred to as "fragmented filamentous nucleoli" or ribbonlike nucleolonema," few intercellular junctions, cytoplasmic glycogen and paucity of cytoplasmic organelles. When glycogen gets depleted during processing glycogen-depleted areas can be made out as shown in Figure 20.25 in the area marked 'G'.
* To differentiate lymphocytic lymphomas from other small round cell tumours as mentioned earlier.
* To differentiate lymphomas from other neoplasms affecting the anterior mediastinum as already mentioned.
* To differentiate between large cell lymphomas and true histiocytic lymphomas. The ultrastructural identification of histiocytic cells in true histiocytic lymphomas is largely based on the presence of lysosomes, phagocytic activity, cytoplasmic flaps and intercellular junctions In Figure 20.26, a mononuclear phagocyte shows many lysosomes in the cytoplasm.

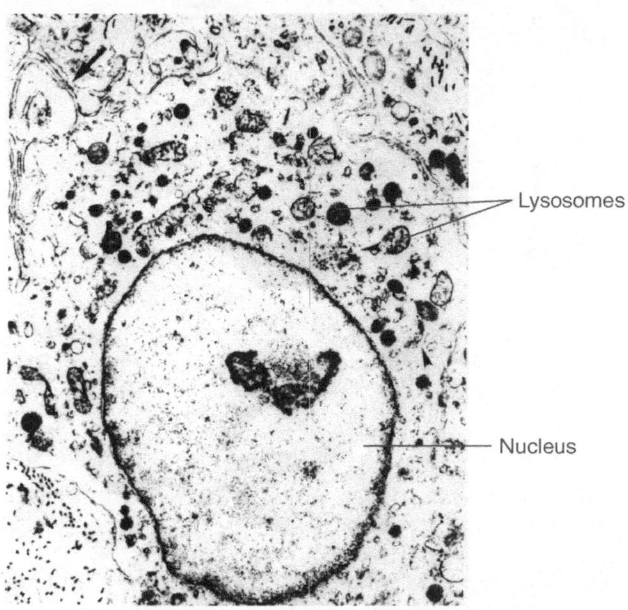

Figure 20.26 True histiocytic lymphoma (9800×). (*Source:* Douglas, W. Henderson, John, M. Papadimitriou & Mark Coleman. (1986). *Ultrastructural Appearances of Tumours*, 2nd edn. Churchill Livingstone, New York. p. 295, Fig. 27.2.)

* Presence of lipid globules, which can be identified with EM is a regular feature of Burkitt's lymphoma where the nucleus is only slightly indented, but the mitochondria, although few, are large and often polarized to one side of the cell.
* Presence of electron-dense dendritic cells of some lymphomas is indicative of follicular centre cell lymphomas.
* In plasmacytomas, the tumour cells show eccentric nucleus where the chromatin deposition exhibits a clock-face or cart-wheel pattern. The cytoplasm is filled with RER, extended cisternae of which are filled with probably proteinaceous material (Figure 20.27).

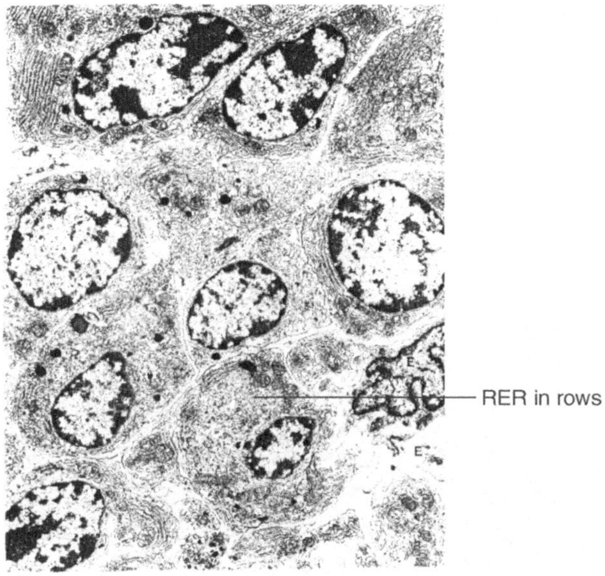

Figure 20.27 Plasmacytoma (10,800×). (*Source:* Douglas, W. Henderson, John, M. Papadimitriou & Mark Coleman. (1986). *Ultrastructural Appearances of Tumours*, 2nd edn. Churchill Livingstone. p. 296, Fig. 27.24.)

In Figure 20.27, a uniform population of plasmacytoid cells with moderate amounts of marginated heterochromatin depicting the characteristic cart-wheel pattern and abundant RER can be seen. Golgi complex is prominent in a few cells.

* In mycosis fungoides/sezary syndrome, the characteristic cerebriform or serpentine nucleus (Figure 20.28) can be clearly visualized in EM.
* In hairy cell leukaemia the diagnostic elaborate cytoplasmic processes and flaps are typically visible in both SEM and TEM (Figure 20.29). The ribosome–lamellar complexes, though not specific, also aid in the diagnosis.

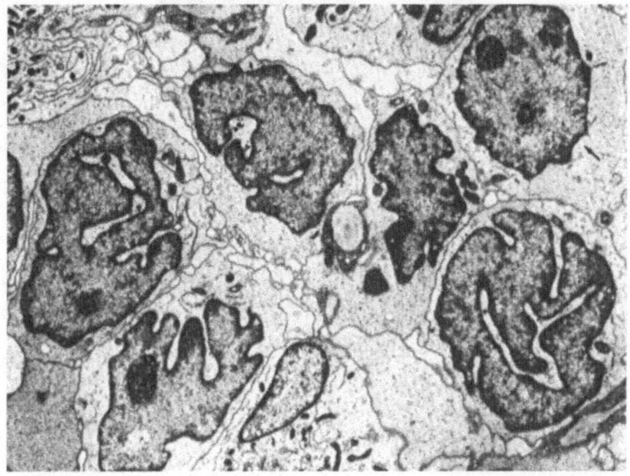

Figure 20.28 Mycosis fungoides—cerebriform or serpentine nucleus (6500×). (*Source:* Douglas, W. Henderson, John, M. Papadimitriou & Mark Coleman. (1986). *Ultrastructural Appearances of Tumours,* 2nd edn. Churchill Livingstone, New York. p. 286, Fig. 27.6.)

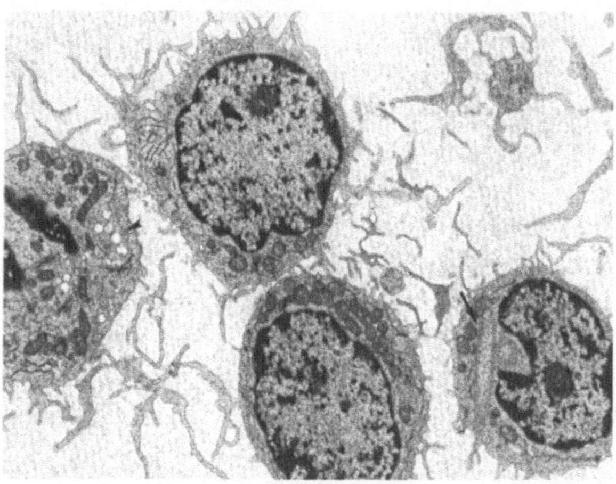

Figure 20.29 Cytoplasmic flaps as multiple hairlike projections of hairy cell leukaemia (8600×). (*Source:* Douglas, W. Henderson, John, M. Papadimitriou & Mark Coleman. (1986). *Ultrastructural Appearances of Tumours,* 2nd edn. Churchill Livingstone, New York. p. 311, Fig. 28.15.)

- ✳ The phenomenon of **Emperipolesis** in RDD (Rosai–Dorfmann disease) also known as "sinus histiocytosis with massive lymphanopathy" is evident at higher magnifications of EM.

In this phenomenon, many lymphocytes are seen inside the histiocyte cytoplasm (Figure 20.30). They are not engulfed by the histiocytes. They are free to enter and leave the histiocytes. This is evident by the fact that each lymphocyte is surrounded by a vacuole separating it from the histiocyte cytoplasm.

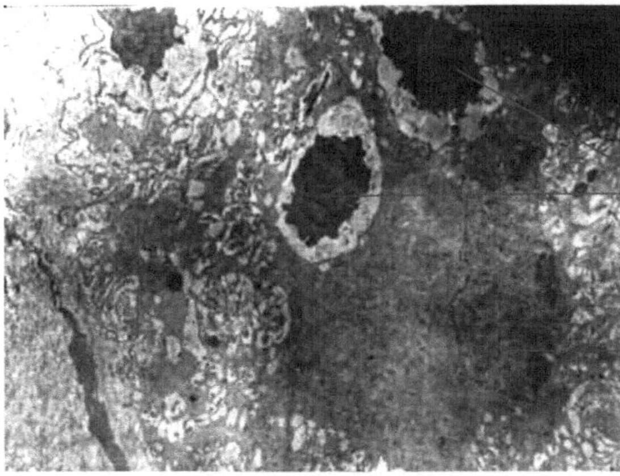

Figure 20.30 Emperipolesis in RDD (2000×)

* In histiocytosis X, EM is confirmatory by identifying the Birbeck granules also called Langerhans cell granules.

  In histiocytosis X, otherwise known as Langerhans cell histiocytosis, the histiolytic cell shows the typical irregular deeply indented nucleus, cytoplasmic flaps, abundant cytoplasm with organelles. The characteristic granules known as Langerhans cell granules or Birbeck granules seem to arise from the membrane and grow inwards and open into the exterior. The granules usually bulge at one end, though bulging can occur anywhere. The bulging at the end gives the typical tennis-racket shape to these granules. The granules get detached from the cell membrane and are seen in the cytoplasm. The typical specific trilaminar structure of the granule can be readily made out in EM. Figure 20.31 shows one cell with a highly indented nucleus. Figure 20.32 shows the granule arising from the cell membrane growing inwards and bulging at one end. Figure 20.33 shows a few typical tennis-racket-shaped granules.

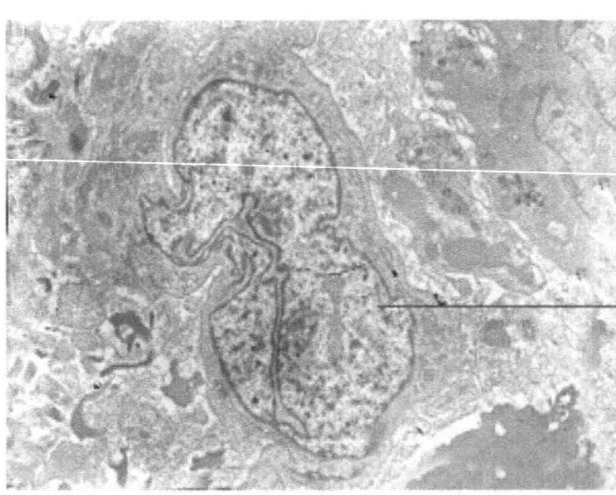

Figure 20.31 Histiocytosis cell showing highly irregular (grooved) nucleus (5000×)

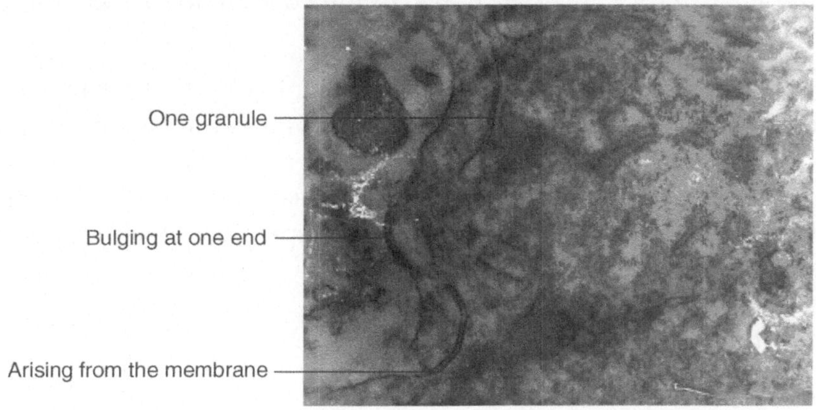

Figure 20.32 Birbeck granules arising from the membrane and growing inwards with bulging at the end (30,000×)

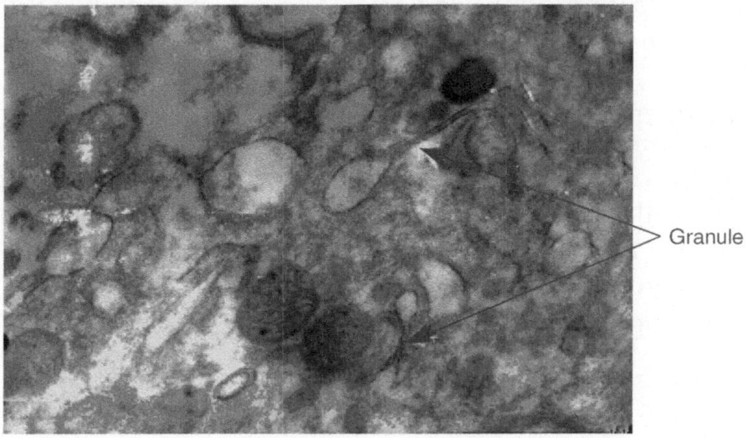

Figure 20.33 The cytoplasmic granule showing characteristic tennis racket shape (50,000×)

BONE AND SOFT-TISSUE TUMOURS

BONE AND CARTILAGINOUS TUMOURS

Bone tumours are diverse in their gross and morphological features and range in their biological potential from the innocuous to the extremely fatal. Most bone tumours are classified according to the normal cell or tissue of origin. Overall, matrix-producing tumours and fibrous tumours are the most common, and among the benign tumours, osteochondroma and fibrous cortical defect are the most frequent. Excluding malignant neoplasms of marrow origin (myeloma, lymphoma and leukaemia), osteosarcoma is the most common primary cancer of bone, followed by chondrosarcoma and Ewings sarcoma.

In addition to the fine structure of the component cells, the stromal morphology of chondroid and osteoblastic tumours is of diagnostic value. Cartilaginous matrix typically contains a loose pattern of proteoglycan granules together with collagen fibres and filaments. In contrast, the stroma of osteogenic sarcoma is characterized by focal hydroxyapatite deposition which appears as dense spicules closely apposed to collagen fibres.

BONE-FORMING TUMOURS

Osteoma

Osteomas, arising from the sub-periosteal or endosteal surfaces of the cortex, may be solitary or multiple. Histologically, they are composed of woven and lamellar bone that is deposited in a cortical pattern with Haversian-like system. The histological differential diagnoses include reactive bone induced by infections, trauma or hemangiomas.

Osteoid Osteoma and Osteoblastoma

Histologically they are composed of a morass of randomly interconnecting trabeculae of woven bone prominently rimmed by osteoblasts. The stroma surrounding the tumour bone consists of loose connective tissue that contains many dilated and congested capillaries. The osteoid osteomas elicit a heavy amount of reactive new bone formation that surrounds the tumour.

Osteosarcoma

Osteosarcomas are the mesenchymal malignant tumours in which the cancerous cells produce bone matrix. Several subtypes of osteosarcomas are recognized and grouped according to (i) The anatomical portion of the bone from which they arise (ii) Multicentricity (iii) Primary or secondary and (iv) Histologic variants (osteoblastic, chondroblastic, fibroblastic, telangiectatic, small cell and giant cell variants). A small cell variant of osteosarcoma is also reported.

The tumours frequently destroy the surrounding cortices and produce soft tissue masses and also spread extensively into the medullary canal. The tumour cells frequently have hyperchromatic nuclei. Bizarre tumour giant cells are common, as are mitoses. The formation of bone by tumour cells is characteristic of osteosarcoma. The neoplastic bone has a coarse, lacelike architecture but is also deposited in broad sheets or as primitive trabeculae. Other matrices including cartilage or fibrous tissue may be present in varying amounts. Vascular invasion is usually conspicuous.

Ultrastructurally osteosarcomas have been reported to contain pleomorphic fibroblastoid cells with copious RER but other mesenchymal cells such as osteoblasts, osteocytes, myofibroblasts, chondroblasts, etc. are also seen. Distinctive features include malignant osteoblasts with abundant dilated RER in annular or lamellar arrangements and undifferentiated cells. Cytoplasmic lipid droplets and glycogen are also reported.

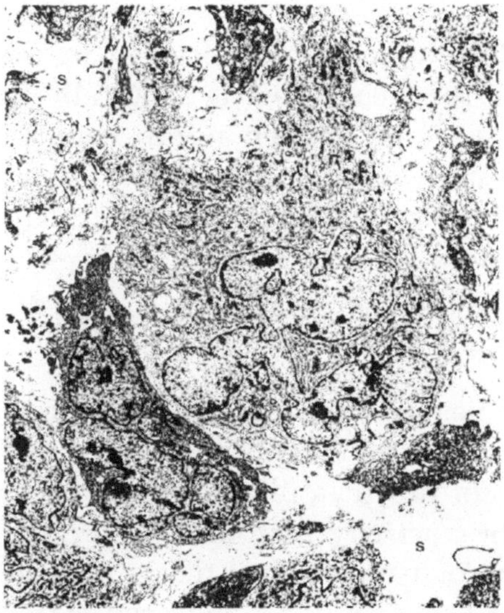

Figure 20.34 A cell from osteosarcoma (5100×). (*Source:* Douglas, W. Henderson, John, M. Papadimitriou & Mark Coleman. (1986). *Ultrastructural Appearances of Tumours*, 2nd edn. Churchill Livingstone, New York. p. 267, Fig. 25.4.)

In Figure 20.34, osteoblastic cells with plentiful RER are seen. Cells have highly irregular nuclear profiles. Mitochondria are numerous and may contain calcific inclusions. Plasma membranes are also irregular often forming ruffles. Collagen bundles inhabit the stroma (S).

CARTILAGE-FORMING TUMOURS

In mesenchymal tumours, the essential ultrastructural features of chondrocytic differentiation include numerous dilated strands of RER (sometimes containing finely granular intracisternal material resembling the extracellular matrix), as well as lipid droplets and glycogen rosettes.

Osteochondroma

Osteochondromas are, usually, solitary benign lesions composed of benign hyaline cartilage varying in thickness and covered by perichondrium. The cartilage has an appearance of a disorganized growth plate and undergoes endochondral ossification, with the newly made bone forming the inner portion of the head and stalk.

Chondroma

Chondromas are benign tumours of hyaline cartilage. They may arise in the medullary canal as in enchondromas, or on the surface of bones when they are called juxtacortical chondromas. Microscopically, the nodules of cartilage are metaplastic and the neoplastic chondrocytes, that reside in the lacunae are cytologically benign. At the periphery, the cartilage cells mature and undergo endochondral ossification. At the centre, the cells calcify and die. Ultrastructurally, in chondromas the chondroid nature is indicated by the arrangements of the matrical components, including a lattice of filaments and granules within the cartilaginous matrix.

Chondroblastoma

Chondroblastomas are rare benign cartilaginous neoplasms. The tumour is cellular and is composed of compact polyhedral chondroblasts that have well-defined cytoplasmic borders, moderate amounts of pink cytoplasm and nuclei that are hyperlobulated, with longitudinal grooves. Mitotic activity and necrosis are present. The tumour cells are surrounded by scant amounts of hyaline matrix that is deposited in a lacelike configuration. When the matrix matures and calcifies, it produces the characteristic chicken-wire pattern of mineralization. Scattered through the matrix are non-neoplastic giant cells.

Chondromyxoid Fibroma

They are rare tumours, microscopically, composed of nodules of poorly formed hyaline cartilage and myxoid tissue delineated by fibrous septae. In contrast to other benign cartilaginous tumours, the neoplastic cells in chondromyxoid fibroma show varying degrees of cytological atypia. Small areas of calcification and scattered non-neoplastic giant cells may also be found.

Chondrosarcoma

Conventional chondrosarcomas are composed of malignant hyaline and myxoid cartilage. The tumours may vary in their cellularity, cytological atypia or their mitotic activity. Approximately 10% of conventional low-grade chondrosarcomas have a second high-grade component that has the morphology of a poorly differentiated sarcoma, such as malignant fibrous histiocytoma, fibrosarcoma or osteosarcoma. This combination is identified as the dedifferentiated variant of chondrosaroma. Clear cell chondrosarcomas consist of large malignant chondrocytes with abundant, clear cytoplasm, numerous osteoclast-like giant cells and intralesional reactive new bone formation. Mesenchymal variants are composed of islands of well-differentiated hyaline cartilage surrounded by sheets of small round cells, which can mimic Ewing's sarcoma.

On ultrastructural examination, chondrocytes are characterized by their typical spiked or scalloping cell membrane. Cytoplasm is abundant with nucleus normally displaced to one end. Extended cisternae of RER and particulate glycogen occupy the major portion of the cytoplasm. The matrix is plentiful and pericellular depletion of filaments corresponds to the lacunae seen by LM. The neoplastic cells have scalloped margins, with small processes protruding into the interstitium. Glycogen is often abundant, particularly in clear cell and myxoid tumours. Lipidic globules vary in both number and size, and are not usually prominent. The amount of RER also varies, but the Golgi apparatus is moderately developed. Figure 20.35 shows a few chondrosarcoma cells and Figure 20.36, the typical spiked cell membrane.

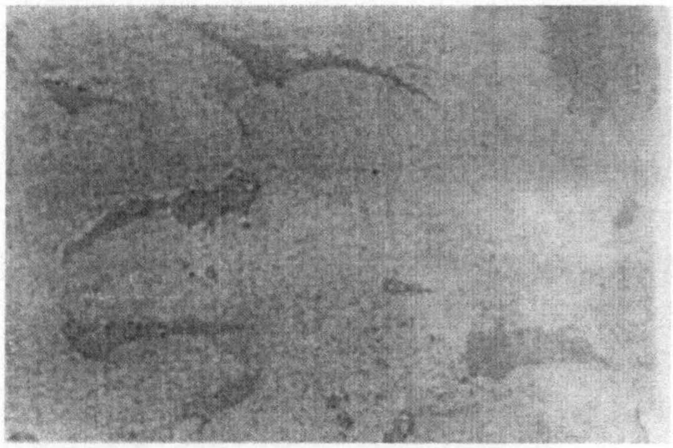

Figure 20.35 A few tumour cells in chondrosarcoma (1000×)

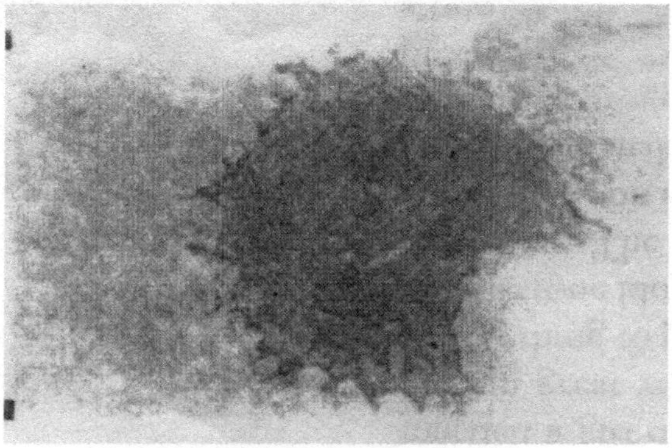

Figure 20.36 One tumour cell showing the characteristic cell membrane (3000×)

In this figure, dilated cisternae of RER inhabit the cytoplasm. The characteristically scalloped periphery of chondrocytes is maintained, with flaps and spikes protruding into the interstitium.

BONE TUMOURS

Ewings and PNET

Arising in the medullary cavity, Ewing's sarcoma and PNET usually invade the cortex, periosteum and soft tissue. It is composed of uniform small, round cells that have scant cytoplasm that appear clear due to rich glycogen. The presence of Homer–Wright rosettes (where the tumour cells are arranged in a circle about a central fibrillary space) is indicative of neural differentiation. Ultrastructurally, close apposition of the rounded tumour cells is characteristic of Ewings. Large lakes of glycogen rosettes are seen. Small, tenuous intercellular junctions have been reported but are generally not prominent.

Ultrastructural assessment of Ewings sarcoma, particularly in suspected cases of malignant round cell tumours, requires optimal preservation of glycogen. En bloc staining with uranyl acetate results in considerable extraction of glycogen. Lead staining is mandatory. Fixation with Karnovsky's osmium tetroxide–potassium ferricyanide fixative enhances glycogen preservation and demonstration. Although glycogen is demonstrable in many other tumours, it typically occurs as rosettes in Ewings. In other round cell tumours, like RMS, it is often seen in monoparticulate form. A few paranuclear glycogen depletion pads which arise due to depletion of glycogen during processing are shown in Ewings sarcoma cells in Figure 20.37.

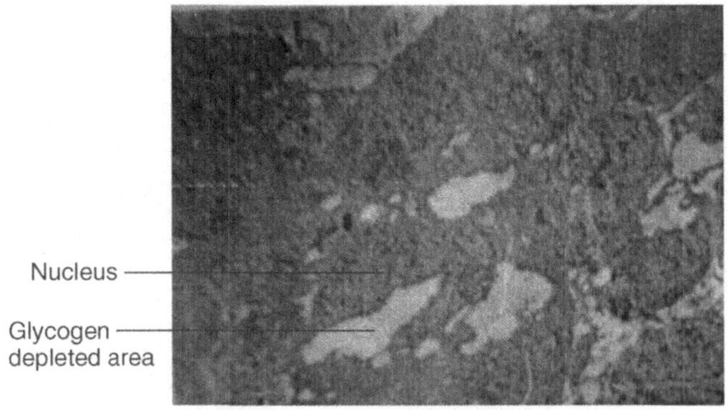

Figure 20.37 Ewings sarcoma (3000×)

OTHER TUMOURS OF THE BONE

Giant Cell Tumour

These tumours are large, and frequently undergo cystic changes. They are composed of uniform, oval, mononuclear cells that have indistinct cell membranes and appear to grow in a syncitium. The mononuclear cells are the proliferating component of the tumour.

Scattered within this background are numerous osteoclast-like giant cells with 100 or more nuclei and having features similar to the mononuclear cells. Necrosis, haemorrhage, haemosiderin and reactive new bone formation are common secondary features.

Ultrastructural examination reveals fibrohistiocytic properties with multiple cell types. The dominant stromal cell is an ovoid to bipolar fibroblastoid cell with abundant RER (stromal cell type I). The next most frequent cell is histiocytic, with rich acid phosphatase activity and a well-developed Golgi apparatus (stromal cell type II). A third stromal cell has lymphocytoid appearance.

The giant cells, ultrastructurally, resemble multinucleated histiocytic giant cells, containing about 20 nuclear profiles. The RER is abundant, and there are numerous mitochondria and sporadic lysosomes, intranuclear filaments. The periphery of the giant cells may form short of microvillous processes. The ruffled borders typical of osteoclasts are generally absent. A part of a giant cell showing 8–9 pleomorphic nuclei is shown in Figure 20.38.

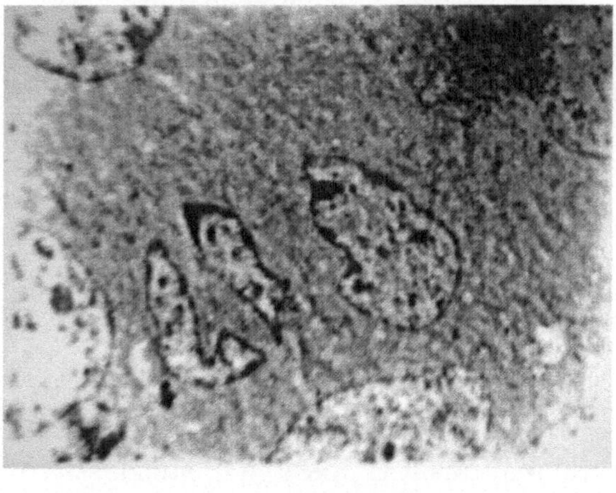

Figure 20.38 Giant cells with pleomorphic nuclei in giant cell tumour (3000×)

SOFT TISSUE TUMOURS/SARCOMA (STS)

STS/STT are defined as mesenchymal proliferations that occur in the extraskeletal, non-epithelial tissues of the body excluding the viscera, coverings of the brain and lymphoreticular system. Benign tumours outnumber their malignant counterparts by a ratio of at least 100 : 1.

These tumours are classified on the basis of tissue of origin:

Tumours of adipose tissue	Lipoma Liposarcoma
Tumours and tumour-like lesions of fibrous tissue	Fibromatoses Fibrosarcoma
Fibrohistiocytic tumours	Fibrous histiocytoma Malignant fibrous histiocytoma
Tumours of skeletal muscle	Rhabdomyoma Rhabdomyosarcoma
Tumours of smooth muscle	Leiomyoma Leiomyosarcoma
Vascular tumours	Hemangioma Lymphangioma hemangiiopericytoma Angiosarcoma
Peripheral nerve tumours	Neurofibroma Schwanoma Granular cell tumour Malignant peripheral nerve sheath tumour
Tumours of uncertain histogenesis	Synovial sarcoma Alveolar soft part sarcoma Epitheloid sarcoma

With the ultrastructural features it is possible to subclassify these tumours accurately. Though the treatment modality is not greatly influenced by subtyping, except, probably in rhabdomyosarcomas, this may help in overall management. The mesenchymal cells are generally elongated, slender to plump with intermediate and microfilaments. The intercellular space is significant and contains mostly collagen.

TUMOURS OF ADIPOSE TISSUE (LIPOMA AND LIPOSARCOMA)

Benign tumours, lipomas, are the most common soft tissue tumours of childhood. Histologically they consist of mature white fat cells with no pleomorphism. Liposarcomas, the most common sarcomas of adulthood, arise in the deep soft tissues of the proximal extremities and retroperitoneum. Histologically they can be divided into a few variants but all variants typically consist of some cells indicative of fatty differentiation which are known as lipoblasts. They mimic foetal fat cells and contain round, clear cytoplasmic vacoules of lipid that scallop the nucleus.

Adipose tissue develops from precursor cells ultrastructurally indistinguishable from fibroblasts. With the onset of lipoblastic differentiation, the cells evolve through a series of stages, designated arbitrarily as early, intermediate and late lipoblasts with eventual development into mature lipocytes. This line of differentiation is marked by a gradual increase in cytoplasmic lipid droplets together with increasing amounts of external lamina and both smooth-surfaced and micropinocytotic vesicles, with corresponding reciprocal diminution of RER and Golgi complexes. The mitochondrial morphology is also altered from spherical to elongated and filamentous. The cell shape also changes from the elongated profiles of the fibroblastoid precursors, to ovoid late lipoblasts and lipocytes. Thus in liposarcomas, undifferentiated and fibroblastoid cells, as well as early, intermediate and late lipoblasts and mature lipocytes are seen in varying proportions, depending on the tumour type and degree of differentiation. In the tumour cells, lipid droplets of various diameters can be seen many of which generally enlarge or coalesce, culminating in a signet-ring configuration where they displace the nucleus and other organelles to the periphery. Figure 20.39 shows the lipid droplets coalescing to form larger lipid vacuoles in a liposarcoma cell.

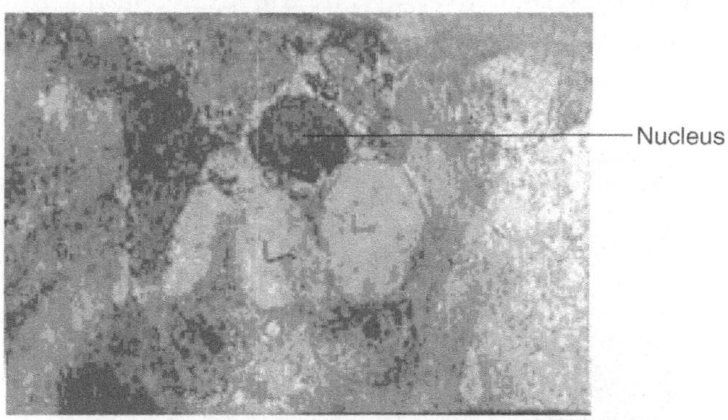

Figure 20.39 Liposarcoma (3000×)

TUMOURS OF FIBROUS TISSUE

Myositis Ossificans

These are tumourlike proliferations that typically occur in traumatized and inflammed tissues, as result of osseous metaplasia of non-osseous tissues. It is distinguished from the other fibroblastic proliferations by the presence of metaplastic bone and histologically based on stage of lesion. In the earliest phase, the lesion is most cellular and consists of plump, elongated fibroblast-like cells. Later on, mineralized trabeculae that closely resemble cancellous bone can be seen in the peripheral zone.

Low-grade fibroproliferative disorders, notably fibromatoses, appear to consist largely of myofibroblasts.

Fibrosarcoma

This occurs mostly in the retroperitoneum, the thigh, the knee and the distal extremities. Histologically they have spindled cells growing in a herringbone fashion and highly cellular neoplasms dominated by architectural disarray, pleomorphism, frequent mitoses and areas of necrosis.

These neoplasms consist of irregular spindle-shaped or stellate fibroblastoid cells with large irregular nuclei containing much euchromatin, and cytoplasmic extensions intruding into the collagenous interstitium. Thick, branching lamellae of RER are the characteristic feature of fibroblastoid cells. The cytoplasm of the elongated cells is filled with extensive ribosome-studded endoplasmic reticulum, forming slender or dilated cisternae containing probably proteinaceous material. Nonspecific intermediate filaments may also be seen in the cytoplasm. Intracellular collagen may also be seen. In the benign condition, "fibromatosis" strands of filaments anchoring the cells to the surrounding collagen and also myofibroblasts are seen which are rare in the malignant tumour, "fibrosarcoma". Figure 20.40 shows a few elongated tumour cells, with cytoplasm filled with cisternae of RER. Copious intercellular matrix is also evident.

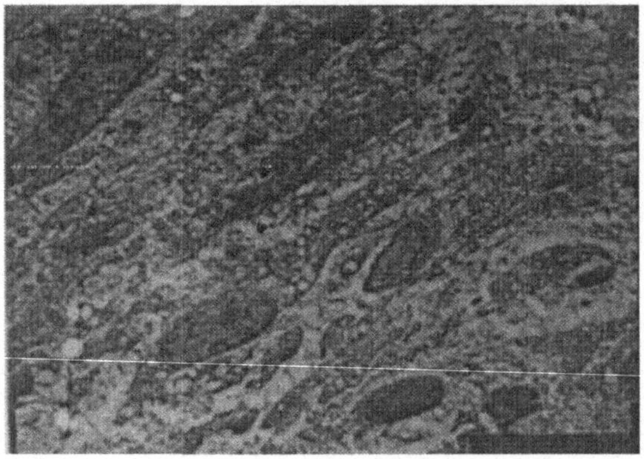

Figure 20.40 Fibrosarcoma—elongated cells with abundant intercellular spaces (2000×)

FIBROHISTIOCYTIC TUMOURS

These tumours contain cellular elements that resemble both fibroblasts and histiocytes. Benign tumours are relatively common and consist of a proliferation of spindle cells arranged in a storiform pattern and show also foam cells, haemosiderin deposits, multinucleated giant cells and hyperplasia of the overlying epidermis. They most frequently arise in the dermis where they are known as **dermatofibrosarcoma.**

Malignant Fibrous Histiocytoma

Characterized by cytologic pleomorphism, presence of bizarre multinucleate cells and storiform architecture, this includes a group of related **STS** which are highly aggressive occurring in the musculature of proximal extremities and retroperitoneum.

EM can contribute to recognition of the fibrohistiocytic characters of a particular tumour. Fibroblastoid, myofibroblastic, histiocytoid, giant cells all can be seen, fibroblastoid cells have irregular nuclear profiles, mildly dilated cisternae of RER and infrequent lysosomes. Histiocytoid cells show numerous membrane bound electron dense lysosomes.

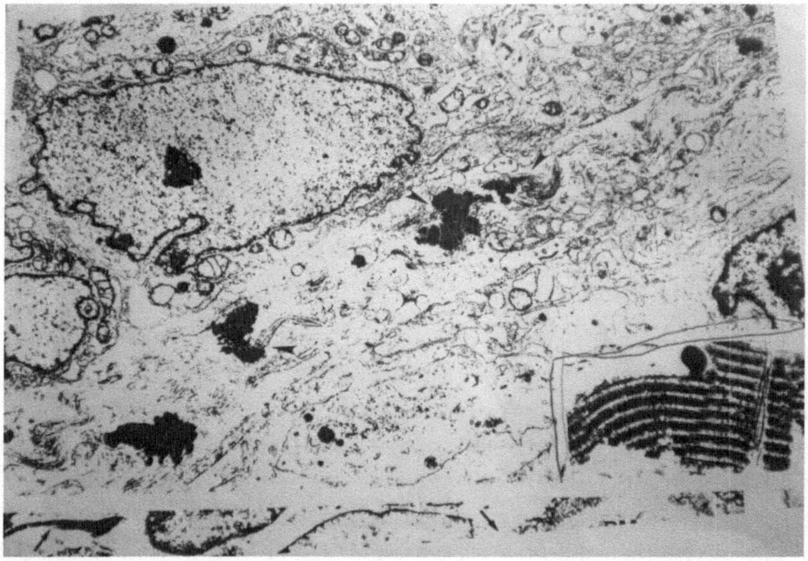

Figure 20.41 Malignant fibrous histiocytoma (8600×). (*Source:* Douglas, W. Henderson, John, M. Papadimitriou & Mark Coleman. (1986). *Ultrastructural Appearances of Tumours,* 2nd edn. Churchill Livingstone, New York. p. 226, Fig. 20.6.)

In Figure 20.41, the malignant cell shows slightly irregular nuclear profile. The cisternae of the RER are mildly dilated and lysosomes, though only a few, could be identified. In addition to native collagen fibres, the extracellular matrix contains sharply circumscribed electron-dense collections of collagen with regular periodicity, which are continuous with collagen fibres in some areas (arrowheads) and may be allied to fibrous long-spacing collagen.

MYOGENIC TUMOURS (SMOOTH MUSCLE AND STRIATED MUSCLE TUMOURS)

EM is useful in establishing or confirming a diagnosis of either smooth muscle tumour (leiomyosarcoma) or striated muscle tumour (rhabdomyosarcoma). But difficulty may be encountered when filaments are present in scattered cells in small numbers. Immunohistochemistry (IHC) is essential. When IHC staining is strong, showing strong positivity, correspondingly abundant filaments may be seen in EM. But even when staining is equivocal or negative, filaments sufficient for making a reliable diagnosis may be identified in EM.

TUMOURS OF SKELETAL MUSCLE

These are almost always malignant. The benign type, rhabdomyoma is rare.

Rhabdomyosarcoma (RMS)

This is the most common STS of childhood and adolescence and mostly occur in head and neck or genitourinary tract. Histologically subclassified into embryonal, alveolar and pleomorphic variants. In all types the diagnostic cell is the rhabdomyoblast which contains eccentric, eosinophilic granular cytoplasms rich in thick and thin filaments. The elongate rhabdomyoblasts are known as "tadpole" or "strap cells" and may contain cross-striations visible in LM. Ultrastructurally, the rhabdomyoblasts contain sarcomeres. Figure 20.42 shows the myofilaments with formation of sarcomeres.

The EM criteria for recognition of striated muscle differentiation in neoplasms include aggregates of elongated thick myosinlike filaments and nonspecific thin actinlike filaments with the formation of abortive sarcomeres. The simple presence of thin and thick filaments is inadequate, and ideally, organization with sarcomeric structures should also be sought. However the thick myosin filaments rarely found in smooth muscle tumours do not form banded clusters with the thin filaments. In many instances smooth muscle differentiation can be distinguished by LM itself. Therefore, the aggregated parallel thick and thin filaments in the small round cell tumours indicate rhabdomyoblastic differentiation, even in the absence of definite sarcomeric arrangements. In EM, in transverse sections a regular hexagonal spacing of thin filaments around myosin filaments may be seen.

The number of myofilaments in the malignant cells varies greatly. Some cells may display highly developed transversely aligned sarcomeres while others contain only nonspecific thin filaments. In poorly differentiated tumours, prolonged search may be required which has been reported to lead to definitive diagnosis in more than 50% of cases.

Depending on the degree of differentiation, the filaments may appear differently as mentioned below:

- abundant skeletal muscle myofilaments several sarcomeres in length
- long, slender fragments of myofibrils in haphazard distribution
- co-presence of actin–myosin filaments in unique spatial distribution with Z band material
- even in absence of an obvious banding pattern, the filamentous internal structure of Z band reveals its identity
- when only myosin filaments are formed, they are recognizable from their calibre and arrangement in small bundles with ribosomes in short rows between the ends of the parallel filaments
- myofilaments may create ladderlike "zebra bodies" or form dense inclusions.

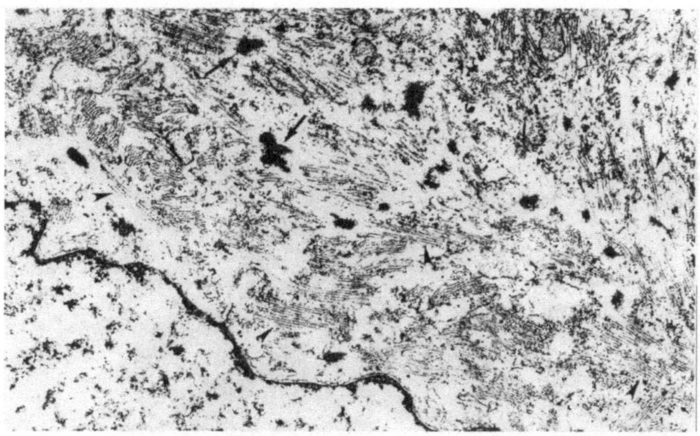

Figure 20.42 Rhabdomyosarcoma—Myofilaments and sarcomere formation (24,500×). (*Source:* Douglas, W. Henderson, John, M. Papadimitriou & Mark Coleman. (1986). *Ultrastructural Appearances of Tumours*, 2nd edn. Churchill Livingstone, New York. p. 244, Fig. 22.6.)

In Figure 20.42, parallel arrays of both microfilaments and thick filaments (arrowheads) can be seen, the former inserting into Z-disclike material (arrows). Ribosomes are closely associated with individual myofilaments.

TUMOURS OF SMOOTH MUSCLE

The benign tumours, leiomyoma often arise in the uterus. They are composed of fascicles of spindle cells that tend to intersect each other at right angles. The tumour cells have blunt-ended, elongated nuclei and show minimal atypia and few mitotic figures.

Leiomyosarcoma (LMS)

These mostly arise in the deep soft tissues of the extremities and retroperitoneum. Histologically they are characterized by malignant spindle cells that have cigar-shaped nuclei arranged in interweaving fascicles.

Ultrastructurally, malignant smooth muscle cells contain bundles of thin filaments, with dense bodies and pinocytotic vesicles and individual cells are surrounded by BL. Longitudinally orientated thin cytoplasmic filaments with focal densities along the course of the filamentous aggregates and subplasmalemmal dense plaques (attachment zones/ hemidesmosomes), intercellular junctions and external lamina are also seen. Figure 20.43 shows these features. The nucleus is mostly cigar-shaped with transverse folds. Thus the minimal criteria are a few longitudinally oriented microfilaments associated with dense bodies together with paucity of RER. Star-shaped configuration of RER can also be observed at times. Such profiles are claimed to be characteristic of the neoplastic cells in LMS.

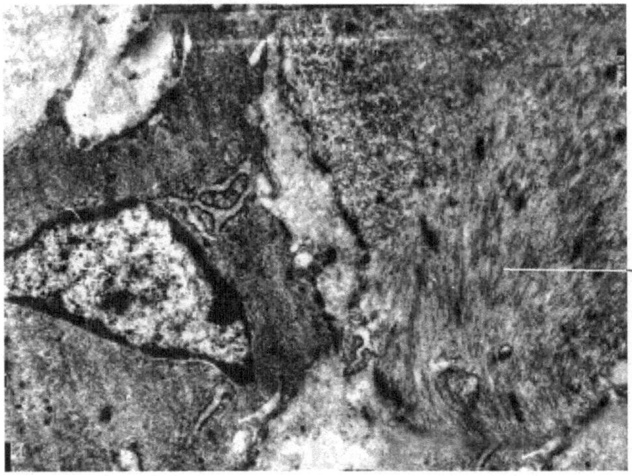

Figure 20.43 Leiomyosarcoma (Hemidesmosomes, filaments with focal densities) (15,000×)

In this figure, hemidesmosomes, filaments with focal densities, nucleus and a few strands of RER could be seen in neoplastic cells.

PERIPHERAL NEURAL TUMOURS

Tumours of peripheral nerves may consist of an admixture of cells resembling Schwann cells, fibroblasts or perineurial cells, where one type tends to predominate. EM may be a definitive means of morphologically typing the predominant cell.

In schwannomas, the component cells have a distinct external lamina which is typically continuous and multilayered. There are also cytoplasmic extensions of varying complexity, as well as infoldings of the plasmalemma. The cell junctions are rare. Figure 20.44 shows a few cells with adjacent processes while Figure 20.45 shows at higher magnifications the processes surrounded by continuous multilayered basal lamina individually.

Figure 20.44 Schwannoma—cells with processes (3000×)

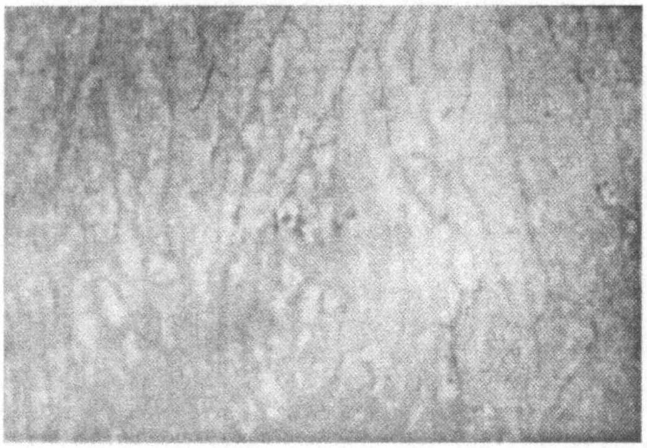

Figure 20.45 Schwannoma—processes surrounded by multilayered basal lamina (6000×)

In melanotic Schwannomas, melanosomes in all stages of evolution can be found within the Schwann cell cytoplasm indicating melanin synthesis by these cells and not simply transfer of melanin from other cells.

In **neurofibromas** the predominant cells usually resemble fibroblasts, lacking a distinct external lamina, junctional contacts or infoldings of the cell membrane.

VASOFORMATIVE TUMOURS

Angiosarcoma

The predominant neoplastic cell is the endothelial cell. But pericytic, myoid and undifferentiated cells as well as sporadic extravascular fibroblasts and histiocytes are often present. The diagnosis is based on the detection of abortive vasoformative structure but the lumina of the neoplastic vessels are often small in poorly differentiates tumours. Figure 20.46 shows the neoplastic endothelium surrounding a small channel containing an erythrocyte (E).

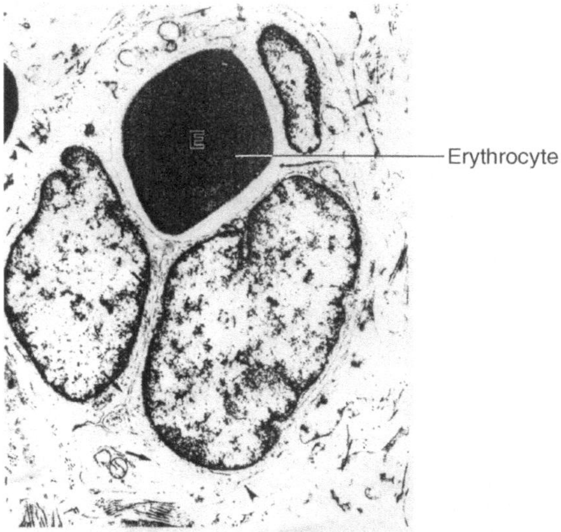

Figure 20.46 Angiosarcoma (1,400×). (*Source:* Douglas, W. Henderson, John, M. Papadimitriou & Mark Coleman. (1986). *Ultrastructural Appearances of Tumours*, 2nd edn. Churchill Livingstone, New York. p. 248, Fig. 23.1.)

Hemangiopericytoma

In all types of hemangiopericytomas, irrespective of their location and age group of the patient, the constituent cells can be identified as pericytes with the help of the ultrastructural features. The major diagnostic criteria are, (i) spindle, oval or polygonal cells in a somewhat palisaded arrangement around capillaries (ii) cytoplasmic filaments (iii) pinocytotic vesicles (iv) focal attachments and junctions between cells.

TUMOURS OF UNCERTAIN HISTOGENESIS

Synovial sarcoma (SS)

SS is a misnomer because the actual cells of origin of this tumour are still unclear with <10% of lesions occurring truly intra-articular. 70% of these lesions involve large joints of lower extremities.

Histologic hallmark of biphasic SS is the dual line of differentiation of the tumour cells, i.e., epithelial-like and spindle cells. The tumour cells do not have the features of synoviocytes. Monophasic tumours are composed of only spindled cells or rarely, epithelial cells.

Ultrastructurally, fibroblastoid cells, mast cells and collagenous connective tissue are seen in the stromal component. The tubular component in biphasic synovial sarcoma strikingly resembles glandular epithelium and consists of cuboidal cells resting on a basal lamina; microvilli and occasional cilia project into the luminal space and junctional complexes occur between adjoining cells.

Alveolar Soft Part Sarcoma

The fine structure is typified by cytoplasmic aggregates of membrane-bound crystalloidal material and numerous mitochondria. Cytoplasmic filaments are absent. One tumour cell showing these granules is shown in Figure 20.47.

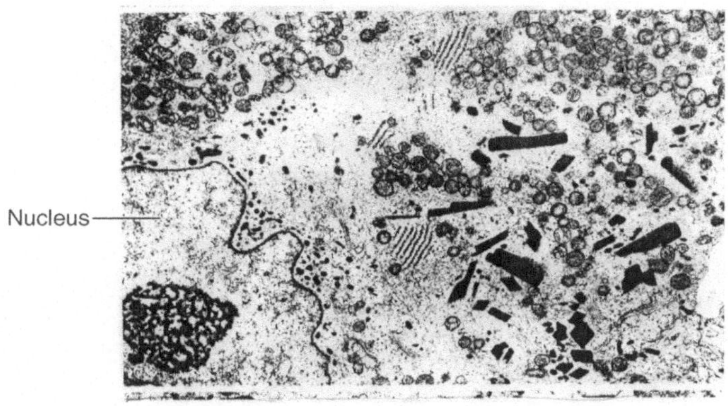

Figure 20.47 Alveolar soft part sarcoma—cell morphology (8800×). (*Source:* Douglas, W. Henderson, John, M. Papadimitriou & Mark Coleman. (1986). *Ultrastructural Appearances of Tumours*, 2nd edn. Churchill Livingstone, New York. p. 258, Fig. 24.1.)

In Figure 20.47, nucleus is slightly irregular and contains abundant euchromatin. The nucleolonema forms a skein-like pattern. Numerous mitochondria are evident in the cytoplasm as are stacked lamellae of RER and prominent Golgi complexes. Rounded and elliptical granules are concentrated in the juxtanuclear region of one cell, while the cytoplasm of the adjacent cell contains numerous crystalloids including elongated and rhomboidal forms.

21

EM IN DIAGNOSIS OF DISEASES OTHER THAN TUMOURS

As already mentioned in the introduction, EM is useful in diagnosis of certain diseases apart from its most important application in tumour diagnosis.

* Certain genetic and metabolic diseases, especially storage diseases.
* Certain renal glomerular diseases
* Certain skeletal muscle diseases
* Certain peripheral nerve diseases
* Certain infectious diseases, particularly viruses.

The salient features of some of these are discussed below:

GENETIC AND METABOLIC DISEASES

This group of genetically transmitted diseases includes several disorders. Most of them are lysosomal and result from the absence or deficiency of particular lysosomal enzyme. Because of the enzyme deficiency the engulfed endogenous or exogenous material cannot be hydrolysed and hence get accumulated in secondary lysosomes. Cellular injury and cell death may occur subsequent to this accumulation. Various organs may be affected in this group of diseases. In all the types, there is increase in the number and size of the secondary lysosomes, which contain accumulated substrate. The appearance of the substrate in the lysosomes often is specific for one disease or for several diseases of one subtype.

In Tay–Sachs disease of the liver (Figure 21.1), the affected kupfer cell cytoplasm is filled with distended and confluent secondary lysosomes (Ly) that contain concentric (ganglioside-rich) membranes and other dense osmiophilic material. The portions of adjacent hepatocytes (H) show less involvement.

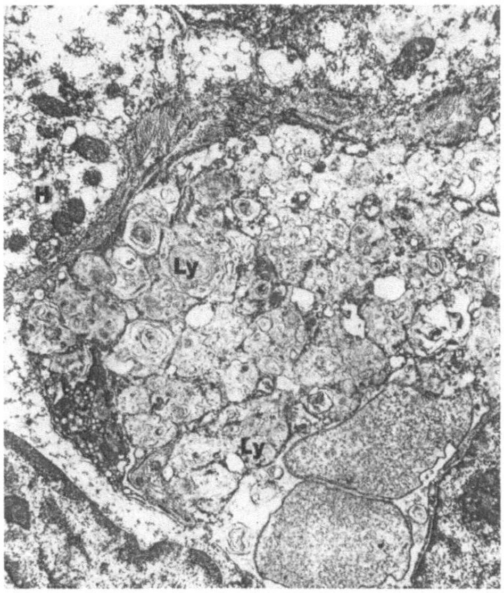

Figure 21.1 Tay-Sachs disease (Liver) (Ly—Lysosomes, H—Hepatocyte) (13,000×). (*Source:* Richard, G. Dickersin. (2000). *Diagnostic Electron Microscopy—A Text/Atlas,* 2nd edn. Springer-Verlag Inc., New York. p. 711, Fig. 11.1.)

In Fabry's disease (Figure 21.2), smooth muscle cells of the bowel wall contain many electron-dense, lamellar and solid bodies.

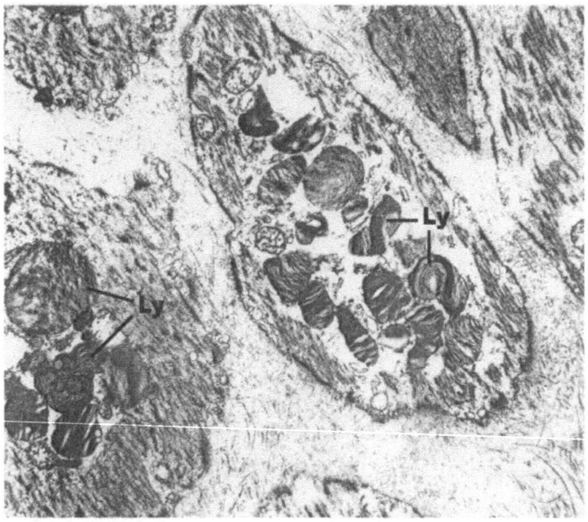

Figure 21.2 Fabry's disease (Jejunum) (Ly—Lysosomes) (5720×). (*Source:* Richard Dickersin, G. (2000). *Diagnostic Electron Microscopy—A Text/Atlas,* 2nd edn. Springer-Verlag Inc., New York. p. 714, Fig. 11.4.)

In Gaucher's disease (Figure 21.3), the cell has copious cytoplasm containing innumerable, elongated lysosomes with substrate which consist of sheaves of tubules. In this figure, the tubular character of the lysosomal contents (Ly) is clearly visible.

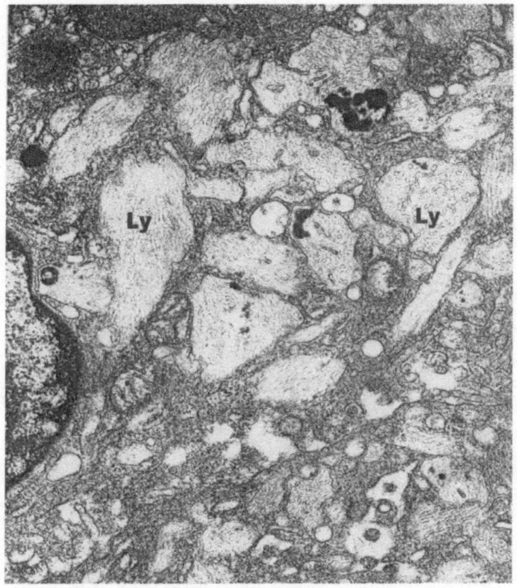

Figure 21.3 Gaucher's disease (Liver) (Ly—Lysosomes) (38,000×). (*Source:* Richard Dickersin, G. (2000). *Diagnostic Electron Microscopy—A Text/Atlas,* 2nd edn. Springer-Verlag Inc., New York. p. 718, Fig. 11.8.)

In Niemann–Pick disease (Figure 21.4), mixed membranous and more solid osmiophilic inclusions occupy most of the cytoplasm of the hepatocyte.

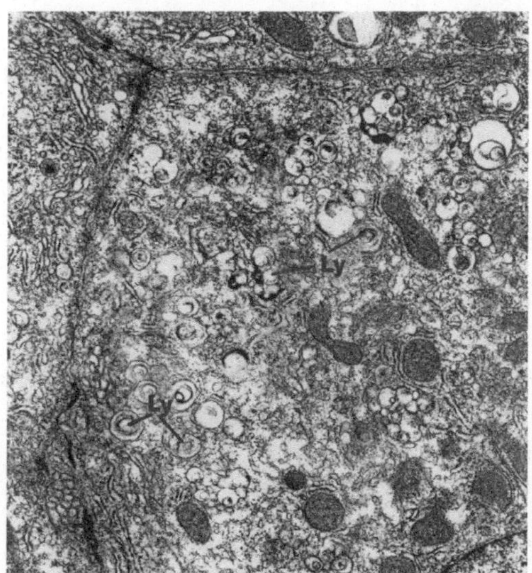

Figure 21.4 Niemann-Pick disease (Liver) (Ly—Lysosomes) (20,770×). (*Source:* Richard Dickersin, G. (2000). *Diagnostic Electron Microscopy—A Text/Atlas,* 2nd edn. Springer-Verlag Inc., New York. p. 719, Fig. 11.9.)

In cholesterol ester storage disease (Figure 21.5), the hepatocyte cytoplasm contains numerous membrane-bound vesicles that have a medium-dense, crystal-like material (arrows) within them.

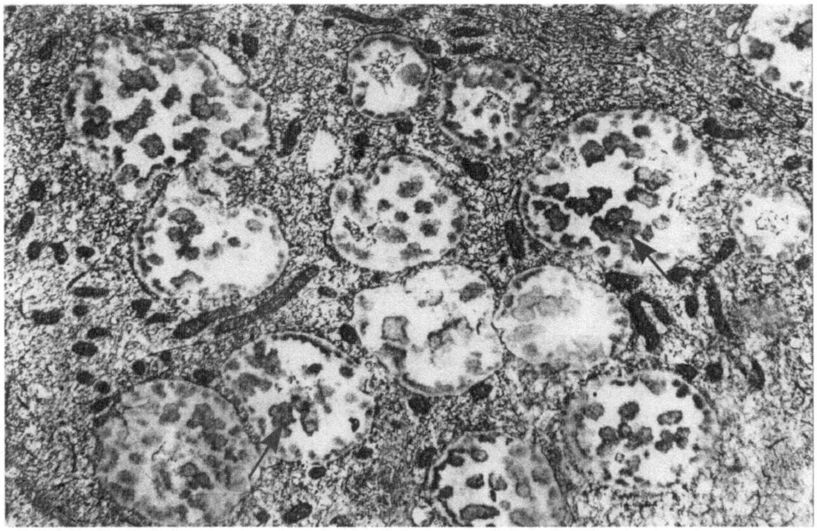

Figure 21.5 Cholesterol ester storage disease (Liver) (19,000×). (*Source:* Richard Dickersin, G. (2000). *Diagnostic Electron Microscopy—A Text/Atlas*, 2nd edn. Springer-Verlag Inc., New York. p. 732, Fig. 11.22.)

Apart from the lysosomal storage diseases, there are other non-lysosomal storage diseases where the materials are stored in the cytoplasm. A few examples are Wilson's disease, glycogen storage disease, alpha-1-antitrypsin deficiency, mannosidosis, melanosis, etc. In drug toxicity conditions also, cytoplasmic accumulations can be seen.

RENAL GLOMERULAR DISEASE

EM has significant contribution in this field. EM can offer diagnosis, confirmation or additional diagnostic information as required. Ultrastructural examination is necessary chiefly for glomerular diseases. Usually a combination of EM and immunofluorescence is optimal to arrive at a confident diagnosis. In some, scant or no glomerular dense deposits are seen ultrastructurally while in others the deposits also aid as diagnostic criteria.

In immunoglobulin M (IgM) nephropathy (Figure 21.6) extensive effacement of podocyte foot processes (arrow) and villous hypertrophy of podocyte (p) are present, similar to minimal change disease. In addition, amorphous mesangial deposits (*) are present. The glomerular basement membrane (GBM) is normal.

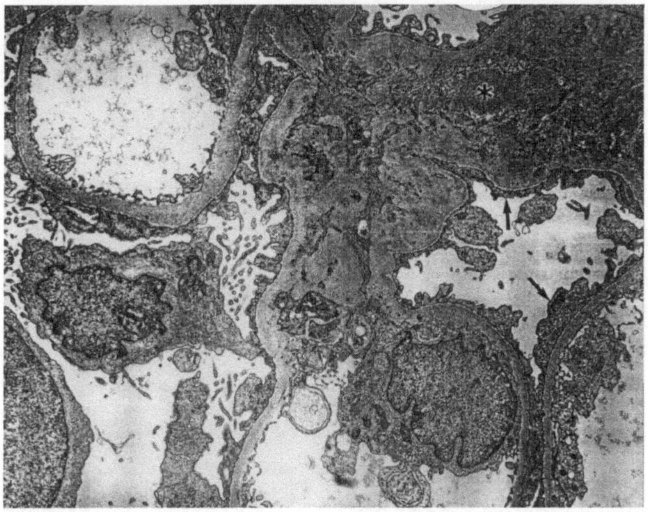

Figure 21.6 IgM nephropathy (P—podocyte, *—Mesangial deposit, Arrow—effacement of foot processes) (5400×). (*Source:* Richard Dickersin, G. (2000). *Diagnostic Electron Microscopy—A Text/Atlas,* 2nd edn. Springer-Verlag Inc., New York. p. 789, Fig. 12.6.)

In hereditary nephritis (Alport's syndrome), (Figure 21.7) the glomerular basement membrane (GBM) has variable thickening and thinning with lamina dense laminations and lucency. Scattered granular microparticles are present in the GBM.

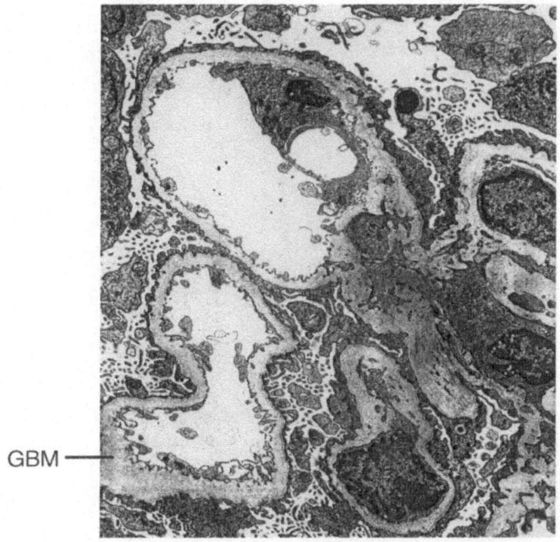

Figure 21.7 Alport's syndrome (Hereditary nephritis) (9500×). (*Source:* Richard Dickersin, G. (2000). *Diagnostic Electron Microscopy—A Text/Atlas,* 2nd edn. Springer-Verlag Inc., New York. p. 804, Fig. 12.20.)

In membranous glomerulonephritis (Figure 21.8) the subepithelial deposits are embedded in the thickened GBM.

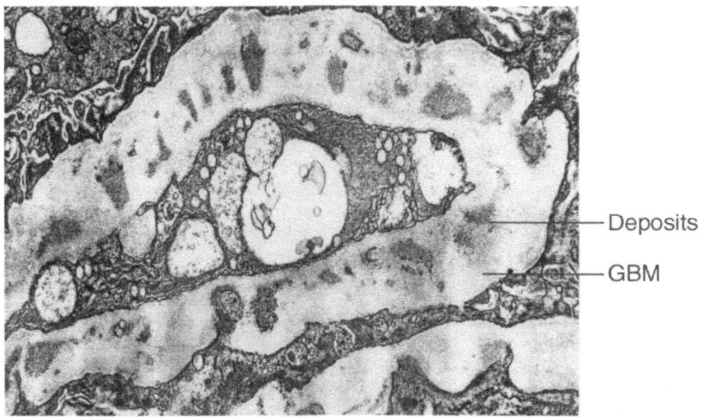

Figure 21.8 Membranous glomerulonephritis (19,000×). (*Source:* Richard Dickersin, G. (2000). *Diagnostic Electron Microscopy—A Text/Atlas*, 2nd edn. Springer-Verlag Inc., New York. p. 815, Fig. 12.29.)

In Nail–patella syndrome (Figure 21.9), the GBM shows subepithelial deposits of collagen fibres (arrows).

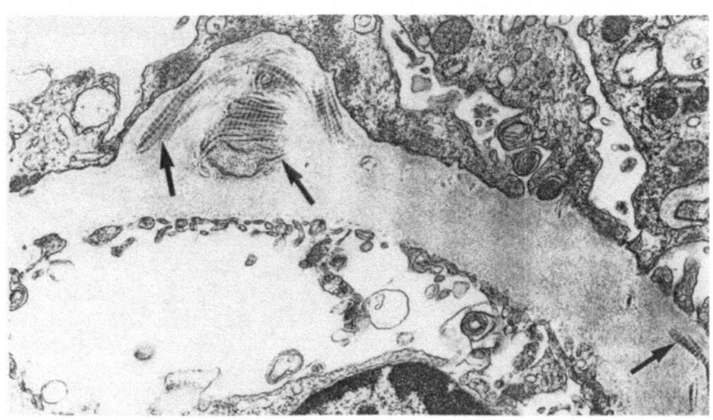

Figure 21.9 Nail-patella syndrome (29,000×). Arrows—collagen, C—capillary lumen (29,000×). (*Source:* Richard Dickersin, G. (2000). *Diagnostic Electron Microscopy—A Text/Atlas*, 2nd edn. Springer-Verlag Inc., New York. p. 860, Fig. 12.67.)

DISEASES OF SKELETAL MUSCLE

Muscular dystrophy As muscle fibre degeneration advances, several different patterns of ultrastructural changes become evident. Fibres in various stages of degeneration and regeneration can be seen. There may be complete disorganization of thick and thin myofilaments and Z bands. The hypercontracted fibres with abnormally structured and shortened sarcomeres are shown in Figure 21.10.

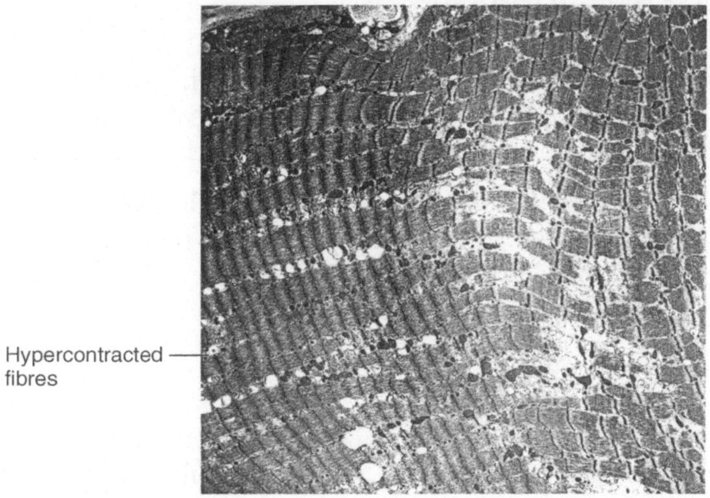

Figure 21.10 Muscular dystrophy (3800×). (*Source:* Richard Dickersin, G. (2000). *Diagnostic Electron Microscopy—A Text/Atlas*, 2nd edn. Springer-Verlag Inc., New York. p. 922, Fig. 13.12.)

Inflammatory myopathy (Polymyositis) Necrotic muscle fibres invaded by phagocytic inflammatory cells are seen light-microscopically and in the EM mononuclear phagocytes can be observed within the cytoplasm of the necrotic muscle fibre.

Toxic myopathy Focal destruction of myofibrillar organization and membranous whorls can be seen as in Figure 21.11, in the example of colchicine myopathy.

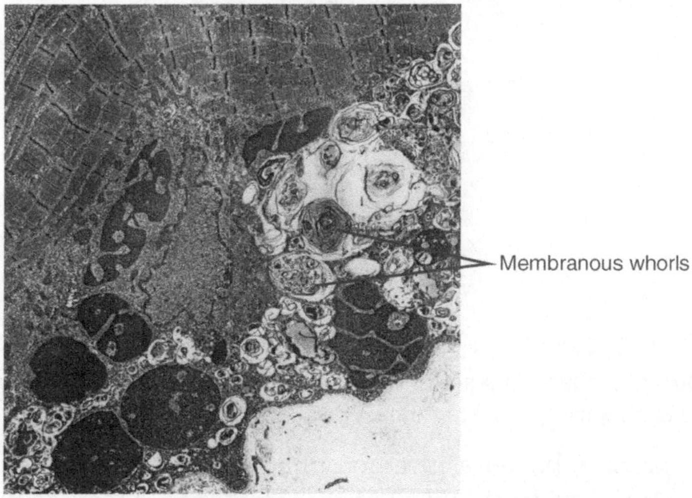

Figure 21.11 Toxic myopathy (Colchicine myopathy) (5800×). (*Source:* Richard Dickersin, G. (2000). *Diagnostic Electron Microscopy—A Text/Atlas*, 2nd edn. Springer-Verlag Inc., New York. p. 947, Fig. 13.14.)

PERIPHERAL NERVE DISEASE

Diabetic and other neuropathies Diminution of myelinated and unmyelinated nerve fibres and a microangiopathy with thickening of the basal laminae of blood vessels and similar thickening of basal laminae of perineurial cells characterize neuropathies.

In infectious neuropathies like herpes infections, leprosy, Lyme disease, etc. the infectious agents can be identified in the affected nerve fibres.

INFECTIOUS DISEASES

Bacterial Infections

The diagnostic criteria of bacteria rods and cocci in general, include, the outer cell walls, inner cell membrane, extracellular capsule, flagella or pili, central nuclear region, dense cytoplasm composed mostly of ribosomes but also a varying number of vesicles.

The Figure 21.12 shows many degenerating bacteria in a case of bacterial pneumonia caused by Legionella. Completely necrotic organisms are represented by osmiophilic debris.

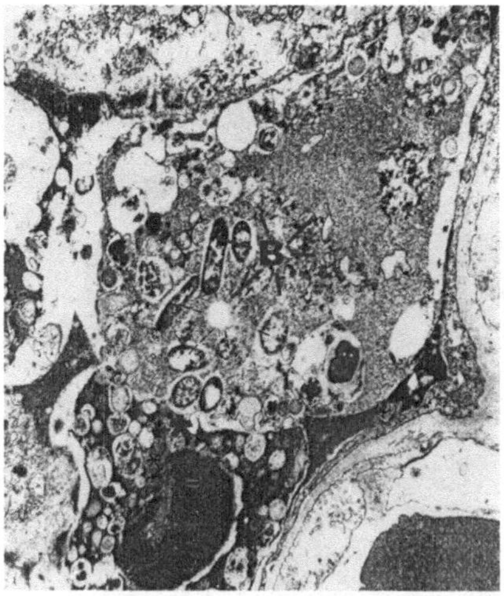

Figure 21.12 Bacterial pneumonia—Legionella (B—Bacteria, ∗—osmiophilic debris) (19,240×). (*Source:* Richard Dickersin, G. (2000). *Diagnostic Electron Microscopy—A Text/Atlas*, 2nd edn. Springer-Verlag Inc., New York. p. 650, Fig. 10.2.)

In Whipple's disease, bacteria in macrophages can be identified as in Figure 21.13, which shows bacterial rods (B) in macrophages (M). Most of the intact bacilli are extracellular, whereas those in the macrophage are in various stages of degeneration, including the end stage of serpiginous membranes (∗).

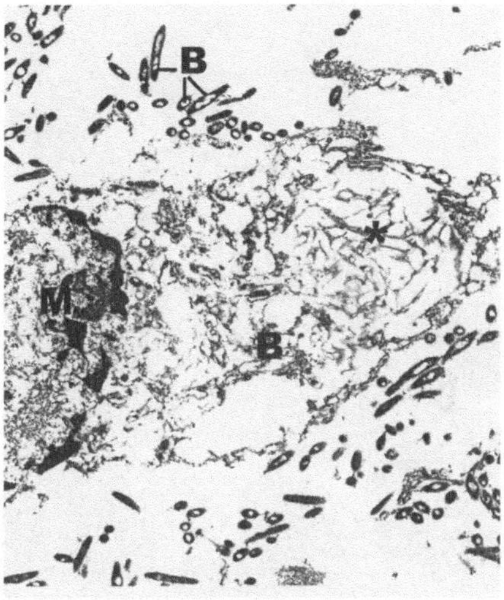

Figure 21.13 Whipple's disease B—Bacterial rods, *—serpiginous membranes (16,500×). (*Source:* Richard Dickersin, G. (2000). *Diagnostic Electron Microscopy—A Text/Atlas,* 2nd edn. Springer-Verlag Inc., New York. p. 652, Fig. 10.4.)

In *Mycobacterium* infection, the bacteria (B) can be seen within lysosomes in the histiocytes (H) which possess copious cytoplasm as in Figure 21.14.

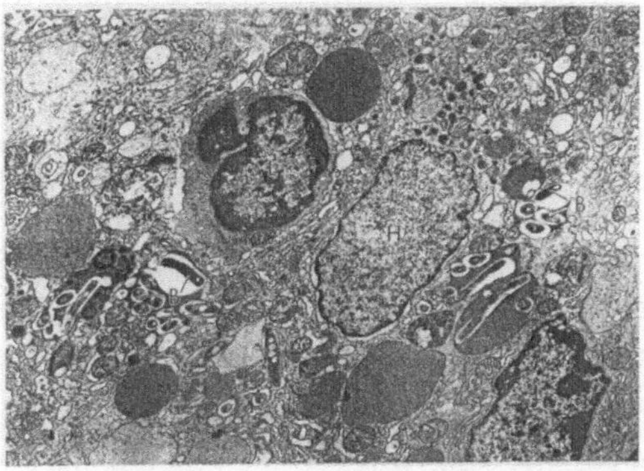

Figure 21.14 *Mycobacterium* infection (H—Histiocytes, L—Lymphocytes, B—Bacteria) (9300×). (*Source:* Richard Dickersin, G. (2000). *Diagnostic Electron Microscopy—A Text/Atlas,* 2nd edn. Springer-Verlag Inc., New York. p. 657, Fig. 10.9.)

In *Chlamydia trachomatis* infection, both elementary bodies with dense cores and more diffusely granular reticulate bodies are identified as shown in Figure 21.15. The elementary bodies (E) represent the infectious phase of the life cycle of *Chlamydia* and they enter the cell by endocytosis and phagocytosis. Reticulate bodies (R) are the metabolic and replicating phase of the cycle. They replicate by binary fission and then condense into elementary bodies.

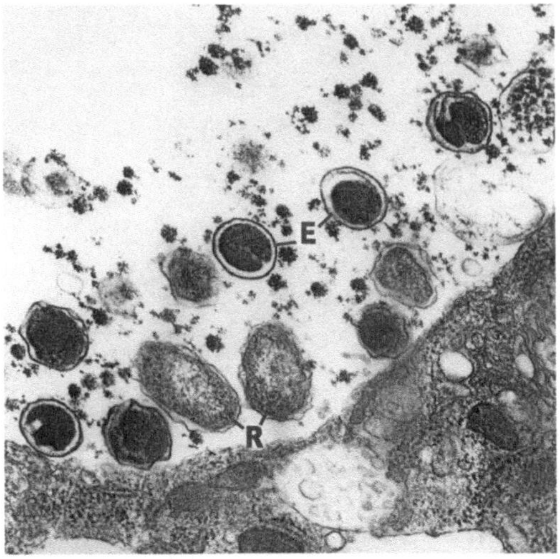

Figure 21.15 Chlamydial infection (E—Elementary bodies, R—Reticulate bodies) (71,000×). (*Source:* Richard Dickersin, G. (2000). *Diagnostic Electron Microscopy—A Text/Atlas,* 2nd edn. Springer-Verlag Inc., New York. p. 660, Fig. 10.12.)

Viral Infections

Viruses can be visualized only at electron microscopic magnifications. Though viral DNA can be identified and characterized using molecular techniques and other techniques, their presence within the cell can be ascertained only by ultrastructural studies. Some of the viruses and their characteristic morphology are given in Figures 21.16 to 21.21. They typically show a central core of nucleic acid and the capsid containing protein.

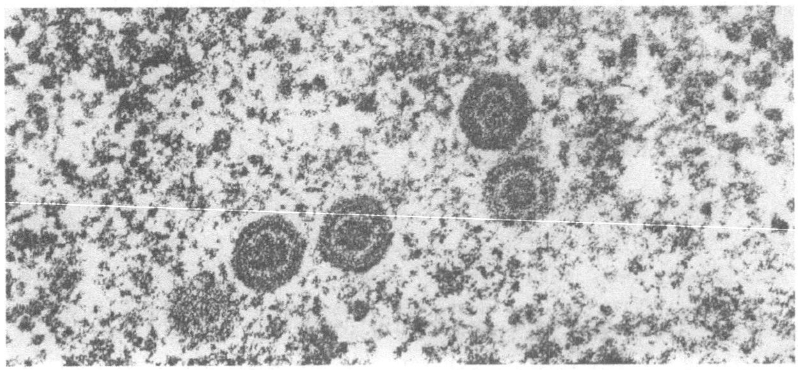

Figure 21.16 Herpes simplex (core and capsid). Virions at high magnification show their basic nucleocapsid structure and hexagonal shape (1,55,000×). (*Source:* Richard Dickersin, G. (2000). *Diagnostic Electron Microscopy—A Text/Atlas,* 2nd edn. Springer-Verlag Inc., New York. p. 664, Fig. 10.15.)

In Figure 21.17 groups of virions (V) are been in the nucleus of the infected cerebral cell.

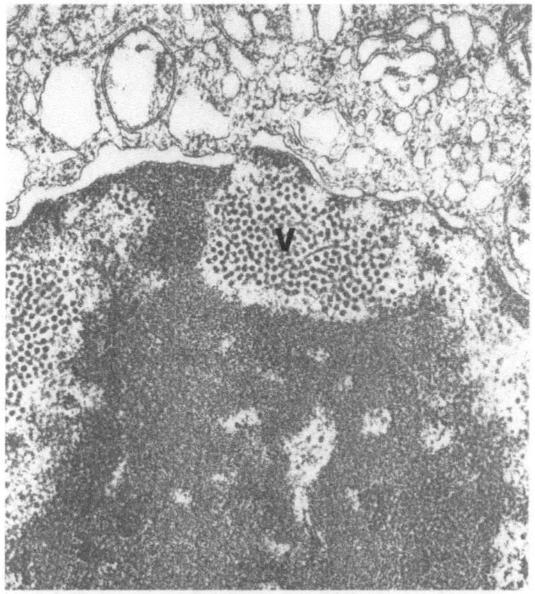

Figure 21.17 Papovavirus (V—Virions) (62,120×). (*Source:* Richard Dickersin, G. (2000). *Diagnostic Electron Microscopy—A Text/Atlas,* 2nd edn. Springer-Verlag Inc., New York. p. 668, Fig. 10.19.)

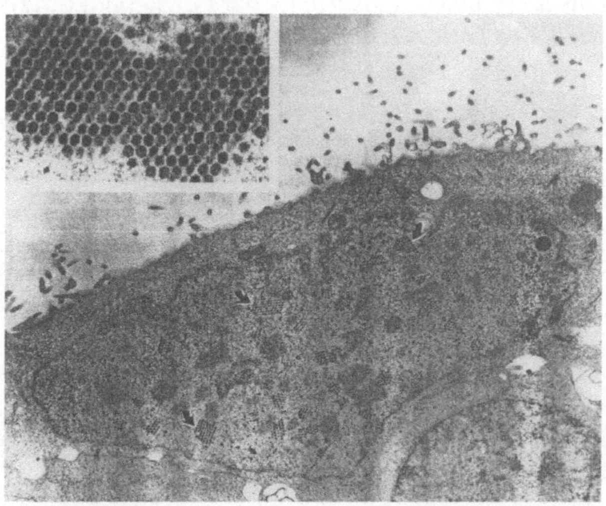

Figure 21.18 Adenovirus (16,000×). (Inset—57,000×). (*Source:* Richard Dickersin, G. (2000). *Diagnostic Electron Microscopy—A Text/Atlas,* 2nd edn. Springer-Verlag Inc., New York. p. 669, Fig. 10.20.)

The nucleus of the adenovirus-infected cell contains numerous packets of virions (arrows) in a regular, lattice-like arrangement. The inset shows virions at higher magnification (Figure 21.18).

In influenza viral infections, round and elongated virions at the cell surface are seen as shown in Figure 21.19.

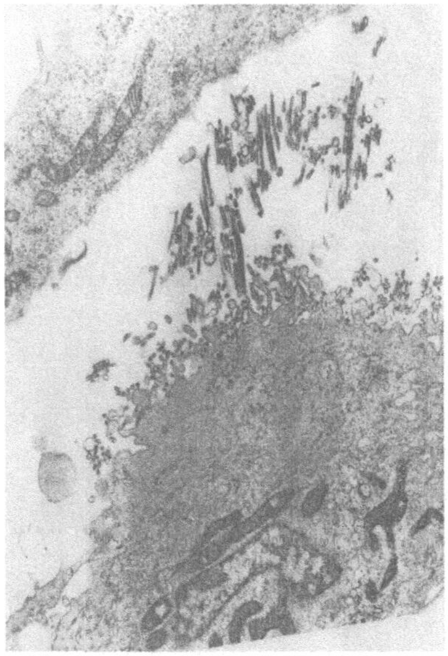

Figure 21.19 Influenza virus (16,000×). (*Source:* Richard Dickersin, G. (2000). *Diagnostic Electron Microscopy—A Text/Atlas,* 2nd edn. Springer-Verlag Inc., New York. p. 670, Fig. 10.21.)

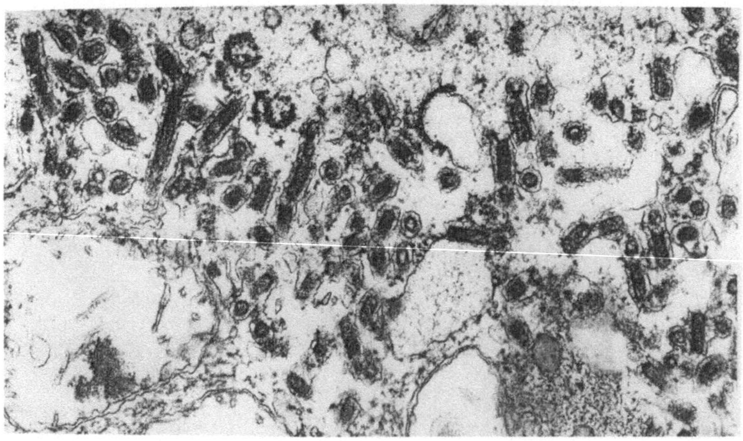

Figure 21.20 Rabies virus V—Virions exhibiting bullet shape when captured in longitudinal axis (57,000×). (*Source:* Richard Dickersin, G. (2000). *Diagnostic Electron Microscopy—A Text/Atlas,* 2nd edn. Springer-Verlag Inc., New York. p. 679, Fig. 10.30.)

Figure 21.20 shows a few rabies viral particles. These virions when captured in longitudinal axis appear typically bullet-shaped.

In HIV infection, a vacuole in the cytoplasm of a macrophage contains several ring-shaped, immature HIV-1 particles (arrows) and numerous mature particles with conical nucleoids (arrowheads) (Figure 21.21). The appearance of the asymmetric conical nucleoids varies with their orientation and plane of sectioning.

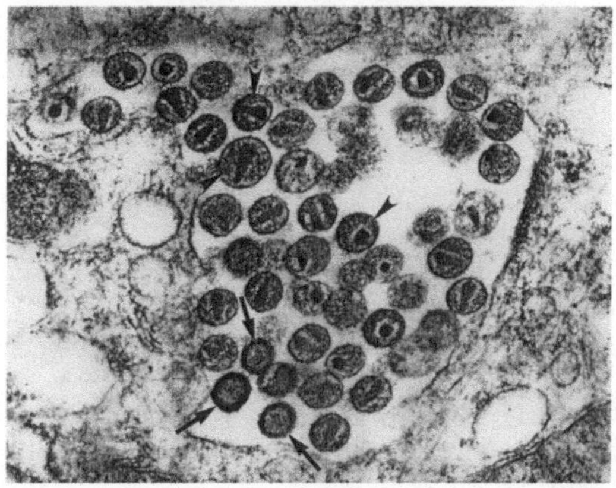

Figure 21.21 HIV particles (1,04,000×). (*Source:* Richard Dickersin, G. (2000). *Diagnostic Electron Microscopy—A Text/Atlas*, 2nd edn. Springer-Verlag Inc., New York. p. 676, Fig. 10.27.)

Parasitic Infections

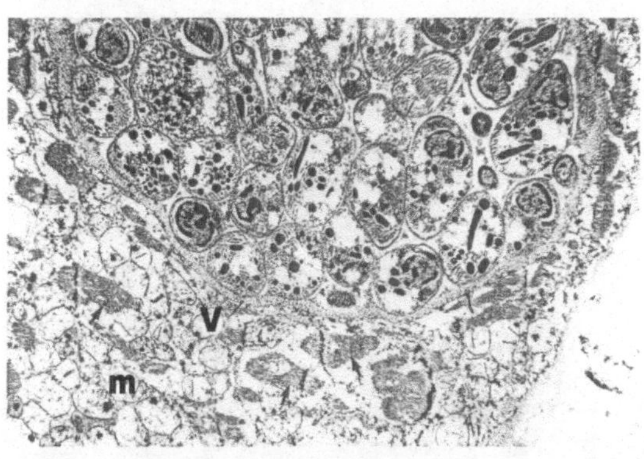

Figure 21.22 *Toxoplasma gondii* (skeletal muscle) (M—Mitochondria, V—Wall of parasitophorus vacuole, Arrow—sarcomeres in muscle cell) (11,000×). (*Source:* Richard Dickersin, G. (2000). *Diagnostic Electron Microscopy—A Text/Atlas*, 2nd edn. Springer-Verlag Inc., New York. p. 686, Fig. 10.36.)

An intramuscular cyst (parasitophorous vacuole) contains numerous sporozoites (S). The daughter cells (Merozoites, M) are present in some of the parasites as shown in Figure 21.22.

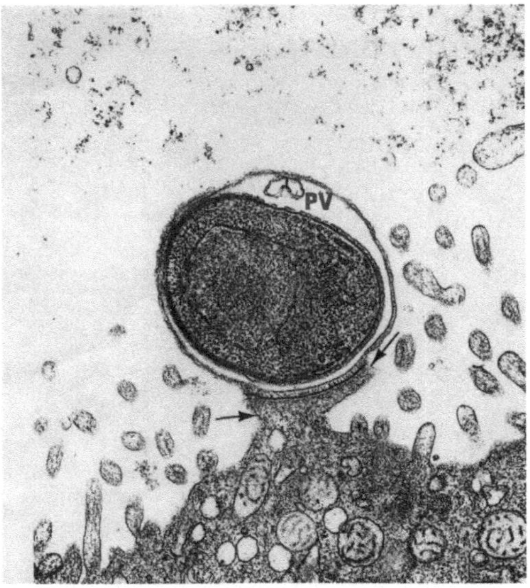

Figure 21.23 *Cryptosporidium* (jejunum) (48,400×). (*Source:* Richard Dickersin, G. (2000). *Diagnostic Electron Microscopy—A Text/Atlas,* 2nd edn. Springer-Verlag Inc., New York. p. 689, Fig. 10.39.)

In cryptosporidium, a trophozoite (T) is attached to a raised, villous portion (between arrows) of the epithelial cell. The edges of the attachment site seemingly extend around the trophozoite, forming a parasitophorous vacuole (PV) as shown in Figure 21.23.

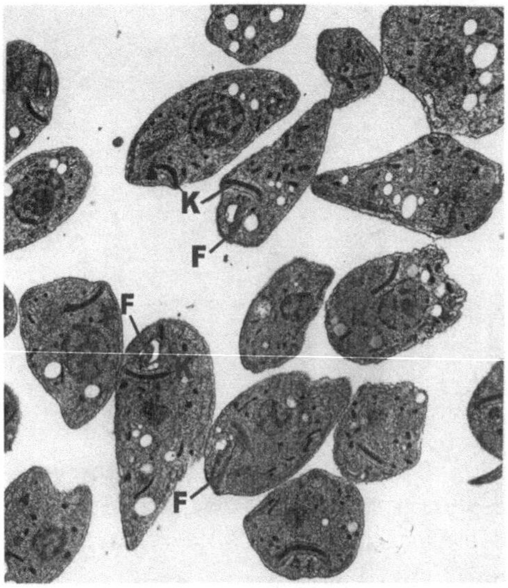

Figure 21.24 *Tryponosoma cruzi* (12,500×). (K—Kinetoplasts, F—Flagella) (*Source:* Richard Dickersin, G. (2000). *Diagnostic Electron Microscopy—A Text/Atlas,* 2nd edn. Springer-Verlag Inc., New York. p. 695, Fig. 10.44.)

Tryponosomes are seen with their oval and elliptical (and teardrop) shapes and the presence of a nucleus and various cytoplasmic organelles (Figure 21.24). Most characteristic of the organisms are the kinetoplasts (K) and flagella (F).

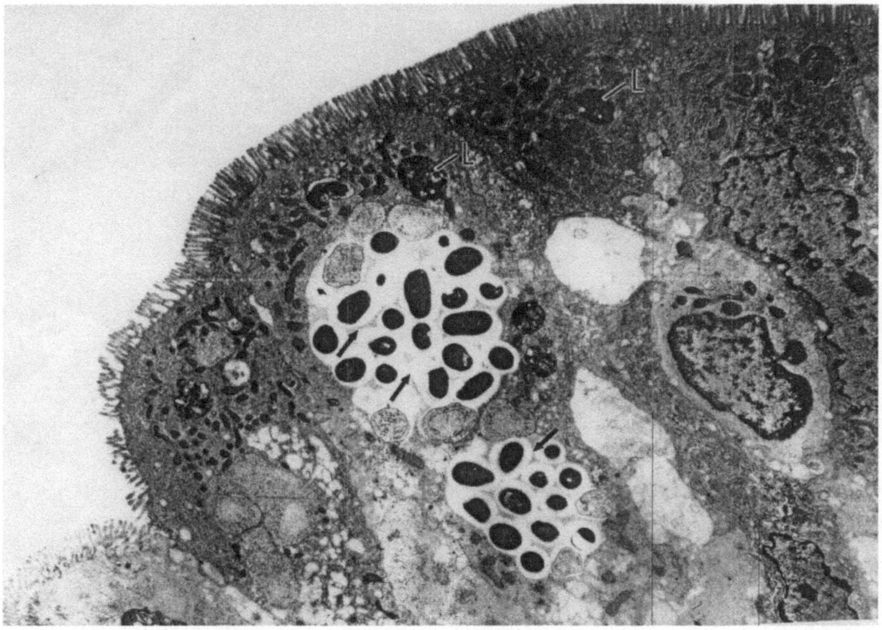

Figure 21.25 *Microspora* (duodenum) (Arrows—Septae) (5800×). (*Source:* Richard Dickersin, G. (2000). *Diagnostic Electron Microscopy—A Text/Atlas,* 2nd edn. Springer-Verlag Inc., New York. p. 699, Fig. 10.48.)

In *Microspora,* spores are grouped in parasitophorous vacuoles in the supranuclear cytoplasm of an enterocyte. Septae (arrows) are faintly visible surrounding individual spores. Secondary lysosomes (L) occupy the subluminal cyst as shown in Figure 21.25.

22

EM IN MICROBIOLOGY

Various types of microscopes are used in microbiological studies. A few are listed in Table 22.1.

Table 22.1 Types of microscopes

Source of illumination	Type of microscope
Visible light	Optical microscope
λ —0.4 μ: Res. 0.2–1 μ	Polaroid microscope
	Dark ground microscope
	Phase-contrast microscope
	Interference microscope
UV	Fluorescence microscope
λ —0.2 μ	Confocal microscope
Electron beam	TEM
λ —0.005 μ (5 nm)	SEM
Res.—0.01 μ (10 nm)	
Magnetic beam	NMR

λ—Wavelength; Res.—Resolution

Depending on the size, various biological organisms can be visualized using different types of microscopes as given in Table 22.2.

Table 22. 2 Microbes and microscopes

Microbes	Size	Microscopy
Helminths	mm–cms	Light microscope
Helminth egg	≥ 50 μ	
Fungi	≥ 5 μ	Light microscope
Protozoa	≥ 2 μ	Light microscope and electron microscope
Bacteria	≥ 0.2 μ	Light microscope and electron microscope
Spirochaete	0.2 × 5–15 μ	Light microscope and electron microscope
Chlamydia	0.3–1 μ	Light microscope and electron microscope
Rickettsia	0.3–0.6 μ	Light microscope and electron microscope
Virus*	0.01–0.4 μ (10–400 nm)	Electron microscope

* Viruses in animal tissues, secretions, excretions and also plant tissues

The science of Virology evolved mainly after the use of electron microscopy, because viruses can be morphologically identified only at electron microscopic magnifications. Only viral inclusion bodies can be seen with the optical microscope.

EM plays a significant role in the field of microbiology in both diagnostic and research aspects. It helps in identifying the causative organisms, especially viruses, in various diseases.

ROLE OF EM IN DIAGNOSTIC MICROBIOLOGY

EM is used as a diagnostic tool. It means EM is used in determining the microbial cause of a disease in man, animal, etc., as in

1. Diagnosis of acute viral infections like,

   Diarrhoea (rota virus)

   Chickenpox (varicella)

   Conjunctivitis (adenovirus)

   Genital infection (HSV)

2. Epidemiological study—diarrhoea due to rota and adenovirus

3. Monitoring disease activity—rotavirus and poliovirus excretion
4. Vaccine trial—drop in carrier rate, duration of polio excretion
5. Other microbial infections:

   Protozoa—Cryptosporidium and microsporidium, etc.

   Bacteria—*Legionella*, etc.

COLLECTION OF SAMPLES FOR DIFFERENT DISEASES

For the diagnostic use of EM, the appropriate samples must be collected in proper sterile conditions and processed suitably. The materials to be collected for the diagnostic possibilities of various diseases is listed in Table 22.3.

Table 22.3 Collection of samples

Disease	Virus	Sample
Diarrhoea, Gastroenteritis	Rotavirus, Calcivirus, Adenovirus, Astrovirus Norwalk virus	i. Stool collected in sterile container; uncontaminated with urine. ii. Rectal swab
Respiratory infections	Rhinovirus, Influenza virus, Corona virus CMV, Legionella, etc.	i. Throat swab; ii. Tracheal secretions; iii. Bronchial lavage fluid.
Genital infections	Herpesvirus, CMV	i. Cervical secretions: ii. Genital secretions
Meningitis, Brain infections	CMV, Herpesvirus, Mumpsvirus, Paramyxovirus	CSF–culture fluid
Conjunctivitis	Adenovirus, Herpesvirus	Conjunctival swab containing infected cells

(CMV—Cytomegalovirus; CSF—Cerebrospinal fluid)

Biopsy/Autopsy Specimens

Meningitis, brain infections	Brain tissue
Respiratory infections	Lung tissue, tracheal tissue
Heart ailments	Heart tissue,
Warts, etc.	Vesical fluid, serum, etc.

If the organisms collected are still infectious, they are fixed immediately in EM fixative, which is buffered glutaraldehyde. Or the specimen can also be exposed to UV radiation. The UV acts on DNA only and so the morphology is preserved.

TRANSPORTATION OF THE COLLECTED SAMPLES

Transportation of the samples to the EM laboratory has to be done as quickly as possible.

STORAGE OF SAMPLES

1. If further processing can be done within 12 hours of collection of the sample, it can be stored at 4°C. Transport medium like Hanks BSS with antibiotics can be used. Drying of the sample and pH changes during storage and transportation should be avoided.
2. If further processing requires more than 12 hours of collection, then the sample can be stored and transported in 'dry ice' and frozen. Storage at –70°C in liquid nitrogen with added cryoprotectant is also advisable. But repeated freezing and thawing should be avoided.
3. 2–4% glutaraldehyde, the primary EM fixative, can be used for storage and transportation. By fixing the cells, the users are protected from infection.

PRECAUTIONS

There are certain precautions to be undertaken during collection, storage and processing of the samples containing the microorganisms.

- collect samples during early illness
- avoid fluctuations in temperature and pH during storage and transportation
- sterile collection
- avoid bacterial proliferation by adding antibiotics
- avoid formalin
- fix immediately; remove fat and take universal precautions while handling microorganisms.

PROCESSING OF THE SAMPLES FOR EM

1. Solid samples like tissue samples, are ground well and then centrifuged to remove the debris. The supernatant containing the organisms, can be processed further.
2. The semi-solid samples, like stool samples are mixed with phosphate-buffered saline (PBS) at a ratio of 1 : 4 (i.e., 1 part faeces and 4 parts PBS) and clarified. Slow speed

spinning of the mixture at 400 to 500 *g* for 10 minutes leads to sedimentation of the solids. The supernatant containing the organisms is collected.

The supernatant collected from the above two procedures, for solids and semi-solids, with sufficient concentration of the organisms can be directly dropped on coated grids, followed by negative staining with PTA (phosphotungstic acid) or UA (uranyl acetate) as discussed in Chapter 14 "negative staining".

If the concentration of the organisms is not sufficient, then the supernatant solution has to be further concentrated using any one of the following methods.

- ✳ Ultracentrifugation at 50 to 1,00,000 *g*, and sediment dropped on carbon-coated grid, stained and screened in the EM.
- ✳ Add PEG, polyethylene glycol, which is a mild precipitant of protein. After spinning, sediment deposited on grid and viewed after staining.
- ✳ Add specific antibody. After spinning, the deposit can be used for EM studies, particularly, SEM studies.
- ✳ Coat the carbon-coated grid with the required antibody or Fab fragment (depending on the protein targeted in the organism) probably tagged with colloidal gold, then add the viral extract on this. Allow for adsorption. Screen in the EM.

3. For liquids such as urine, the suspension is purified, concentrated by spinning and dropped on coated grid, stained with PTA and then scanned in EM.

   The concentration of organisms, of 10,000/ml is sufficient for both LM and EM for larger organisms, but for viruses generally, 10^6 /ml is suitable.

   Infectious organisms should be disinfected by any one of the following methods as soon as the samples are collected:

   - ✳ exposure to UV, which does not usually alter the morphology,
   - ✳ treating with disinfectants like, sodium hypochlorite,
   - ✳ fixing the organisms with EM primary fixative, i.e., glutaraldehyde, which by killing the cells and arresting enzymatic reactions by binding with proteins, preserves morphology.

STAINING

1. Negative staining is done using 4% PTA at pH 6.3. Ammonium molybdate and uranyl acetate can also be used for this purpose.

   Empty viral particles contain no nucleus and when the stain penetrates inside it appears dark even in negative staining.

2. Positive staining

   - ✳ Ruthenium red stains mucopolysaccharides,

* Colloidal thorium stains bacterial capsule
* Platinum blue stains cell wall (peptidoglycan)

3. For intracellular localization of proteins, immunological techniques can be employed as explained in Chapter 17.

CONFIRMATION AS CAUSATIVE AGENT

The causative agent of a disease can be confirmed only if, the organisms are

* demonstrated in large numbers and correlated clinically.
* demonstrated on repeated samples also.
* But in the case of a normally sterile specimen, like CSF, single demonstration is sufficient for diagnosis.

Mixed infections, two viruses in the same sample, can be easily identified with EM, which is not possible with serology, and even with fluorescent microscopy. The size, shape and topography of the viruses can be visualized in EM, with which mixed infections can be identified.

But pleomorphic viruses may pose difficulty in identification.

Pathogenesis of the disease can also be studied using EM. Apart from morphology and structure of organisms, including viruses, their entry and route within the cell can also be studied using EM and hence pathogenesis can be identified which may help in vaccine preparations.

The intracellular as well as surface antigens of different organisms can also be traced and characterized using immunoelectron microscopic techniques as mentioned in chapter 17. Viral nucleic acid studies can also be carried out with EM, which is actually *in situ* hybridization studies at the ultrastructural levels.

EM APPLICATIONS IN MICROBIOLOGY

Thus the applications of EM in Microbiology can be enumerated as follows:

* Identification of morphology
* Quantification of organisms, especially viruses
* Serotyping of viruses, adenovirus, rotavirus, etc.
* Size determination
* Classification of organisms—spirochaetes, viruses, etc.
* Insight into structure–function relationships:

* Giardia damages intestinal villi, resulting in malabsorption. The structural damage is correlated with the functional disability.
* Flagella and mechanism of bacterial motility can be interrelated. The presence and identification of flagella indicates the functional motility.
* Fimbriae and bacterial adhesion can be easily correlated with the pathogenicity.
* Spore structure and classification—may indicate the nature of their spread and germination.
* Study of intracellular structures like mitochondria, ribosomes and nuclear material can also be identified with the varied functions of the organism.

ADVANTAGES OF EM

Thus the advantages of using EM in the study of organisms can be summarized as follows:

1. Study of organisms and their characterization
2. Is a sensitive technique
3. Specific technique
4. Rapid technique
5. Multiple infections can be demonstrated
6. Is a sole diagnostic aid for non-cultivable or difficultly cultivable viruses, such as Norwalk, etc.

LIMITATIONS OF EM APPLICATIONS IN MICROBIOLOGY

There are certain limitations also in the application of EM in the study of microorganisms, including viruses. Some important ones are listed below:

i. The sensitivity is high when the number of organisms is around 10^6 particles per gram or per ml of the specimen. False negatives may occur when their concentration is very low.

ii. Pleomorphic viruses pose difficulty in identification.

iii. Disintegrated viruses are not detectable. Thus, they are not demonstrable in tissue specimens when their DNA gets integrated with the host DNA.

iv. Occasionally it is non-specific, e.g., rotavirus vs. reovirus

v. Problems are encountered in interpretation which include the following.

* demonstration of the presence of the organism does not confirm etiology because, some may be present in normal conditions also. For example, reo-, entero-, and adenoviruses may occur as normal flora in the gut and rhino-, adeno- and herpesviruses in throat.

- ✳ But demonstration of an organism in a normally sterile site is indicative of infection. For example identification of virus in CSF is a highly significant finding in the diagnosis of the disease.
- ✳ Carrier state cannot be distinguished from disease state. Similarly chronic infections, and altered morphology also cannot be accurately determined.

vi. High cost and time are required for EM studies and also the expertise needed for identification and interpretation of the images. Therefore it is not suitable for large-scale study.

vii. Epidemiological studies are more difficult to carry out in practice.

viii. More sensitive molecular techniques are available but even then, the presence of viruses with their size, shape and symmetry can be visualized in the cellular context only with EM.

SOME ILLUSTRATIONS

Some important uses of EM in microbiological research are illustrated in Figures 22.1–22.25. The isolated organisms in suspension are negatively stained with aqueous phosphotungstic acid. A few tissue specimens processed and screened in EM are also included as specified. The specimens received from several students are processed and screened in EM in our laboratory.

For diagnostic applications please also refer Chapter 21.

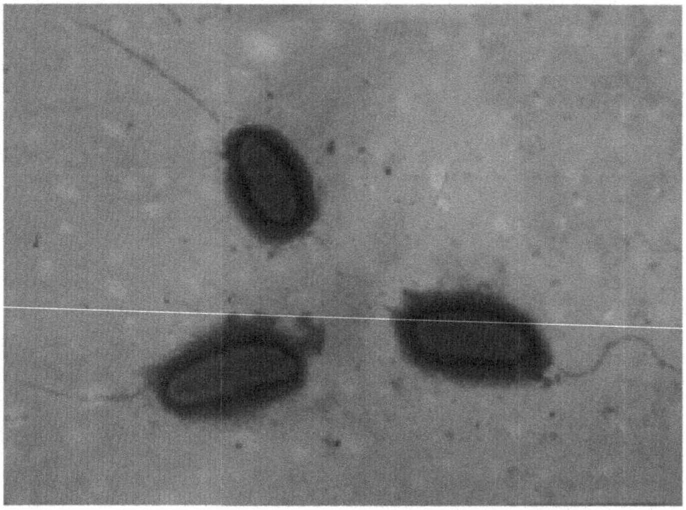

Figure 22.1 Three bacteria with single flagellum each (10,000×)

Figure 22.2 *Aeromonas* sp. isolated from cat fish from Kodiveri dam in Tamil Nadu. The organism causes *Aeromonas* septicaemia, an ulcerative disease in the fish. The isolated organism is shown with its flagellum (5000×).

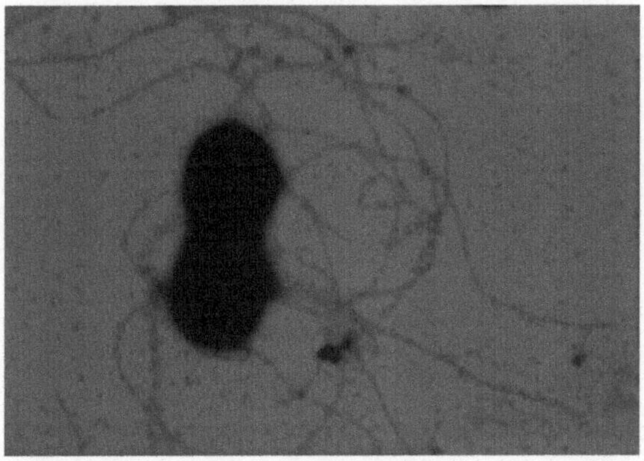

Figure 22.3 A new bacterial species/genus of mangrovibacter is shown to possess peritrichous flagella (15,000×).

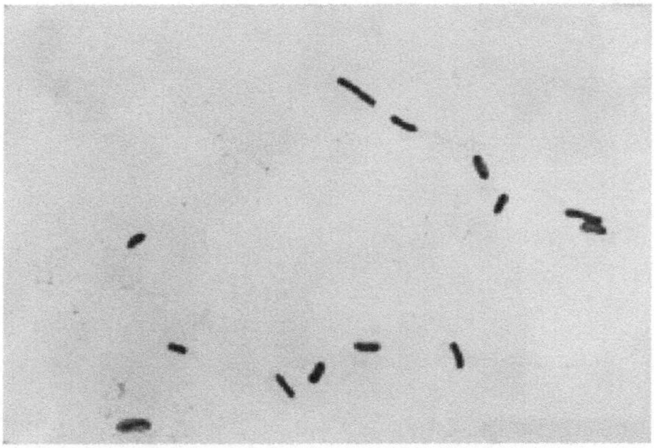

Figure 22.4 Nitrate-oxidizing bacteria isolated from soil—a few bacteria (3000×)

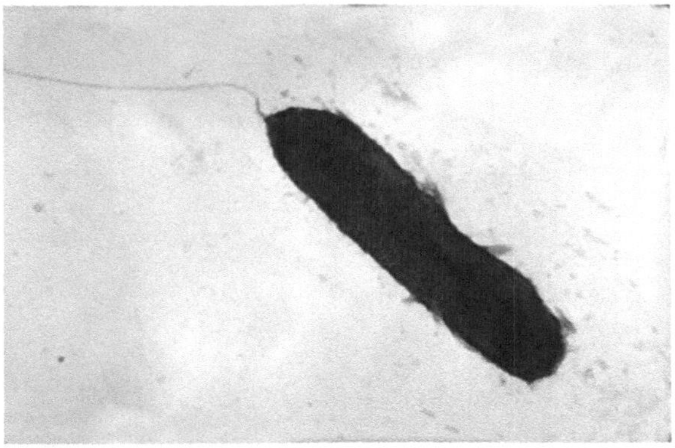

Figure 22.5 Nitrate-oxidizing bacteria isolated from soil, at higher magnification; one bacterium with flagellum (30,000×)

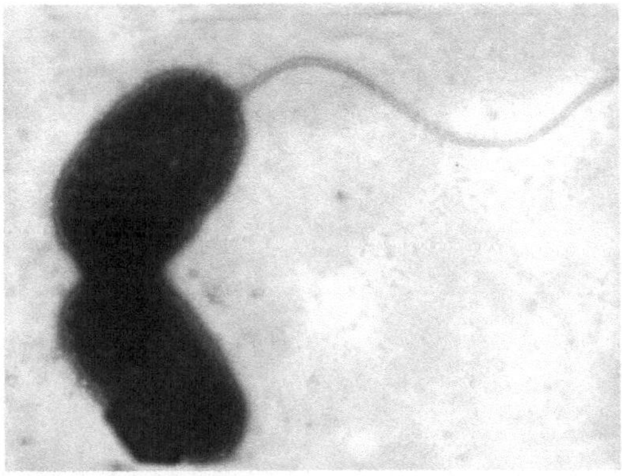

Figure 22.6 Bacteria isolated from post-tsunami estuary waters. Rod-shaped with polar flagella belonging to a species of *Pseudomonas*. One bacterium probably in a dividing stage with a polar flagellum (30,000×)

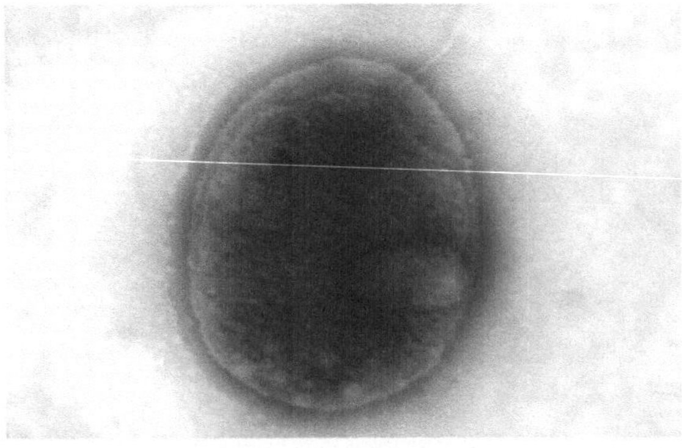

Figure 22.7 One bacterium of the post-tsunamic estuary waters at a higher magnification showing the cellular features (40,000×)

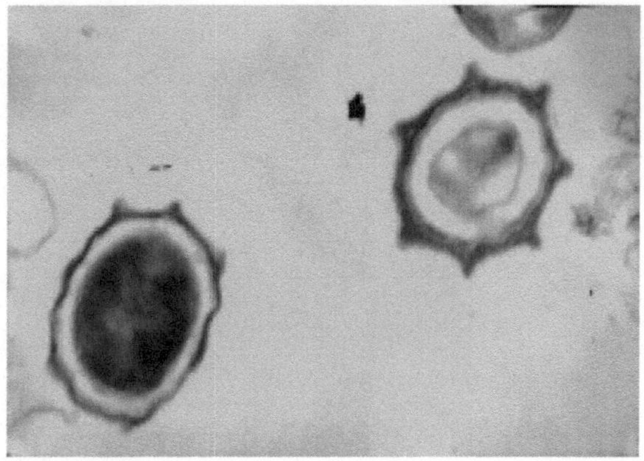

Figure 22.8 Bacteria isolated from termite. *Paenebacillus* grown in nutrient broth. Bacterial spores are shown in the figure (30,000×)

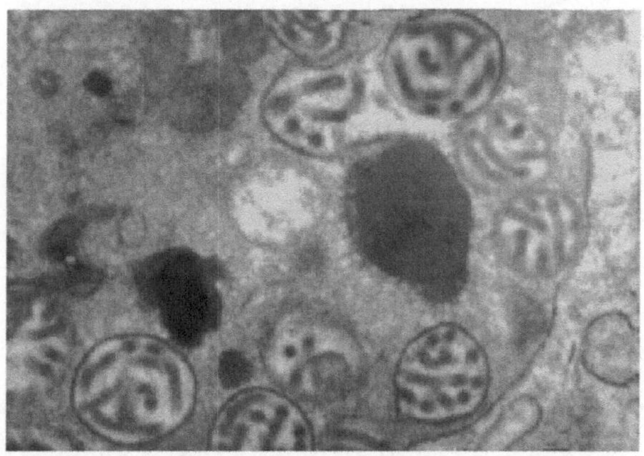

Figure 22.9 Bacteria within a parasite (25,000×)

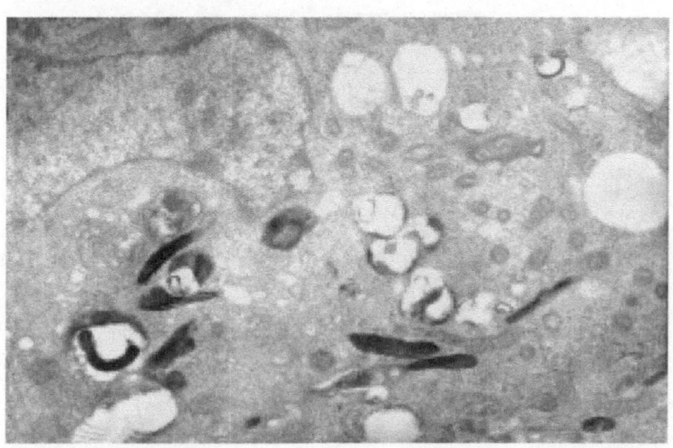

Figure 22.10 *Mycobacterium tuberculosis* in the cytoplasm of a mesothelial cell. Organisms within vacuoles can be seen (15,000×)

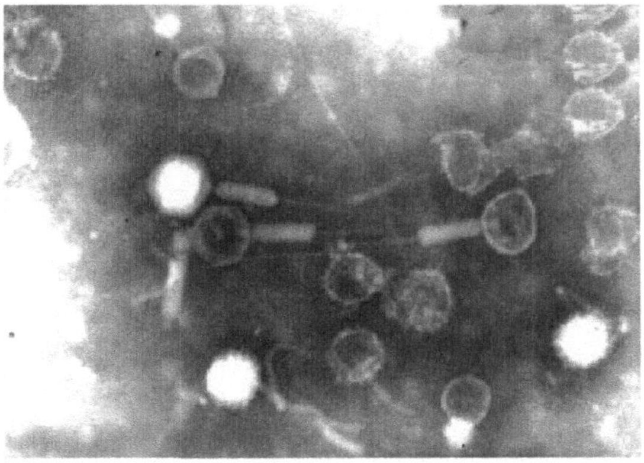

Figure 22.11 A few bacteriophage particles (1,00,000×)

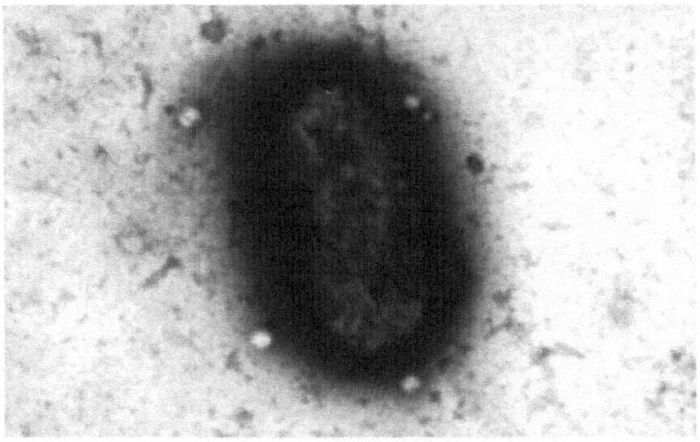

Figure 22.12 *Vibrio* (bacterium) affected by phage. A few phage particles attached to the surface are seen. The phage is isolated from marine sediments and belong to the family Myoviridae (30,000×)

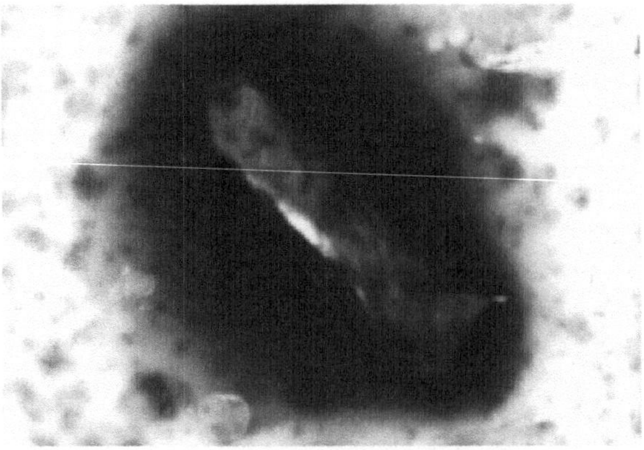

Figure 22.13 The lysed bacterium after the phage attack (30,000×)

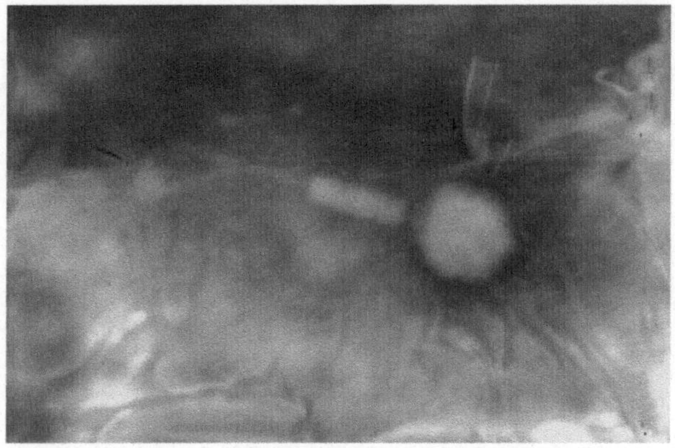

Figure 22.14 One bacteriophage showing the hexagonal head, neck and tail (1,50,000×)

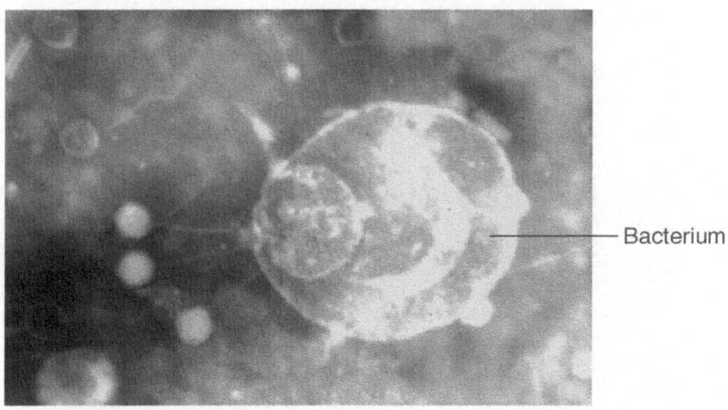

Figure 22.15 One bacterium being attacked by three phage particles getting attached to the bacterium with their tails (80,000×)

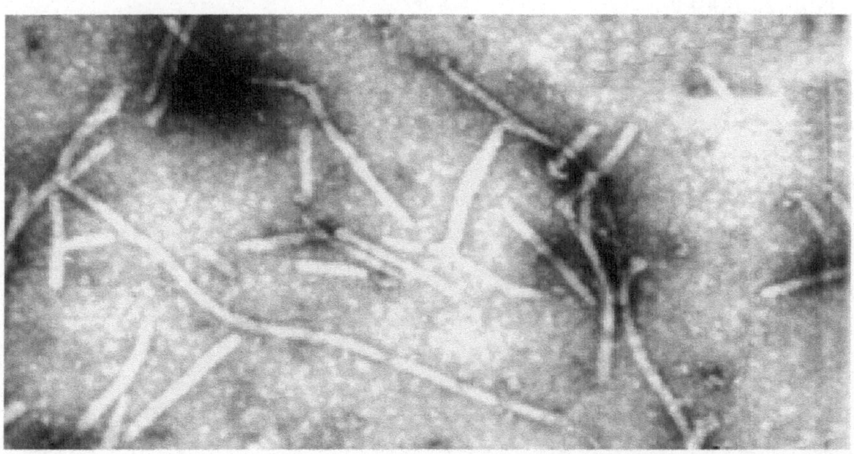

Figure 22.16 Plant virus—a few viral particles (50,000×)

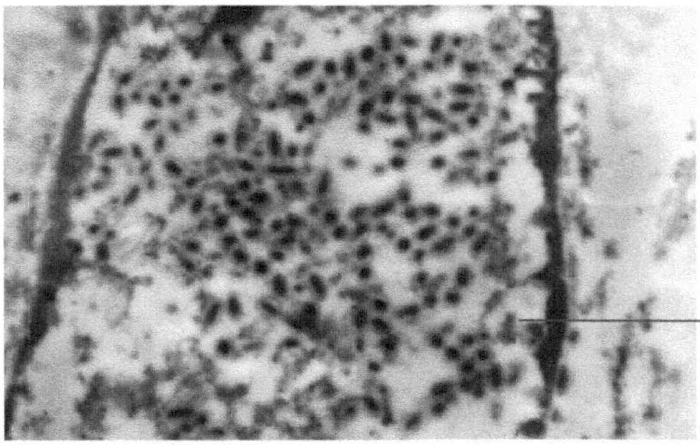

Figure 22.17 Animal virus. A few particles within the nucleus. The infected animal tissue is dissected, processed, sectioned for EM and stained. The nuclear outline is seen. Some particles are circular while some are elongated. The circular ones are cross sections and the elongated ones are longitudinal sections of the particles (20,000×)

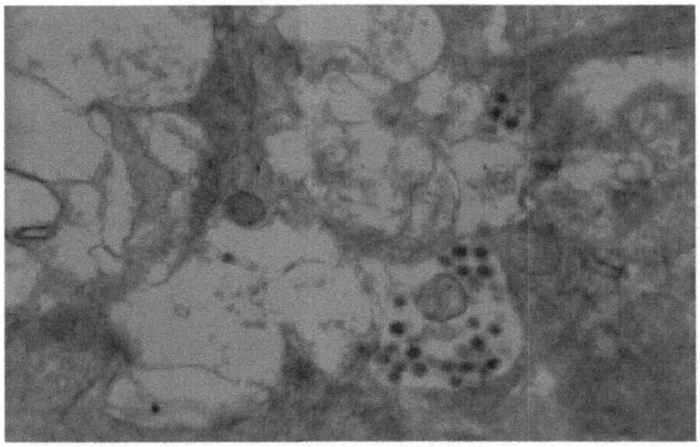

Figure 22.18 Animal virus. Also from tissue sections. Viral particles within vesicles in an affected cell (20,000×)

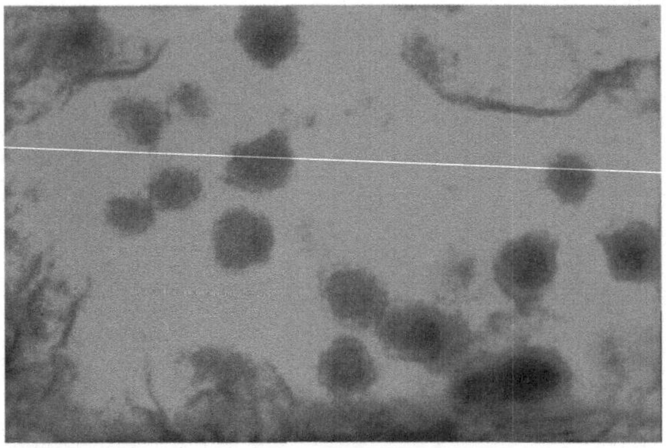

Figure 22.19 The same specimen as Figure 23.18 at higher magnification showing the viral particles with typical morphology (1,00,000×)

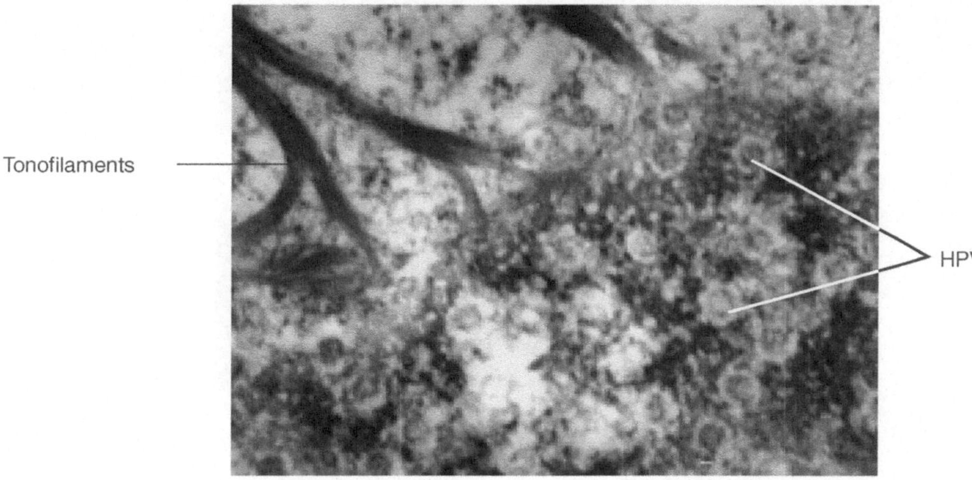

Figure 22.20 Human virus—HPV (Human papilloma virus) in an epithelial cell tumour tissue from cervical cancer patient. The presence of tonofilaments suggest the epithelial nature of the cell (40,000×)

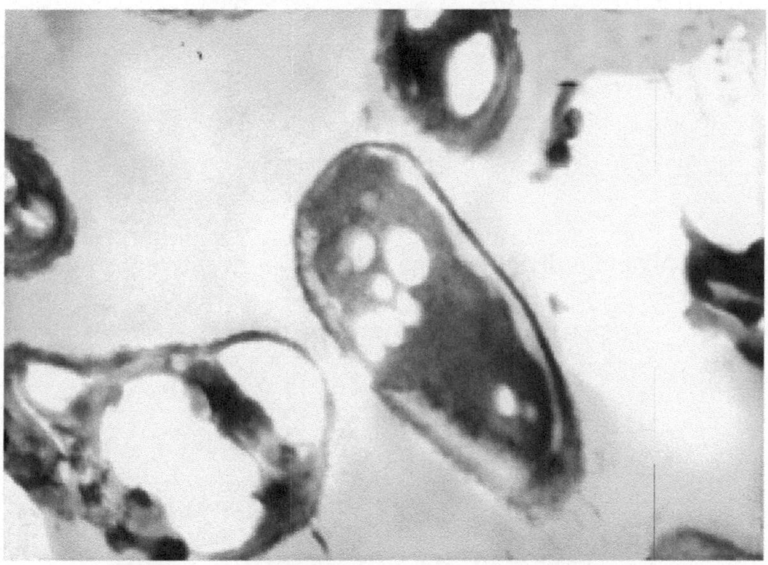

Figure 22.21 Organism isolated from effluents rich in carbohydrate contents. The excess carbon in the environment is converted to polyhydroxybutyrate (PHA) particles which get accumulated within the organism leading to its swelling and subsequent bursting. (30,000×)

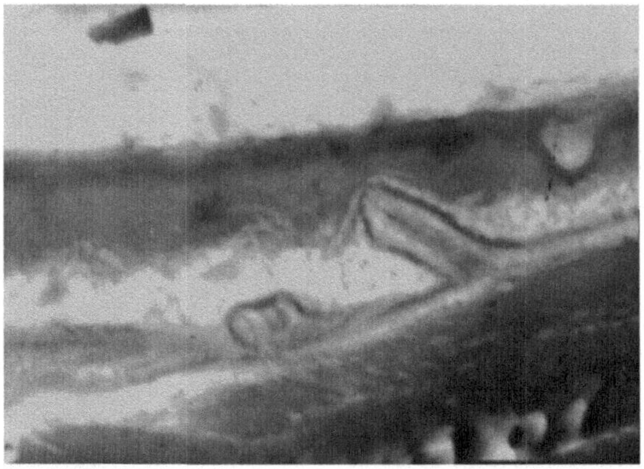

Figure 22.22 Fungal infection of human hair. Two fungal hyphae invading the hair can be seen (50,000×)

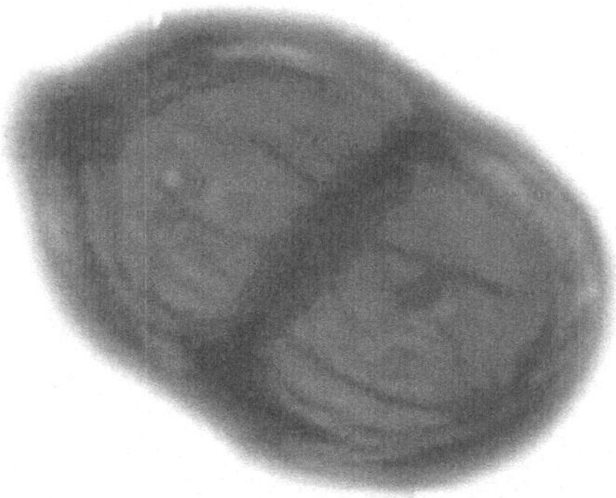

Figure 22.23 Alga in a dividing stage. Wild type, negatively stained (20,000×)

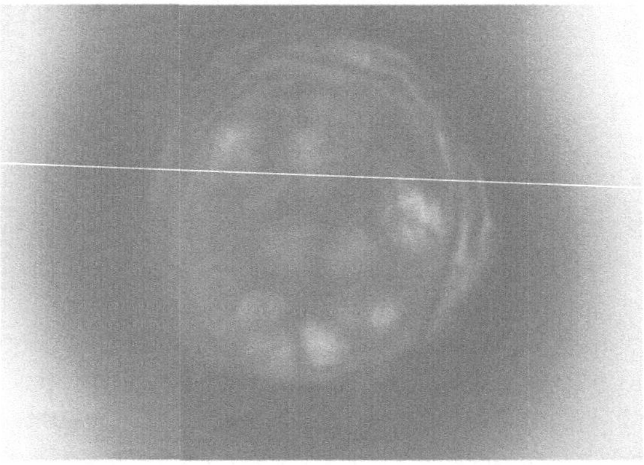

Figure 22.24 Alga—wild type negatively stained (20,000×)

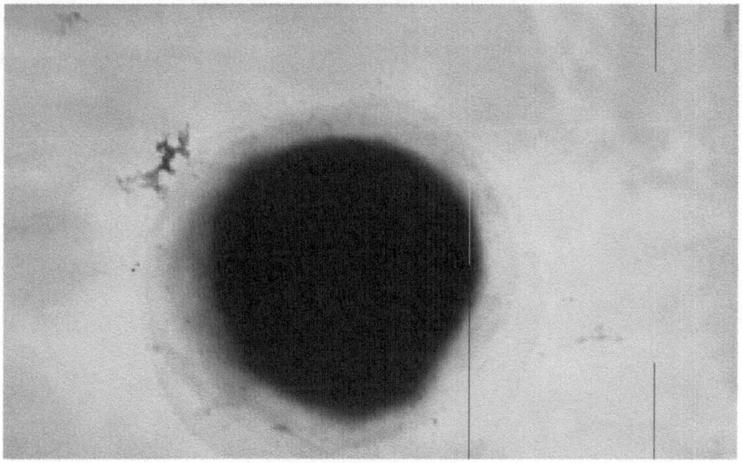

Figure 22.25 Alga—mutated by UV irradiation for 4 hours (Unstained specimen) (20,000×)

23

EM IN ORAL PATHOLOGY

A few pathological conditions including non-malignant, pre-malignant and malignant lesions where EM has a contributory role in diagnosis are discussed in this chapter. When fresh specimens are not available, paraffin block sections have been used for ultrastructural study which has helped in their differential diagnosis as well as in confirmation of the light-microscopic findings. The processing of specimens for EM and the screening of the ultrathin sections using TEM were carried out in the Electron Microscopy Department at Cancer Institute (WIA) Chennai. Many of the specimens, particularly those included under the headings Non-tumour and tumours were received as formalin-fixed, paraffin block materials, probably the need for ultrastructural study was realized only after histopathological diagnosis. Hence only representative paraffin block sections were brought to our laboratory where they were further processed and screened. Though ultrathin sections from these blocks do not reveal all fine structural details, important characteristic features could be delineated as discussed in the following pages.

NON-TUMOUR

ECRM—Epithelial Cell Rests of Malassez

The epithelium of dental lamina is involved in the formation of enamel and maps out the shape of the tooth. On completion of tooth formation, epithelial remnants remain in various areas. The most frequent epithelial rests are the debris of Malassez, which are found in the periodontal ligament as small clusters and strands of non-keratinizing stratified squamous epithelium but occasionally as larger nests.

To assess whether the ECRM is involved in cementum repair and periodontal regeneration, a study was undertaken to identify the presence of ECRM cells in the regenerating periodontal ligament after Emdogain application in surgical defects created in a rodent model. Light microscopy showed presence of ECRM-like cells in the regenerated periodontium. But immunohistochemistry did not reveal positivity with cytokeratin 19.

So, ECRM cluster area was identified using the light microscope and selected for EM study only paraffin block sections were made available for EM study as already mentioned. ECRM cells, being the only odontogenic epithelial cells in the periodontal ligament, have well-characterized ultrastructural features as observed in EM. At lower magnifications in the EM, a cluster of closely packed cells, ovoid in shape with a high nuclear–cytoplasmic ratio and scant cytoplasm are seen in the connective tissue matrix (Figure 23.1).

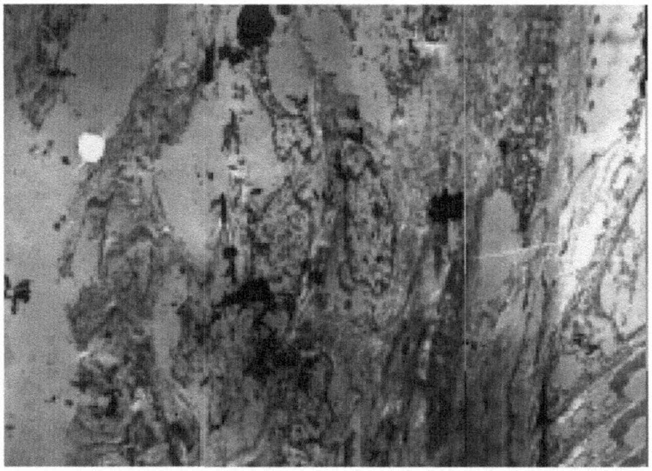

Figure 23.1 A few ECRM cells (2000×)

These cells at higher magnifications show cytoplasmic membrane fusion between adjacent cells in certain areas and also tonofilaments (Figure 23.2)

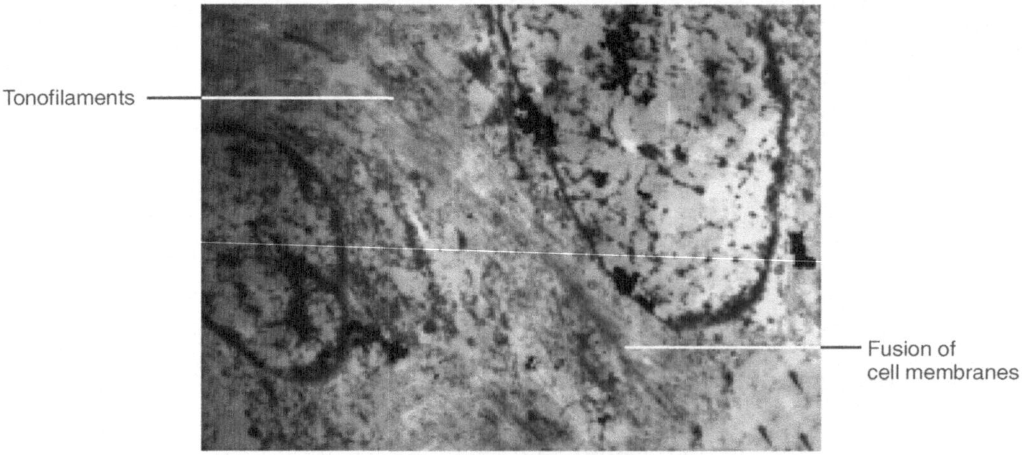

Figure 23.2 Fusion between adjacent cell membranes (8000×)

Small micropodlike projections also are seen as in Figure 23.3.

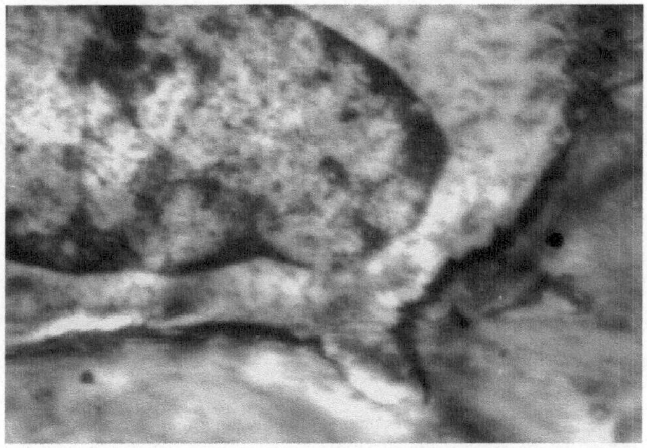

Figure 23.3 Micropod (15,000×)

These ultrastructural features of ECRM support the light-microscopic findings that the ECRM cells are present in the regenerating periodontium.

TUMOURS—BENIGN AND MALIGNANT

Here also, all specimens included were received only as paraffin block sections.

Odontogenic Keratinocyst

This is a cyst which is considered to be a separate entity and these cysts are typically parakeratinized. The ultrastructural features suggest that the basal lamina complex in the odontogenic keratocyst appears to be morphologically normal. The frequently observed splitting of the epithelium from the capsular tissue is marked by the separation of the epithelium and the basal lamina complex from the connective tissue (Figure 23.4).

Figure 23.4 Splitting of the epithelium (80,000×)

Though inflammation is seen, keratocyst specimen also shows juxtaepithelial collagenolysis that is not associated with the degree of inflammation (Figure 23.5).

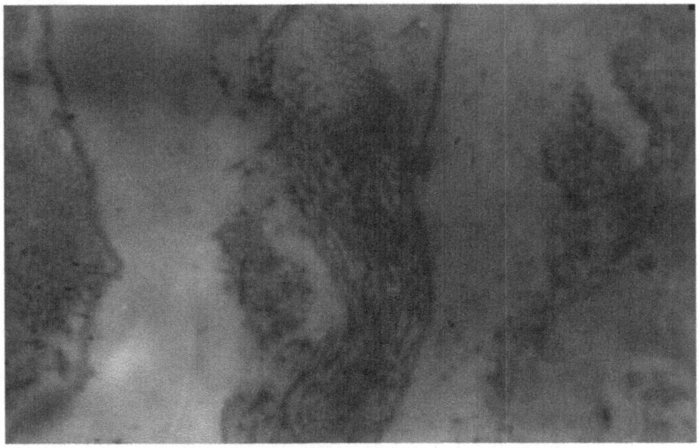

Figure 23.5 Collagenolysis (8,000×)

Fenestrated capillaries are seen in keratocyst, where their presence may indicate rapid transfer of fluid (Figure 23.6).

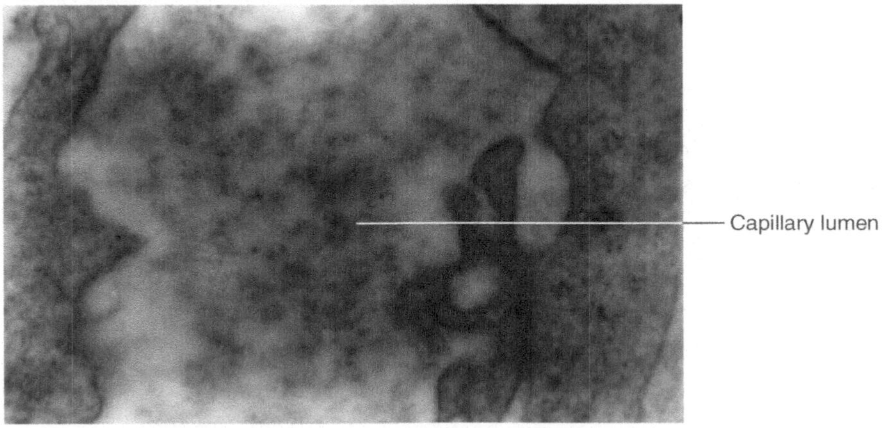

Figure 23.6 Fenestrated capillary (50,000×)

Adenomatoid Odontogenic Tumour (AOT)

It is considered as a benign lesion. It is an epithelial odontogenic tumour making about 3 to 7% of all odontogenic tumours. It is characterized by diverse morphologic features and eosinophilic coagulum. Cuboidal cells with round to oval nuclei and few tonofibrils are shown in Figures 23.7 and 23.8 while different stages of calcifications and amorphous granular material can be seen in Figure 23.9.

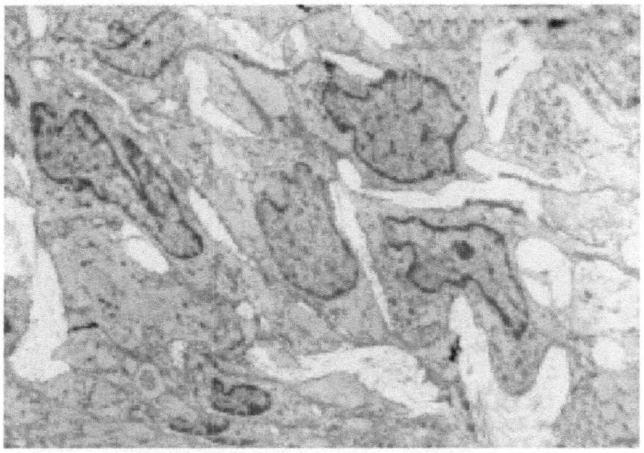

Figure 23.7 A few epithelial cells in AOT (4000×)

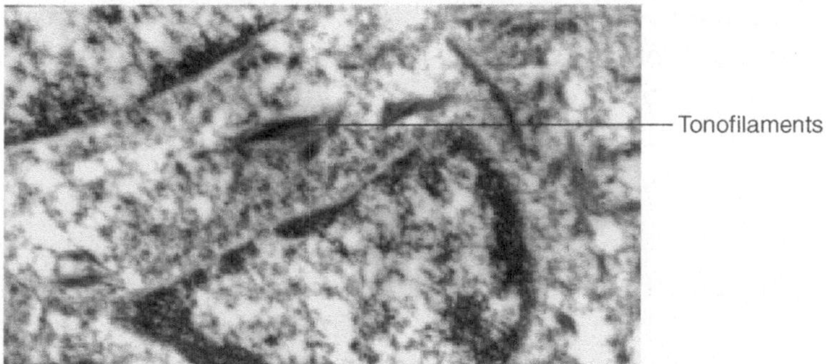

Figure 23.8 One cell with cytoplasmic tonofilaments exhibiting the epithelial nature (25,000×)

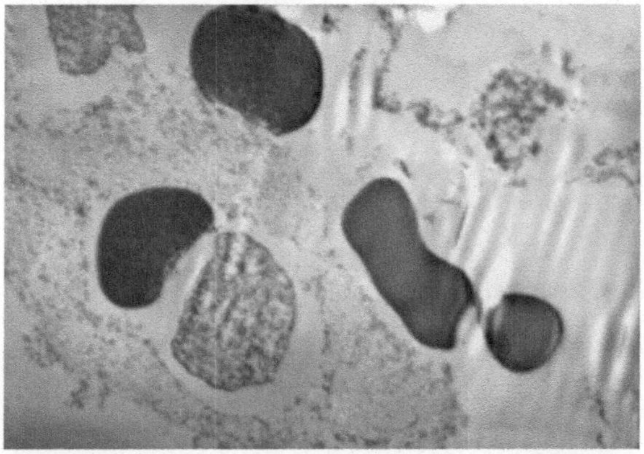

Figure 23.9 Few dense bodies with differing densities (5,000×)

Figure 23.10 shows the granular amorphous material with haphazard fibrils probably indicating amyloid.

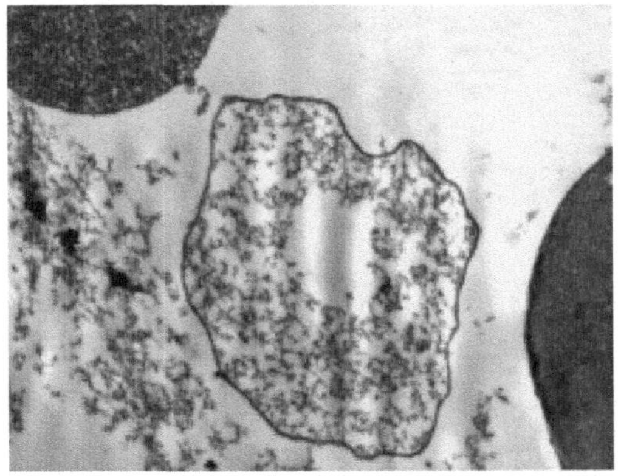

Figure 23.10 Deposits containing haphazard fibrils (15,000×)

The specimen received is only paraffin block sections and not fresh ones and hence the details are not fully explicit but a few characteristic feature are good enough to be made out. The ultrastructural studies suggest that the eosinophilic coagulum secreted by epithelial tumour is probably amyloid.

Desmoplastic RCT

In this rare tumour, usually there is evidence of more than one line of differentiation with epithelial, mesenchymal and neural features. Electron microscopy reveals several lines of differentiation including neuroendocrine, supporting derivation from a multipotential cell. A case of tumour arising from body of mandible extending behind mandible was offered a differential diagnosis of desmoplastic RCT or poorly differentiated carcinoma histopathologically. Ultrastructurally, islands of closely apposed, oval and polygonal cells surrounded by abundant banded collagen (Figure 23.11) and collagen in between cells within groups could be seen (Figure 23.12).

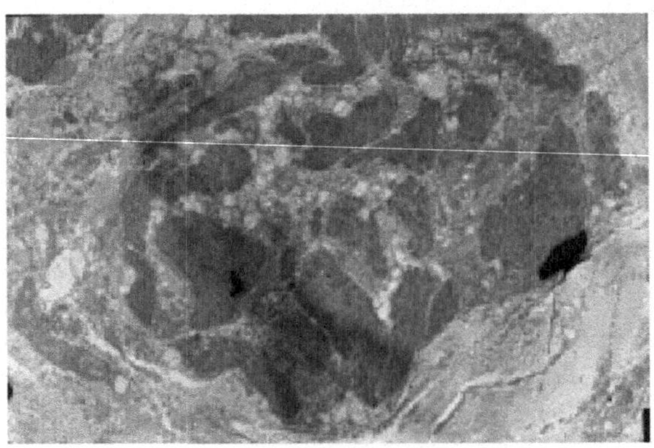

Figure 23.11 A few closely apposed cells surrounded by collagen bundles (2000×)

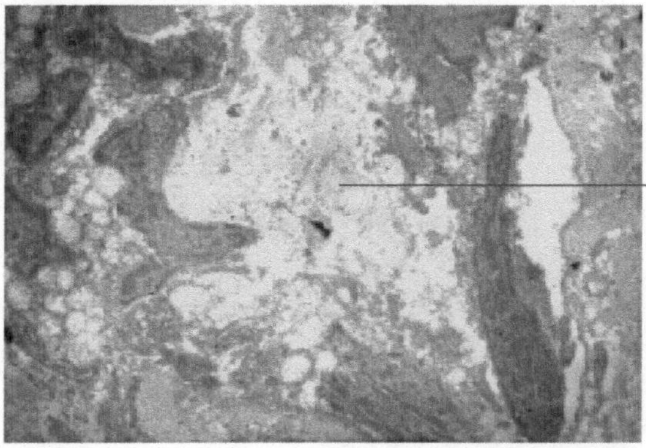

Figure 23.12 Collagen bundles between cells within group (4000×)

Cells have high nuclear cytoplasmic ratio. Nuclei are extremely pleomorphic with condensed chromatin at the nuclear membrane (Figure 23.13).

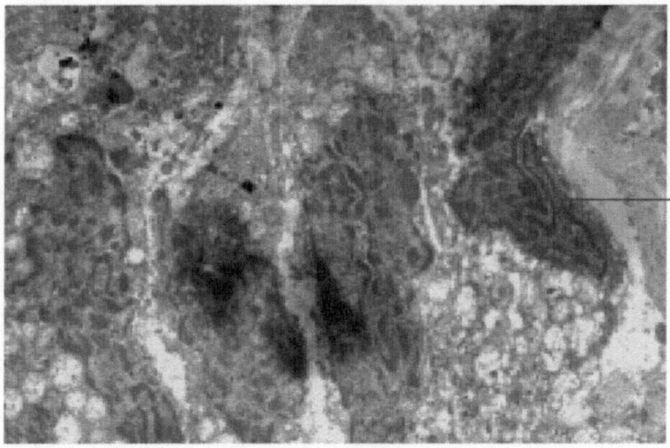

Figure 23.13 Highly indented (pleomorphic) nuclei with peripherally condensed chromatin (5000×)

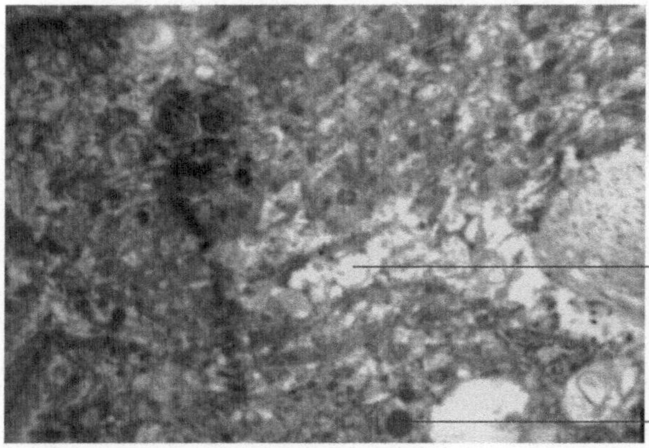

Figure 23.14 Cytoplasmic vesicles and dense granules (10,000×)

Junctions are made out at higher magnifications. Cytoplasm is rich in ribosomes; a few mitochondria; small vesicles; dense granules; (Figure 23.14) and paranuclear filaments (Figure 23.15). Basal lamina surrounding groups of cells rather than individual cells also could be seen.

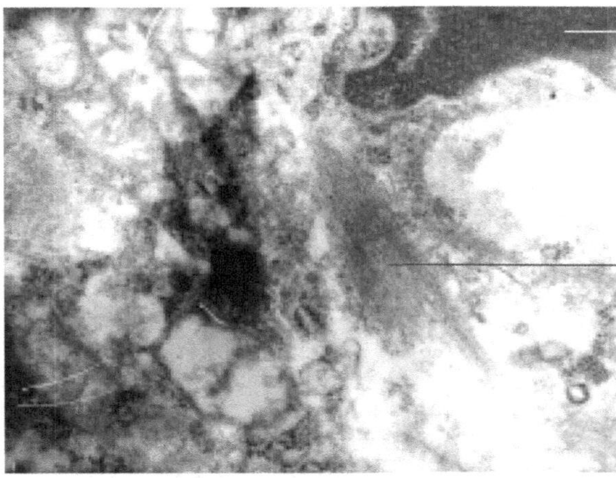

Figure 23.15 Paranuclear filaments (30,000×)

Granular Cell Ameloblastoma

Ameloblastoma is a tumour arising from enamel organ, i.e., bone forming structures and granular cell ameloblastoma is one of its variants. It is a rare tumour. Ultrathin sections of formalin-fixed paraffin-embedded material show epithelial cells containing tonofilaments and lysosomal granules. Figure 23.16 shows a few epithelial cells.

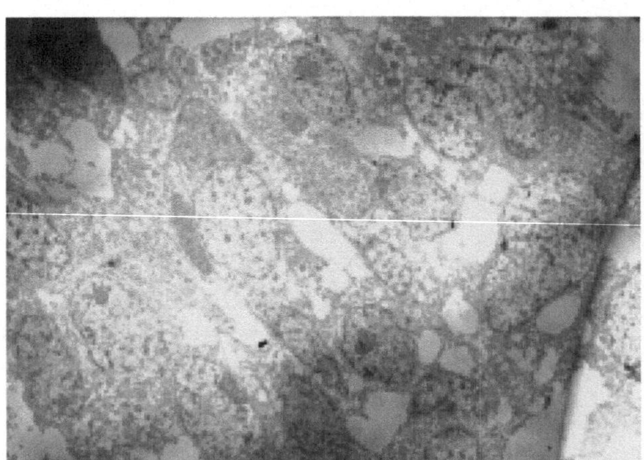

Figure 23.16 A few epithelial cells (2500×)

The presence of tonofilaments is suggestive of the epithelial nature of these cells while the lysosomal granules are responsible for their granular nature. The lysosomal granules have been reported to show apoptotic bodies. This is verified both ultrastructurally as well as using immunofluorescence techniques. In carcinomas, apoptotic bodies from the apoptotic cells appear to be phagocytosed more frequently by neoplastic epithelial cells than by histiocytes. Both tonofilaments and lysosomal granules could be identified in epithelial cells ultrastructurally. The dense bodies are osmiophilic bodies representing phagolysosomes. Heterogeneous character of osmiophilic material as well as the limiting membrane of the phagolysosomes can be easily recognized in Figure 23.17 and 23.18.

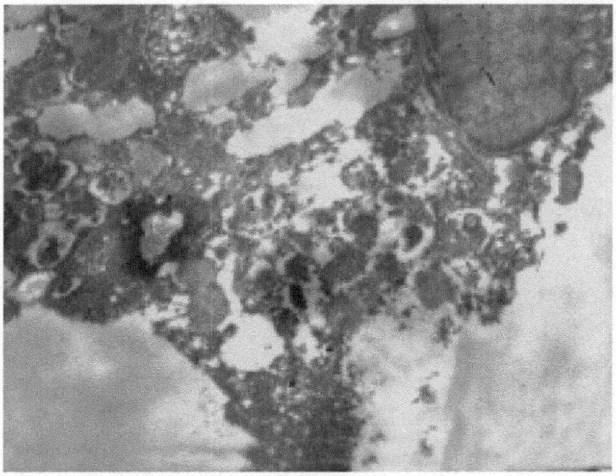

Figure 23.17 Granules containing small globular densities (15,000×)

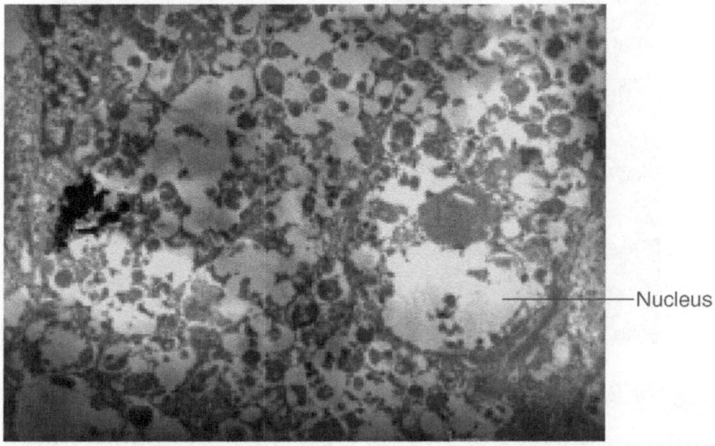

Figure 23.18 Apoptotic bodies in lysosomes (6000×)

Osteosarcoma

A biopsy specimen from a patient with diffuse swelling in the body and ramus of mandible showed histopathologically spindle-shaped and round cells of varying size and shape. Scanty area of osteoid was seen. There was evidence of numerous nerve bundles with mature collagen fibres. Special staining showed dark red zones resembling osteoid. For confirmation, EM was done. EM show pleomorphic cells, spindle, round and oval cells round or highly irregular nucleus with euchromatin, moderate to abundant cytoplasm showing RER with extended cisternae and a few mitochondria (Figure 23.19.)

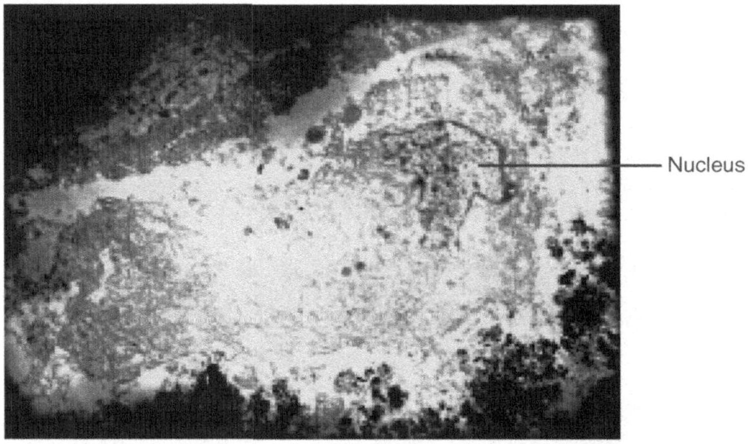

Figure 23.19 One large cell close to osteoid showing RER and mitochondria (4000×)

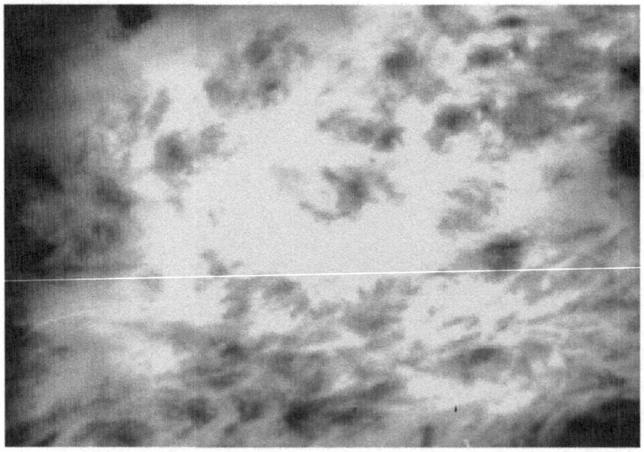

Figure 23.20 Dense deposits on collagen fibres (30,000×)

Extracellular collagen bundles could be seen. Electron-dense deposits on prominent fibres of collagen resembling osteoid could be clearly made out (Figure 23.20) which confirms the clinical and histopathological diagnosis.

Squamous Odontogenic Tumour (SOT)

This is derived from the debris of Malassez and considered to be a benign tumour. Though rare, it is described as a distinct entity. It consists of islands and strands of stratified squamous epithelium lying in a collagenous stroma. Individual cell keratinization and small foci of keratin may be seen. Calcified, crystalloid structures can occur within the epithelium but dental hard tissue or amyloid is not found. Here again only paraffin block sections were made available for EM work. A few closely placed epithelial cells with round to oval nucleus are seen in Figure 23.21.

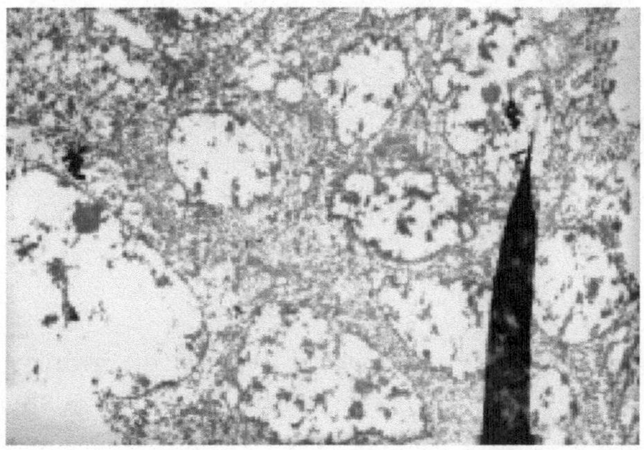

Figure 23.21 A few epithelial cells in SOT (4000×)

One cell with round nucleus, small nucleolus, moderate cytoplasm in which tonofilaments (keratin filaments) could be made out is shown in Figure 23.22. One granular deposit in the same tumour cell can be seen in Figure 23.23.

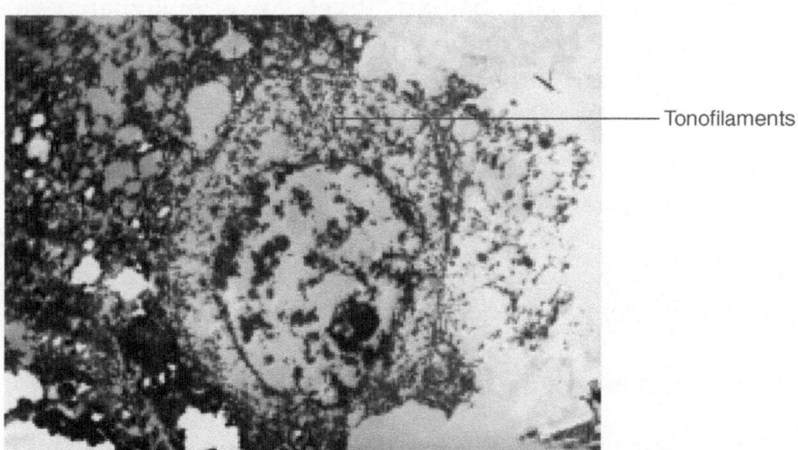

Figure 23.22 One cell with tonofilaments (6000×)

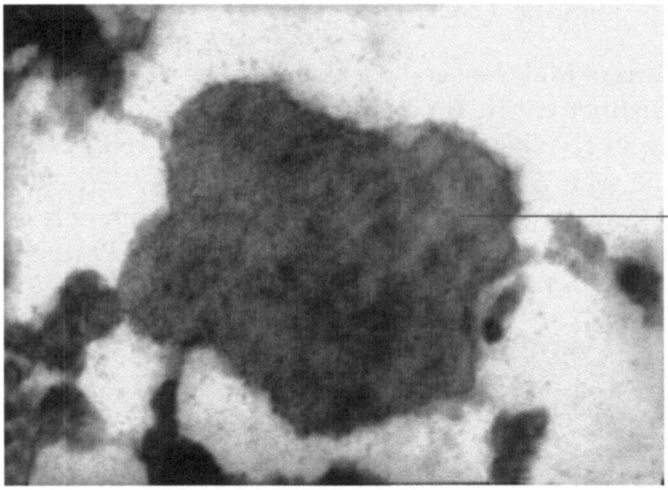

Figure 23.23 Granular deposit (1,00,000×)

Squamous Cell Carcinoma (SCC)

This is the most common malignant neoplasm of the mouth. Oral squamous cell carcinomas show the same range of features and variants as such carcinomas at other sites and hence not discussed any further here.

PRE-MALIGNANT LESIONS

It is known that tobacco smokers develop leukoplakia and chewers submucous fibrosis (SMF) where both are identified as premalignant conditions with malignant potential to develop into carcinoma. Leukoplakia has a relatively lower malignant potential than SMF but in the case of the former, the condition can revert back after cessation of the habit while in SMF the changes are rather permanent. Cellular alterations precede clinical changes.

In SMF, the fibrosis in the submucous tissue induces changes in the epithelium leading to premalignant and malignant changes. Thus the changes in the mucosal epithelium and the underlying connective tissue are interrelated and interdependent. Some of the ultrastructural features of the connective tissues, muscle bundles and pigmentation in chronic smokers and chewers as carried out in our laboratory are discussed below.

Connective Tissue Changes in Smokers and Pan Chewers

The physical state of collagen, overall fibroblastic response and inflammatory response assessed show the following significant features.

In the case of smokers, numerous discontinuous collagen bundles are seen. Three types of fibroblasts, namely, epithelioid, stellate and spindle-shaped are seen. Occasional elongated fibroblasts actively secreting collagen could also be made out. Increased fibrosis in smokers similar to SMF rather than leukoplakia as seen in our study is also reported in literature (Figures 23.24 to 23.27).

In chewers, increased collagen synthesis is significant though the fibroblast proliferation is insignificant indicating that the fibrosis is mainly due to increased collagen synthesis by the fibroblasts. The collagen bundles are wavy, and appear fragmented with frayed and bent ends. In some areas apparent partial degeneration of the fibres into amorphous material are also seen. Inflammatory cell infiltration is also evident (Figure 23.28 to 23.30).

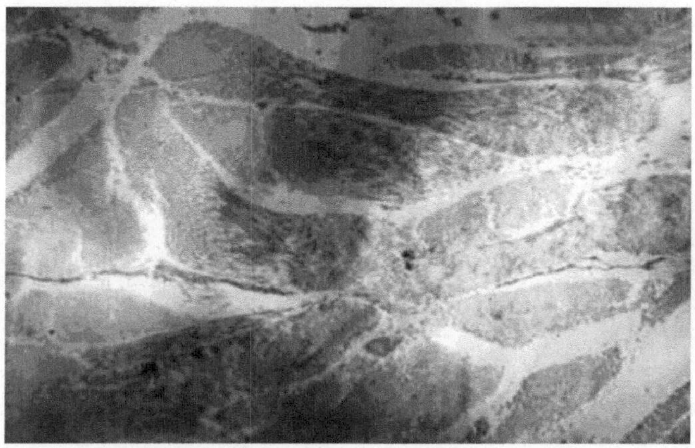

Figure 23.24 Proliferation of collagen bundles in smoker (4000×)

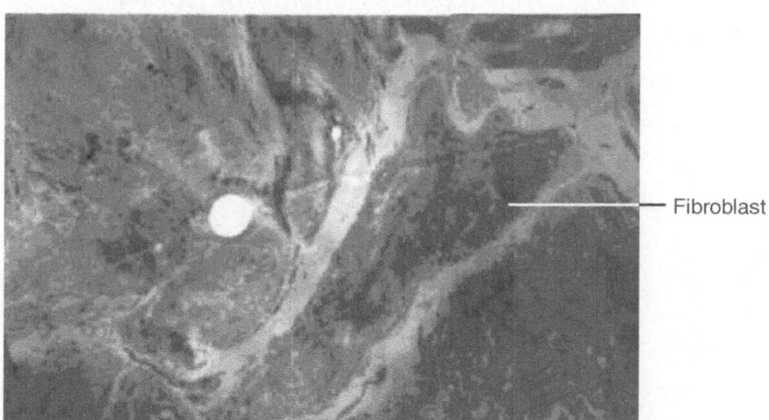

Figure 23.25 Spindle-shaped fibroblast (6000×)

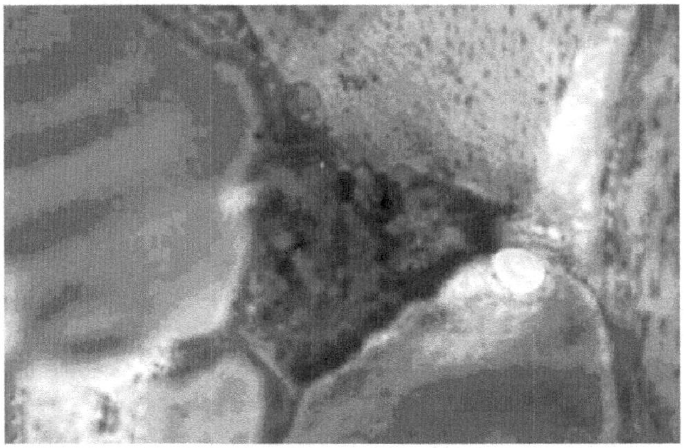

Figure 23.26 Stellate-shaped fibroblast (5000×)

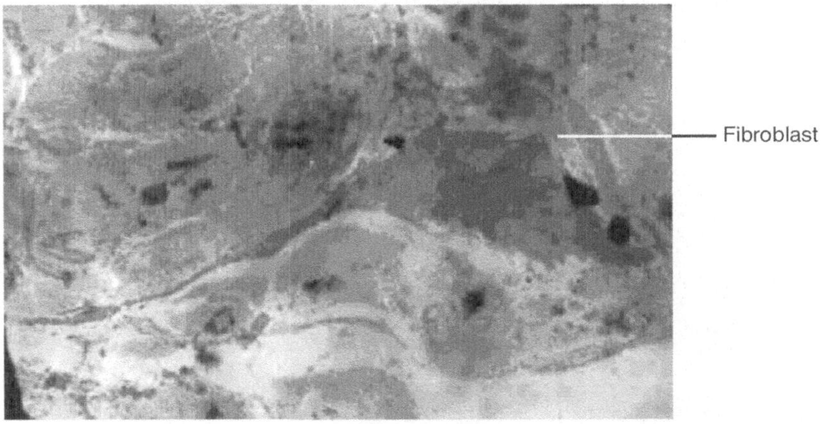

Figure 23.27 Epithelioid fibroblast (5000×)

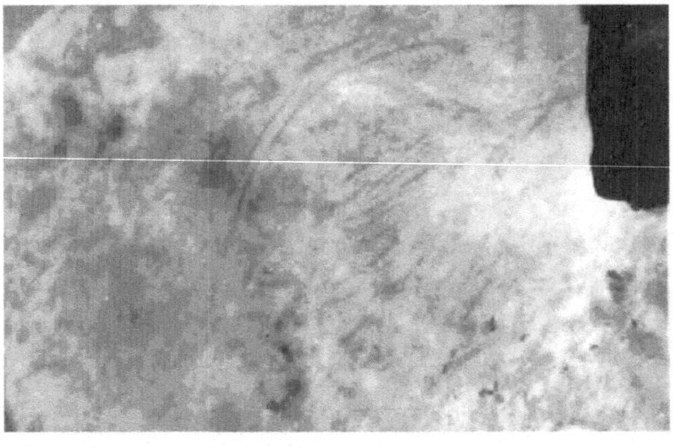

Figure 23.28 Collagen filaments in chewers (15,000×)

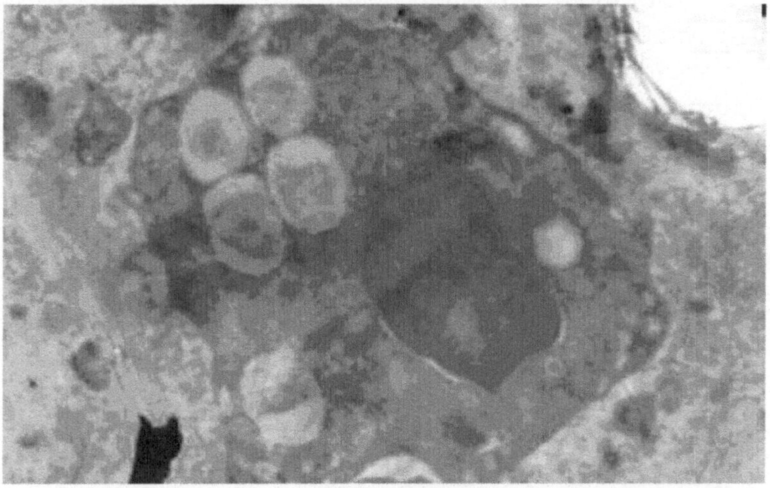

Figure 23.29 Vacuolated cell in chewers (8000×)

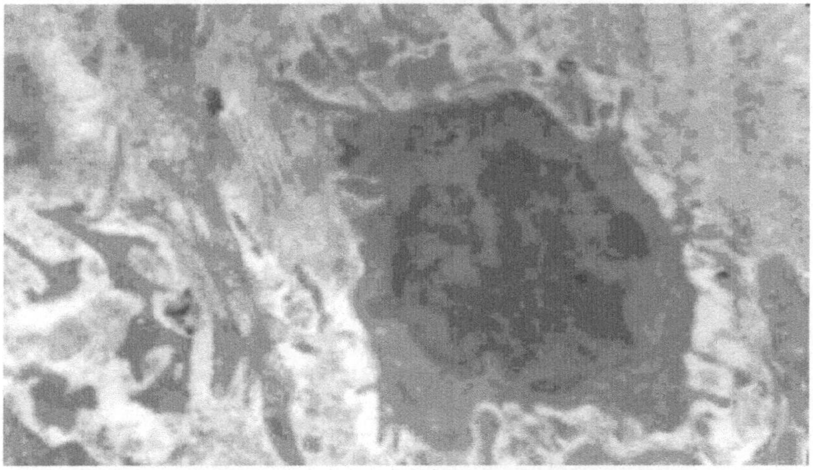

Figure 23.30 Lymphocyte and collagen bundles in chewers (10,000×)

Involvement of Muscle in SMF

Submucous fibrosis is a premalignant condition for which the major etiological factor has been identified as tobacco chewing. It is a chronic, slow progressive disease leading to severe restriction of movement of mouth including the tongue. SMF involves the mucosal epithelium, subepithelial connective tissue and the deeper muscle fibres. The ultrastructure of muscle fibres at different stages of restriction of mouth opening has been studied, where the following characteristic features have been noted (Figures 23.31 to 23.36).

With no restriction of mouth opening, majority of the fibres appear normal. Though some hypercontracted and stretched fibres are noted, the organelles appear normal. When there is restriction of mouth opening, severe degenerative changes in high proportion of fibres are seen. Splitting of muscle fibres and Z-line abnormalities are some of the constant features observed. The lost fibres are replaced by homogeneous material with different degrees of electron density and also by autophagic vacuoles.

Common irregularity of surface and foldings of sarcolemma, an indicator of atrophy is often seen. The plasma membrane disappears but the fibres are maintained by intact basement membrane which is also replicated.

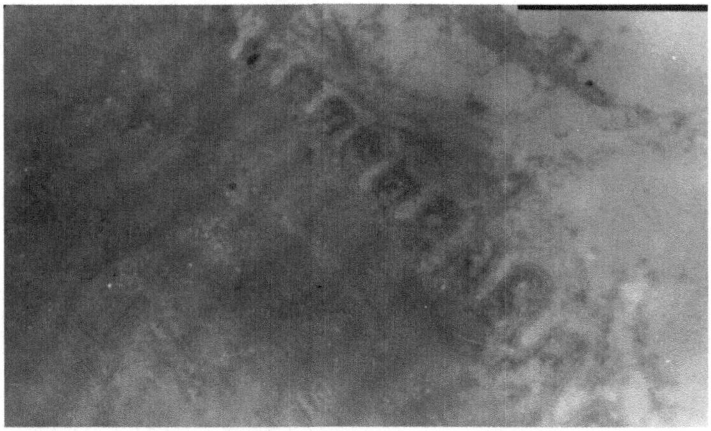

Figure 23.31 Exaggerated villi-like projections into extracellular space exhibited by the sarcolemma (10,000×)

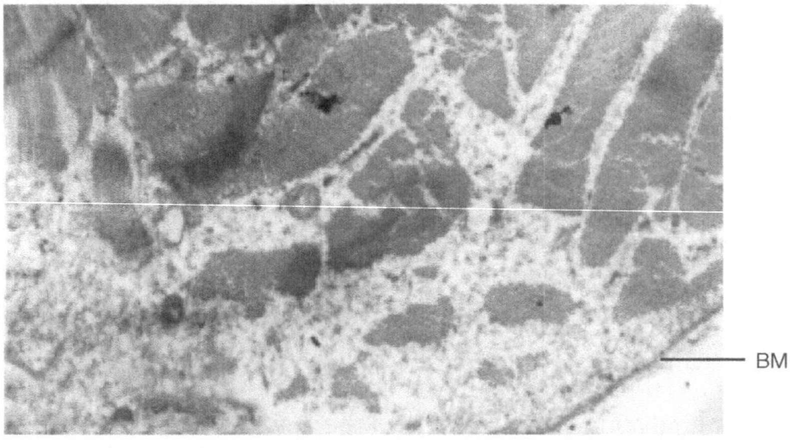

Figure 23.32 Loss of plasma membrane and fibres maintained by intact basement membrane (BM) (10,000×)

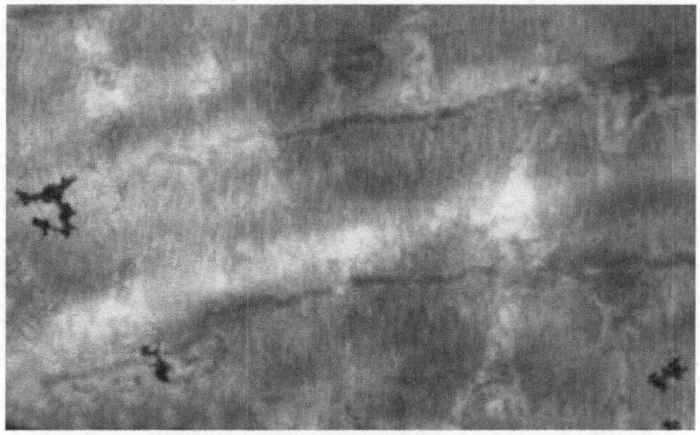

Figure 23.33 Splitting of bundles (15,000×)

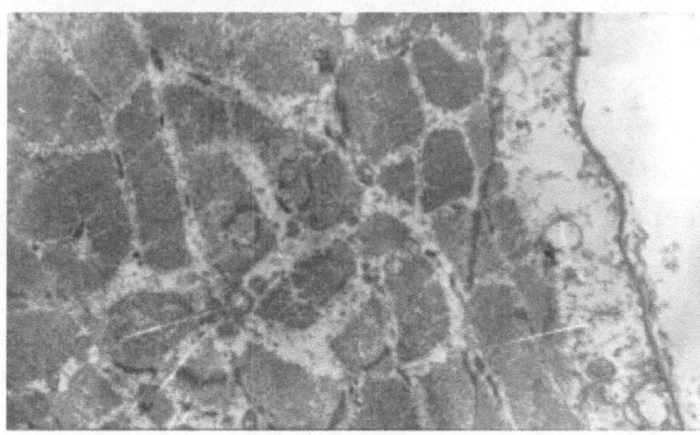

Figure 23.34 Z-line abnormalities (8000×)

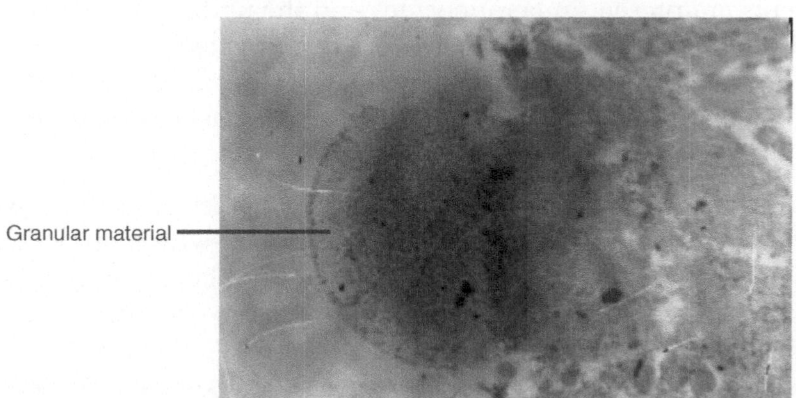

Figure 23.35 Loss of myofilaments replaced by granular material (15,000×)

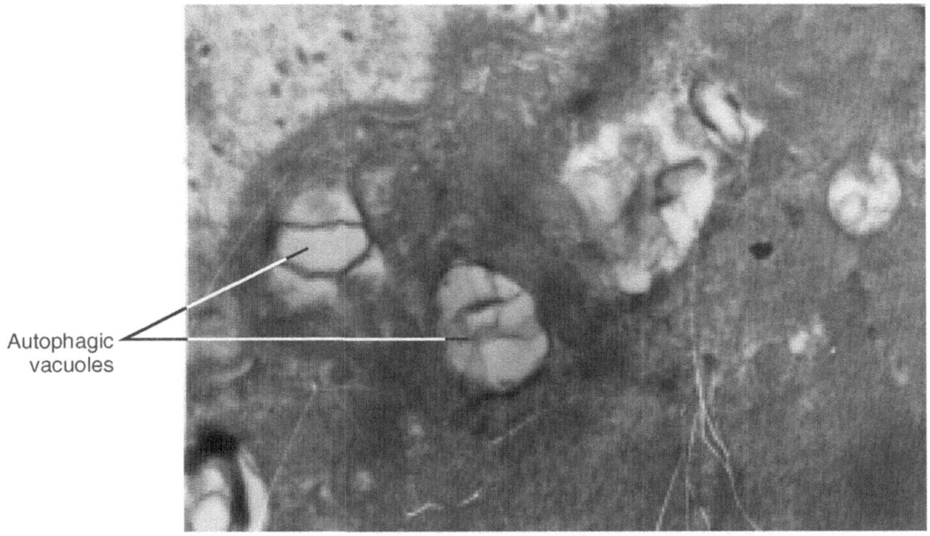

Figure 23.36 Lost filaments replaced by autophagic vacuoles (25,000×)

Thus the results indicate that restriction of mouth opening may depend not only on fibrosis but also on deeper muscle degeneration.

Oral Pigmentation in Smokers and Chewers

Increased pigmentation known as smoker's melanosis is a reversible condition. The melanin is shown to have a protective role. But in advanced stages, depigmentation occurs which leads to loss of protection for the oral mucosa.

Smokers as expected, show increased pigmentation in both labial and buccal mucosa. Melanin pigments are present in keratinocytes and the underlying connective tissue also with increased deposition in basal layer of mucosa. Activated melanocytes, characterized by highly elaborate dendrites, abundant melanosomes and complex melanosomes, dispersed dense aggregates of melanosomes along with melanosomes at different stages of maturation could be made out.

Chewers show keratinocytes, lymphocytes and melanocytes with a few melanosomes in the keratinocytes. Melanocytes are less in number which show well-melanized melanosomes as aggregates. Large intracellular multilocular vesicles, some showing dispersed dense particles, probably, chew particles are also observed. A few keratinocytes in non-keratinized normal epithelium is shown in Figure 23.37. The various ultrastructural features as described with regard to pigmentation in smokers and chewers are depicted in Figures 23.38 to 23.45.

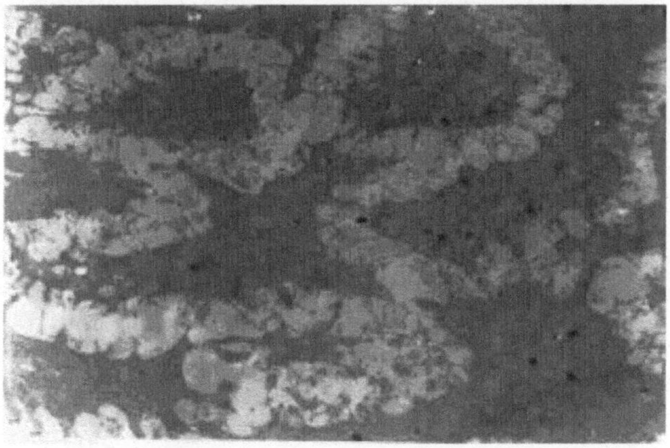

Figure 23.37 Keratinocytes in non-keratinized epithelium of normal control (4000×)

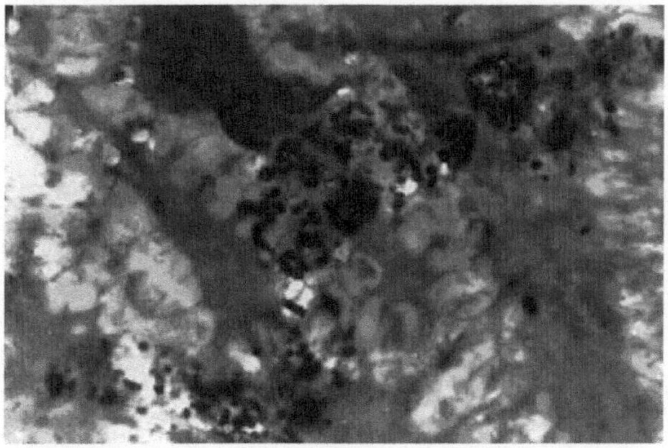

Figure 23.38 Keratinocytes in non-keratinized epithelium of smoker (6000×)

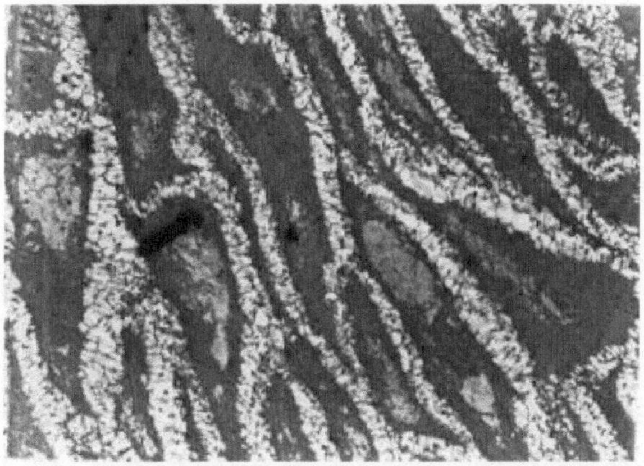

Figure 23.39 Keratinocytes in non-keratinized epithelium of tobacco chewer (4000×)

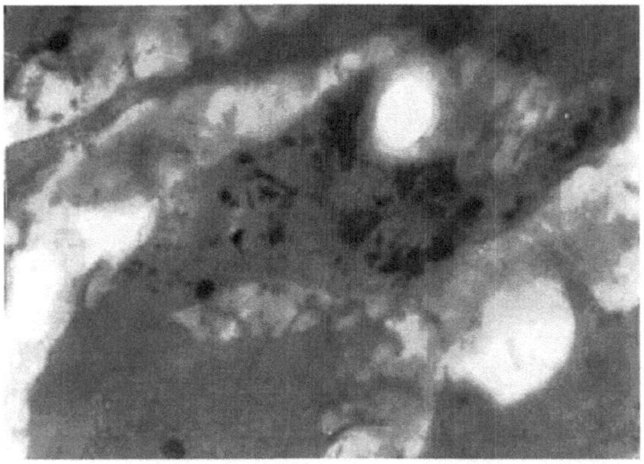

Figure 23.40 Melanocyte in labial mucosa of smoker (10,000×)

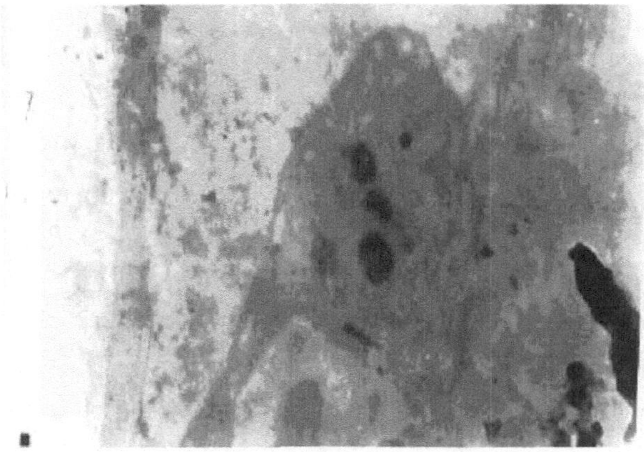

Figure 23.41 Melanocyte in buccal mucosa of chewer (20,000×)

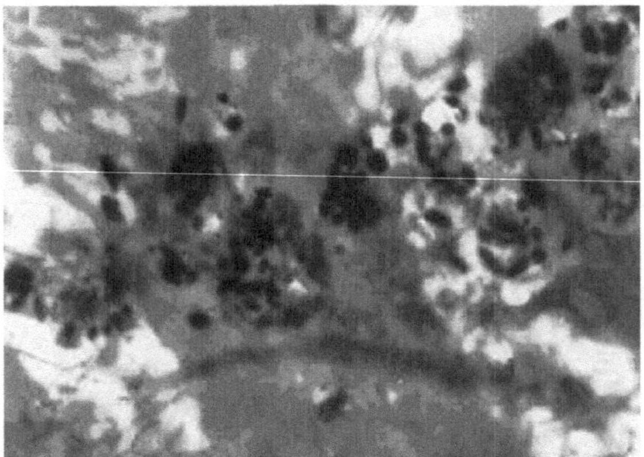

Figure 23.42 Melanosome within keratinocytes of smoker (10,000×)

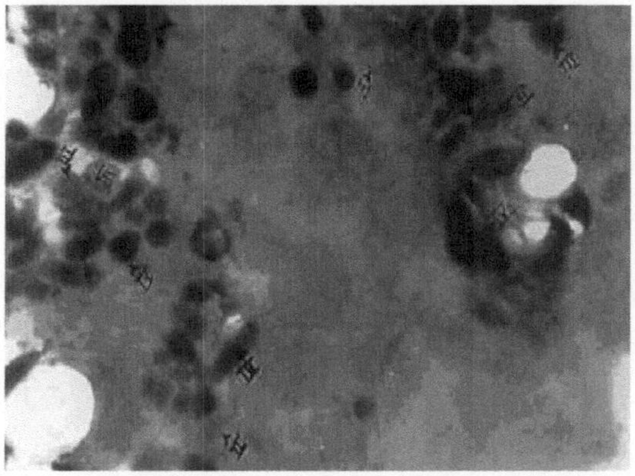

Figure 23.43 Melanosomes in different stages (20,000×)

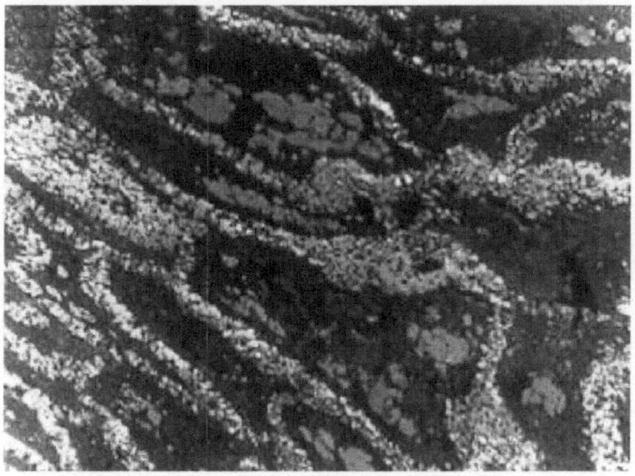

Figure 23.44 Keratinocytes of buccal mucosa of chewer showing vesicles (1000×)

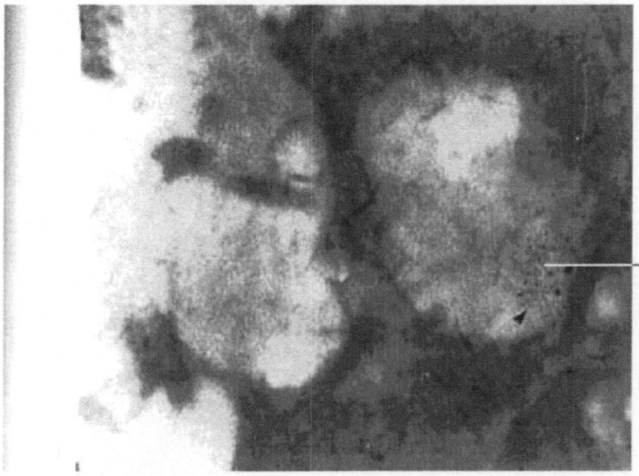

Figure 23.45 Intracellular vesicles showing the presence of electron-dense particles of size 0.025 to 0.075 (10,000×)

24

APPLICATIONS IN NANOSCIENCE

"Nano" is a prefix which means one billionth of a metre (10^{-9} m). "Nanotechnology" covers engineering, research and development on nanometre-sized substances. It creates the smallest functional unit (about 1 nm in size) by building groups of atoms.

Nanoscience is defined as an area of science devoted to the manipulation of atoms and molecules leading to the construction of structures on the nanometer scale. The size of the nanoparticles range from 1 nm to 100 nm though generally nanoscale indicates size less than 1 micron. The importance of size can be realized with the fact that the particles acquire unique properties depending on the size. For example, gold nanoparticles of different sizes produce different colours in transmitted light.

Size	**Colour**
300 nm	Red
100 nm	Bluish red
60 nm	Blue

Nanotechnology depends on the unique properties and behaviour of nanophase systems and the nanoparticles making up such systems often have properties that are significantly different from bulk materials. The behaviour of the system may be strongly influenced by particle size, shape and the interactions between particles. The configuration and evolution of the system will also be influenced by temperature, ambient atmosphere and associated gas–solid interactions.

Nanoparticles are unique moieties with properties entirely different from their bulk masses. Silver nanoparticles exhibit antimicrobial effect when the size is around 50 nm. The nanoparticles can be prepared from the corresponding salt solutions using chemical or biological methods.

Usually metal nanoparticles are chemically synthesized by reduction of the ions by a reducing agent. In the biological method, the reducing agent is an organism or the biological molecule responsible for the reduction. For example, it has been reported that the enzyme α-amylase is involved in the preparation of gold nanoparticles and the enzyme nitrate reductase in the preparation of silver nanoparticles. The organisms possessing the respective enzymes will produce the corresponding nanoparticles. The size and shape depend very much on the conditions used for their preparation.

Applications of nanoscience and nanotechnology are multiple and in the ever-increasing trend. In all fields including physics, chemistry, biology, medicine it has several applications in the form of nanoparticles, nanotubes, nanowires, nanorods, etc.

Silver nanoparticles have been shown to have anti-platelet activity preventing their adhesion to form internal clots in the bloodstream. The silver and gold nanoparticles have medicinal properties and are used in the traditional medicinal preparations as in Siddha, Ayurvedha, etc. They have been shown to have wound-healing properties also. The research on their potential for use as vehicles for targeted drug delivery is progressing in many laboratories.

Since the properties of the nanosystems are highly influenced by their environments, it is essential to characterize them after preparation before proceeding with the functional research. One important tool in characterization of the nanosystems is the electron microscope, because the particles can be visualized and size, shape, uniformity determined with the help of EM, both SEM and TEM.

The TEM can be used as a **nanobalance** when the size of the particle is in nanometre range and the weight in femto gram range. This is achievable by use of multilayered nanotubes of 5–50 nm diameter and 20 micron length. With application of external voltage to the tube, the tube bends. With variation of voltage, the tube vibrates at a particular frequency. A mass attached to the tube would change this frequency and generate information about its weight.

Though dynamic light scattering (DLS) method is used to measure the size of these particles, they can be easily characterized using modern microscopy and microanalysis tools. The most important ones are:

1. **Confocal scanning optical microscopy** makes a great contribution to nanotechnology with rapid, non-destructive three-dimensional nanoscale localization of the sample area generating a given detected photon and the analysis (spectroscopy) of that photon. This localization property of the confocal allows real-time spectroscopy of individual nanoscale objects. Thus, confocal plays an important role in the investigation of structure and dynamics of systems relevant to nanotechnology, complementing the other techniques employed.

The main contribution of confocal microscopy has not been to image nanoscale objects with light, but rather to use the capability to analyse the photons that have interacted with nanoscale structures, looking at their energies, temporal distribution, polarization, etc. Therefore it gathers information which cannot be obtained with other techniques. It characterizes, three-dimensional systems such as **nanoemulsions, nanocapsules**; two-dimensional systems like **nanopores, nanoholes, nanomembranes,** etc., and one-dimensional systems like **nanotubes, nanowires,** etc.

2. **XCT** (X-ray computed tomography system) enables the non-invasive, non-destructive visualization of buried, internal structures at the nanoscale. It provides the ability to image thick sections of low Z materials without the addition of contrast-enhancing agents (stains). The use of X-ray phase optimized sample evaluation and multi-keV X-rays provides deep penetration, as well as high contrast and resolution down to 30 nanometres. With little or no sample preparation, the nano-XCT system provides advanced imaging capability to see inside the sample to view critical internal structures in a 2D/3D CT format.
3. **TEM** allows direct observation of substances at the nanometre level for basic research.
4. **SEM** allows observation of surfaces at the nanometre level.
5. **SPM (Scanning probe microscope)** allows direct observation of atoms, enables to directly observe and investigate extremely small objects that cannot be seen with the naked eye.

Other techniques which can be employed in nanoscience are:

6. Scanning transmission electron microscopy (STEM)
7. Focused ion beam microscopy (FIB)
8. Scanning near-field optical microscopy
9. Fluorescence microscopy
10. Raman microscopy
11. Scanning tunnelling microscopy (STM)
12. Atomic force microscopy (AFM)
13. Scanning thermal and thermoelectric microscopy
14. Imaging secondary ion mass spectrometry (SIMS)
15. Atom probe tomography
16. Electron beam lithography
17. High resolution scanning electron microscopy

18. High spatial resolution quantitative electron beam microanalysis
19. High resolution TEM (HRTEM)
20. ***In situ* electron microscopy** *In situ* microscopy refers to the techniques that allow direct observation of the dynamic properties at the nanoscale through imaging and diffraction.

Nanoparticles prepared elsewhere both by biological and chemical methods which have been dropped on coated grids screened and photographed in our EM are given below as examples in Figures 24.1 to 24.20.

The metallic preparations like silver, gold, etc. can be directly dropped on carbon-coated grids, air-dried and screened. Non-metallic preparations like PHA, etc. have to be stained (negative staining) with phosphotungstate suitably to increase contrast.

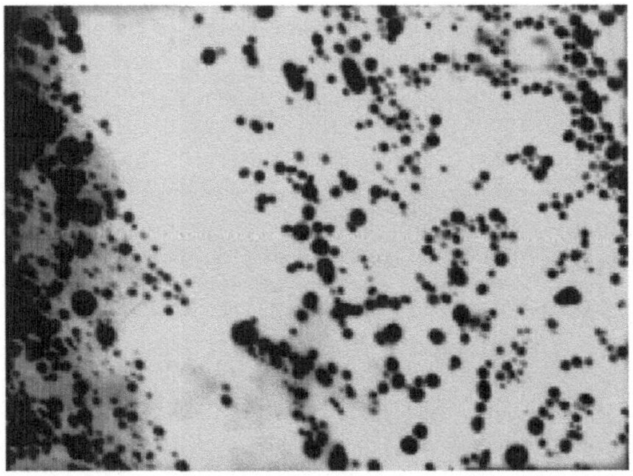

Figure 24.1 Silver nanoparticles prepared with plant extract (80,000×)

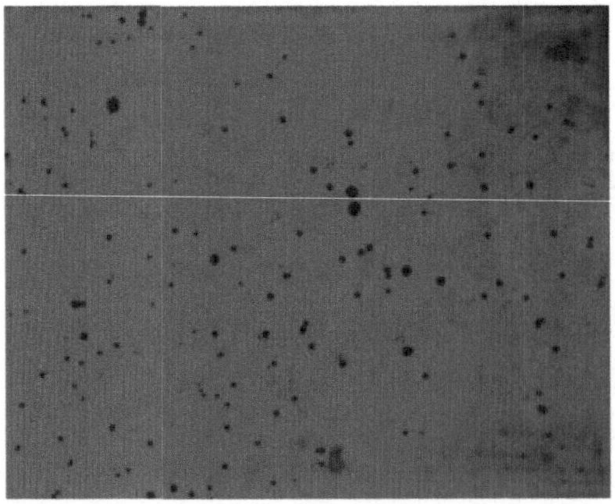

Figure 24.2 Silver nanoparticles prepared with starch (80,000×)

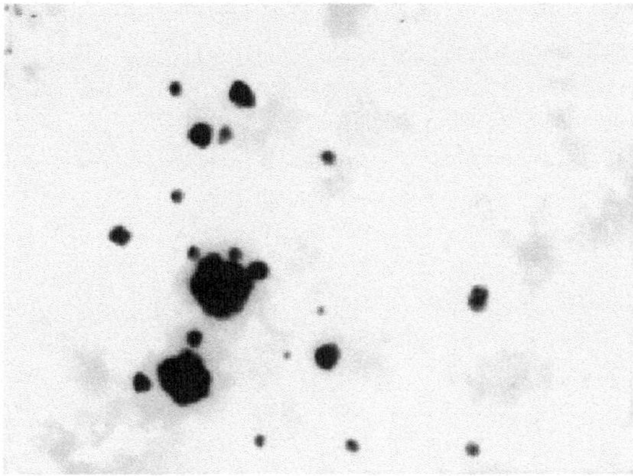

Figure 24.3 Silver nanoparticles on clay (1,00,000×)

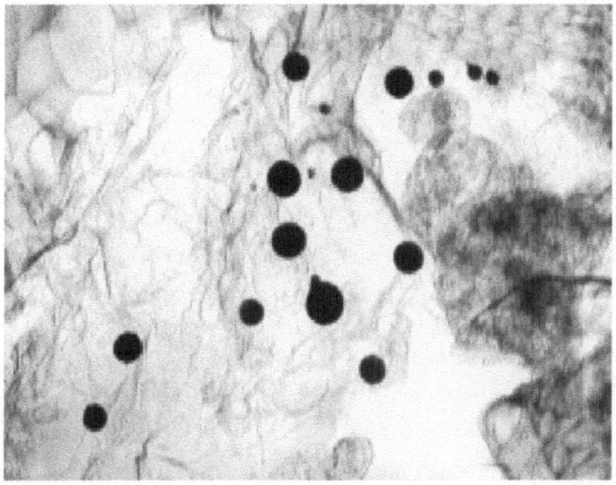

Figure 24.4 Silver nanoparticles stabilized with guargum (50,000×)

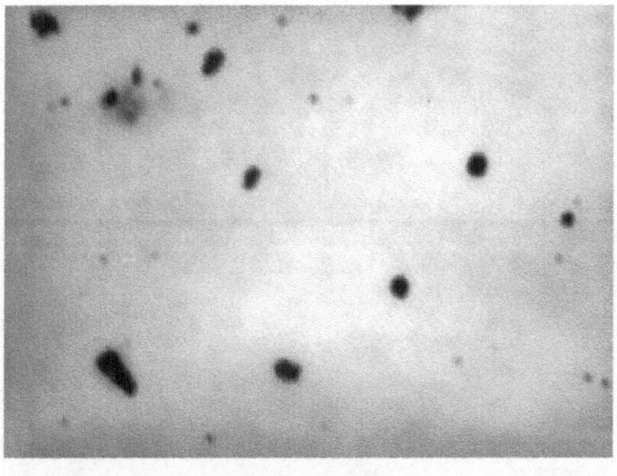

Figure 24.5 Silver nanoparticles stabilized with gum-arabic (80,000×)

Figure 24.6 Silver nanoparticles stabilized with neem (80,000×)

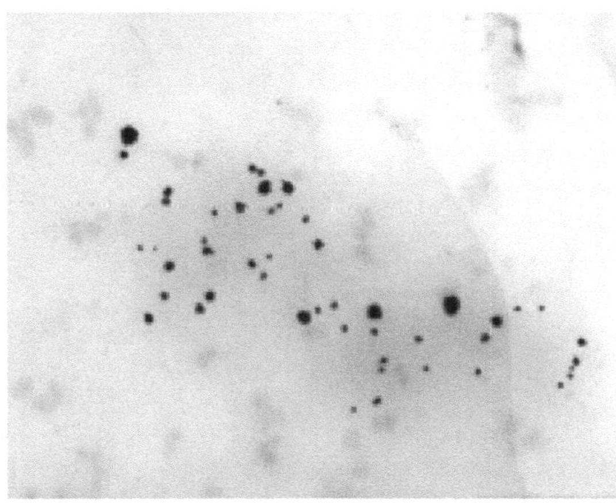

Figure 24.7 Gold nanoparticles prepared with *Bacillus licheniformis* (40,000×)

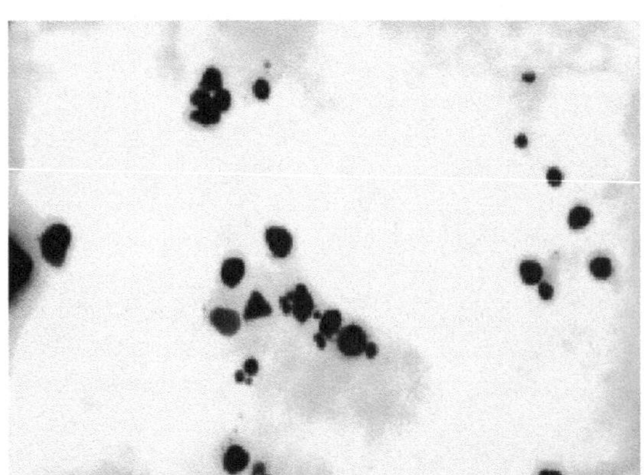

Figure 24.8 Triangular and spherical gold nanoparticles prepared with *Brevibacterium* (1,00,000×)

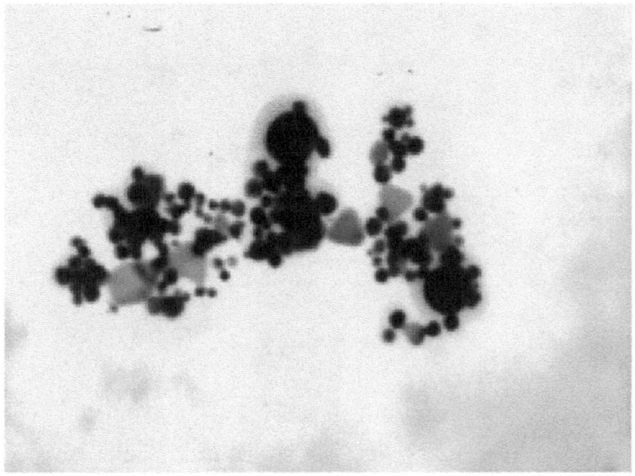

Figure 24.9 Gold nanoparticles prepared with *Hypnea* sp. (spherical and triangular particles) (1,00,000×)

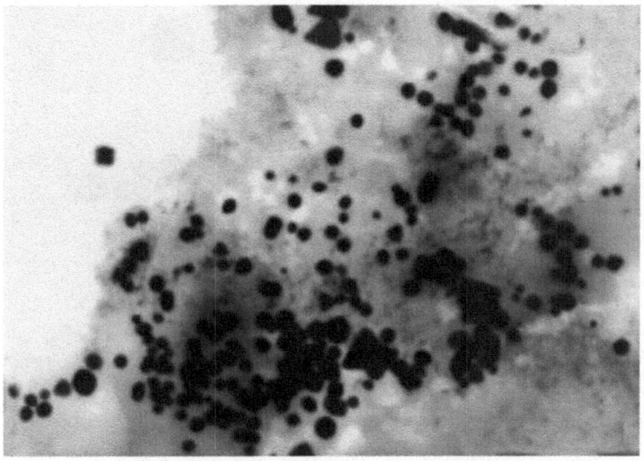

Figure 24.10 Gold nanoparticles prepared with Kappa picus (spherical and triangular particles) (80,000×)

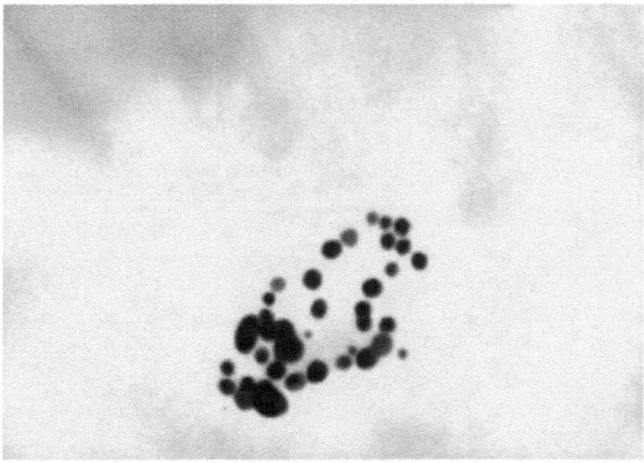

Figure 24.11 Gold nanoparticles prepared with macroalgae (1,50,000×)

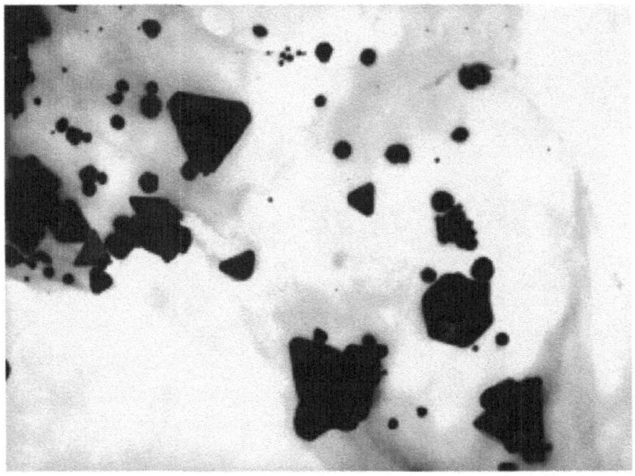

Figure 24.12 Hexagonal and triangular gold nanoparticles (30,000×)

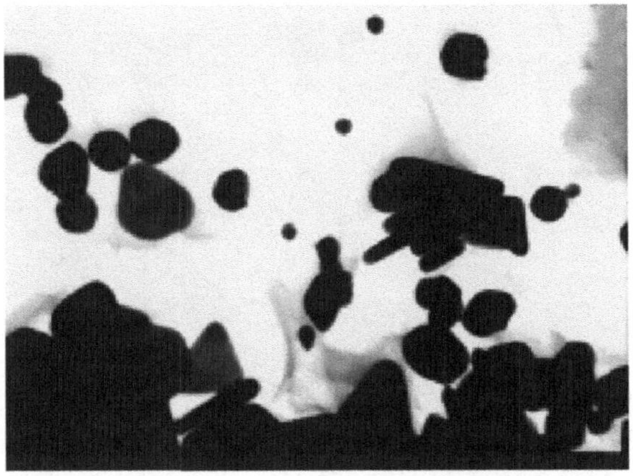

Figure 24.13 Gold nanoparticles—rods and hexagons (80,000×)

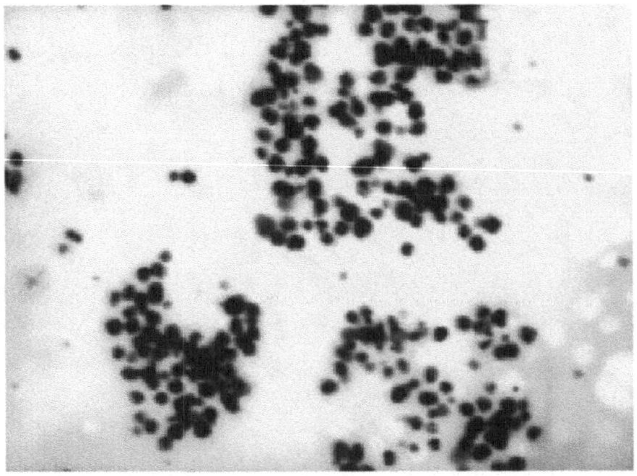

Figure 24.14 Ag—Au bimetal nanoparticles (80,000×)

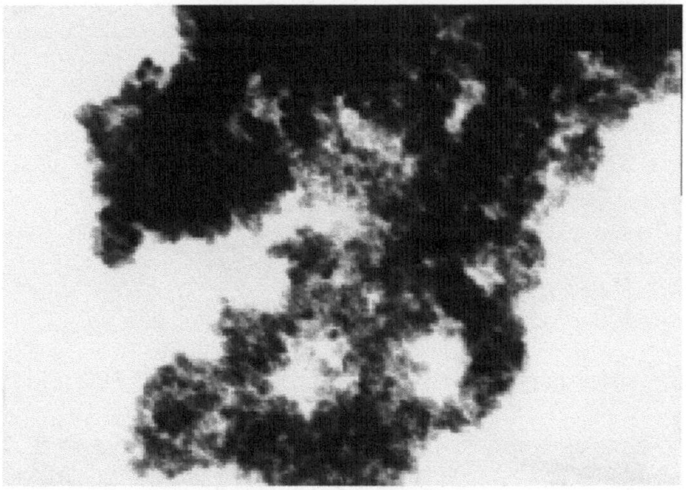

Figure 24.15 Fe_3O_4 + Au particles prepared chemically (1,00,000×)

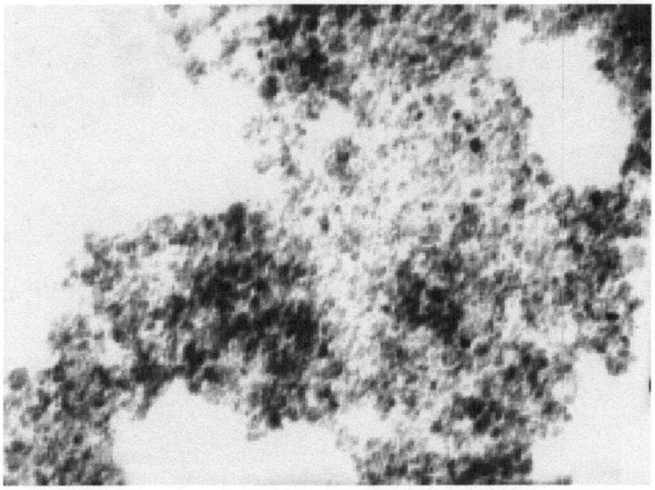

Figure 24.16 Fe_3O_4 nanoparticles prepared chemically (1,00,000×)

The formation of nanoparticles is highly influenced by their environmental conditions. Silver nanoparticles prepared from silver nitrate ($AgNO_3$) using polyethylene glycol (PEG) and polyvinyl pyrrolidine (PVP) are shown in Figure 24.17. After addition of potassium chloride (KCl) to the above mixture consisting of $AgNO_3$, PEG and PVP, the change in size and shape of the particles can be seen in Figure 24.18.

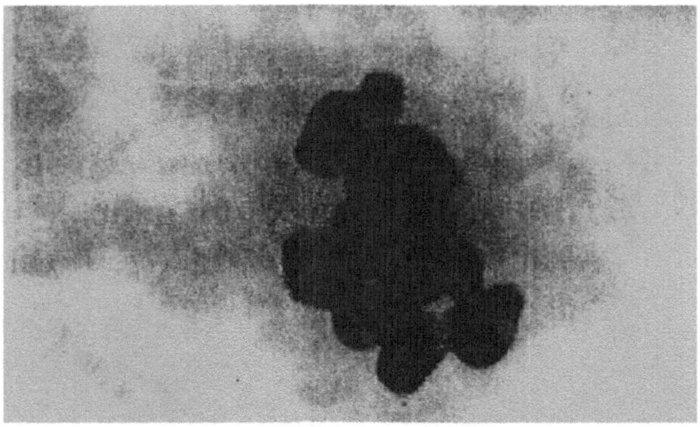

Figure 24.17 Silver nanoparticles prepared using $AgNO_3$, PEG and PVP (50,000×)

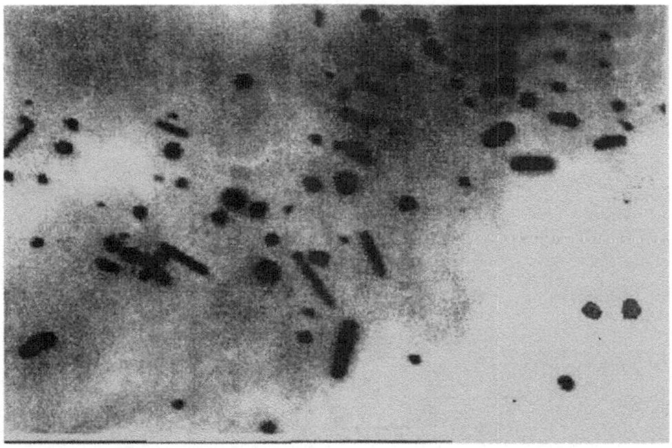

Figure 24.18 Silver nanoparticles prepared using $AgNO_3$, PEG, PVP and KCl (80,000×)

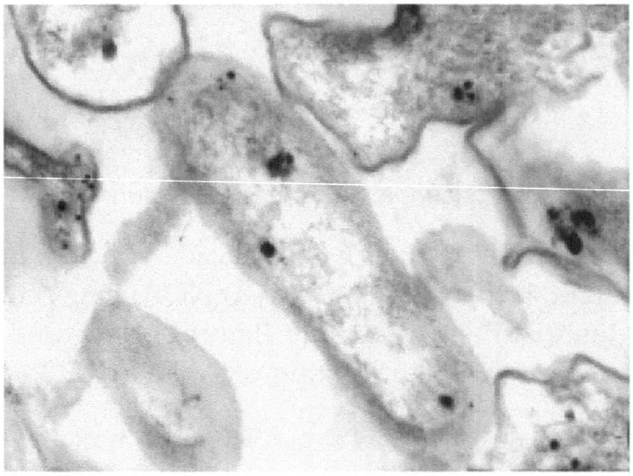

Figure 24.19 Silver nanoparticles penetrating inside a *Bacillus* (50,000×)

Cultured endothelial cells are treated with silver nanoparticles at specific concentration for a fixed duration to verify the penetration of the particles into the cells. The cells treated thus, are pelletized by centrifugation and the cell pellet is processed further for routine transmission electron microscopy. The ultrathin sections after double staining with lead and uranium are screened in the EM. Figure 24.20 shows the nanoparticles entering the cells probably by endocytosis. The accumulated particles in the cytoplasmic vesicles in one of the cells are seen in Figure 24.21.

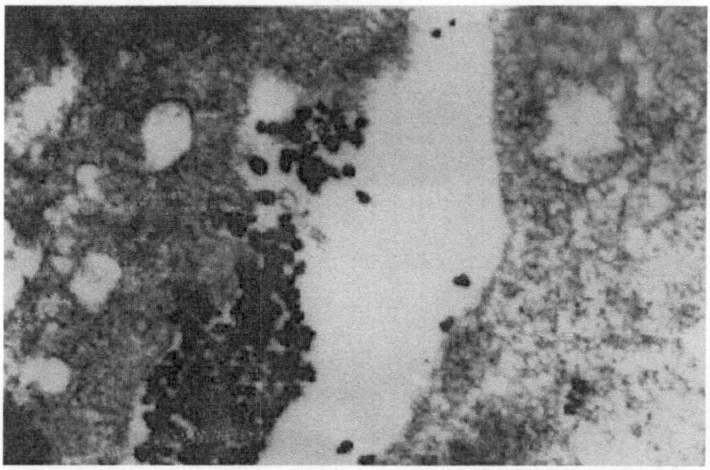

Figure 24.20 Cultured endothelial cells treated with Ag nanoparticles (40,000×)

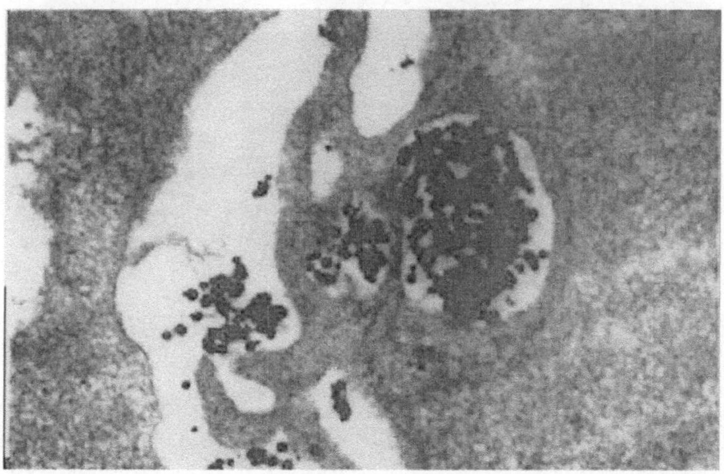

Figure 24.21 Accumulated Ag nanoparticles in cultured endothelial cells (40,000×)

Similarly, retinal epithelial cells treated with gold nanoparticles are shown in Figures 24.22 and 24.23.

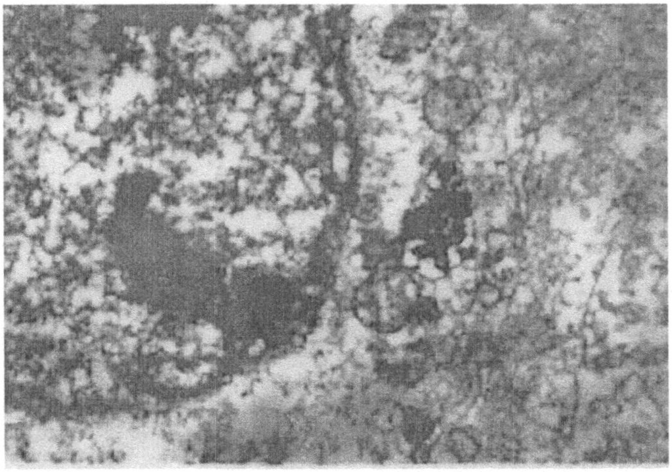

Figure 24.22 Retinal epithelial cells treated with Au nanoparticles. A part of the nucleus is also seen (50,000×)

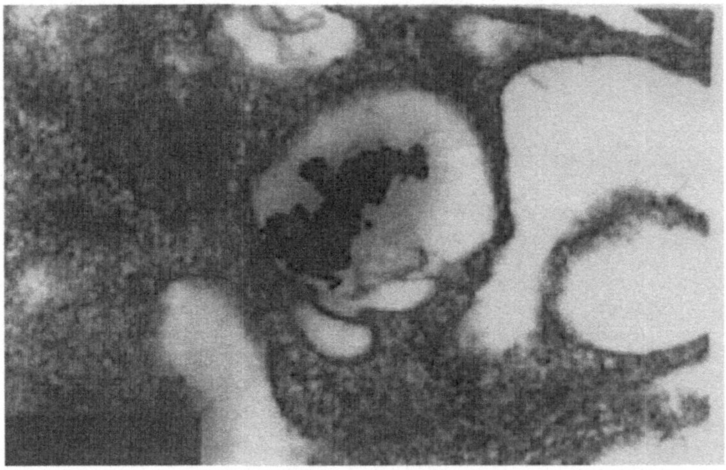

Figure 24.23 Retinal epithelial cells with accumulated Au nanoparticles in cytoplasmic vesicles (40,000×)

Polyhydroxy alkanoate (PHA) is a polymer synthesized by the organism *Bacillus* sp. It is accumulated by the organism during availability of excess carbon source in the environment. (An organism with accumulated PHA is shown in Figure 22.21 in Chapter 22. It is extracted using organic solvents and converted to nanoparticles using solvent displacement technique. Similarly, polyhydroxy butyrate is also synthesized and accumulated by the organism. Figure 24.24 shows the isolated PHA particles and Figure 24.25 the PHB particles.

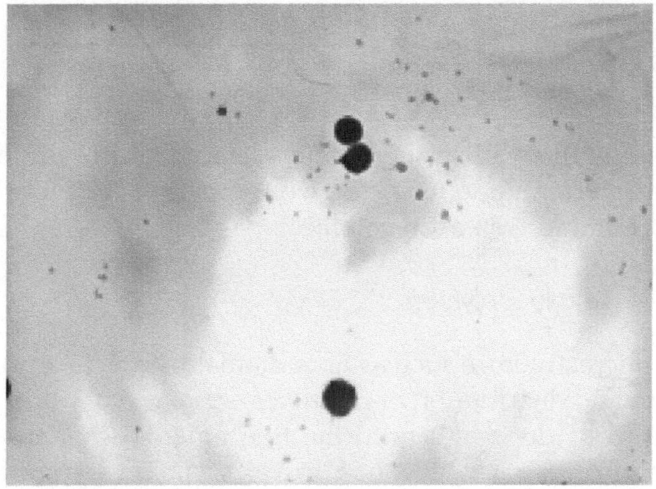

Figure 24.24 PHA (polyhydroxy alkanoate) particles stained with PTA (20,000×)

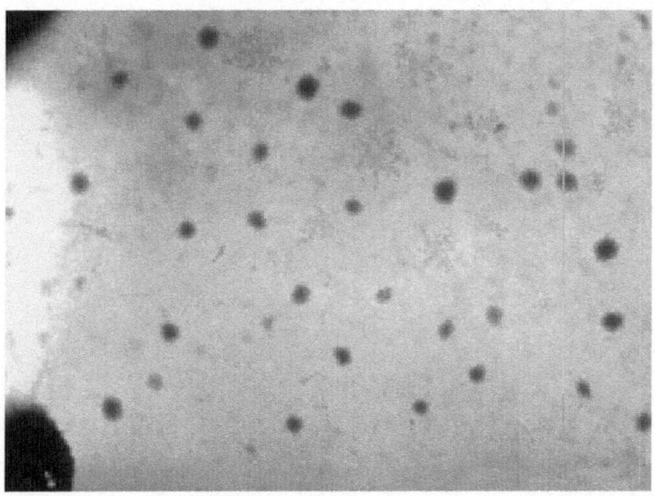

Figure 24.25 PHB (polyhydroxy butyrate) particles stained with PTA (15,000×)

APPLICATIONS OF EM IN OTHER FIELDS

Applications in Earth Sciences

The most common use of electron microscopy in palaentology is the reproduction of the morphological details of microfossils. Some microfossils have small globular forms with fine ornamentation and wall structure which can be visualized only at high magnifications offered by EM. Many of these morphological characteristics help in identifying and classifying the microfossils. The study of spores, pollen and other microscopic organisms like phytoplanktons, diatoms, and nano-fossils is possible using the SEM. Determination of their size, shape, aperture character and surface ornamentation using EM helps in their identification.

Use of SEM to study the wall structure and surface features of very small organisms such as primitive bacteria, algae and fungi have helped in identifying their spatial relationship as individual as well as in colonies.

Application of EM in Textile Polymers

Characterization of microstructure such as fibre diameter, structure, molecular orientation of polymer fibre, size and distribution of additives, structure of fibre assemblages, failure mechanism, etc. can provide insight into the fundamental structures and relationship between structure and properties. SEM has proved to be a very important technique for the study of textile fibre. Fibre finish and cross section, unevenness and lack of uniformity of fibre coatings, etc. can be directly seen in the SEM. The mechanism of fibre failure can be determined by fractography studies in the SEM.

Application in pharmaceutical industry, forensic and agriculture fields are very many and very essential. With advancements in technology and instrumentation, the applications in varied fields are still in the progressive trend.

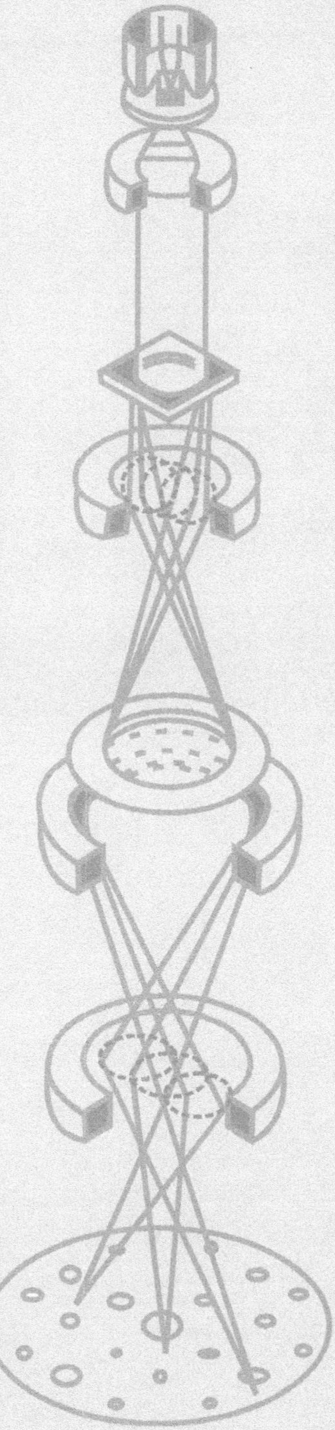

PART IV

RECENT ADVANCES

25

AN INTRODUCTION TO MODERN MICROSCOPES

Evergreen technological innovations have paved way for ever-increasing advancements in microscopic instrumentation techniques. Some of the most important ones are listed in this chapter.

STM SCANNING TUNNELLING MICROSCOPY

STM was developed by Binnig and Rohrer in 1982 to image the topography of surfaces, in which electron tunnelling is used as the mechanism to probe a surface. In 1981, Binnig, Rohrer, Gerber and Weibel successfully combined vacuum tunnelling with scanning capability and developed the first STM in the world.

STM works on the principle that electron tunnelling occurs when a voltage is applied across two conductors brought close enough to establish a tunnelling current. The tunnelling current varies exponentially with the height distance. Once the tunnelling phenomenon is established, the instrument can be operated in one of the following modes:

- constant current mode
- constant height mode
- spectroscopy mode

In the constant current mode, the tunnelling current is maintained at a constant value. As the tip scans the surface laterally, the variation in gap width due to surface undulations of the sample would cause a variation in the tunnelling current. This variation is sensed and a feedback loop repositions the tip vertically to maintain a constant gap between the tip and the sample. The probe tip therefore profiles the surface or follows the undulations of the specimen surface and an image of the surface can be easily produced as a simple function of the corrections applied to the tip. This mode is used mostly to topograph imaging of rough surfaces since the distance between the tip and sample is adjusted by the feedback circuit.

The tip is maintained at a constant height in the constant height mode. The resulting variation in gap between the tip and the surface causes a variation in tunnelling current which is recorded and used to image the surface. Here, no feedback loop is required and faster scan speeds are possible. In the spectroscopy mode, the variation of tunnelling current as a function of height and voltage at several discrete points on the specimen surface can be carried out.

As a microscope, STM can provide very-high-resolution images in real space. These images can be used to investigate surface structures, and also surface or even subsurface atomic dynamic processes. In nanoscience, the, STM has been widely used to study the structure and electronic properties of nanotubes and nanowires.

To achieve the atomic resolution, there are many requirements in STM design and instrumentation. Recent developments in STM include the following.

- Spin-polarized STM (SPSTM)
- Ultra-low temperature STM (ULT-STM)
- Dual-tip STM
- Variable temperature fast-scanning STM, etc.

SCANNING PROBE MICROSCOPY

As a sequel to the invention of the STMs, the SPMs were developed, which attracted the biologists for many reasons. The SPMs offer resolutions between 1 and 100 nm, particularly for imaging uncoated samples. These microscopes permit real space, three-dimensional imaging at atomic resolution in air, vacuum or liquid medium. The samples are generally exposed to very low electric fields of a few millivolts. These can be used as local probes to study various physical properties at the nanometre level.

All SPMs use probes which detect the specimen surfaces only at very close proximity. The interaction or the property sensed is strongly dependent on the height between the probe and the specimen surface. This height sensitivity distinguishes SPMs from other microscopes and allows imaging sample topography.

ATOMIC FORCE MICROSCOPY (AFM)

Both STM and AFM have revolutionized surface analysis by providing high-resolution visualization of structures at the atomic and nanometre scales. Both can be used to examine samples not only in an ultrahigh vacuum but also at ambient conditions and even in liquids. In both methods, the localized interaction between a sharp probe and a sample is employed for surface imaging. STM is based on detection of tunnelling current between a sharp metallic tip and conducting surface and it is applied mostly to studies of atomic structures and atomic-scale processes on different conducting and semiconducting samples, primarily in UHV (ultra high vacuum) conditions. But AFM can be applied for characterization of materials of any kind because this method is based on detection of more universal tip–sample mechanical forces.

AFM was invented in 1986 by Binnig, Quate and Gerber to overcome the limitation of STM which can image only conducting surfaces. In the AFM neither the tip nor the sample needs to be conducting. The AFM probes the surface of a sample with a sharp tip, a few micrometres long and often less than 10 nm in diameter at its apex. The tip is located at the free end of a cantilever that is up to a few hundred μm long. Forces between the tip and the sample surface cause the cantilever to bend or deflect. An optical detector measures the cantilever deflection as the tip is scanned over the sample, or the sample is scanned under the tip. The measured cantilever deflections allow a computer to generate a map of surface topography.

The advantages of AFM in the biological field include the following:

* It represents a unique approach to detect structures, chemical interactions and forces in the nanoscale range.
* It can yield high-resolution spatially resolved information on properties exhibited by living tissues under near physiological conditions.
* Due to its ability to measure forces, it is also possible to use AFM to investigate the material properties of biological specimens including cells.

AFM applications include, high-resolution examination of surface topography, compositional mapping of heterogeneous samples and studies of local mechanical, electric, magnetic and thermal properties. These measurements can be performed on scales from hundreds of microns down to nanometers and thus find application in nanoscience and nanotechnology. AFM together with SEM is applied for examination of deep trenches and under-cut profiles with tens and hundreds of nanometres dimensions. AFM can be applied for imaging in various environments and at different temperatures.

AFM is capable of compositional imaging of heterogeneous polymer systems. Electric force microscopy is the most known AFM technique for mapping of conducting regions of various samples and is based on measurements of electric field gradients acting between a metal-coated probe and conducting sample regions. AFM provides the unique capability to visualize single macromolecules and their size, conformation, surface interactions and other phenomena can be studied by analysis of the image. The sample preparation is simple, the only challenge being fixation of individual molecules to the surface.

Examples of common applications in biology include:

- topographic analysis of cell membranes and intracellular structures
- measuring binding forces
- mapping cellular distribution of receptors interacting with cognate ligands in live cells
- nanomanipulation studies.

Combinations of AFM

Confocal and AFM AFM combined with confocal microscopy has been used for studies on:

- Single-cell mechano-transduction and modulation
- Membrane receptor–ligand interactions
- Cellular topography

SEM and AFM Similarly, AFM and SEM can be used as complementary techniques for surface investigations. Both techniques raster a probe across the surface to detect some interaction with the surface to form an image and both have a lateral resolution which is similar in scale. On thin films both produce a similar representation of the sample surface. But the three-dimensional nature of the AFM can be used to calculate changes in roughness and surface area variations due to differences in deposition parameters. Another area of difference between the two techniques is in interpreting subtle differences in height. In the SEM image, changes in slope can result in an increase in electron emission from the sample surface producing a higher intensity in the image. However, it can sometimes be difficult to determine whether the feature is sloping up or down. Since the AFM data contains the height information, a bump or a pit can be easily identified. Though AFM does not provide elemental analysis as SEM does, it can supply compositional information by differentiating materials based on their physical properties.

Another important difference between the two techniques is the environment in which they are performed. SEM is conducted in a vacuum environment and AFM is conducted in an ambient or fluid environment. The two techniques when used in conjunction can definitely provide more information about the specimens than when used singly.

CryoSEM and AFM Cryo-electron microscopy of vitrified specimens is the most useful high-resolution imaging technique for the investigation of cells, particles and biological macromolecules. CryoSEM is the preferred technique for structural studies on fungi because of their tough cell walls and high water content. CryoSEM and environmental SEM have been combined with atomic force microscopy (AFM) and force spectroscopy

(FS) for probing the ultrastructure, chemistry, physical characteristics and motion of fungal cells. CryoSEM offers high-resolution imaging for frozen fungal samples, whereas environmental SEM allows the analysis of robust samples such as spores under ambient conditions. Dual-beam FIB-SEM allows manipulation capabilities along with elemental detection. AFM can image live cells including growing fungal hyphae with similar lateral but better depth resolution compared with SEM. FS can analyse cell wall chemistry, elasticity and dynamic cell characteristics. The integration of AFM with optical microscopy will allow examination of individual molecules or cellular structures in the context of fungal cell architecture. Thus SEM and AFM are complementary techniques that are clarifying our understanding of fungal biology. These techniques can be extended to other systems in biology.

TEM and AFM Investigating systems with both AFM and TEM are gaining importance. In fact, both techniques have been used in conjunction to examine a number of different biological systems like collagen fibrils, proteins, protein crystals, bacteriophages, cells, cartilage, polysomes, etc. By developing a novel sample holder for carbon-coated indexed TEM grids, Lin and Goh have been able to image exactly the same area under both AFM and TEM.

SECONDARY ION MASS SPECTROMETRY (SIMS)

One area of research utilizing high-resolution-imaging SIMS involves the use of stable isotopic tracers to track biological activity within organisms. One study of melanin granules within mammalian hair shafts demonstrates the utility of high-resolution elemental SIMS imaging in these systems. Compositional information from sub-100 nm structures within the hair shaft has been readily obtained using SIMS.

The distribution of iodo-benzamide melanoma markers within the cell ultrastructure has been studied successfully via high-resolution imaging SIMS. Here, the iodo-benzamide marker is used as a diagnostic tool for the detection of cancerous cells. The iodine ion ($^{127}I^{-}$) was detected only in the melanosomes, and not in other structures within the cells. This confirms the suspected preferential uptake of this molecule as a marker for cancerous cells. This also demonstrates the potential utility of imaging SIMS in the development of therapeutic agents to attack specific cancerous structures.

High-resolution SIMS finds applications in nanotechnology where one needs to probe the composition of fine structures, at trace levels, with isotopic specificity and a spatial resolution on the order of 50–1000 nm.

FOCUSED ION BEAM SYSTEM

The two-beam FIB system incorporates both the FIB and SEM into a single machine. Images can be obtained from a secondary electron detector using either the SEM or the

FIB by simply toggling between the two. Each has its advantages depending on the specific needs. For example, using electrons to image, as in the SEM entails much less damage to the material while providing a much better resolution. On the other hand, ion beam imaging, found in the FIB, provides better sensitivity to details because of the depth of penetration from channelling.

In both the FIB and SEM machines, a source emits charged particles which are then focused into a beam and scanned across small areas of the sample using deflection plates or scan coils. In the SEM, the focusing is accomplished using magnetic lenses. Ions, on the other hand, are much heavier and therefore slower than electrons. As a result, magnetic lenses are less effective on ions, so the FIB system is equipped with electrostatic lenses which prove much more effective in focusing the ions into a beam. Both instruments are used for high-resolution imaging by collecting the secondary electrons produced by the beam's interactions with the sample surface. Contrast is formed by the fact that raised areas of the sample (hills) produce more collectable secondary electrons than depressed areas (valleys). For the FIB, imaging resolution below 10 nm is possible. This image will show both topographic information and materials contrast and often provides information complementary to that obtained from a SEM image.

HIGH-RESOLUTION SCANNING ELECTRON MICROSCOPY

SEM is the most widely used of all electron beam instruments. This popularity can be attributed to many factors, such as,

- versatility of its various modes of imaging
- excellent spatial resolution now achievable
- very modest requirement on sample preparation and condition
- relatively straightforward interpretation of the acquired images
- accessibility of associated spectroscopy and diffraction techniques

With recent generation of SEM instruments, a magnification range from 5 × to >1,000,000× can be obtained, i.e., up to nanometre dimensions. Image resolution of about 0.5 nm can also be achieved with the FEG-SEM (Field-emission-gun SEM) nearing that obtained with TEM.

FEG-SEM

The introduction of the use of a field-emission gun (FEG) in a scanning electron microscope made it possible to image individual heavy atoms in the transmission mode by collecting scattered electrons with an annular detector.

The development of low-voltage SEM (LV-SEM) clearly broadened the application of the SEM techniques to non-conducting or delicate specimens.

Nanometre-scale resolution is achievable in the modern FEG-SEM. The significant reduction in the electron–specimen interaction volume clearly makes low-voltage SEM more surface-sensitive; and the charge balance at the specimen surface makes it possible to directly observe non-conducting materials without a conductive coating layer. With further improvement of low-voltage FEG-SEM optical systems, the ultra-low-voltage SEM (ULV-SEM) technique is a powerful tool for studying nanoscale systems.

Almost any sample can be examined in a modern, variable pressure or environmental FEG-SEM—powders, films, pellets, conducting, semiconducting, non-conducting. The modern high-resolution FEG-SEM and associated techniques are suitable for applications to both material sciences and life sciences. Apart from its applications in chemical, metallurgical and semiconductor industries, it is very useful in food industry for particle and morphological applications, in the consumer health industry for nanoparticle applications and in the pharmaceutical industry for drug and drug delivery applications.

HR FEG-SEM gives true three-dimensional images of various samples to provide not only the lateral but also the depth information on a nanometre scale. It integrates the traditional SEM techniques to those of the STEM techniques in which both bulk and thin samples can be examined. These instruments have the highest resolutions (1 nm) and probe currents (200 nA) and with an in-column energy filter, are designed for optimal X-ray microanalysis and ultra-low-voltage imaging of uncoated, beam-sensitive specimens.

TABLETOP EMs

Tabletop Mini SEMs

These are ideal for research laboratories who have neither the space nor an adequate budget for SEM and also for schools and colleges as an entry level SEM. These mini-SEMs are as simple to operate as a digital camera yet feature sub-10-nm resolution and have been designed to bridge the gap between light microscopy and conventional SEM. Among these instruments, those having a solid-state field emission gun permit higher brightness, longer life and better performance at low voltages, but do require a superior vacuum system.

Low Voltage Electron Microscope (Desk-top TEM)

The first idea of increasing contrast by decreasing the energy of electrons was proposed by Nixon in 1958. Subsequently, the first low-voltage TEM was built in 1960 by A.P. Wilska and G. Mollenstedt. This was unsuccessful because of the following reasons:

- low brightness of the electron gun at low voltage
- low efficiency of fluorescent screen
- higher contamination due to bad vacuum
- electrical charges on contaminated surfaces

The modern version of low-voltage TEM is LVEM5 marketed by Delong America in 2005 which is the world's only benchtop TEM with optional SEM and STEM. It is basically a light microscope with EM attachments. It is a compact, truly desktop design which can accommodate up to four imaging modes—TEM, STEM, SEM and EDS—and allows for effortless transitioning between modes. Because of the low accelerating voltage (5 kV) the contrast is high and therefore there is no need for staining or shadowing the samples as required for the conventional TEM. Also no cooling water and no compressed air are required. But because the energy of the electrons is low, the specimen thickness has to be below 50 nm which may be difficult to achieve in some samples. In the conventional TEM, since the voltage can be altered over a range, specimen thickness can also be altered within a range.

HIGH-RESOLUTION TEM (HRTEM)

It is now possible to attain resolving powers at or close to the one Angstrom (0.1 nm) level almost on a routine basis. Using correct operating conditions and well-prepared samples, high-resolution image characteristics are interpretable directly in terms of projections of individual atomic-column positions. It is a powerful and indispensable tool for characterizing nanostructured materials.

The elastically scattered electrons contribute to formation of the high-resolution bright field image. The inelastically scattered electrons can be used for information about the sample composition using the technique of electron-energy-loss spectroscopy (EELS). Electrons scattered at very large angles are utilized for Z-contrast annular-dark-field imaging in the scanning transmission electron microscope (STEM).

The transmission electron microscope with 0.5 Å resolution was first designed in 2008. This breakthrough, which was accomplished by the use of beam monochromators and probe and imaging aberration correctors, now allows electron microscopists to directly image individual nanostructures and perform atomic resolution tomography and single atom spectroscopy to investigate local electronic structure and bonding and the dynamics and mechanisms of reactions *in situ*. All new high-resolution TEMs are now offered with the option of monochromated and/or aberration-corrected electron optics. The new environmental TEM has the ability to image the sample in a controlled gaseous environment to study processes such as nanotube growth, crystal nucleation and heterogeneous catalysis.

HIGH-VOLTAGE ELECTRON MICROSCOPY

For analysing stereological ultrastructure in cells and tissues, high-voltage electron microscopy, operated at accelerating voltages of 400–1000 kV on thick biological specimens either from whole mount cultured cells or thick sections of embedded tissues, can be used.

SCANNING TRANSMISSION ELECTRON MICROSCOPY (STEM)

The same electromagnetic, or electrostatic lenses used in the conventional transmission electron microscopy (TEM) instruments could also be used to form a small electron probe by demagnifying a small bright source, and that probe could be scanned over the specimen in a two-dimensional raster. Albert Crewe realized that the way to produce an electron probe of very small diameter was to make use of a field emission gun (FEG) in which electrons are extracted from a cold, sharply pointed metal tip by use of a high electrical field gradient. He built a small microscope column using electrons accelerated by 30 kilovolts and focused with a strong, short-focus electron lens to form a probe of diameter of approximately 0.4 nm. The electrons scattered out of the incident beam cone were collected by use of an annular detector to form a dark-field image. With this arrangement, the images of individual single heavy atoms, thorium atoms on a very thin amorphous carbon support, were recorded. Successful applications to many areas of non-biological and biological science then followed rapidly.

The principal components of a modified STEM instrument are as follows the source of electrons is a field-emission electron gun (FEG), preferably operating with a cold (i.e., room temperature) sharpened tungsten wire tip in ultra-high vacuum. A positive voltage, of about 3000 volts with respect to the tip, creates a sufficiently high voltage gradient to draw the electrons out of the tip. The effective diameter of this very bright electron source is as small as 4 or 5 nm. The main lens producing the demagnification of the effective source, to form the small-diameter probe at the specimen level, is the objective lens, similar to the objective lenses used in TEM with a high magnetic field and a focal length as small as 1 nm. Deflector coils built into the bore of the objective lens, serve to scan the electron probe over the specimen. For each position of the incident electron beam on the specimen, a diffraction pattern of the illuminated area is formed on a distant fluorescent screen. Since the beam focused by the objective lens is necessarily convergent, the pattern is a convergent-beam electron diffraction (CBED). Since the illuminated region of the specimen has a diameter approximately equal to the resolution limit of the images formed, usually less than 1 nm, the pattern is often referred to as an electron nano-diffraction (END) pattern. The diffraction pattern may be observed and/or recorded by the use of a light-optical system consisting, for example, of a wide-aperture optical lens forming an image of the pattern in a low-light-level TV camera, coupled to a video-tape recorder or on a CCD camera for digital recording. An aperture in the observation screen allows some part of the diffraction pattern, usually the central spot, to be transmitted into an electron energy-loss spectrometer (EELS), so that the EELS spectrum may be recorded, or else an image may be made with electrons having any desired energy loss. The normal bright field STEM image is obtained with the zero-loss electrons from the central beam of the pattern. Dark-field images may be obtained by selecting electrons that have lost the few electron volts. As in TEM, the specimens needed for STEM imaging must usually be very thin with thicknesses less than a few hundred nanometres. It is highly useful in the study of shapes, sizes, crystal structures, surface structures and compositions of nanometre-size regions of small particles, thin films and nanocrystals.

Also possible with STEM are:

- Reflection Electron Microscopy (REM)
- Scanning reflection electron microscopy (SREM)
- STEM holography

FLUCTUATION ELECTRON MICROSCOPY (FEM)

This is a combined imaging and diffraction technique that analyses the spatial fluctuations in diffracted intensity across a disordered sample to detect the medium-range structural order. Because it examines the spatial statistics of electron scattering, it is relatively unaffected by the background averaging that interferes with pure modes of diffraction and imaging.

IN SITU TRANSMISSION ELECTRON MICROSCOPY

It is another modern technique where dynamic *in situ* experiments can be very efficient, and productive of data, in scanning a range of experimental conditions in a single session, saving time and costs.

ETEM (ENVIRONMENTAL TEM) (COUNTERPART OF ENVIRONMENTAL SEM)

Dynamic *in situ* exploration of nanomaterials at the atomic scale with variable gas pressures and temperatures is possible using the ETEM developed by TITAN (FEI).

Along with the advancements in the instrumentation technologies, simultaneous research in the processing techniques and specimen preparation techniques including instruments for processing like ultramicrotome, knives, etc. have also been taking place. This is, of course an on-growing process.

APPENDIX

HANDLING OF CHEMICALS AND SAFETY PRECAUTIONS IN AN EM LABORATORY

Most of the chemicals used for various processing procedures in electron microscopy are generally harmful and toxic. Quite a few of them are carcinogenic also. Hence extreme precautions have to be taken while using as well as disposing these chemicals. Majority of the chemicals used in electron microscopy must be manipulated in a fume cupboard. The chemicals must be handled wearing gloves and safety goggles apart from strictly following the necessary precautions for toxic chemicals.

Aldehyde Fixatives

Glutaraldehyde is a skin contact hazard causing dermatitis, etc. Glutaraldehyde solutions should be stored at 4°C and should be discarded if the pH falls below 3.5. pH of glutaraldehyde may drop slightly during storage and therefore should be tested regularly. The solutions are prepared and stored in clean, glass-stoppered bottles. Formaldehyde is more volatile than glutaraldehyde. The vapours are irritant to lungs and eyes. Formaldehyde is a carcinogen also. Thus with the dilute solutions used for fixation, the main hazard with glutaraldehyde is skin contact, with formaldehyde, inhalation and with acrolein, both. Gloves should be worn at all times.

For disposal, small quantities of formaldehyde and glutaraldehyde solutions can be poured down the sink with plenty of running water. Both these aldehydes and acrolein can be made safe by mixing them with an excess of a 1 M solution of glycine and pouring the resulting solution down a sink with plenty of running water.

For spillages of aldehyde fixatives, the best treatment is to cover the spillage immediately with a liberal amount of milk powder, allow time for reaction to take place and then mop up the powder and place it in a plastic bag for disposal.

The standard latex surgical gloves do not offer adequate skin protection when handling glutaraldehyde. Nitrile or neoprene gloves provide the needed protection. But double-gloving with latex gloves is recommended for delicate work.

Cacodylate Buffer

This contains toxic arsenic and it may act as a fixative. It has an unpleasant smell. It should be always handled in a fume cupboard. Buffer solutions can be poured down the sink accompanied by plenty of running water, but large quantities of cacodylate buffer should not be disposed of at any one time.

Osmium Tetroxide

Both crystals and dilute solutions give off fumes of osmium tetroxide even at 4°C. It volatilizes readily at room temperature. The vapours are very harmful to the eyes, nose and throat. These vapours are irritant to respiratory system and eyes. Continued exposure to vapour causes serious damage to the eyes and may result in disturbance of vision; appearance of haloes around bright light indicates over-exposure to osmium. Osmium tetroxide is usually supplied in the form of yellow crystals in glass ampoules, generally in amounts of 0.1 g to 1.0 g or as an aqueous solution. The ampoules have pre-scored necks so that a cutting tool is not required. The label on the ampoule is removed and the outer surface is cleaned well. Then the ampoule is scratched on the scored line on the neck and dropped into a brown glass bottle with a glass stopper containing the required quantity of buffer and immediately stoppered tightly in a fume cupboard. The ampoule breaks inside and the crystals of osmium tetroxide dissolve in the buffer. The crystals take time to dissolve, so the solution should be prepared at least a day before it is required. The solution is best stored in a shelf in a fume cupboard away from direct sunlight. It is stable for some weeks at room temperature and need not necessarily be stored in a refrigerator.

Osmium vapours readily leak out of many containers as is evidenced by the black precipitates on inside refrigerator walls in many EM laboratories. A convenient way to prevent such leakage is to store the osmium solutions in glass vacuum-type blood collection tubes. The osmium vapours will not penetrate past the rubber stopper and contaminate the surrounding area.

Since even the vapours can fix the epithelium of the eyes, nose, etc., it should be handled only in a fume cupboard with good ventilation. OsO_4 is toxic as well as expensive. So only small quantities must be used. Waste fixative should be added to either aqueous alcohol or ferrous sulphate solution to produce a black precipitate of osmium dioxide, which is the most suitable form for further specialist disposal. A 2% solution of osmium tetroxide can be fully neutralized by twice the volume of oil (corn oil is preferred because of its high percentage of unsaturated bonds). In a fume cupboard, pour twice the volume of corn oil into used osmium tetroxide solution. Wait for the oil to completely turn black which indicates the neutralization of osmium. Dispose in accordance to local regulations.

Another method of disposal is as follows: The used solution should be poured into a brown glass bottle, one-third filled with 50–70% aqueous ethanol and kept tightly stoppered. Or for disposal this fixative may first be reacted with milk. To dispose of osmium down a sink, very large quantities of running water should be used. For the spillage of osmium tetroxide, oleic acid or a polyunsaturated fat is very effective. Or milk powder can be used to cover any spilt fixative.

Both skin contact and inhalation must be rigorously avoided at all times with all these fixatives. More so, because the fixing action is irreversible with these fixatives.

Propylene Oxide

This is corrosive to eyes and skin. It causes burning sensation, coughing, wheezing, laryngitis, nausea, vomiting, dizziness, narcosis, etc. It is shown to cause cancer also. It must be flushed down the sink with a large volume of water after use and care must be taken not to inhale the fumes.

Embedding Media

Many of the components can cause dermatitis. Some resins of the Epon type are classified as mildly carcinogenic. The chemicals of the embedding media should be handled in a fume cupboard because vapours are also toxic. Embedding media should be handled with great care. Most of the components can cause dermatitis if they come into contact with the skin, and consequently all procedures should be carried out as cleanly as possible. In addition, since the vapours of the components are often toxic, particular care should be taken when handling embedding media of low viscosity which tend to be more volatile.

Most resin monomers and other components of embedding media are sensitizing when they are allowed to contact the skin, and so adequate protective clothing, including suitable gloves should be worn. The toxic, irritating and sensitizing powers of resin monomers are greatest when they are diluted with solvents during the infiltration stage. Resins must not be removed from the skin with an organic solvent, since this will aid penetration of toxic components. Wipe the skin immediately with disposable paper towels and then cleanse the affected area with a resin-removing cream, followed by washing with warm soapy water.

For disposal, excess mixtures of resins and solvents should be placed in separate waste bottles. They should not be poured down the sink because they may slowly polymerize and in the long run may block the outlet. Excess embedding media should always be cured or polymerized before disposal, since they are then relatively inert.

Glass Knives

Glass knives have to be handled carefully because of their sharp edge. A further hazard arises during sectioning since dust particles from resin blocks particularly from Spurr's resin are potentially sensitizing.

Stains

Lead is toxic and uranium is even radioactive, in fact the only radioactive chemical used routinely in the EM laboratory. It is only mildly radioactive and can be handled with standard precautions, provided extra care is taken against spillages and with disposal.

Uranyl acetate presents two different hazards. It is radioactive and it is a cumulative poison. The uranyl acetate supplied has been depleted of the 239 isotope and the radiation is blocked by ordinary glassware, but any spillage is very dangerous. Evaporation of the solutions must be prevented since any fine deposit is potentially carcinogenic if it becomes dispersed and inhaled into the lungs. Ingestion of the solution must, of course, be avoided.

For disposal, solutions containing heavy metals such as uranium must not be poured down the sink. Large quantities must be stored and arranged for specialist disposal.

Apart from this, the routine precautionay measures mandatory in any chemical laboratory should also be adopted in the EM laboratory.

GLOSSARY

Absorbed electrons These are electrons that irradiate a specimen, then lose their energy and are absorbed by the specimen. The image obtained from these absorbed electrons is called absorbed electron image.

Accelerating voltage High voltage applied to accelerate the electrons emitted from a cathode. This determines the wavelength of the electrons emitted from the cathode filament.

Accelerator Catalyst added in minute quantities to the embedding medium to accelerate the polymerization/curing of the resins.

Acrylic resins A water-miscible embedding resin which can be polymerized at very low temperatures. But due to its many disadvantages it is used mainly for immuno-electron microscopy.

Adamantinoma/Ameloblastoma Rare slow-growing cancerous bone tumour of odontogenic epithelium, i.e., the epithelium that either has produced or will produce the crown of tooth. It is most often found in mandible (jaw bone). Predominantly arises in bone in subcutaneous location. Histologically, islands of epithelial cells are found in fibrous stroma.

Adenocarcinoma Malignant epithelial tumour with glandular differentiation.

Adenomatoid odontogenic tumour (AOT) An odontogenic tumour arising from the enamel or dental lamina.

A/E ratio The ratio between the anhydride hardener and epoxy resin, which determines the hardness of the block.

Air-lock mechanism Mechanism by which the specimen grids can be removed or introduced into the electron microscope easily and quickly. Using this mechanism the specimen chamber can be isolated from the rest of the column during change of specimens, such that the column vacuum is not disturbed during this procedure.

Analytical SEM In a SEM, where X-ray detectors, which can detect characteristic X-rays emitted from the specimen, are installed, then it is possible to identify and quantitatively analyse the chemical composition in micro areas on the specimen while observing the SEM image. Such a SEM is called analytical SEM.

Antibody An immunoglobulin with ability to react with specific antigen.

Anticontamination device A cooling trap to hold liquid nitrogen surrounding the specimen in the electron microscope in order to minimize the specimen contamination.

Antigen A substance which is able to produce specific detectable immune reaction.

Antigen presenting cells A cell that breaks down antigens and displays their fragments on surface receptors next to major histo-compatibility complex (MHC) molecules.

This presentation is necessary for some T lymphocytes which are unable to recognize soluble antigens. Macrophages are the primary antigen presenting cells but dendritic cells can also act as antigen presenting cells.

Aperture Physical hole introduced in condenser and objective pole pieces in order to reduce the size of the electron beam. In condenser, the smaller aperture reduces beam size as well as the intensity, while in the objective lens the smaller aperture improves the contrast in bright field imaging partly by removing the off-axis electron beam.

Apoptosis Programmed cell death.

Artifacts A structure or substance which is not normally present but produced by some external agency or action, say for example, due to processing of the specimens.

Astigmatism Lens aberration due to the construction of the lens; in electron lenses it arises due to lens inaccuracies in machining the pole-pieces.

Atomic force microscope (AFM) In this microscope, similar to scanning tunnelling microscope (STM), the localization interaction between a sharp probe and a sample is employed for surface imaging. AFM is based on detection of more universal tip-sample mechanical forces and hence, can be applied for characterization of materials of any kind. Thus high-resolution examination of surface topography, compositional mapping of heterogeneous samples and studies of local mechanical, electric, magnetic and thermal properties are possible with this.

Auger electrons When the atoms of a specimen that are excited by electron beam irradiation return to the normal state, the released energy is used mainly to generate X-rays or light. But a portion of the released energy is used to release electrons in the atoms of the specimen. This phenomenon is called Auger effect and the emitted electrons are called Auger electrons.

Autolysis Spontaneous disintegration of cells in tissues by autologous enzymes, as occurs after death and also in some pathologic conditions.

Autophagic vacuoles or autophagosomes Membrane-bound vacuoles within the cell that contain fragments of cell components, e.g., mitochondria, endoplasmic reticulum, etc.

Autophagy In biology, autophagy or autophagocytosis is a catabolic process involving the degradation of a cell's own components through the lysosomal machinery.

Backscattered electrons Primary electrons from the source which are reflected or scattered backwards by the specimen atoms; also referred to as "reflected" electrons. These electrons can be made to form the image of the surface topography of the specimen in SEM.

Balanced break method Method of cutting glass strip for the preparation of glass knife. The long glass strip is cut into equal halves successively such that the stress is equalized on both sides and hence the name. Small square glass bits are obtained finally from which by diagonal cut, glass knives can be made.

Basal lamina or basement membrane Thin layer of delicate non-cellular material of a fine filamentous texture underlying the epithelium. Its principal component is collagen.

Beam current The current of a few amperes applied to the cathode filament directly to heat it.

BEEM capsules BEEM is a registered trademark of "Better Equipment for Electron Microscopy, Inc." It is the trade name for their polymer capsules used for embedding.

Benzo(a)pyrene Carcinogen known to induce lung cancer.

Bias voltage Voltage applied to the cathode shield. The cathode shield is often referred to as Wehnelt cylinder which controls and stabilizes the electron beam emitted from the cathode.

Birbeck granules Also known as Langerhans cell granules. These are characteristic cytoplasmic granules in the Langerhans cells. These granules arise from the cell membrane, grow inwards and bulge most often at one end giving the typical "tennis racket" shape to these granules. They slowly get detached from the membrane and remain free in the cytoplasm. They are pentalaminar, rodlike tubular structures.

Blocker To prevent non-specific binding of antibodies to "reactive groups" of antigens within the specimen, the specimen is incubated with either high ionic strength buffer or competing inert protein which are referred to as "blockers".

Block face The surface of the trimmed block from which sections are cut in the ultramicrotome.

Bone marrow Soft tissue found within bone cavities and which is the site of blood cell production.

Buffer Solution which can maintain the pH of the system constant even after addition of small quantities of acids or alkalies.

Cacodylate buffer Buffer which can maintain pH around physiological pH and hence used for preparation of fixatives for EM. This contains arsenic and therefore prevents growth of organisms.

Capsaicin Natural antioxidant derived from capsicum fruit extract.

Carbon-coated grid Grid coated with a thin film of carbon. These grids are commercially available and used for dispersions/ suspensions.

Cathodoluminescence When a specimen is irradiated with an electron beam, the electrons in the valence electron bands of the atoms comprising the specimen are excited. Light emitted during the process in which the hole created in the valence electron bands recombine with the electrons is called cathodoluminescence.

Carcinoma A malignant tumour arising from cells in the surface layer or lining membrane of an organ , i.e., the epithelium.

Cardiomyocyte Heart muscle cell.

Carrier protein Probes are attached to certain proteins at one particular site either by chemical fixation or physical binding. These proteins which recognize either the antigen or antigen–antibody complex are called carrier proteins.

Cell cultures Cells grown in controlled conditions which can be used as specimen for EM for their ultrastructural study.

Cell monolayer Thin layers of cultured cells attached to the substrate on which the cells are cultured.

Cerebrospinal fluid The fluid flowing inside and around the brain and spinal cord that provides support and nutrition.

Characteristic X-rays X-rays generated by the interaction between the electron probe and the specimen atoms, from which the elemental composition can be characterized both qualitatively and quantitatively.

Charge-up Build-up of charges due to electron accumulation.

Chatter Effect of low or high frequency vibrations during ultrathin sectioning which appear as wavy parallel lines in the EM. It is an artifact appearing as periodic wavy changes in thickness.

Chondroblastoma Benign tumour of cartilage tissue.

Chondrocytes Cells of cartilage tissue.

Chondrosarcoma A malignant growth of cartilage that usually affects the long bones.

Chromatic aberration Lens aberration due to the inability of the lens to bring rays of different wavelengths in a beam, light or electron, to a focus. Though there is no colour possible in EM due to the very low wavelength of electrons, the aberration arising due to wavelength difference of the source electrons is still referred to as "chromatic aberration".

Chromatin The material inside cells consisting of DNA and protein from which chromosomes are formed.

Clearance angle Angle that the knife edge makes with the block face in an ultramicrotome to facilitate easy sectioning.

Clearing Addition of intermediary solvents like propylene oxide, xylene, toluene, etc. for replacement of dehydrating agent in the tissue specimens facilitating easy penetration of resins into tissues.

Collagen A structural protein present in connective tissue and ECM (extracellular matrix).

Collagenolysis Lysis of collagen fibres.

Colloidal gold marker One of the smallest electron-dense markers for cytochemistry. Colloidal gold is a negatively charged hydrophobic sol, formed by electron-dense metallic particles. It can be tagged to macromolecules and the gold probe-complexed antibody facilitates easy identification of the antigen–antibody complex.

COMPO image (composition image) A backscattered electron image where the signal represents the composition differences in the specimen.

Compression Shortening of the length of the ultrathin sections cut in an ultramicrotome which are floating on water.

Condenser lens A type of lens which functions to converge the electron beam from the source onto the specimen in an electron microscope.

Confocal microscope An optical imaging technique used to increase optical resolution and contrast of a micrograph by using point illumination and a spatial pinhole to eliminate out-of-focus light in specimens that are thicker than the focal plane. It enables the reconstruction of 3D structures from the obtained images. It uses point illumination and a pinhole in an optically conjugate plane in front of the detector to eliminate out-of-focus signal. The name "confocal" stems from this configuration. The increased resolution is at the cost of decreased signal intensity and so long exposures are often required. Three types of confocal microscopes are commercially available: i) confocal laser scanning microscope ii) spinning-disk confocal microscope iii) programmable array microscope.

Conjugation Probes and carrier proteins may be conjugated using glutaraldehyde which can bridge them together for immunoelectron microscopy.

Contrast Phenomenon by which different features are distinguishable clearly in the magnified image; usually introduced through staining.

Copper complex stain A new Cu(II)–UA (Copper (II)–Uranyl acetate) complex stain which can be used for staining ultrathin sections on grids instead of the routine double staining with uranyl acetate followed by lead citrate.

Critical point drying (CPD) A method of drying a tissue specimen by adjusting the pressure and temperature so that the distinction between liquid and gas phases disappears. This is the most commonly used technique to dry biological samples for SEM.

Cross-over point The electrons from the hot cathode filament are concentrated to a point by the bias potential. This electron converging point is called the "crossover point" and this is considered as the size of the source.

Cryoelectron microscopy Electron microscopic techniques carried out at low temperatures, say, liquid nitrogen temperatures. The cryosections obtained with cryoultra-microtome are transferred via cryotransfer system to the EM where the specimen is cooled by liquid nitrogen.

Cryofixation Fixation (preservation) of biological specimens by using low-temperature methods.

Cryogen Cooling agent like liquid nitrogen or Freon.

Cryoimmunoelectron microscopy In cryotechniques, immunocytochemical reactions are performed on cryoultrathin sections or freeze-fractured surfaces.

Cryonegative staining Negative staining carried out at low temperatures.

Cryoprotectant Chemicals added to protect the specimen from freezing damage to living cells during cryofixation.

Cryosectioning Cryofixed tissue could be sectioned in specially designed cryoattachments to microtomes. The sections obtained are transferred to cryo-EM using cryoholders.

Cryo-SEM SEM provided with a temperature-controlled cryostage in it, with which the specimens can be examined at very low temperatures.

CUM (cryoultramicrotomy) Sectioning at low temperatures. This technique does not require dehydration and embedding of specimens in resins and hence the antigenicity is well-preserved, hence is most suited for immunoelectron microscopy.

Curing Addition reaction by which the epoxy resins are hardened. This method is used for preparing resin blocks of tissue specimens for EM.

Cytoplasmic neolumina An aberrant feature seen in adenocarcinoma (glandular cell tumour). The feature actually appears as a membrane-bound cytoplasmic collection of microvilli.

DDSA (Dodecenyl succinic anhydride) A hardener used in Epoxy resin formulations. It is a low temperature hardener and confers plasticity on the cured resin.

Dehydration Removal of water. But in EM processing, it is a misnomer because during tissue processing addition of alcohol/acetone is not for removal of water but only for its replacement.

Dendritic cells Cells of mononuclear phagocytic system located in the skin, lymph nodes and spleen.

Depth of field Range over which the recording plate could be displaced without producing a loss of focus. This is large in EM such that though the photographic plate is well below the fluorescent screen, the image focused on the screen is sufficient for photography.

Depth of focus Range within which the object point could be moved over the axial distance above and below its true position, without changing the size of the corresponding image point. In EM it is high.

Dermatitis Inflammation of the skin.

Desmosomes A type of intercellular junction characteristically seen in epithelial tissue, particularly squamous epithelium.

Detector Positioned in the specimen chamber of SEM, it detects and collects the signal for further amplification. Separate detector is used for different signals.

Developer Chemical preparation to develop the image captured on the negative of the photographic film.

Diamond knife Knife for sectioning in an ultramicrotome where the knife edge is coated with diamond. This has preformed boat fixed to it.

Diatom Aquatic organism, the finely structured shells of which can be used as a magnification calibration standard.

Diffraction grating Carbon replica of a ruled diffraction grating having specified number of lines per inch is used as a magnification calibration standard.

DMP 30 The catalyst used in resin formulations, to hasten polymerization or curing. It is added in minute quantities during polymerization or curing to reduce the temperature during the embedding procedure.

DNA Deoxyribonucleic acid. The hereditary material.

Easymould Brand name for a particular type of mould for embedding in EM for resin block preparations.

Edge effect Effect of edge of the features in the specimen on electron interactions and thus the image.

EDS (Energy Dispersive X-ray Spectrometer) An X-ray spectrometer which uses a semiconductor detector to detect the energies of X-rays emitted from the specimen surface and convert them into electric pulse heights. Characteristic X-rays emitted from the specimen have energies characteristic of individual elements. EDS can be attached to both SEM and TEM.

Elastically scattered electrons Electrons scattered by the specimen atoms which undergo a directional change due to atomic collision but retain their energy.

Electron Extranuclear negatively charged particles in an atom of any element, which has negligible mass and which could be considered to have the properties of a light wave. Hence can be substituted for light in a microscope thus forming the electron microscope.

Electron diffraction Similar to X-rays, electron beam is reflected diffracted at the net plane of a crystal lattice, which can be used for the study of crystal structures. The advantage of electron diffraction is that the structure of the specimen can be examined by correlating the diffraction pattern and the microscope images at the same time using the EM.

Electron gun The part of the electron microscope which houses the cathode filament, the source of electron beam as well as the anode and the Wehnelt cylinder.

Electron lens Electromagnetic or electrostatic or permanent magnet lenses used in the electron microscopes.

Electron micrograph Photograph taken with an electron microscope.

Electron microscope A microscope where a beam of electrons is used as the source of illumination instead of light.

Electron noise The graininess in an enlarged micrograph due to the random arrival of electrons at the photographic film.

Electron optics Science on the motion of free electrons under the influence of electrostatic and electromagnetic fields.

Electron probe Electron beam that is focused finely on the surface of a specimen.

Electron tomography Constructs 3D models from the two dimensional information contained in a series of projected images acquired over a range of perspectives as the sample rotates in incremental steps about an axis perpendicular to the viewing direction.

Embedding A processing step for biological specimens by which a hard block of the specimen is obtained for further sectioning.

Embedding mould Moulds in which embedding is carried out to prepare specimen blocks which have definite shapes and consistency for further sectioning.

Emperipolesis Phenomenon seen in RDD (Rosai Dorfmann Disease)/Sinus histiocytosis with massive lymphadenopathy (SHML) wherein the lymphocytes enter and leave the histiocytic cytoplasm freely.

En bloc staining Staining of small tissue pieces as such and not thin sections as done routinely. This staining is carried out during fixation itself when the fixative contains heavy metal atoms which can act as electron stains.

Endothelial cells Cells of the single layer of flattened cells lining the cavities of the heart and of the blood and lymphatic vessels and the serous cavities of the body.

Enzyme cytochemistry Subcellular localization of enzymes using their specific substrates.

Epithelial cells Cells of the epithelium which is the covering of the internal and external surfaces of the body.

EPON 812 Epoxide embedding resin for EM processing. Most commonly used resin.

ESEM (Environmental scanning electron microscope) A SEM in which the specimen is placed in an internal chamber at higher pressure than the vacuum in the electron optical column. The pressure of gas in the chamber can be controlled. It is especially useful for non-metallic and biological samples because specimens can be viewed directly without metal coating.

Euchromatin Chromosomal material that is genetically active and stains lightly with basic dyes. It is the loosely coiled DNA that is transcriptionally active.

Ewings sarcoma A bone tumour affecting children.

Exposure time Duration of exposure of specimen to the electron beam.

Exothermic reaction A chemical reaction in which heat is liberated.

External lamina Same as basal lamina.

Extracellular matrix (ECM) Acellular material around cells is called ECM. It is the extracellular part of animal tissue that usually provides structural support to the animal cells in addition to performing various other important functions. It is the defining feature of connective tissue in animals. It includes interstitial matrix and the basement membrane.

Eyepiece Eye lens through which the image can be viewed directly by the human eye in the light microscope. But in an electron microscope the image on the screen can be further magnified for fine-focusing the image.

Fenestrated capillaries Blood capillary found in renal glomeruli, intestinal villi and some glands in which ultrastructurally pores of variable size can be seen.

Ferritin Iron-containing protein used for magnification calibration, as well as for immunostaining.

Fibroblast Any cell or corpuscle from which connective tissue is developed; it produces collagen, elastin and reticular protein fibres.

Field emission gun SEM (FEG-SEM) FEG-SEM is actually used to image individual heavy atoms by collecting scattered electrons with an annular detector. This low voltage SEM makes it possible to observe directly non-conducting materials without a conductive coating. This SEM is more surface-sensitive. Variable pressure FEG-SEMs can be used for examining powders, films, pellets, conducting, semiconducting and non-conducting materials.

Fine needle aspirates Aspirated cellular material from a tumour using a fine needle which can be used for both cytological and ultrastructural diagnosis.

Fixation Preservation of cells and tissues by chemical or physical methods.

Fluorescent screen Screen coated with fluorescent material on which the magnified image falls and which converts electron image into optical image for direct visualization by the naked eye.

Focal length Distance between the lens and the point at which the image is focused.

Focused ion beam system This incorporates both FIB (Focused ion beam) and SEM into a single machine. The SEM image is obtained from secondary electron detector which gives much less damage to the specimen while providing a better resolution, while the ion beam imaging provides better sensitivity to details because of the depth of penetration from channeling. In both, a source emits charged particles which are then focused into a beam and scanned across small areas of the sample using deflection plates or scan coils. In SEM, the focusing is accomplished using magnetic lenses, while in FIB, electrostatic lenses focus the ions into a beam. The image will show both topographic information and materials contrast.

Focusing Act of converging at a point.

Follicular dendritic cell sarcoma A very rare type of soft tissue sarcoma. It develops from specialized cells in the lymph nodes. Follicular dendritic cells support lymphocytes by recognizing cells that are foreign or unwanted in the body.

Formalin Commercial formaldehyde solution used as a routine fixative for light microscopy.

FORMVAR Material used for the making of plastic films both for the preparation of replicas and also support films on grids.

Freeze-drying This method depends on molecular distillation to achieve drying. This is a method of drying a specimen containing water by freezing it, using for example, liquid nitrogen, then sublimating the water in a vacuum.

Freeze-etching Similar to freeze-fracture technique. In this, after fracturing, the specimen is elevated to higher temperatures and then replicated. This could be done in specimens which are not cryoprotected or are cryoprotected in organic solvents like methanol.

Freeze-fracture A cryotechnique used to study the macromolecular architecture of biological membranes. When biomembranes are frozen to low temperatures, they can be fractured with a fine knife. The frozen and fractured specimen is immediately replicated and the replica is studied in TEM.

Freeze-substitution A method of dehydrating the tissue specimen, rather, replacement of water by an organic solvent at temperatures of –80 to –90°C. The specimens can be dehydrated and infiltrated with an embedding medium at a low temperature without the presence of a gas phase by using suitable intermediates. The solidified fluid of the quenched tissue is slowly dissolved or substituted at low temperatures in fluid solvents.

Freeze-thawing In this, immediately following quenching (rapid cooling), the tissue specimen is rapidly warmed by being immersed in a chemical fixative at room temperature (RT). The specimen is then embedded in the conventional manner. Thus the tissue is subjected to two rapid changes of temperatures.

Fresnel fringes Diffraction bands observed in through-focus images.

Gap junction A type of intercellular junction.

Germinoma Any neoplasm with morphological features of a germ cell.

Glass knife Knife made of special glass for the purpose of knifemaking. The knife has a sharp edge with which sections can be cut from resin blocks using the ultramicrotome.

Glomerular basement membrane (GBM) Basement membrane surrounding glomerulus which is a tuft of convoluted mass of capillaries enclosed in a Bowman's capsule of the kidney.

Glutaraldehyde A pentacarbon dialdehyde used as a routine primary fixative for EM.

Glycocalceal bodies/glycocalyx A thin layer of glycoprotein and oligosaccharides on the outer surface of cell membrane that contributes to cell adhesion and forms antigens involved in the recognition of "self".

Glycogen Polymer of glucose present in a biological cell.

Gold probes Colloidal gold giving gold nanoparticles which can be used for immunolabelling for EM. (*See* Colloidal gold marker.)

Golgi complex A cytoplasmic organelle in a biological cell where lipid synthesis and protein modifications occur.

Granular cell ameloblastoma An unusual variant of ameloblastoma, which shows granular transformation of neoplastic cells.

Granulation tissue Young vascular fibrous tissue developing in healing wounds and organizing fibrin.

Grids Metallic discs on which the specimens are placed for TEM. The most commonly used grids are made of copper, but grids made of other metals like gold, nickel, platinum or alloys like platinum–palladium, etc. are also used. Only these grids can be inserted into the specimen chamber of EM.

Gun assembly The assembly of parts of the electron gun, namely, the cathode filament, anode and the Wehnelt cylinder.

Hairy cell leukaemia Multiple hairlike projections on the surface of the leukemic (blood cancer) cell.

Hardeners Chemicals, most commonly anhydrides, added to the resin for hardening the blocks.

Hemidesmosomes Structure representing half of the desmosome found on the basal surface of some epithelial cells, forming the site of attachment between the basal surface of the cell and the basement membrane.

Hepatocyte A liver cell.

Histiocytic lymphoma Lymphoid cell neoplasm where the tumour cell shows histiocytic differentiation.

Histiocytosis X or Langerhans cell histiocytosis Tumour of the Langerhans cell, a specialized histiocyte.

Holey carbon film A standard grid coated with a thin film of carbon with numerous round holes of various sizes. This standard is useful for objective lens astigmatism correction.

HPV (Human Papilloma Virus) A virus whose infection has been shown to lead to uterine cervical cancer.

Hybrid antibody An antibody that contains two different specificities united into one molecule. For example, an antibody in which one specificity is directed to ferritin and the other to the antigen to be marked.

Ice crystal formation During freezing, formation of ice crystals takes place, which is a major disadvantage in cryotechniques. Rapid freezing is carried out to minimize this hazard.

Immersion fixation Method of fixation where the specimen is completely immersed into the fixative. This is the only method suitable for human samples.

Immunocytochemistry Subcellular localization of antigens using corresponding specific antibodies.

Immunoelectron microscopy Immunological techniques at the ultrastructural level.

Immunogold silver staining (IGSS) A method in which gold probes are further enhanced by metallic silver.

Immunohistochemistry (IHC) Refers to the process of detecting antigens, for example, proteins, in cells of a tissue by employing the principle of antibodies binding specifically to antigens. It is the localization of antigens or proteins in tissue sections by the use of labelled antibodies as specific reagents through antigen–antibody interactions that are visualized by a marker such as a fluorescent dye, enzyme or colloidal gold.

Immunonegative staining A combination of negative staining with immunolabelling.

Inelastically scattered electrons Electrons scattered by specimen atoms which undergo both a directional change and a partial energy loss.

Infiltration Processing step for penetration of embedding resins into tissue specimens such that hard specimen blocks can be prepared subsequently.

Inherent artifact Artifacts which occur due to the type of technique employed for processing the tissue specimens.

Interaction volume Volume of the specimen surface in which the primary electron beam interacts.

Interference colours Apparent colours of the floating sections which are imparted due to the light interference phenomenon. Using these interference colours, the range of thickness of the floating sections can be ascertained.

Intermediate image The image formed by the objective lens which is further magnified by magnifying lenses.

Intracisternal tubules Tubular structures within the cisternae of RER/SER (rough or smooth endoplasmic reticulum).

Intracytoplasmic neolumina or crypts Same as cytoplasmic neolumina.

Introduced artifact Artifact which arises due to the actual way of carrying out the chosen technique for processing tissue specimens.

Inverted capsule technique A technique for embedding monolayer of cultured cells. The gelatin or BEEM capsule (embedding moulds) is inverted over selected areas of the monolayer. After curing, the capsule is removed from the substrate, leaving the monolayer embedded in the surface of the block.

Isoproterenol (ISO) An inducer of myocardial infarction.

Keratin A scleroprotein consisting of sulphur that is a major component of epidermis, hair, nail, horny tissues and the organic matrix of the enamel of the teeth.

Keratin filaments Intermediary type of filaments in the cytoplasm of epithelial cells which normally are grouped in bundles referred to as "tonofibrils".

Keratinocyte The epidermal cell that synthesizes keratin.

Knife angle Angle of the knife.

Knife edge The sharp edge of the glass or diamond knife which can be employed for sectioning the specimen.

Knifemaker Instrument for preparing glass knives from special glass.

Langerhans cell A macrophage that presents antigen to lymphocytes in the upper dermis which sends dendritic processes.

Langerhans cell histiocytosis or Histiocytosis X Proliferation of special type of immature dendritic cells called the Langerhans cells.

Lead citrate One of the routine electron stains for ultrastructural study. It preferentially stains many proteins, glycogen, endoplasmic reticulum vesicles, mitotic spindles, etc.

Leiomyosarcoma Malignant tumour of smooth muscle.

Lens Part of the microscope which forms as well as magnifies the image of the specimen. In the light/optical microscope optical lenses are used while in electron microscopes electron lenses—electromagnetic or electrostatic or permanent magnetic lenses—are used.

Lens aberrations Deviation from the normal functioning of the lens/unequal refraction or focalization of a lens. The major types of aberrations are spherical and chromatic aberrations and astigmatism.

Leukaemic cell Tumour cell in blood cancer.

Light microscope A microscope where light is used as the source of illumination.

Lipase Enzyme that metabolizes lipids.

Lipoblast An immature fat cell.

Lipocyte A fat cell.

Liposomes Synthetic spherical vesicle with a lipid bilayer which functions as artificial membrane systems to deliver DNA, etc. into cells and as a drug delivery system of toxic drugs.

Loose shell syndrome (LSS) An emerging infectious disease in cultured shrimp, in which the affected shrimp shows muscular dystrophy and dysfunction of hepatopancreas leading to slow progressive mortality.

Lowicryl Acrylic resin suitable for low-temperature embedding. It is mainly used for immunoelectron microscopy.

LR gold, LR White and Lowicryls Commonly used acrylic resins for immunoelectron microscopy.

LVSEM SEM in which the specimen chamber is maintained at relatively lower vacuum such that even live specimens can be viewed.

Lymphocyte A cell present in the blood and lymphatic tissue. There are many types of lymphocytes with respect to function but cannot be differentiated by microscopic examination.

Lysosomes Cytoplasmic organelle which is mainly concerned with removal of waste material.

Macrophage Cells of reticuloendothelial system possessing the ability to phagocytoze bacteria and other foreign particles.

Magnification calibration Standards used for calibrating the magnification in the microscope for quantitative measurements.

Malignant Condition of cancerous growth, resisting treatment. It is a tumour that tends to invade and metastasize.

Melanocyte Cells producing the pigment, melanin.

Melanoma Tumour of melanocytes, which are melanin pigment-producing cells.

Melanosomes Cell organelle of the melanocyte, in which melanin pigment is formed.

Merkel cell carcinoma A type of neuroendocrine carcinoma. Merkel cells are tactile receptor cells present in deeper layers of the epidermis.

Mesenchymal cells Cells of the connective tissue.

Mesothelioma A rare form of cancer that develops from the protective lining, the mesothelium, that covers many of the body's internal organs. It is usually caused in lungs by exposure to asbestos.

Mesothelium Layer of cells derived from the mesoderm lining the primitive body cavity. In the adult, it becomes the epithelium covering the serous membranes.

Methacrylate A type of resin used for embedding tissue specimens for EM.

Methyl nadic anhydride (MNA) Second hardener used in epoxy resin formulations. Nadic is a registered trademark of the Allied Chemical Corporation of USA. The unsaturated "Nadic" ring of MNA enhances beam stability of the cured resin.

Microscope An instrument which magnifies the image of a specimen.

Microtome An instrument for preparing thin sections from tissue blocks.

Microvilli Tiny hairlike folds in the plasma membrane that extend from the surface of many absorptive or secretory cells. They are most clearly visible with an electron microscope but may be seen as "brush border" with an optical microscope.

Microwave fixation A rapid method of fixation of biological specimens using microwave energy.

Microwave radiation Non-ionizing radiation that interacts with dipolar molecules like water and polar side chains of proteins. This can be used for all steps of processing, including, fixation, embedding, staining, etc.

Mitochondria Cytoplasmic organelle which is the storehouse of energy.

Mitosis A type of biological cell division as a result of which two identical daughter cells are produced.

Mottling A smudging effect seen at high magnifications in EM in Epon sections cut with glass knives but not seen in blocks cut with diamond knives.

Multiplate heater An accessory for the ultramicrotome. In this, three different temperatures can be maintained at three levels. One is a wax bath for fixing the turf to the knife edge. The second one is for drying and staining of semithin sections and the third one for prewarming the glass knife.

Mycosis fungoides / Sezary syndrome A type of skin tumour; T-cell lymphomas involving skin.

Myelin Covering of nerve fibres (myelinated nerve fibres) formed by the successive layering of cell membranes of Schwaan cells.

Myelinated nerve Nerve fibre covered by myelin sheath.

Myelin figures A rolled up or scroll-like arrangement of a lipid bilayer within a cell, superficially resembling the myelin sheath of nerves. Myelin figures are obtained with the EM in the cytoplasm or as inclusion in mitochondria and autpophagic vacuoles where they may represent artifacts of lipid fixation in normal cells. They can be observed in reversibly or irreversibly injured cells. They are concentric, membranous, lamellar formulations related with drug administration, hormone-exposure, wide variety of pathological processes, tumour cells, potent inducers of apoptotic cell death, etc.

Myofibrils Slender, longitudinal fibrils occurring in a skeletal or cardiac muscle fibre.

Myofibroblast Contractile fibroblasts that help in contraction of healing wounds.

Nanoparticles Particles of "nano" dimensions.

Nanoscience "Nano" is a prefix which means one billionth of a metre, i.e., 10^{-9} m. Science dealing with particles at those dimensions is referred to as nanoscience.

Negative immunoelectron microscopy In this method, particulate samples deposited on coated nickel grids are treated with specific antibodies and then negatively stained for screening in the EM.

Negative staining A staining technique for EM by which particulate structures are stained externally such that the particles appear as electron-lucent structures in an electron-dense environment similar to a photographic negative and hence the name. Here the stains do not penetrate the specimen.

Nephroblastoma (Wilms tumour) Rapidly growing tumour of the kidney that occurs in children.

Neuritic processes The cytoplasmic processes of a neuron, the nerve cell.

Neuroblastoma Malignant tumour with neuroblasts that usually originates in the adrenal medulla. It may also originate in any part of the sympathetic nervous system.

Neuroendocrine carcinoma Growth with malignant potential that secretes one or more hormones with neuroendocrine properties. They are derived from embryonic neural crest cells.

Neurosecretory granules (NSGs) Granules with typical morphology—having a limiting membrane, dense core and a halo in between. These granules are seen in nerve cells and cells secreting certain hormones and their tumours.

Nuclear arm The nuclear region of the spermatocyte projecting by putting forward armlike structures.

Nucleolus A small spherical mass within the cell nucleus concerned with the production of nucleoprotein.

Nucleolonema A network of strands formed by organization of a finely granular substance, perhaps containing RNA, in the nucleolus of a cell and seen characteristically in a few tumours, e.g., Seminomas.

Nucleus Membrane-bound structure that contains the cell's hereditary information and controls the cell's growth and reproduction.

It contains chromatin and the nucleolus and acts as the central unit of a living cell.

Objective lens Lens which is placed close to the object thus forming the first small image of the specimen. In the scanning electron microscope this demagnifies the beam forming an electron probe.

Odontogenic keratocyst A cyst that originates around the crown of an unerupted tooth and has aggressive growth potential.

Odontogenic tumour Tumour derived from odontogenic epithelium, i.e., the epithelium that either has produced or will produce the crown of tooth.

Osmiophilic Having affinity for osmium.

Osmium tetroxide Secondary fixative for EM which fixes as well stains membranes. Staining action is because osmium is a heavy metal.

Osteochondroma/chondroma Benign tumour originating from the growing epiphyseal cartilage.

Osteoblastoma/Osteoma Benign tumour of bone.

Osteosarcoma A malignant bone tumour of childhood or early adult life. It arises from primitive bone-forming cells.

Paraffin blocks Specimen blocks prepared using paraffin wax as the embedding material for routine histopathology, which can be used for diagnostic EM also if fresh specimens are not available.

Paraformaldehyde A polymer of formaldehyde from which methanol-free formaldehyde can be prepared instantaneously for use as EM fixative.

Partial dehydration So-called dehydration step carried out partially, where the tissue is transferred from 70% ethanol directly to 1 : 1 mixture of ethanol and resin.

Pepper effect Deposition of black granular precipitates of lead arising out of lead stains after prolonged bombardment with the electron beam.

Perfect loop Small loop of platinum or other wire useful for picking up floating ultrathin sections on grids.

Perfusion fixation A type of fixation where the fixative is perfused into the anaesthetized animal through the bloodstream. Not applicable for human samples.

Phase contrast microscope An optical microscopy illumination technique in which small phase shifts in the light passing through a transparent specimen are converted into amplitude or contrast changes in the image. Thus phase differences caused by variations in the specimen appear as intensity variations in the image.

Pheochromocytoma A benign tumour of the adrenal medulla which secretes the hormones adrenaline and nor-adrenaline. Characteristic neurosecretory granules with large haloes called "halo granules" can be seen electron microscopically.

Phosphotungstic acid (PTA) An EM stain mostly used for negative staining.

Pinocytotic vesicles Pinocytosis is a process by which cells ingest liquid to form small vesicles and these vesicles are referred to as pinocytotic vesicles.

Plasmacytoma Tumour of plasma cells.

Plasticizer Chemical used to decrease the viscosity of the embedding medium and to improve sectioning properties of the final blocks.

Pleomorphic Varying in size and shape.

Polarizing microscope A special microscope which uses polarized light for investigating the optical properties of the specimen. There are two types of polarizing microscopes—transmitted light models and incident light models.

Polymerization Chemical process by which the resin monomers harden to form hard blocks of the specimen for further sectioning.

Polymerization damage Damage to the specimen details which specifically occur during polymerization of methacrylate embedding.

Polyhydroxy alkanoate (PHA) and Polyhydroxy butyrate (PHB) Polyhydroxy products that get accumulated in the cytoplasm of some organisms when they assimilate the excess carbon source in the environment and convert it into such products.

Poorly differentiated tumour A fast growing aggressive tumour, where the tumour cells show very little differentiation.

Post-embedding immunoelectron microscopy Immunolabelling on ultrathin sections obtained after routine processing and sectioning of tissue specimens.

Post-fixation Secondary fixation with osmium tetroxide.

Pre-embedding immunoelectron microscopy Technique where the immunolabelling is carried out prior to the routine processing, i.e., embedding, sectioning, etc.

Premalignant Any lesion with a tendency to undergo malignant degeneration.

Premelanosomes Organelles characteristic of melanocytic differentiation. They consist of membrane-bound vesicles with a distinctive internal structure, oval structure with a dense grainy matrix and cross striations. Depending on the stage of development, these may be parallel lamellar, helical or zigzag structures.

Primary antibody Antibody raised against a particular antigen or its epitope.

Probe current In the SEM, the total amount of current to be irradiated on the specimen.

Probes Substances attached to antibodies for ready visualization of the antigen–antibody reaction product.

Projector lens An electron lens in the transmission electron microscope which magnifies and projects the final magnified image on the fluorescent screen.

Propylene oxide Intermediary solvent used for facilitating easy penetration of resin mixtures into tissue specimens.

Radiation damage Damage to the specimen due to bombardment of the electron beam in the EM.

Rapid freezing/plunge freezing Freezing the tissue specimen by plunging it into liquid nitrogen at a fast rate, such that ice-crystal formation is minimized. This initial rapid cooling is also called "quenching".

Rapid processing Fast processing method in electron microscopy for diagnostic purposes.

Replica Replicas of surfaces quite faithfully reproduce the smallest details of the specimen surface and are electron-transparent.

Replica techniques/replication Technique by which replicas of surface can be prepared for study of surface topography using TEM.

RER (Rough Endoplasmic Reticulum) Cytoplasmic organelle where the tubules of endoplasmic reticulum are studded with ribosomes on the outer side where synthesis of proteins takes place.

Resins Viscous liquids used for embedding the tissue specimens for EM for making hard blocks of the specimen.

Resolution / Resolving power of a microscope Shortest distance between two points or two lines in the specimen image which can be recognized as two different images. Resolving power of human eye is 0.2 mm, that of light microscope is 0.2 μm and electron microscope is 0.2 nm.

Reynolds lead citrate Commonly used lead stain formulation which can be freshly prepared.

Rhabdomyosarcoma Malignant tumour of skeletal muscle cells.

Ribosomes Organelles which may be seen as dispersed in the cytoplasm or as attached to the endoplasmic reticulum forming the so-called RER (rough endoplasmic reticulum). These are the organelles responsible for protein synthesis.

Sarcolemma The cell membrane of the muscle cell (skeletal, cardiac and smooth muscle). It consists of a true cell membrane, called the plasma membrane and an outer coat made up of a thin layer of polysaccharide material that contains numerous thin collagen fibrils.

Sarcoma A cancer arising from any tissue including muscle, fat, bone or blood vessels other than skin and mucous membrane.

Sarcomere The basic unit of a muscle's cross-striated myofibril. The sarcomeres give the typical striated appearance to skeletal and cardiac muscles. The myofibrils of smooth muscle cells are not arranged into sarcomeres.

Saturation point When the beam current is slowly increased to heat the cathode filament, at a particular point the quantity of emitted electrons reaches its limit which is called the saturation state. Increasing the beam current beyond this point overheats the filament and thus shortens the filament life.

Saw-teeth Appearance seen on the knife edge end which denotes the non-usable portion of the knife.

Scanning electron microscope An electron microscope where the electron beam, called an "electron probe" scans the surface of the specimens in a raster pattern and the electrical signals are amplified and fed into a CRT. 3D image of the surface topography can be visualized using this.

Scanning immunoelectron microscopy Immunological methods for antigen detection carried out using SEM.

Scanning probe microscopy (SPM) A sequel to STMs. It uses electron probes, which detect specimen surfaces only at a very close proximity. The interaction or the property sensed is strongly dependent on the height between the probe and the specimen surface. This height sensitivity distinguishes SPMs from other microscopes and allows imaging sample topography. Uncoated samples can be imaged at a resolution of 1 to 100 nm.

Scanning transmission electron microscopy (STEM) Highly useful in the study of shapes, sizes, crystal structures, surface structures and compositions of nanometre-size regions of small particles, thin films and nanocrystals. The electromagnetic lenses in the conventional TEM are used to form a small electron probe by demagnifying a small bright source and that probe is scanned over the specimen in a 2D raster.

Scanning tunnelling microscope (STM) A microscope which uses the tunnelling current which is established when a voltage is applied across two conductors brought close enough. This tunnelling current varies exponentially with the height distance and hence the tunnelling phenomenon can be used to study the surface topography of the specimens.

Schwaanoma Tumour of Schwaan cells.

Secondary antibody Antibody raised against a primary antibody or a portion of it. This secondary antibody generally serves as carrier for the probe and is used in indirect methods for localization of antigens.

Secondary electrons The primary electrons from the source are absorbed by the specimen, which instead emits its own electrons from its surface. These secondary electrons arise out of the interaction between the source electrons in the electron beam and the specimen atoms. These can contribute to the study of surface topography of specimen in the SEM.

Secondary Ion Mass Spectrometry (SIMS) Uses stable isotopic tracers to track biological activity within organisms.

Secretory granule Specialized intracellular organelles that serve as a storage pool for selected secretory products.

Section ribbons Floating sections that form a ribbon, with sections sticking edge to edge with one another on the water surface.

Seminoma A type of germ cell tumour. It is a testicular tumour presenting in the fourth decade of life as chronic painless enlargement of the testicles.

Semithin sections Tissue sections cut in the ultramicrotome of thickness around 1μ which are thinner than the routine histopathology sections but thicker than useful EM sections. Also called "survey sections".

Shadow-casting Evaporating atoms of a dense material, usually a metal from a point source at an oblique angle to the specimen. When the specimen is contoured, metal piles up on surfaces which face the source, but surfaces facing away from the source are shielded and receive no deposit. This is the shadow and from the length and angle of the shadow, the height of the specimen contour can be measured.

SHADOW image Stereoscopic image. This is obtained by mixing TOPO and COMPO signals, where an enhanced 3D effect is obtained.

Signal-to-noise ratio (S/N) Signal is the result of the beam interacting with the specimen. Noise arises as a result of imperfections in the electronics of the detector and display system, which should naturally be minimized.

Single-stage replica In this method, a thin film of replicating material is deposited on the specimen surface which can be detached for EM observation.

Specimen chamber Part of the EM where specimen is inserted.

Specimen stage Stage on which specimen is placed in a microscope for viewing its image.

Specimen tilt or tilt angle The range through which the specimen can be tilted for better view of features, 3D tomography, etc. It is useful in both SEM and TEM for obtaining stereomicrographs, observing topographic features, observing the sides of the specimen, and for improving image quality of secondary electron images.

Spermatocyte A cell developing from a spermatogonium that gives rise to spermatozoa.

Spherical aberration Lens aberration which arises due to the inability to bring central and peripheral rays in a beam, light or electron, to a single focus.

Spot size The diameter of the initial focused spot of electrons.

Spurr Low-viscosity resin used mainly for embedding plant specimens. It is a water-immiscible epoxy resin.

Sputter coating Uniform coating on the specimen surface including the parts which are not directly facing the metal to be evaporated. It is the most acceptable technique, but not suitable for carbon coating.

Squamous cell A type of epithelial cells.

Squamous odontogenic tumour (SOT) A rare benign odontogenic tumour found in any area of the alveolar bone. It is thought to originate from remnants of the dental lamina.

Stain exclusion Stain artifact caused by bulky polymeric complexes at cross-linking sites.

Stain pad/grid mat Silicone rubber mats for staining ultrathin sections collected on grids.

Stains Usually coloured dyes which are added to the sections to impart contrast to various features in the specimen. This contrast facilitates easy identification of the features. Since no colour is possible in EM due to its very low wavelengths, "electron stains", i.e., heavy metals are used in EM to stain the sections on grids.

Sticky side The side of the grid on which the ultrathin sections have to be collected. It is the duller side of the grid.

Stigmator A correction lens in the electron microscope to correct the astigmatism in the condenser and objective lenses.

Submucous fibrosis Premalignant lesion of the oral cavity which arises mostly due to tobacco chewing.

Support film Thin film of plastic or carbon deposited on the grid to support the specimen placed over it.

Surface coating Coating of the surface with a metal.

Survey sections Semithin sections with thickness in between the routine histopathology sections and the routine ultrathin sections for TEM. The thickness of these sections are around 1μ. These sections are cut using the ultramicrotome, collected on glass slides and seen in the optical microscope to ascertain the usefulness of the specimen in that particular block from which these sections are cut, for further ultrastructural study using the EM.

Thermionic electrons Electrons emitted from a hot cathode filament.

Thymoma Tumour of thymic epithelial cells.

Through-focus series Micrographs taken from under-focus to over-focus images to identify the true-focus image.

Tight junction A type of intercellular junction seen between epithelial cells.

Toluidine blue Stain for resin-embedded semithin sections for screening in the conventional light microscope.

Tonofibrils Bundle of cytoplasmic cytokeratin filaments seen in epithelial cells, particularly, the squamous type.

TOPO image Backscattered electron image, where the signal represents the **topo**graphy of the specimen surface.

Transmission electron microscope Electron microscope where the electrons are transmitted through the specimen such that a magnified image of the internal details can be seen.

Transmitted electrons When the specimen is sufficiently thin, the electrons irradiating the specimen at high energy can pass through the specimen, with almost no energy loss. Such electrons are called transmitted electrons which help in forming the image of the internal features of the specimen.

Trapezoid/pyramid The preferred shape of the block face to be maintained during trimming the block. This shape is most suitable for obtaining continuous ribbon of sections.

Trehalose A disaccharide sugar, used as a biological protectant during negative staining.

Trimming Shaping of the specimen embedded in the block such that the specimen is exposed and the block face obtains a pyramidal or trapezoidal shape.

Trufs Brand name of boat or trough which is fixed to knife and filled with water on which cut sections float.

Trypsinized cells Cells removed from culture flask by using trypsin, which is a protein-digesting enzyme.

Tumour/cancer Neoplastic growth with uncontrolled proliferation of cells.

Tumour marker Non-specific substance which under certain circumstances can be used to detect tumour and its recurrence.

Two-stage replica The primary replica is commonly made of plastics like FORMVAR. The secondary replica is evaporated carbon. After dissolution of the primary replica, the secondary replica can be screened in EM. It is most suitable for soft surfaces.

Ultramicrotome Instrument used for ultrathin sectioning of tissue specimens for TEM.

Ultrathin sections Tissue sections cut in the ultramicrotome using glass or diamond knives with thickness less than 100 nm, which alone can be used for TEM.

Uranyl acetate One of the routine electron stains for ultrastructure. It preferentially stains structures containing nucleic acid.

Usable edge or useful edge The central part of the knife edge free from stress marks and saw-teeth. This portion is most useful for ultrathin sectioning. This part is longer in knives with small counterpieces.

Vacuum coating (thermal evaporation) Coating the specimen surface with heavy metal by heating the metal in vacuum. Carbon coating is done using this technique only.

Vasoformative tumour Tumour of the blood-vessel-forming cells.

Viruses One of the smallest infectious agents which can be visualized only at electron microscopic magnifications.

Wallner line Also called "stress line", produced in a knife cut using the knifemaker. This line usually starts from the left hand corner of the cutting edge. It indicates the side where the break started and the irregularities increase in magnitude towards the opposite side.

WDS (Wavelength Dispersive X-ray Spectrometer) Technique for the elemental analysis of the specimen using the characteristic X-rays. X-rays emitted from the specimen are dispersed in their wavelength by the analysing crystal and the X-rays having specified wavelength are detected by a proportional counter.

Wehnelt cylinder The control electrode which controls and stabilizes the electron beam emitted from the cathode when a bias voltage is applied to it. It is sometimes called a "grid".

Wiebel–Palade bodies Electron-dense granules seen in the cytoplasm of endothelial cells.

Wilms tumour *See* Nephroblastoma.

Working distance Distance from the lower face of the lower pole piece of the objective lens to the surface of the specimen in the SEM.

Xylene Organic solvent to remove wax from paraffin block sections.

REFERENCES

Alvin C. Lin and Cynthia Goh, M. (2002). Investigating the ultrastructure of fibrous long spacing collagen by parallel atomic force and transmission electron microscopy. *Proteins: Structure, Function and Genetics.* 49**:** 378–384.

Audrey, M. Glauert and Peter, R. Lewis. (1988). *Biological Specimen Preparation for Transmission Electron Microscopy*. Vol. 17. Princeton University Press, Princeton/Portland Press Ltd., London.

Audrey, M. Glauert. (Ed.). (1974). *Practical Methods in Electron Miocroscopy*. Vol. 3. American Elsevier, North Holland. p. 290.

Boyd, J.D., Johnson, F.R. and Lever, J.D. (Eds.). (1961). *Electron Microscopy in Anatomy*. Edward Arnold Publishers Ltd., London. p. 34.

Connie, R. Mahon, Donald, C. Lehman and George Manuselis. (2007). *Textbook of Diagnostic Microbiology*, 3rd edn. Elsevier, New Delhi, India. p.251.

Cosslett, V.E. (1951). *Practical Electron Microscopy*. Butterworths Scientific Publications. p.74.

Daniel, C. Pease. (1964). *Histological Techniques for Electron Microscopy*, 2nd edn. Academic Press, New York and London. p. 290.

Douglas, W. Hendenson and John, M. Papadimitriou and Mark Coleman. (1986). *Ultrastructural Appearances of Tumours*, 2nd edn. Churchill Livingstone. p. 22.

Gaurav Sharma, Sharma, M.L., Sharma, S.K. and Kang, H.P.S. "Single step surface replication of flower petals for SEM". *Asia-Pacific Microscopy Analysis*, p. 4.

Hagelquist, E. and Engstrom, B. (1973). Electron microscopy studies on cytologic material acquired by fine-needle biopsy. **Ups. J. Med. Sci. 78:** 153–159.

Henry, A. Azar. (Ed.). (1988). *Pathology of Human Neoplasms. An Atlas of Diagnostic Electron Microscopy and Immunohistochemistry*. Raven Press, New York.

Idelman, S. (1964). Modification de la technique de Luft en vue de la conservation des lipids en microscopie electronique. *Journal de Microscopie.* 3, 715–718.

John, E. Johnson *et al*. (1988). "New perpectives of electron microscopy in medicine." In: Jose Russo and Sheldon, C. Sommers (Eds.). *Tumour Diagnosis by Electron Microscopy*, Vol.2. Field and Wood Medical Publishers Inc., USA.

Keiji Kawamoto, Asao Hirano and Fritz Herz. (1980). "Simplified *in situ* preparation of cultured cell monolayers for electron microscopy. *J. Histochem.Cytochem.* **28**(2): 178–180.

Lars-Gunnar Kindblom. (1983). "Light and electron microscopic examination of embedded fine-needle aspiration biopsy specimens in the pre-operative diagnosis of soft tissue and bone tumors". *Cancer* 51: 2264–2277.

Login and Dvorak. (1994). "Methods of MW fixation technology". *Prog. Histochem. Cytochem.* **27**(4): 72–94.

"Microwave Protocols" prepared for the workshop at the Pathology Dept. of the Tata Memorial Hospital, Mumbai, by PELCO International, USA. p.23.

Mohammed Akhtar, M. Ashraf Ali, Rajeh Sabbah, Mohammed Bakry and Joan E. Nash. (1985). "Fine-needle aspiration biopsy diagnosis of round cell malignant tumors of childhood". *Cancer*. 55: 1805–1817.

Monica, A. Abrew *et al.* (1993). "A toluidine blue/basic fuchsin stain for GMA embedded tissues." *J. of Histotechnology*, Vol.16, No.2.

Richard, G. Dickersin. (2000). *Diagnostic Electron Microscopy—A Text/Atlas*, 2nd edn. Springer-Verlag Inc., New York.

Richard, T. Giberson and Richard, S. Demaree, Jr., (1995). "Microwave fixation: Understanding the variable to achieve rapid reproducible results". *Microscopy Res. and Technique.* **32**: 246–254.

Robin Harris, J. David Bhella and Marc Adrian. (2006). "Recent developments in negative staining for transmission electron microscopy. *Asia/Pacific Microscopy and Analysis*. pp. 5–8.

Stein, O. and Stein, Y. (1971). "Light and electron microspicial radioautography of lipids: techniques and biological applications". *Advances in Lipid Research*. 9, 1–72.

TEM Instructions Manual for JEM 100 SX JEOL Transmission Electron Microscope, JEOL Ltd., Tokyo, Japan. pp. 1–5.

Virendra Kumar. (2008) "High resolution structural charcterizations of *Mycobacterium leprae* by shadow-casting technique". *EMSI Bulletin.* **7**(1). 7–12.

INDEX

C

N

O

P

Q

R

S